Final Take Home 5/1
        Return 5/8

        4/17 => 4/24 => 5/1 Groups
        chapts 4, 5, & 6

Call Joe Smith.

ENGINEERING ECONOMY

FOURTH EDITION

Leland Blank, P. E.
Texas A&M University

Anthony Tarquin, P. E.
University of Texas at El Paso

*We dedicate this book to our parents,
for all they have given to us and to others.*

WCB/McGraw-Hill

A Division of The **McGraw·Hill** *Companies*

ENGINEERING ECONOMY

Copyright © 1998 by The McGraw-Hill Companies, Inc. All rights reserved. Previous editions © 1976, 1983, and 1989. Printed in the United States of America. Except as permitted under the United States Copyright Act of 1976, no part of this publication may be reproduced or distributed in any form or by any means, or stored in a data base or retrieval system, without the prior written permission of the publisher.

This book is printed on acid-free paper.

1 2 3 4 5 6 7 8 9 0 DOC/DOC 9 0 9 8 7

ISBN 0-07-063110-7

Editorial director: *Kevin T. Kane*
Publisher: *Tom Casson*
Executive editor: *Eric Munson*
Marketing manager: *John T. Wannemacher*
Project manager: *Alisa Watson*
Production supervisor: *Heather D. Burbridge*
Senior designer: *Laurie J. Entringer*
Cover designer: *Ellen Pettengell*
Cover illustration: *Judy Rush*
Compositor: *Interactive Composition Corporation*
Typeface: *10/12 Times Roman*
Printer: *R. R. Donnelley & Sons Company*

Library of Congress Cataloging-in-Publication Data

Blank, Leland T.
    Engineering economy / Leland T. Blank, Anthony J. Tarquin.—4th ed.
        p.   cm.—(McGraw-Hill series in industrial engineering and management science)
    Includes index.
    ISBN 0-07-063110-7
    1. Engineering economy. I. Tarquin, Anthony J. II. Series.
TA177.4.B58   1998
658.1′52—dc21   97-28589

http://www.mhhe.com

McGraw-Hill Series in Industrial Engineering and Management Science

Consulting Editors

Kenneth E. Case, Department of Industrial Engineering and Management, Oklahoma State University

Philip M. Wolfe, Department of Industrial and Management Systems Engineering, Arizona State University

Barnes: *Statistical Analysis for Engineers and Scientists: A Computer-Based Approach*
Bedworth, Henderson, and Wolfe: *Computer-Integrated Design and Manufacturing*
Black: *The Design of the Factory with a Future*
Blank and Tarquin: *Engineering Economy*
Blank: *Statistical Procedures for Engineering, Management, and Science*
Bridger: *Introduction to Ergonomics*
Denton: *Safety Management: Inproving Performance*
Ebeling: *An Introduction to Reliability and Maintainability Engineering*
Grant and Leavenworth: *Statistical Quality Control*
Hicks: *Industrial Engineering and Management: A New Perspective*
Hillier and Lieberman: *Introduction to Mathematical Programming*
Hillier and Lieberman: *Introduction to Operations Research*
Huchingson: *New Horizons for Human Factors in Design*
Juran and Gryna: *Quality Planning and Analysis: From Product Development through Use*
Kelton, Sadowski, and Sadowski: *Simulation with ARENA*
Khoshnevis: *Discrete Systems Simulation*
Kolarik: *Creating Quality: Concepts, Systems, Strategies, and Tools*
Law and Kelton: *Simulation Modeling and Analysis*
Marshall and Oliver: *Decision Making and Forecasting*
Moen, Nolan, and Provost: *Improving Quality through Planned Experimentation*
Nash and Sofer: *Linear and Nonlinear Programming*
Nelson: *Stochastic Modeling: Analysis and Simulation*
Niebel, Draper, and Wysk: *Modern Manufacturing Process Engineering*
Pegden: *Introduction to Simulation Using SIMAN*
Riggs, Bedworth, and Randhawa: *Engineering Economics*
Steiner: *Engineering Economic Principles*
Taguchi, Elsayed, and Hsiang: *Quality Engineering in Production Systems*
Wu and Coppins: *Linear Programming Extensions*

CONTENTS

PREFACE

The first three editions of this text presented, in a clearly written fashion, the basic principles of economic analysis for application in the decision-making process. Our objective has always been to present the material in the clearest, most concise fashion possible without sacrificing coverage or true understanding on the part of the learner. In this fourth edition, we have made every effort to do the same thing, while retaining the basic structure of the text developed in previous editions.

MATERIAL AND ORGANIZATIONAL CHANGES

In addition to the rewording and restructuring that always takes place in new editions of textbooks, some new topics have been added and considerable material has been updated to render this edition an even more valuable resource. One of the most exciting changes to this edition is the introduction of spreadsheets in a way that makes their use natural and optimal throughout the text. In keeping with the easy-to-learn format of this book, spreadsheets have been introduced in a way that even learners totally unfamiliar with them will be able to "Excel" in a mattter of minutes. An appendix has been added which leads the student step-by-step through the process of setting up and executing a simple spreadsheet. From this beginning, students and instructors learn to use the built-in functions and many of the powerful features of spreadsheets that render them ideally suited to engineering economic analyses. All spreadsheet-based examples are clearly marked, and a manual solution is included immediately prior to the spreadsheet solution, thus allowing both manual and spreadsheet options for the course.

The second major addition to this text is the incorporation of case studies at the end of several chapters. These real-world, more in-depth treatments underscore the importance of economic analysis to the engineering profession.

A Monte Carlo simulation section has been added to Chapter 20 which introduces in a clear and simple way the use of probability in engineering economic decision making. As in previous editions, each chapter contains many problems, over 75% of which are new and representative of real-world situations. Finally, the latest U.S. depreciation and tax law information has been incorporated into Chapters 13 through 15.

In making any changes, the overriding consideration has been the preservation of the free-flowing, easy-to-understand format which characterizes previous editions. We are confident that this text represents an up-to-date, well-balanced presentation of economic analysis at the undergraduate level with coverage that is particularly relevant to engineers and other decision makers.

USE OF TEXT

This text has been prepared in an easy-to-read fashion for use in learning and teaching and as a reference book for the basic computations used in an engineering economic analysis. It is best suited for use in a one-semester or one-quarter undergraduate course in engineering economic analysis, project analysis, or engineering cost analysis. The students should have at least a sophomore, and preferably a junior, standing. A background in calculus is not necessary to understand the material, but a basic understanding of economics and accounting (especially from the cost viewpoint) is helpful. However, the building-block approach used in the text's design allows a practitioner unacquainted with economics and engineering principles to use the text to learn, understand, and correctly apply the techniques in the process of decision making.

COMPOSITION OF TEXT MATERIAL

Each chapter contains a chart of progressive learning objectives followed by the study material with section headings that correspond to the learning objectives. Section 5.1, for example, contains the material which pertains to the first learning objective of the chapter. Sections contain one or more illustrative solved examples, which are separated from the textual material and include comments about the solution and any pertinent relations to other topics in the book. Many sections make reference to additional solved examples at the end of the chapter's sections. Sections also include a cross reference to the end-of-chapter unsolved problems, which the learner should now be able to understand and solve. This approach allows the opportunity to apply material on a section-by-section basis or to wait until a chapter is completed. As a new feature in each chapter, a short summary of the concepts and major topics covered is located immediately prior to the case study or the end-of-chapter problems.

Appendixes A through C contain supplementary information: a basic introduction to the use of spreadsheets and Microsoft Excel for readers unfamiliar with them; the basics of accounting reports; and the final answers to selected problems arranged by chapter. Interest factor tables are located at the very back of the text for easy access. Finally, the inside covers contain easy reference to factor notation, formulas, and graphs (front cover) and to common terms and symbols used in engineering economy (back cover).

OVERVIEW OF TEXT

The text is comprised of 20 chapters clustered into five levels as indicated by the flowchart. Coverage of the material should approximate the flow in this chart to ensure its understanding. The material in Level One emphasizes basic computational skills. Level Two discusses the four most widely used techniques to evaluate alternatives, and Level Three extends these analysis techniques to other methods commonly needed by the engineering economist. Level Four introduces the important topics of depreciation and taxation (corporate and individual), while Level Five treats supplementary and advanced analysis procedures, especially variation, sensitivity analysis, and decision making under risk. Much of the Level Five material may be considered optional by those requiring a shortened version of the text. We thank Peter Chan for his excellent work in preparing the spreadsheets, appendix A material, and clipart layout for the text. Also, we thank Corel for the permission to use the clipart presented throughout this book. Both of these additions make the book easier to use and very current.

Lee Blank
Tony Tarquin

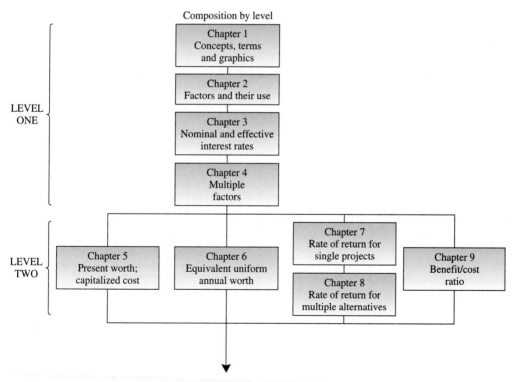

(*flowchart continues*)

(*flowchart concluded*)

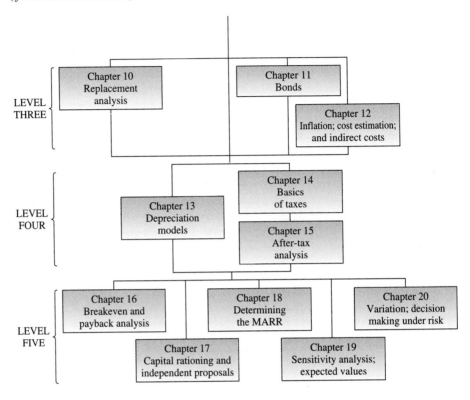

LEVEL

ONE

*T*he first four chapters will help you learn to construct and use cash-flow diagrams and to correctly account for the time value of money. The concept that money has different values over different time intervals is called equivalence. The movement of revenues and disbursements through time (with equivalence) requires the understanding and use of engineering economy factors which greatly simplify otherwise complicated computations.

There are different ways of stating and using interest rates in economic computations. The explanation of nominal and effective rates is included in this level so you can correctly use the factors tabulated at the end of this text. The effective and nominal rates, and the factors, are directly applicable to individual, business, industry, and government economic and investment evaluations.

CHAPTER

Basic Concepts, Terms, and Graphics

This chapter provides you with an understanding of the basic concepts and terminology necessary to perform an engineering economy analysis. It explains the role of engineering economy in the decision-making process and describes the major elements of an engineering economy study. Finally, a basic graphical approach—the cash-flow diagram—is introduced.

LEARNING OBJECTIVES

Purpose: Understand the meaning, role, approach, and basic concepts of engineering economy.

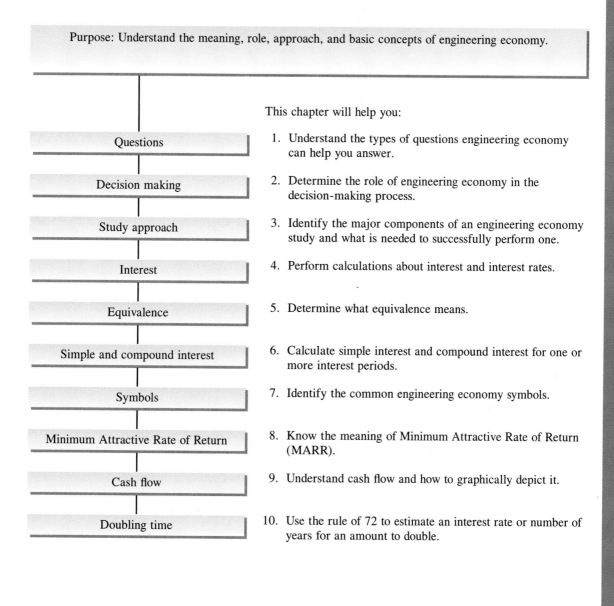

	This chapter will help you:
Questions	1. Understand the types of questions engineering economy can help you answer.
Decision making	2. Determine the role of engineering economy in the decision-making process.
Study approach	3. Identify the major components of an engineering economy study and what is needed to successfully perform one.
Interest	4. Perform calculations about interest and interest rates.
Equivalence	5. Determine what equivalence means.
Simple and compound interest	6. Calculate simple interest and compound interest for one or more interest periods.
Symbols	7. Identify the common engineering economy symbols.
Minimum Attractive Rate of Return	8. Know the meaning of Minimum Attractive Rate of Return (MARR).
Cash flow	9. Understand cash flow and how to graphically depict it.
Doubling time	10. Use the rule of 72 to estimate an interest rate or number of years for an amount to double.

1.1 WHY IS ENGINEERING ECONOMY SO IMPORTANT?

On virtually a daily basis, we all make decisions which affect our future. The selections we make change our lives in small, and sometimes large, ways. For example, buying a new shirt for cash increases our fashion choices when we dress each day, and it decreases the amount of money we have with us at the moment. On the other hand, purchasing a new car and assuming a car loan gives us newfound transportation options, but it may cause a significant reduction in available cash as we make the monthly payments. In both cases, *economic* and *noneconomic factors,* as well as *tangible* and *intangible factors,* are important in the decision to purchase the shirt or the car.

Individuals, small-business owners, large-corporation presidents, and government agency heads are routinely faced with the challenge of making significant decisions when selecting one alternative over another. These are decisions of how to best invest the funds, or *capital,* of the company and its owners. The amount of capital is always limited, just as the cash available to an individual is usually limited. These business decisions will invariably change the future, hopefully for the better. The usual factors considered, may be, once again, economic and noneconomic, as well as tangible and intangible. However, when corporations and public agencies select one alternative over another, the financial aspects, return on invested capital, societal considerations, and time frames often increase substantially over those for an individual's selection.

Engineering economy, quite simply, is about determining the economic factors and the economic criteria utilized when one or more alternatives are considered for selection. Another way to define engineering economy is a collection of mathematical techniques which simplify economic comparisons. With these techniques, a rational, meaningful approach to evaluating the economic aspects of different methods (alternatives) of accomplishing a given objective can be developed. The techniques work equally well for an individual or a corporation faced with an economic-based decision. Some of the typical questions which can be methodically addressed by individuals, businesses and corporations, and public (government) agencies using the material in this book are posed here.

Individuals

- Should I pay off my credit card balance with borrowed money?
- What are graduate studies worth financially over my professional career?
- Are federal income tax deductions for my home mortgage a good deal or should I accelerate my mortgage payments?
- Exactly what rate of return did we make on this stock investment?
- Should I buy or lease my next car or keep the one I have now and continue to pay on the loan?

→ *Corporations and Businesses*

- Will we make the required return if we install this new manufacturing technology in the plant?
- Do we build or lease space for the new branch in Asia?
- Is it economically better to make in-house or buy outside a solid-state component part on a new product line?

→ *Government Units Which Serve the Public*

- How much new tax revenue does the city need to generate to pay for the school bond issue up for vote?
- Do the benefits outweigh the costs of a bridge over the intracoastal waterway at this point?
- Is it cost-effective for the city to construct a covered dome for major sports events?
- Should the state university contract with a local community college to teach foundation-level undergraduate courses or have university faculty teach them?

Example 1.1

The presidents of two small businesses play racquetball each week. After several conversations, they have decided that, due to their frequent commercial-airline travel around the region, they should evaluate the purchase of a plane jointly owned by the two companies. What are some of the typical economic-based questions the two presidents should answer as they evaluate the alternatives to (1) co-own a plane or (2) continue as is?

Solution

Some questions (and what is needed to respond to them) might be

- How much will it cost each year? (Estimates needed here)
- How do we pay for it? (A financing plan needed here)
- Are there tax advantages? (Tax law information needed here)
- Which alternative is more cost-effective? (Selection criteria needed here)
- What is the expected rate of return? (Equations needed here)
- What happens if we use different amounts each year than we estimated? (Sensitivity analysis needed here)

1.2 ROLE OF ENGINEERING ECONOMY IN DECISION MAKING

People make decisions; computers, methodologies, and other tools do not. The techniques and models of engineering economy *assist people in making decisions*. Since decisions affect what will be done, the time frame of engineering economy usually is *the future.* Therefore, numbers used in an engineering economic analysis are *best estimates of what is expected to occur.*

It is common to include outcomes in an analysis of observed facts. This utilizes engineering economic methods to analyze *the past,* since a decision to select one (future) alternative over another is not made. Rather, the analysis explains or characterizes the outcomes. For example, a corporation may have initiated a mail-order division 5 years ago. Now it wants to know the actual return on investment (ROI) or rate of return (ROR) experienced by this division. Both of these—outcomes analysis and future alternative decision—are considered the domain of engineering economy.

There is a very popular procedure used to address the development and selection of alternatives. Common names for this procedure are the *problem-solving approach* or the *decision-making process.* Typically the steps in the approach are as follows:

Steps to Problem Solving

1.   Understand the problem and goal.

2.   Collect relevant information.

3.   Define the alternative solutions.

4.   Evaluate each alternative.

5.   Select the best alternative using some criteria.

6.   Implement the solution and monitor the results.

Engineering economy has a major role in steps 2, 3, and 5, and it is the primary technique in step 4 to perform the economic-based analysis of each alternative. Steps 2 and 3 set up the alternatives, and engineering economy helps structure the estimates of each one. Step 4 utilizes one or more engineering economy models discussed in this book to complete the economic analysis upon which a decision is made.

Example 1.2

Reconsider the questions presented in the previous example about jointly purchasing a corporate airplane. State some ways in which engineering economy may contribute to the decision-making process of selecting between the two alternatives.

Solution

Assume that the problem and goal are the same for each president—available, reliable transportation which minimizes total cost. Engineering economy assists in several ways. Use the problem-solving approach as a framework.

Steps 2 and 3: The framework of estimates necessary for an engineering economy analysis assists in structuring what data should be estimated and collected. For example, for alternative 1 (buy the plane), these include estimated purchase cost, financing methods and interest rates, annual operating costs, possible increase in annual sales revenue, and income tax deductions. For alternative 2 (maintain the status quo), these include observed and estimated commercial transportation costs, annual sales revenue, and other relevant data. Note that engineering economy does not specifically include the estimation; it helps determine what estimates and data are needed for the analysis (step 4) and decision (step 5).

Step 4: This is the heart of engineering economy. The techniques result in numerical values called *measures of worth,* which inherently consider the *time value of money.* Some common measures of worth are

Present worth (PW)	Future worth (FW)
Annual worth (AW)	Rate of return (ROR)
Benefit/cost ratio (B/C)	Capitalized cost (CC)

In all these cases, the fact that money today is worth a different amount in the future is considered.

Step 5: For the economic portion of the decision, some criterion based on one of the measures of worth is used to select only one of the alternatives. Additionally, there are so many noneconomic factors—social, environmental, legal, political, personal, to name a few—that the result of the engineering economy analysis may seem, at times, to be used less than the engineer may wish. But this is exactly why the decision maker must have adequate information for all factors—economic and noneconomic—to make an informed selection. In our case, the economic analysis may significantly favor the co-owned plane (alternative 1). But because of noneconomic factors, one or both presidents may decide to remain with the current situation by selecting alternative 2.

The concept of *time value of money* was mentioned in the solution of Example 1.2. For alternative aspects which can be quantified in terms of dollars, it is vitally important to recognize this concept. It is often said that money makes money. The statement is indeed true, for if we elect to invest money today (for example, in a bank, a business, or a stock mutual fund), we inherently expect to have more money in the future. This change in the amount of money over a given time period is called the *time value of money;* it is the most important concept in engineering economy. You should also realize that if a person or company finds it necessary to borrow money today, by tomorrow more than the original loan principal will be owed. This fact is also explained by the time value of money.

1.3　PERFORMING AN ENGINEERING ECONOMY STUDY

We use the terms engineering economy, engineering economic analysis, economic decision making, capital allocation study, economic analysis, and similar terms synonymously throughout this book. There is a generally accepted approach, called the *Engineering Economy Study Approach,* used to perform an engineering economic analysis. It is outlined in Figure 1–1 for two alternatives, labeled (alternative 1) New Plant Design and (alternative 2) Upgrade Old Plant for illustration purposes.

Once the alternatives are identified and relevant information (meaning estimates) is available, the flow of the economic analysis generally follows the problem-solving steps 3 through 5 described in the previous section. Let us track these steps through the major sections identified in the Engineering Economy Study Approach (Figure 1–1).

Alternative Description   The result of problem-solving approach step 1 (understand the problem and goal) is a general description of how the solution may be

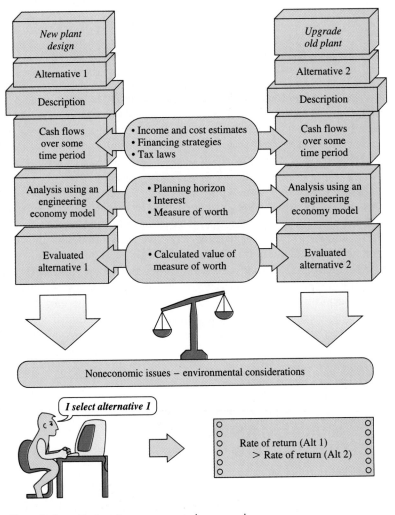

Figure 1–1   Engineering economy study approach.

approached. There may initially be many alternatives described, but only a few will be viable and actually prepared for evaluation. Alternatives are stand-alone options.

Alternatives usually involve information such as *first cost* (including purchase price, construction, installation, and delivery costs), *expected life, estimated annual incomes and expenses* of the alternative (including annual maintenance and upkeep costs), *projected salvage value* (resale or trade-in value), an appropriate *interest rate* (rate of return), and possibly *income-tax effects*. The annual expenses are usually lumped together and called annual operating costs (AOC).

Cash Flows over Some Time Period   The actual inflows and outflows of money are called cash flows. To perform the economic analysis, estimates may be needed on financing, interest rates, life of assets, revenues, costs, tax effects, etc. The relevant estimates (step 2) are gathered and formatted to specify the economic aspects of the alternatives (step 3). Without cash-flow estimates over a stated time period, no engineering economy study can be conducted.

Analysis Using an Engineering Economy Model   Computations of the type in this book are performed on the cash flows of each alternative to obtain one or more of the measures of worth. This phase is the essence of engineering economy. However, the procedures enable you to make economic decisions only about those alternatives which have been recognized as alternatives; the procedures do not assist in identifying the alternatives themselves. If only alternatives *A*, *B*, and *C* have been identified and defined for analysis, when method *D*, though not recognized as an alternative, is the most attractive, the wrong decision is certain to be made. The importance of alternative identification and definition in the decision-making process cannot be overemphasized, since this step (step 4 in the Problem-Solving Approach) makes the result of an economic analysis of real value.

Evaluated Alternative   The measure of worth is stated for each alternative. This is the result of the engineering economy analysis. For example, the result of a rate-of-return analysis of the two alternatives may be

Select alternative 1 where the rate of return is estimated at 18.4% per year over a 20-year life.

Noneconomic Issues   As mentioned earlier, there are many other factors— social, environmental, legal, political, personal, to name a few—that must be considered before a selection is made. Some of these are tangible (quantifiable), while others are not.

Evaluation or Selection Criteria   Some combination of economic criteria using the measure of worth, and the noneconomic and intangible factors, is applied by a decision maker to select one alternative. If only one alternative is defined, a second is always present in the form of the *do-nothing alternative*. We also call this the *as-is* or *status quo* alternative. We will discuss this option throughout the book, but, in short, it means that the current approach is maintained.

Whether we are aware of it or not, we use criteria every day to choose between alternatives. For example, when you drive to campus, you decide to take the "best" route. But how did you define best? Was the best route the safest, shortest, fastest, cheapest, most scenic, or what? Obviously, depending upon which criterion or combination of criteria is used to identify the best, a different route might be selected each time. In economic analysis, *financial units* (dollars) are generally used as the tangible basis for evaluation. Thus, when there are several ways of accomplishing a given objective, the alternative with the lowest overall cost or highest overall net income is selected.

In most cases alternatives involve *intangible factors,* which cannot readily be expressed in terms of dollars. When the alternatives available have approximately the same equivalent cost, the noneconomic and intangible factors may be used as the basis for selecting the best alternative.

Alternative Selected   Once selected, it is expected that implementation and continued monitoring (step 6) will take place. Usually the monitoring prompts new alternatives as technology changes, markets change, and assets deteriorate.

Now, let's learn some of the basics about engineering economy that you can apply in everyday life, when you go to work professionally, or if you start your own business.

Problems 1.1 to 1.11

1.4   INTEREST CALCULATIONS

The manifestation of the time value of money is termed *interest,* which is the increase between an original sum of money borrowed and the final amount owed, or the original amount owned (or invested) and the final amount accrued. The original investment or loan amount is referred to as *principal*. If you invested money at some time in the past, the interest would be

$$\text{Interest} = \text{total amount now} - \text{original principal} \qquad [1.1]$$

If the result is negative, you have lost money and there is no interest. On the other hand, if you borrowed money at some time in the past, the interest would be

$$\text{Interest} = \text{amount owed now} - \text{original principal} \qquad [1.2]$$

In either case, there is an increase in the amount of money that was originally invested or borrowed, and the increase over the original amount is the interest.

When interest is expressed as a percentage of the original *amount per time unit,* the result is an *interest rate.* This rate is calculated as

$$\text{Percent interest rate} = \frac{\text{interest accrued per time unit}}{\text{original amount}} \times 100\% \qquad \text{[1.3]}$$

By far the most common time period in which to express an interest rate is 1 year. However, since interest rates may be expressed over periods of time shorter than 1 year, for example, 1% per month, the time unit used in expressing an interest rate must also be identified. This period is called the *interest period.* The following examples illustrate interest calculations.

Example 1.3

The Oracle Investment Group invested $100,000 on May 1 and withdrew a total of $106,000 exactly 1 year later. Compute (*a*) the interest gained and (*b*) the interest rate on the investment.

Solution

(*a*) Applying Equation [1.1],

$$\text{Interest} = \$106,000 - 100,000 = \$6000$$

(*b*) Equation [1.3] determines the interest rate over the 1-year interest period.

$$\text{Interest rate} = \frac{\$6000 \text{ per year}}{\$100,000} \times 100\% = 6\% \text{ per year}$$

Comment

For borrowed money, computations are similar to those shown above except interest is computed by Equation [1.2]. For example, if Oracle had borrowed $100,000 now and repaid $110,000 after 1 year, the interest is $10,000 and the interest rate via Equation [1.3] is $10,000/$100,000 × 100% = 10% per year.

Example 1.4

Stereophonics, Inc. plans to borrow $20,000 from a bank for 1 year at 9% interest for new recording equipment. Compute (*a*) the interest and (*b*) the total amount due after 1 year. (*c*) Construct a graph which shows the numbers that would be used to compute the loan interest rate of 9% per year.

Solution

(*a*) Compute the total interest accrued by Equation [1.3].

$$\text{Interest} = \$20,000(0.09) = \$1800$$

(*b*) The total amount due is the sum of principal and interest.

$$\text{Total due} = \$20,000 + 1800 = \$21,800$$

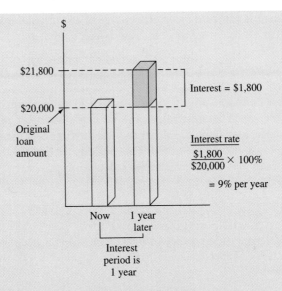

$21,800

Interest = $1,800

$20,000

Original
loan
amount

Interest rate

$\dfrac{\$1,800}{\$20,000} \times 100\%$

$= 9\%$ per year

Now   1 year
       later

Interest
period is
1 year

Figure 1–2   Values used to compute an interest rate
of 9% per year, Example 1.4.

(*c*) Figure 1–2 shows the values needed for Equation [1.3]: $1800 interest, $20,000
original loan principal, 1-year interest period.

Comment

Note that in part (*b*), the total amount due may also be computed as

$$\text{Total} = \text{principal}\,(1 + \text{interest rate}) = \$20{,}000(1.09) = \$21{,}800$$

Later we will routinely use this method to compute future amounts.

Example 1.5

(*a*) Calculate the amount of money that must have been deposited 1 year ago to have
$1000 now at an interest rate of 5% per year.

(*b*) Calculate the amount of interest earned during this time period.

Solution

(*a*) The total amount accrued is the sum of the original deposit and the earned
interest. If X is the original deposit,

$$\text{Total accrued} = \text{original} + \text{original}(\text{interest rate})$$

$$\$1000 = X + X(0.05) = X(1 + 0.05) = 1.05X$$

The original deposit is

$$X = \frac{1000}{1.05} = \$952.38$$

(b) Apply Equation [1.1] to determine interest earned.

$$\text{Interest} = \$1000 - 952.38 = \$47.62$$

In Examples 1.3 to 1.5 the interest period was 1 year and the interest amount was calculated at the end of one period. When more than one interest period is involved (for example, if we had wanted to know the amount of interest owed after 3 years in Example 1.4), it is necessary to state if the interest is accrued on a *simple* or *compound* basis. These concepts are discussed in Section 1.6.

Additional Example 1.18
Problems 1.12 to 1.17

1.5 EQUIVALENCE

When considered together, time value of money and interest rate help us develop the concept of *equivalence*, which means that different sums of money at different times are equal in economic value. For example, if the interest rate is 6% per year, $100 today (present time) would be equivalent to $106 one year from today.

Amount accrued $= 100 + 100(0.06) = 100(1 + 0.06) = \$106$

So, if someone offered you a gift of $100 today or $106 one year from today, it would make no difference which offer you accepted. In either case you have $106 one year from today. The two sums of money are equivalent to each other when the interest rate is 6% per year. However, at a higher or lower interest rate $100 today is not equivalent to $106 one year from today.

In addition to future equivalence, we can apply the same logic to determine equivalence for previous years. If you have $100 now, it is equivalent to $100/1.06 = $94.34 one year ago at an interest rate of 6% per year. From these illustrations, we can state the following: $94.34 last year, $100 now, and $106 one year from now are equivalent at an interest rate of 6% per year. The fact that these sums are equivalent can be established by computing the two interest rates for 1-year interest periods.

$$\frac{\$106}{\$100} = 1.06 \qquad (6\% \text{ per year})$$

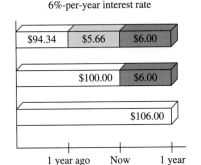

6%-per-year interest rate

Figure 1-3   Equivalence of three amounts at a 6%-per-year interest rate and 1 year apart.

and

$$\frac{\$100}{\$94.34} = 1.06 \qquad (6\% \text{ per year})$$

Figure 1–3 indicates the amount of interest each year necessary to make these three different amounts equivalent at 6% per year.

Example 1.6

Make the calculations necessary at an interest rate of 5% per year to show which of the following statements are true and which are false.

(a) $98 now is equivalent to $105.60 one year from now.

(b) $200 one year ago is equivalent to $205 now.

(c) $3000 now is equivalent to $3150 one year from now.

(d) $3000 now is equivalent to $2887.14 one year ago.

(e) Interest accumulated in 1 year on an investment of $2000 is $100.

Solution

(a) Total amount accrued = 98(1.05) = $102.90 ≠ $105.60; therefore, it is false. Another way to solve this is as follows: Required investment 105.60/1.05 = $100.57 ≠ $98.

(b) Required investment = 205.00/1.05 = $195.24 ≠ $200; therefore, it is false.

(c) Total amount accrued = 3000(1.05) = $3150; true.

(d) Total amount accrued = 2887.14(1.05) = $3031.50 ≠ $3000; false.

(e) Interest = 2000(0.05) = $100; true.

Problems 1.18 to 1.21

1.6   SIMPLE AND COMPOUND INTEREST

The terms interest, interest period, and interest rate (introduced in Section 1.4) are useful in calculating equivalent sums of money for one interest period in the past and one period in the future. However, for more than one interest period, the terms *simple interest* and *compound interest* become important.

Simple interest is calculated using the principal only, ignoring any interest accrued in preceding interest periods. The total simple interest over several periods is computed as

$$\text{Interest} = (\text{principal})(\text{number of periods})(\text{interest rate}) \qquad \textbf{[1.4]}$$

where the interest rate is expressed in decimal form.

Example 1.7

If Jonathan borrows $1000 from his older sister for 3 years at 5%-per-year simple interest, how much money will he repay at the end of 3 years? Tabulate the results.

Solution

The interest for each of the 3 years is

$$\text{Interest per year} = 1000(0.05) = \$50$$

Total interest for 3 years from Equation [1.4] is

$$\text{Total interest} = 1000(3)(0.05) = \$150$$

The amount due after 3 years is

$$\$1000 + 150 = \$1150$$

The $50 interest accrued in the first year and the $50 accrued in the second year do not earn interest. The interest due each year is calculated only on the $1000 principal.

The details of this loan repayment are tabulated in Table 1–1. The end-of-year figure of zero represents the present, that is, when the money is borrowed. No payment is made by the borrower until the end of year 3, so the amount owed each year increases uniformly by $50, since simple interest is figured only on the loan principal.

Table 1–1	Simple-interest computations			
(1) End of Year	(2) Amount Borrowed	(3) Interest	(4) Amount Owed	(5) Amount Paid
0	$1000			
1	—	$50	$1050	$ 0
2	—	50	1100	0
3	—	50	1150	1150

For *compound interest* the interest accrued for each interest period is calculated on the *principal plus the total amount of interest accumulated in all previous periods*. Thus, compound interest means interest on top of interest. Compound interest reflects the effect of the time value of money on the interest also. Now the interest for one period is calculated as

$$\text{Interest} = (\text{principal} + \text{all accrued interest})(\text{interest rate}) \qquad \textbf{[1.5]}$$

Example 1.8

If your friend Jonathan borrows $1000 at 5%-per-year compound interest from his sister, instead of simple interest as in the preceding example, compute the total amount due after 3 years. Graph and compare the results of this and the previous example.

Solution

The interest and total amount due each year is computed separately using Equation [1.5].

$$\text{Year 1 interest: } \$1000(0.05) = \$50.00$$

$$\text{Total amount due after year 1: } \$1000 + 50.00 = \$1050.00$$

$$\text{Year 2 interest: } \$1050(0.05) = \$52.50$$

$$\text{Total amount due after year 2: } \$1050 + 52.50 = \$1102.50$$

$$\text{Year 3 interest: } \$1102.50(0.05) = \$55.13$$

$$\text{Total amount due after year 3: } \$1102.50 + 55.13 = \$1157.63$$

The details are shown in Table 1–2. The repayment plan is the same as that for the simple-interest example—no payment until the principal plus accrued interest is due at the end of year 3.

Figure 1–4 shows the amount owed at the end of each year for 3 years. The difference due to the time value of money is recognized for the compound interest case. An extra $1157.63 − $1150 = $7.63 of interest is paid compared with simple

Table 1–2	Compound-interest computations			
(1) End of Year	(2) Amount Borrowed	(3) Interest	(4) Amount Owed	(5) Amount Paid
0	$1000			
1	—	$50.00	$1050.00	0
2	—	52.50	1102.50	0
3	—	55.13	1157.63	$1157.63

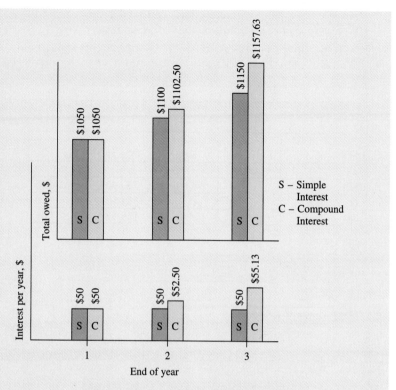

Figure 1–4   Comparison of simple- and compound-interest calculations for Examples 1.7 and 1.8.

interest over the 3-year period. The difference between simple and compound interest grows each year. If the computations are continued for more years, for example, 10 years, the difference is $128.90; after 20 years compound interest is $653.30 more than simple interest.

If $7.63 does not seem like a significant difference in only 3 years, remember that the beginning amount here is $1000. If we make these same calculations for an initial amount of $100,000 or $1 million we are talking real money. This all indicates that the power of compounding is vitally important in all economic-based analyses.

Example 1.8 (Spreadsheet)

Develop a spreadsheet to compute the compound interest and loan balance each year for the $1000 Jonathan borrowed at 5% per year. Graphically compare the results for compound interest and simple interest from Example 1.7.

Solution

Figure 1–5a presents a spreadsheet to compute annual compound interest and loan balance. Figure 1–5b compares simple and compound interest in tabular and

(a)

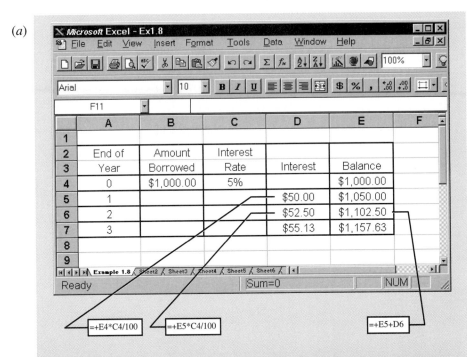

	A	B	C	D	E	F
1						
2	End of	Amount	Interest			
3	Year	Borrowed	Rate	Interest	Balance	
4	0	$1,000.00	5%		$1,000.00	
5	1			$50.00	$1,050.00	
6	2			$52.50	$1,102.50	
7	3			$55.13	$1,157.63	
8						
9						

=+E4*C4/100   =+E5*C4/100   =+E5+D6

(b)

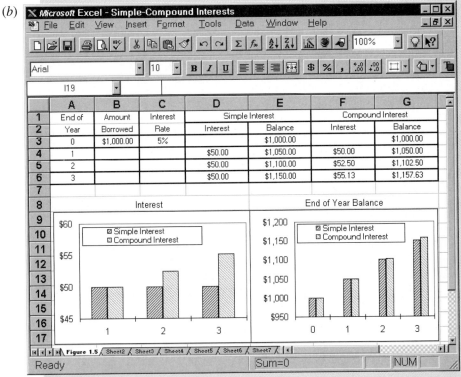

	A	B	C	D	E	F	G
1	End of	Amount	Interest	Simple Interest		Compound Interest	
2	Year	Borrowed	Rate	Interest	Balance	Interest	Balance
3	0	$1,000.00	5%		$1,000.00		$1,000.00
4	1			$50.00	$1,050.00	$50.00	$1,050.00
5	2			$50.00	$1,100.00	$52.50	$1,102.50
6	3			$50.00	$1,150.00	$55.13	$1,157.63
7							

Figure 1-5 Spreadsheet and graph, Example 1.8 (spreadsheet).

graphical formats. The spreadsheet relation used to determine the value in selected cells is shown in Figure 1–5a. For example, the D6 cell value is

$$\text{Interest} = (\$1050)(0.05) = \$52.50$$

In cell notation,

$$D6 = E5 \times C4 \text{ in decimal form}$$

Appendix A includes an introduction to understanding and using Microsoft Excel and spreadsheets to solve engineering economy problems. It is there for reference throughout your use of this text.

We combine the concepts of simple interest, compound interest, and equivalence to demonstrate that different loan repayment plans, or investment plans, are equivalent but differ substantially in monetary amounts from one year to another. This also shows how many ways there are to take into account the time value of money. The following example illustrates equivalence for five different loan repayment plans.

Example 1.9

Demonstrate the concepts of equivalence using the different loan-repayment plans described below. Each plan repays a $5000 loan in 5 years at 8% interest per year using simple or compound interest.

- **Plan 1: Simple interest, pay all at end.**   No interest or principal is paid until the end of year 5. Interest accumulates each year on the principal only.

- **Plan 2: Compound interest, pay all at end.**   No interest or principal is paid until the end of year 5. Interest accumulates each year on the total of principal and all accrued interest.

- **Plan 3: Simple interest paid as accrued; principal paid at end.**   The accrued interest is paid each year and the entire principal is repaid at the end of year 5.

- **Plan 4: Compound interest and portion of principal paid annually.**   The accrued interest and one-fifth of the principal (or $1000) is paid each year. The outstanding loan balance decreases each year, so the interest for each year decreases.

- **Plan 5: Equal payments paid annually of compound interest and principal.**   Equal payments are made each year with a portion going toward principal repayment and the remainder covering the accrued interest. Since the loan balance decreases at a rate slower than in plan 4, due to the equal end-of-year payments, the interest decreases but at a slower rate.

Solution

Table 1–3 presents the interest, payment amount, total owed at the end of each year, and the total amount paid over the 5-year period (column D totals). We will not

A End of Year	B Interest Owed for Year	C Total Owed at End of Year	D End-of-Year Payment	E Total Owed after Payment
Table 1–3 Different repayment schedules for $5000 borrowed or lent at time 0 over 5 years at 8%-per-year interest				
Plan 1: Simple Interest, Pay All at End				
0				$5000.00
1	$400.00	$5400.00	—	5400.00
2	400.00	5800.00	—	5800.00
3	400.00	6200.00	—	6200.00
4	400.00	6600.00	—	6600.00
5	400.00	7000.00	$7000.00	
Totals			$7000.00	
Plan 2: Compound Interest, Pay All at End				
0				$5000.00
1	$400.00	$5400.00	—	5400.00
2	432.00	5832.00	—	5832.00
3	466.56	6298.56	—	6298.56
4	503.88	6802.44	—	6802.44
5	544.20	7346.64	$7346.64	
Totals			$7346.64	
Plan 3: Simple Interest Paid as Accrued; Principal Paid at End				
0				$5000.00
1	$400.00	$5400.00	$400.00	5000.00
2	400.00	5400.00	400.00	5000.00
3	400.00	5400.00	400.00	5000.00
4	400.00	5400.00	400.00	5000.00
5	400.00	5400.00	5400.00	
Totals			$7000.00	
Plan 4: Compound Interest and Portion of Principal Paid Annually				
0				$5000.00
1	$400.00	$5400.00	$1400.00	4000.00
2	320.00	4320.00	1320.00	3000.00
3	240.00	3240.00	1240.00	2000.00
4	160.00	2160.00	1160.00	1000.00
5	80.00	1080.00	1080.00	
Totals			$6200.00	
Plan 5: Equal Payments Paid Annually of Compound Interest and Principal				
0				$5000.00
1	$400.00	$5400.00	$1252.28	4147.72
2	331.82	4479.54	1252.28	3227.25
3	258.18	3485.43	1252.28	2233.15
4	178.65	2411.80	1252.28	1159.52
5	92.76	1252.28	1252.28	
Totals			$6261.41	

concentrate now upon how all the column amounts are calculated; however, the interest values (column B) are determined as follows:

- **Plan 1.** Simple interest = (original principal)(0.08)
- **Plan 2.** Compound interest = (total owed previous year)(0.08)
- **Plan 3.** Simple interest = (original principal)(0.08)
- **Plan 4.** Compound interest = (total owed previous year)(0.08)
- **Plan 5.** Compound interest = (total owed previous year)(0.08)

Note that the amounts of the annual payments are different for each repayment schedule and that the total amounts repaid for most plans are different, even though each repayment plan requires exactly 5 years. The difference in the total amounts repaid can be explained (1) by the time value of money, (2) by simple or compound interest, and (3) by the partial repayment of principal prior to year 5.

Plans 1 and 3 have total amounts paid of $7000, because simple interest does not accrue interest on previous interest, as is the case in plans using compound interest.

What can be stated about equivalence for these plans at 8%-per-year interest? We can state: Table 1–3 shows that $5000 at time 0 is equivalent to each of the following:

- **Plan 1.** $7000 (simple interest) at the end of year 5.
- **Plan 2.** $7346.64 (compound interest) at the end of year 5.
- **Plan 3.** $400 per year (simple interest) for 4 years and $5400 at the end of year 5.
- **Plan 4.** Decreasing, compound-interest payments of interest and partial principal in years 1 ($1400) through 5 ($1080).
- **Plan 5.** $1252.28 per-year payments for 5 years.

Problems 1.22 to 1.34

1.7   SYMBOLS AND THEIR MEANING

The relations in engineering economy commonly include the following symbols and their (sample) units:

P = value or amount of money at a time denoted as the present, called the *present worth* or *present value;* currency, dollars

F = value or amount of money at some future time, called *future worth* or *future value;* dollars

A = series of consecutive, equal, end-of-period amounts of money, called the *equivalent value per period* or *annual worth;* dollars per year, dollars per month

n = number of interest periods; years, months, days

i = interest rate per interest period; percent per year, percent per month

t = time stated in periods; years, months, days

The symbols P and F represent one-time occurrences: A occurs with the same value once each interest period for a specified number of periods. It should be clear that a present value P represents a single sum of money at some time prior to a future value F or equivalent series amount A.

It is important to note that the symbol A always represents a uniform amount (i.e., the amount must be the same each period), which must extend through *consecutive* interest periods. Both conditions must exist before the dollar value can be represented by A.

The compound-interest rate i is expressed in percent per interest period, for example, 12% per year. Unless stated otherwise, assume that the rate applies throughout the entire n years or interest periods. The decimal equivalent for i is always used in engineering economy computations.

All engineering economy problems involve the element of time and, therefore, the symbol t. Of the other five symbols, every problem will involve at least four of the symbols P, F, A, n, and i, with at least three of them known. The following examples illustrate the use of the symbols.

Example 1.10

A college student about to graduate plans to borrow $2000 now and repay the entire loan principal plus accrued interest at 10% per year in 5 years. List the engineering economy symbols involved and their values if the student wants to know the total amount owed after 5 years.

Solution

In this case P and F, but not A, are involved, since all transactions are single payments. Time t is expressed in years.

$$P = \$2000 \quad i = 10\% \text{ per year} \quad n = 5 \text{ years} \quad F = ?$$

The future amount F is the unknown amount.

Example 1.11

Assume you borrow $2000 now at 12% per year for 5 years and must repay the loan in equal yearly payments. Determine the symbols involved and their values.

Solution

Time t is in years.

$P = \$2000$
$A = ?$ per year for 5 years
$i = 12\%$ per year
$n = 5$ years

There is no future value F involved.

In both Examples 1.10 and 1.11, the $2000 *P* value is a receipt (inflow to the borrower) and *F* or *A* is a disbursement (outflow from the borrower). It is equally correct to use these symbols in the reverse roles.

Example 1.12

On May 1, 1998, you deposited $500 into an account which paid interest at 10% per year and withdrew an equal annual amount for the following 10 years. List the symbols and their values.

Solution

Time *t* is in years; the *P* (deposit) and *A* amounts (ten withdrawals) are

$P = \$500$
$A = ?$ per year
$i = 10\%$ per year
$n = 10$ years

Comment

The value for the $500 deposit *P* and receipt *A* are indicated by the same symbol names as above, but they are considered in a different context.

Example 1.13

Carli deposited $100 each month for 7 years at an interest rate of 7% per year compounded monthly and withdrew a single amount after the 7 years. Define the symbols and their amounts.

Solution

The equal monthly deposits are in a series *A* and the withdrawal is a future sum or *F* value. The time periods *t* are in months.

$A = \$100$ per month for 84 months (7 years)
$F = ?$ after 84 months
$i = 7\%$ per year
$n = 84$ months

In Chapter 3 we will learn to compute interest when the compounding period is other than a year.

Example 1.14

Assume that you plan to make a lump-sum deposit of $5000 now into an account that pays 6% per year, and you plan to withdraw an equal end-of-year amount of $1000 for 5 years starting next year. At the end of the sixth year, you plan to close your account by withdrawing the remaining money. Define the engineering economy symbols involved.

Solution

Time t is expressed in years.

$P = \$5000$
$A = \$1000$ per year for 5 years
$F = ?$ at end of year 6
$i = 6\%$ per year
$n = 5$ years for the A series and 6 for the F value

Problems 1.35 to 1.37

1.8   MINIMUM ATTRACTIVE RATE OF RETURN

For any investment to be profitable the investor (corporate or individual) must expect to receive more money than the amount invested. In other words, a fair *rate of return,* or *return on investment* must be realizable. Over a stated period of time, the rate of return (ROR) is calculated as

$$\text{ROR} = \frac{\text{Current amount} - \text{original investment}}{\text{original investment}} \times 100\% \qquad \textbf{[1.6]}$$

The numerator may be called *profit, net income,* or a variety of other terms. Note that this computation is essentially the same as that for the interest rate in Equation [1.3]. The two terms may be used interchangeably, depending upon the vantage point—interest rate is used for the borrower's vantage, when money has been borrowed, or when a fixed interest is established. The term *rate of return* is commonly used when estimating the profitability of a proposed alternative or when evaluating the results of a completed project or investment. Both are represented by the symbol i.

Investment alternatives are evaluated upon the prognosis that a reasonable ROR can be expected. Some reasonable rate must, therefore, be stated and utilized in the selection criteria phase of the engineering economy study approach (refer back to Figure 1–1). The reasonable rate is called the *Minimum Attractive Rate of Return (MARR)* and is higher than the rate expected from a bank or some safe investment which involves minimal investment risk. Figure 1–6 indicates the relations between different rate-of-return values. In the United States, the current U.S. Treasury bill return is sometimes used as the benchmark safe rate. The MARR is also referred to as the *hurdle rate* for projects; that is, to be considered financially viable the expected ROR must meet or exceed the MARR or hurdle rate.

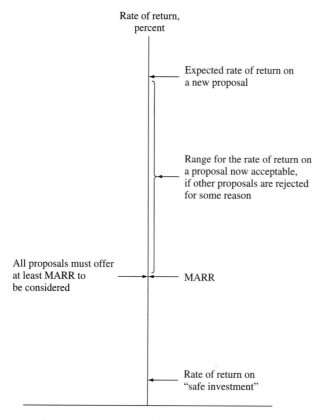

Figure 1–6   Location of the MARR relative to other rate-of-return values.

The terms *capital, capital funds,* and *capital investment money* all refer to funds available for investment to help the company generate business and revenue. The term *capital* is most frequently used. For most industrial and business organizations, capital is a limiting resource. Even though there are many alternatives that may yield a ROR which exceeds the MARR as indicated in Figure 1–6, there will likely not be sufficient capital available for all, or the project's risk may be estimated as too high to take the investment chance. New projects that are undertaken because they have an expected ROR > MARR are usually those projects which have an expected return that is at least as great as the return on another alternative not yet funded. Such a selected new project would be a proposal represented by the top ROR arrow in Figure 1–6. For example, if MARR = 12% and proposal 1 with an expected ROR = 13% cannot be funded due to a lack of capital funds, while proposal 2 has an estimated ROR = 14.5% and can be funded from available capital, only proposal 2 is undertaken. Since proposal 1 is not pursued due to the lack of capital, its estimated ROR of 13% is referred to as the *opportunity cost;* that is, the opportunity to make an additional 13% return must be forgone. Limited capital and opportunity costs are discussed further in Chapter 17.

We will use the concept of MARR throughout. The important points now are (1) that a MARR or hurdle rate must be stated or established to evaluate a single proposal or compare alternatives, and (2) that estimated project RORs less than the MARR should be regarded as economically unacceptable. Of course, if it is decreed that one alternative will be selected, and all RORs are less than the MARR, the alternative closest to the MARR should be chosen.

As you might expect, individuals use much the same logic as that presented above, but with less specificity and structure in setting the MARR and in the definition, evaluation, and selection of an alternative. Also, individuals have substantially different dimensions than corporations for risk and uncertainty when alternatives are evaluated using ROR estimates and associated intangible factors.

Problems 1.38 to 1.41

1.9   CASH FLOWS: THEIR ESTIMATION AND DIAGRAMMING

In this first chapter, we also want to discuss one of the fundamental elements of engineering economy—*cash flows*. In Section 1.3 cash flows are described as the actual inflows and outflows of money. Every person or company has cash receipts—revenue and income (inflows); and cash disbursements—expenses, and costs (outflows). These receipts and disbursements are the cash flows, with a positive sign usually representing cash inflows and a negative sign representing cash outflows. Cash flows occur during specified periods of time, such as 1 month or 1 year.

Of all the elements of the engineering economy study approach (Figure 1–1), cash-flow estimation is likely the most difficult and inexact. Cash-flow estimates are just that—estimates about an uncertain future. Once estimated, the techniques of this book can guide you in decision making. But the time-proven accuracy of an alternative's estimated cash inflows and outflows clearly dictates the quality of the economic analysis and conclusion. Ostwald's book *Engineering Cost Estimating* (see Bibliography) is an excellent source for cost estimation.

Cash inflows, or receipts, may be comprised of the following, depending upon the nature of the proposed activity and the type of business involved. (Commonly used inflows and outflows in the initial chapters of this book are printed in bold.)

Samples of Cash Inflows

- **Revenues** (usually *incremental* due to the alternative).
- **Operating cost reductions** (due to the alternative).

- **Asset salvage value.**
- **Receipt of loan principal.**
- Income-tax savings.
- Receipts from stock and bond sales.
- Construction and facility cost savings.
- Savings or return of corporate capital funds.

Cash outflows, or disbursements, may be comprised of the following, again depending upon the nature of the activity and type of business.

Samples of Cash Outflows
- **First cost of assets** (with installation and delivery).
- **Operating costs** (annual and incremental).
- **Periodic maintenance and rebuild costs.**
- **Loan interest and principal payments.**
- Major, expected upgrade costs.
- Income taxes.
- Bond dividends and bond payment.
- Expenditure of corporate capital funds.

Background information for estimates is provided by departments such as marketing, sales, engineering, design, manufacturing, production, field services, finance, accounting, and computer services. The accuracy of estimates is largely dependent upon the experiences of the person making the estimate with similar situations. Usually *point estimates* are made; that is, a single-value estimate is developed for each economic element of an alternative. If a statistical approach to the engineering economy study is to be taken, a *range estimate* or *distribution estimate* may be developed. We will use point estimates throughout most of this book. Advanced chapters will introduce distribution estimates.

Once the cash inflow and outflow estimates are developed, the net cash flow over a stated period of time may be represented as

$$\text{Net cash flow} = \text{receipts} - \text{disbursements} \qquad \textbf{[1.7]}$$

$$= \text{cash inflows} - \text{cash outflows}$$

Since cash flows normally take place at frequent and varying time intervals within an interest period, a simplifying assumption is made that all cash flow occurs at the end of an interest period. This is known as the *end-of-period convention.* When several receipts and disbursements occur within a given interest period, the net cash flow is assumed to occur at the end of the interest period. However, it should be understood that, although F or A amounts are located at the end of the interest period by convention, the end of the period is not necessarily December 31. In Example 1.12, since investment took place on May 1, 1998, the withdrawals will take place on May 1 of each succeeding year for 10 years. *Thus, end of the period means one time period from the date of the transaction.*

A *cash-flow diagram* is simply a graphical representation of cash flows drawn on a time scale. The diagram, which represents a restatement of the situation, includes what is known and what is needed. That is, once the cash-flow diagram is complete, another person should be able to essentially work the problem by looking at the diagram. So, it is important that you understand the meaning and construction of the cash-flow diagram, since it is a valuable tool in problem solution.

Cash-flow diagram time $t = 0$ is the present, and $t = 1$ is the end of time period 1. (We will assume that the periods are in years until Chapter 3.) The time scale of Figure 1–7 is set up for 5 years. Since the end-of-year convention places cash flows at the end of years, the '1' marks the end of year one.

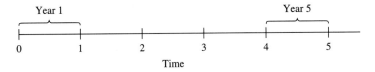

Figure 1–7   A typical cash-flow time scale for 5 years.

While it is not necessary to use an exact scale on the cash-flow diagram, you will probably avoid many errors if you make a neat diagram to approximate scale for both time and cash-flow magnitude.

The direction of the arrows on the cash-flow diagram is important. Throughout this text, a vertical arrow pointing up will indicate a positive cash flow. Conversely, an arrow pointing down will indicate a negative cash flow. The cash-flow diagram in Figure 1–8 illustrates a receipt (cash inflow) at the end of year 1 and a disbursement (cash outflow) at the end of year 2.

The perspective or vantage point must be determined prior to placing a sign on each cash flow and diagramming it. As an illustration, if you borrow $2500 from a credit union to buy a $2000 car for cash, and use the remaining $500 for

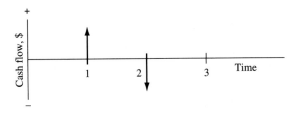

Figure 1–8   Example of positive and negative cash flows.

a paint job, there may be several different perspectives taken. The perspectives, cash-flow signs, and amounts are as follows.

Perspective	Cash Flow, $
Credit union	−2500
You as borrower	+2500
You as car purchaser, and as paint customer	−2000 −500
Used car dealer	+2000
Paint shop owner	+500

As we begin our study of engineering economy, we will routinely take the perspective of the middle entry, you as car purchaser and as paint customer. Therefore, the two negative cash flows would be constructed. When other perspectives are taken, it will be made clear.

Example 1.15

Reconsider Example 1.10, where $P = \$2000$ is borrowed and F is sought after 5 years. Construct the cash-flow diagram for this case, assuming an interest rate of 10% per year.

Solution

Figure 1–9 presents the cash-flow diagram from the vantage point of the borrower. The present sum P is a cash inflow of the loan principal at year 0, and the future sum F is the cash outflow of the principal repayment at the end of year 5.

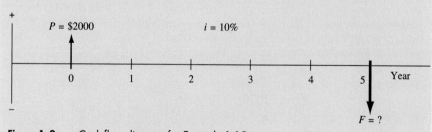

Figure 1–9   Cash-flow diagram for Example 1.15.

Example 1.16

present

Assume that Mr. Ramos starts now and makes five equal deposits of $A = \$1000$ per year into a 17%-per-year investment and withdraws the accumulated total immediately after the last deposit. Construct the cash-flow diagram.

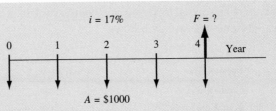

Figure 1-10   Cash-flow diagram for Example 1.16.

Solution

The cash flows are shown in Figure 1–10 from the perspective of Mr. Ramos. Since he starts now, the first deposit (negative cash flow) is at year 0. The fifth deposit and withdrawal both occur at the end of year 4 but are in opposite directions. The amount accumulated immediately after the fifth deposit is to be computed, so the future amount F is indicated by a question mark.

Example 1.17

Assume that you want to deposit an unknown amount into an investment opportunity 2 years from now that is large enough to withdraw $400 per year for 5 years starting 3 years from now. If the rate of return you expect is 15.5% per year, construct the cash-flow diagram.

Solution

Figure 1–11 presents the cash flows from your vantage point. The present value P is a cash outflow 2 years hence and is to be determined ($P = ?$). Note that this present value does not occur at time $t = 0$, but it does occur one period prior to the first A value of $400, which is your cash inflow or receipt.

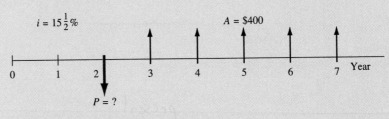

Figure 1-11   Cash-flow diagram for Example 1.17.

Additional Examples 1.19 to 1.21
Problems 1.43 to 1.51

1.10 RULE OF 72: ESTIMATING DOUBLING TIME AND INTEREST RATE

Sometimes it is important to estimate the number of years n, or the rate of return i, that is required for a single cash-flow amount to double in size. The *rule of 72* for compound-interest rates can be used to estimate i or n, given the other value. The estimation is simple; the time required for an initial single amount to double in size with compound interest is approximately equal to 72 divided by the rate of return value (in percent).

$$\text{Estimated } n = \frac{72}{i}$$

For example, at a rate of 5% per year, it would take approximately $72/5 =$ 14.4 years for a current amount to double. (The actual time required is 14.3 years, as will be shown in Chapter 2.) Table 1–4 compares the times estimated from the rule of 72 to the actual times required for doubling at several compounded rates. As you can see, very good estimates are obtained.

Alternatively, the compound rate i in percent required for money to double in a specified period of time n can be estimated by dividing 72 by the specified n value.

$$\text{Estimated } i = \frac{72}{n}$$

In order for money to double in a time period of 12 years, for example, a compound rate of return of approximately $72/12 = 6\%$ per year would be required. The exact answer is 5.946% per year.

Table 1–4	Doubling time estimates using the rule of 72 and the actual time using compound-interest calculations	
	Doubling Time, Years	
Rate of Return, % per Year	**Rule-of-72 Estimate**	**Actual Years**
1	72	70
2	36	35.3
5	14.4	14.3
10	7.2	7.5
20	3.6	3.9
40	1.8	2.0

If the interest rate is simple, a rule of 100 may be used in the same way. In this case the answers obtained will always be exactly correct. As illustrations, money doubles in exactly 12 years at $100/12 = 8.33\%$ simple interest. Or, at 5% simple interest it takes exactly $100/5 = 20$ years to double.

Problems 1.52 to 1.55

ADDITIONAL EXAMPLES

Example 1.18

INTEREST CALCULATIONS, SECTION 1.4   Last year Jane's grandmother offered to put enough money into a separate savings account to generate $1000 per year to help pay Jane's expenses at college. Calculate the amount that had to be deposited exactly 1 year ago to earn $1000 in interest now if the rate is 6% per year.

Solution

Using the logic of Equations [1.1] and [1.3], let F = total amount accrued and P = original deposit. We know that $F - P$ is the accrued interest and that $F - P = \$1000$ in this case. Now we can determine P for Jane and her grandmother.

$$F = P + P(\text{interest rate in decimal form})$$

The $1000 interest can be expressed as

$$\text{Interest} = F - P = [P + P(\text{interest rate})] - P$$

$$= P(\text{interest rate})$$

$$\$1000 = P(0.06)$$

$$P = \frac{1000}{0.06} = \$16,666.67$$

Example 1.19

CASH-FLOW DIAGRAMS, SECTION 1.9   The Hot-Air Balloon Company invested $2500 in a new air compressor 7 years ago. The annual income from the compressor has been $750. Additionally, the $100 spent on maintenance during the first year has increased each year by $25. The company plans to sell the compressor for salvage at the end of next year for $150. Construct the cash-flow diagram from the company's perspective.

Solution

Use time now as $t = 0$. The incomes and costs for years -7 through 1 (next year) are tabulated below with net cash flow computed using Equation [1.7]. The net cash flows (1 negative, 8 positive) are diagrammed in Figure 1–12.

End of Year	Income	Cost	Net Cash Flow
-7	$ 0	$2500	$-2500
-6	750	100	650
-5	750	125	625
-4	750	150	600
-3	750	175	575
-2	750	200	550
-1	750	225	525
0	750	250	500
1	750 + 150	275	625

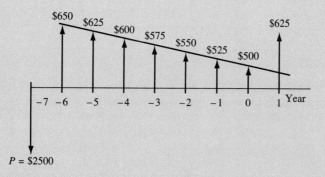

$P = \$2500$

Figure 1–12   Cash-flow diagram for Example 1.19.

Example 1.20

CASH-FLOW DIAGRAMS, SECTION 1.9   Suppose you hope to make an investment of amount P now such that you can withdraw an equal annual amount of $A_1 = \$200$ per year for the first 5 years starting 1 year after your deposit and a different annual withdrawal of $A_2 = \$300$ per year for the following 3 years. How would the cash-flow diagram appear if $i = 14.5\%$ per year?

Solution

The cash flows are shown in Figure 1–13. Your negative cash outflow P occurs now. The first withdrawal (positive cash inflow) for the A_1 series occurs at the end of year 1, and A_2 occurs in years 6 through 8.

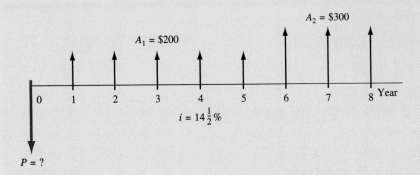

Figure 1–13   Cash-flow diagram with two different A series, Example 1.20.

Example 1.21

CASH-FLOW DIAGRAMS, SECTION 1.9   If Melissa buys a big-screen TV with Internet connection for $900 one year from now, maintains it for 3 years at a cost of $50 per year, and then sells it for $200, diagram the cash flows. Label each cash flow as F or A with its respective dollar value. Also, locate and label the present-worth amount P that is equivalent to all the cash flows shown. Assume an interest rate of 12% per year.

Solution

Figure 1–14 presents the cash-flow diagram. The two $50 negative cash flows form a series of two equal end-of-year A values. As long as the amounts are equal and in two or more consecutive periods, they may be an A series, regardless of where they begin or end. However, the $150 positive cash flow ($200 − 50) in year 4 is a different amount and is labeled as a single-occurrence future F value.

Comment

It is equally correct to diagram all the individual cash flows in years 1 through 4 as separate F values, as shown in Figure 1–15. Usually, if two or more equal end-of-period amounts occur consecutively, they should be labeled A values because, as you will discover in Chapter 2, the use of A values wherever possible simplifies calculations considerably. Therefore, the interpretation in Figure 1–15 is discouraged and is not generally used.

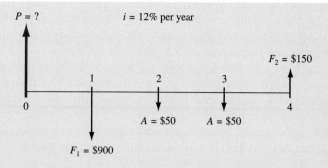

Figure 1-14   Cash-flow diagram for Example 1.21.

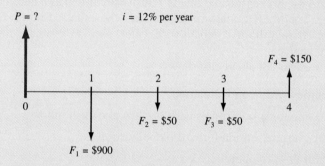

Figure 1-15   An alternate cash-flow diagram for Example 1-21 considering all values as future amounts.

CHAPTER SUMMARY

We have covered a variety of fundamental engineering economy topics in this initial chapter. We have learned that the engineering economy approach and mathematical computations are helpful to individuals, as well as business, industry, and government, in evaluating alternatives described in economic terms.

Engineering economy is the application of economic factors and criteria to evaluate alternatives considering the time value of money by computing a specific economic measure of worth of estimated cash flows over a specific period of time.

The engineering economy study approach (Figure 1–1) plays an important role in the problem-solving process. The study approach is the primary method by which economic-based analysis is accomplished leading to the selection and implementation of a recognized and detailed alternative.

The very basic concept of *equivalence* helps us understand how different sums of money at different times are equal in economic terms at an interest rate that is compounded each interest period. And the differences between simple interest (based on principal only) and compound interest (based on principal and interest upon interest) have been described in formulas, tables, and graphs. This power of compounding is very noticeable, especially over long periods of time.

Basic engineering economy computations involve only five symbols: P, A, F, n, and i. We will develop relations between these to account for the time value of money as we move into the next chapter.

The MARR is a reasonable rate of return established as a hurdle rate for an alternative. The MARR is always higher than the expected return from a safe investment. It may be a companywide value used to evaluate most capital investment projects.

Also, we learned about cash flows:

- Difficulties with their estimation.
- What commonly constitutes cash outflow and cash inflow.
- End-of-year convention for cash-flow location.
- Net cash-flow computation.
- Different perspectives when determining the cash-flow sign.
- How to construct a correct cash-flow diagram.

The rule of 72 is used in two ways: to estimate the years to double an amount for a stated compound interest rate ($72/i$) or to estimate the rate at which the amount is doubled after a stated number of years ($72/n$).

CASE STUDY #1 CHAPTER 1:   CANNED SOUP DISTRIBUTION

Branndon and Julia are relatively new employees of the Service Quality Division of the management consulting firm Qupelt, Inc. They have been teamed with a senior consultant, Xavier, and the three have just exited a briefing by First Choice Soups' Vice President for Distribution and two regional distribution managers. The team has been asked to assist First Choice in an economic analysis of distribution system improvements in association with a nationwide TV campaign to boost sales this coming winter season.

The briefing was about several concepts that the soup company's distribution managers have discussed to significantly reduce the time it takes to deliver soup cans from the end of the canning line to the grocery store shelf for display. This reduction is becoming much more important because First Choice President Jane Hinwhin has told marketing to start an aggressive advertising program this next winter (at high soup season) that promotes a particular flavor of soup on TV one evening with the soup on the shelves in adequate supply throughout the country the next morning for shoppers. This is all to be accomplished, says the Distribution Vice President, without an increase in the performance measure of delivery and storage costs as a percentage of sales revenue, thus improving the total revenue of the company for the advertised flavors. This performance measure is currently collected and reported monthly on a nationwide basis for all the company's major market areas.

The basic concepts described to the consulting team have been recorded by Xavier in rough outline form as follows:

Option 1: Temporary space. First Choice leases temporary storage space in selected, high-sales cities in order to meet projected demand.

Option 2: Contractor revenue sharing. Offer contract truckers and warehousers a percentage of the increased sales revenue for them to make the increased supply available at the supermarkets they currently supply.

Option 3: Supermarket takes lead. Inform the supermarkets of which flavors are to be advertised well in advance and make them responsible for stocking and storing through regular First Choice distribution channels the increased amount they expect to sell.

Even though precise project alternatives are difficult to develop due to the sketchy information provided by the Vice President and managers, the team is now expected to develop the alternatives and perform an economic analysis, taking into account to a reasonable degree intangible factors which may make this campaign a success.

Branndon, Julia, and Xavier must initially prepare a plan for First Choice of how to formulate the alternatives and how to use the problem-solving process and economic analysis to approach the analysis. Prepare this report. Consider elements such as alternative description, the various estimates which the team must develop or obtain from First Choice, intangible factors which are pertinent, and use of the engineering economy approach. Also, be prepared to make a brief verbal report of the recommended approach.

PROBLEMS

1.1   What is meant by the term *time value of money*?

1.2   You meet a friend on a bus to the beach and tell her you are taking a course in engineering economy. She asks what it is. What do you answer?

1.3   List at least three criteria which might be used besides money for evaluating each of the following: (*a*) quality of the service and food in a neighborhood restaurant; (*b*) flight on a commercial aircraft; (*c*) an apartment where you might sign a 1-year lease.

1.4   Describe the concept of equivalence in a way that your brother-in-law can understand it. He has a degree in psychology and works as a personnel counselor in the human resources department of a large corporation.

1.5   Write one-half to one page on your current understanding of how engineering economy is best used in a decision-making process which usually involves both economic and noneconomic factors.

1.6   Describe two separate situations in your life in which you have made a decision and a significant amount of money was involved. To the degree possible, make one situation an analysis of observed outcomes and the other one about deciding on a future action, as described in Section 1.2.

1.7   Explain how you would use the problem-solving approach to consider economic and noneconomic factors for the following: You and three friends want to go on a long trip for the next spring break. There are currently three alternatives: a cruise to the Caribbean, a ski trip to a new mountain resort, and a camping trip to a desert wilderness park.

1.8   Assume you are president of the student chapter of your professional society this year. List what you think are the most important economic and intangible factors to apply in deciding between two alternatives recommended by your executive committee: (1) the traditional, semester-end banquet for members and invited faculty guests or (2) a more informal, end-of-semester evening meeting with the faculty to discuss and evaluate the quality of education in the department. Snacks will be served at the chapter's expense at the informal discussion. You do not plan to hold both events; only one or neither.

1.9   Consider the following situations and determine if each one is suitable or not suitable for solution using the engineering economy study approach. Explain your answer.

(*a*) Decide if two new machines should be leased to replace five currently owned machines. Current employees can work on either machine.

(*b*) Determine whether to live on-campus in a residence hall with a friend from high school or live off-campus with three new friends.

(c) Decide between two different mortgage strategies for your first house: 15-year mortgage or 30-year mortgage, if the 15-year interest rate is 1% lower.

(d) Decide to keep your major in Engineering or switch to Business.

(e) Lease or buy an automobile.

(f) Pay off a student credit card balance which has a special low 14% per year interest rate, or pay the minimum and vow to invest the remaining amount each month with an expected return between 10% and 15% per year.

1.10 Explain the term *measure of worth,* and discuss its role in an engineering economy study.

1.11 Assume you are the president of an international manufacturing corporation and have just been presented with a proposal to subcontract all the engineering design work in the U.S. and European plants. Currently the corporation does all the design work using its own engineering and technology personnel. What are five of the major elements and factors that you would use as the bases for the decision? Assume that additional information can be generated based upon the factors you identify here before you make the final decision.

1.12 Jules borrowed $1000 from a bank and paid 12%-per-year compounded semiannually. He repaid the loan in six equal payments of $203.36 each. Determine the total dollar amount of interest he paid, and determine what percent of the original loan this interest represents.

1.13 Cheryl collected advertised loan rates from three places. They are 10% per year compounded semiannually, 11% per year compounded quarterly, and 11.5% per year. State the interest period in months for each rate.

1.14 Explain the terms *interest, interest rate,* and *interest period.*

1.15 Calculate the amount of interest payable after 1 year on a loan of $5000 if the interest is 8% per year. What is the interest period?

1.16 What was the loan amount if the interest rate is 1.5% per month payable monthly and the borrower just made the first monthly payment of $25 in interest?

1.17 Which of the following has a better rate of return: $200 invested for 1 year with $6.25 paid in interest or $500 invested for 1 year with $18 paid in interest?

1.18 Give a simple numerical example which demonstrates the concept of equivalence that is understandable to someone who is your age but does not know the principles of engineering economy or finance.

1.19 At what annual interest rate are $450 a year ago and $550 one year from now equivalent?

1.20 University tuition and fees can be paid using one of three plans:

> **On-time.**   Total amount on the first day of a semester
>
> **Pay-later.**   Total amount plus 2% two weeks after classes start
>
> **Early-bird.**   Get a 2% discount to register and pay 2 weeks before classes start

(*a*) If a student's on-time bill is $1200, determine the equivalent pay-later and early-bird amounts. (*b*) What total difference in dollars can a student experience between the early-bird and pay-later plans for a $1200 bill? What percentage of the $1200 does this represent? Why is the percentage not exactly 4%?

1.21 Joan bought a compact disc (CD) player for $399 two years ago, John bought the same model last year on sale for $438, and Carol wants to buy one this year for an equivalent amount. (*a*) What should Carol pay? (*b*) If the percentage increase is an estimate of the annual inflation rate in CD player prices, what is this estimated inflation rate?

1.22 Starburst, Inc. invested $50,000 in a foreign co-venture just 1 year ago and has reported a profit of $7500. What is the annual rate that the investment is returning?

1.23 Which is a better investment opportunity: $1000 at 7%-per-year simple interest for 3 years, or $1000 at 6%-per-year compound interest for 3 years?

1.24 (*a*) How much total interest would you pay if you borrowed $1500 for 3 months at a rate of ¾% per month compounded monthly? (*b*) What percentage of the original loan is this interest amount?

1.25 Resolve Problem 1.24 for ¾%-per-month simple interest.

1.26 A newly married couple and the groom's parents each bought new furniture which they won't have to pay for with interest for some months. The newlyweds' purchase price is $3000 with simple interest at 12% per year, and one delayed payment of principal and interest is due in 6 months. The parents' purchase price is also $3000 with interest at 9% per year compounded monthly, and one delayed payment is due in 13 months. Determine the accumulating interest by month and determine the total payment for each couple. Who paid more and by how much?

1.27 How much money will your friend have after 4 years if she saves $1000 now at 7% per year simple interest?

1.28 How much can you borrow today if you pay $850 two years from now at an interest rate of 6% per year compounded yearly?

1.29 If you borrow $1500 now and must repay $1850 two years from now, what is the loan's annual interest rate?

1.30 You have just invested $10,000 in a friend's business venture that promises to return $15,000 or more at some time in the future. What is the minimum number of years (whole number) that you can wait to receive the $15,000 in order to make 10% or more per year compounded yearly?

1.31 If you invest $3500 now in return for a guaranteed $5000 income at a later date, when must you receive your money in order to earn exactly 8% per year simple interest?

1.32 A colleague tells you that she just repaid principal and interest for a loan she took out 3 years ago at 10%-per-year simple interest. If her payment was $1950, determine the principal.

1.33 At 9% per year simple interest $1000 is equivalent to $1270 in 3 years. Find the compound-interest rate per year for this equivalence to be correct.

1.34 Compute, graph, and compare the annual interest and total interest amount for 10 years on a million dollars under two different scenarios. First, the $1 million is borrowed by a company at 6%-per-year simple interest. Second, the $1 million is invested in a company at 6% per year compounded annually.

1.35 Write the relevant engineering economy symbols and values for the following: A total of five deposits of $2000 each are made every 2 years starting next year at 10% per year. What is the total accrued amount to be withdrawn exactly when the last deposit is made?

1.36 Describe the economy symbols and their values for the following plan: Dr. Raffle hopes to borrow $800 now and repay it at $100 per year for the next 5 years and $200 per year for the following 2 years. What is the interest rate?

1.37 Define the economy symbols for a problem to determine how many years it would take for $5000 to double in size at a compound-interest rate of 5.5% per year.

1.38 An engineering graduate bought 200 shares of a company's common stock at $52.00 each and sold it 4 years later, after commissions, for a total of $15,010. (*a*) What was the 4-year rate of return? (*b*) What was the annual, simple-interest rate of return?

1.39 Why is it commonly accepted that the MARR for a corporation will be greater than the rate of return obtainable from a bank or other safe investment such as U.S. Treasury bills?

1.40 Assume you want a MARR of 5% per year compounded annually on your investment in a college education and that you expect to receive an annual salary of at least $60,000 ten years after the year you graduate. What is the equivalent amount you could have invested in a college education during just your last year in college?

1.41 Explain the term *capital* (or investment capital) and give two examples from your own experiences in which capital had to be found. Explain how the capital funds were actually generated.

1.42 Determine your own cash inflows and outflows for a 3-month period and keep track of their size. Plot them on a monthly cash-flow diagram. (Instructor: This is an excellent semester-long assignment to discuss in general during class, but it should not be a homework assignment to be submitted for grading.)

1.43 Construct the cash-flow diagrams for Examples 1.8, 1.10, and 1.12.

1.44 Construct the cash-flow diagram of yearly net-cash flows for Ms. Jameson, an investment manager, who developed this plan for a client: Invest $5000 immediately and every other year through year 10 from now. Then, plan to withdraw $3000 every year starting 5 years from now and continuing for 8 additional years.

1.45 Construct individual cash-flow diagrams for the five end-of-year payment plans presented in Table 1–3.

1.46 Your uncle has offered to make five $700-per-year deposits into an account in your name starting now. You have agreed not to withdraw any money until the end of year 9, at which time you plan to remove $3000. Further, you plan to withdraw the remaining amount in three equal year-end installments after the initial withdrawal to zero-out the account. Indicate the cash flows for your uncle and yourself on one diagram.

1.47 Jaime wishes to invest at an 8%-per-year return so that 6 years from now he can withdraw an amount F in a lump sum. He has developed the following alternative plans. (*a*) Deposit $350 now and again 3 years from now. (*b*) Deposit $125 per year starting next year and ending in year 6. Draw the cash-flow diagram for each plan if F is to be determined in year 6.

1.48 Draw a cash-flow diagram for the following: Deposit $100 per year starting 1 year from now. Withdraw the entire amount 15 years from now. The expected earning rate is 10% per year.

1.49 Construct a cash-flow diagram which will assist in calculating the current equivalent value of an expenditure of $850 per year for 6 years that starts 3 years from now, if the interest rate is 13% per year.

1.50 Define the economy symbols and draw the cash-flow diagram for the following: Invest $100,000 now in a real-estate venture, sell the property 10 years from now, and make a 12%-per-year return on your investment.

1.51 Develop a cash-flow diagram for the following situation: Equal-amount payments for 4 years beginning 1 year from now are equivalent to spending $4500 now, $3300 three years from now, and $6800 five years from now if the interest rate is 8% per year.

1.52 At a rate of 8.5% per year, estimate the time it takes to double $500 if interest is (*a*) compounded and (*b*) not compounded. (*c*) How many years will it take to double $1000 at 8.5% per year compounded annually?

1.53 Joey hopes to purchase a boat in 5 years and believes he must double the size of the stock portfolio set aside as his "boat kitty." Estimate the rate of return at which his portfolio must grow if interest is (*a*) simple and (*b*) compound.

1.54 Clarissa went to work several years ago and has placed all the employer's contributions to her retirement fund into an investment that is currently returning exactly 12% per year. All interest is reinvested into the retirement account.

(*a*) What is the doubling time for each dollar contributed to Clarissa's retirement fund?

(*b*) A special program allows employees to borrow against the current value of their retirement fund, but inflation is considered and reduces the fund's value for loan purposes. Clarissa has a value of $30,000 now. If she applies for a loan when the amount is doubled to $60,000, what is the maximum amount she can borrow against her retirement fund? Assume the current 12% compounded return continues and that inflation is estimated to be 4% per year compounded annually.

1.55 Choose one or more things that you have learned in this chapter and set up, and solve, an engineering economy problem of your own.

Factors and Their Use

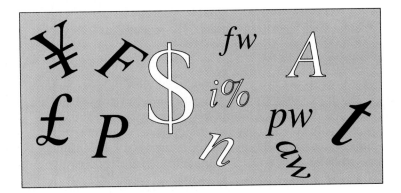

In this chapter we learn the derivation of the engineering economy factors and the use of these basic factors in economic computations. This chapter is one of the most important, since the concepts presented here are relied upon throughout the text.

Purpose: Understand the derivation of the engineering economy formulas and how they are used.

This chapter will help you:

P/F and *F/P* factors	1. Derive the single-payment compound-amount and present-worth factors.
P/A and *A/P* factors	2. Derive the uniform-series present-worth and capital-recovery factors.
F/A and *A/F* factors	3. Derive the uniform-series compound-amount and sinking-fund factors.
Use tables	4. Find the correct factor value in a table.
P/G and *A/G* factors	5. Derive the uniform-gradient present-worth and annual-series factors.
Geometric gradient	6. Derive the geometric (escalating) gradient formula.
Interpolate	7. Linearly interpolate to find a factor value.
P, *F*, or *A* of cash flows	8. Calculate the present worth, future worth, or annual worth of various cash flows.
P, *F*, or *A* of a uniform gradient	9. Calculate the present worth, future worth, or annual worth of cash flows involving a uniform gradient.
P, *F*, or *A* of a geometric gradient	10. Calculate the present worth, future worth, or annual worth of cash flows involving a geometric gradient.
Calculate *i*	11. Calculate the interest rate (rate of return) of a sequence of cash flows.
Calculate *n*	12. Determine the number of years *n* required for equivalence for a sequence of cash flows.

2.1   DERIVATION OF SINGLE-PAYMENT FACTORS (*F/P* AND *P/F*)

In this section, a formula is developed which allows determination of the future amount of money F that is accumulated after n years (or periods) from a *single* investment P when interest is compounded one time per year (or period). As in Chapter 1, an interest period of 1 year will be assumed. However, you should recognize that the i and n symbols in the formulas developed here apply to *interest periods,* not just years, as will be discussed in Chapter 3.

You will recall from Chapter 1 that compound interest refers to interest paid on top of interest. Therefore, if an amount of money P is invested at some time $t = 0$, the amount of money F_1 that will be accumulated 1 year hence at an interest rate of i percent per year will be

$$F_1 = P + Pi$$
$$= P(1 + i)$$

At the end of the second year, the amount of money accumulated F_2 is the amount that accumulated after year 1 plus the interest from the end of year 1 to the end of year 2. Thus,

$$F_2 = F_1 + F_1 i \tag{2.1}$$
$$= P(1 + i) + P(1 + i)i$$

This may be written as

$$F_2 = P(1 + i + i + i^2)$$
$$= P(1 + 2i + i^2)$$
$$= P(1 + i)^2$$

Similarly, the amount of money accumulated at the end of year 3, using Equation [2.1], will be

$$F_3 = F_2 + F_2 i$$

Substituting $P(1 + i)^2$ for F_2 and simplifying,

$$F_3 = P(1 + i)^3$$

From the preceding values, it is evident by mathematical induction that the formula can be generalized for n years to

$$F = P(1 + i)^n \tag{2.2}$$

The factor $(1 + i)^n$ is called the single-payment compound-amount factor (SPCAF), but it is usually referred to as the *F/P factor.* When the factor is multiplied by P, it yields the future amount F of an initial investment P after n years at interest rate i. Solving for P in Equation [2.2] in terms of F results

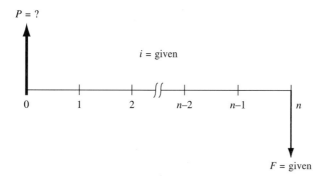

Figure 2–1   Cash-flow diagram to determine P, given F.

in the expression

$$P = F\left[\frac{1}{(1 + i)^n}\right] \qquad [2.3]$$

The expression in brackets is known as the *single-payment present-worth factor* (SPPWF), or the *P/F factor*. This expression determines the present worth P of a given future amount F after n years at interest rate i. The cash-flow diagram for this formula is shown in Figure 2–1. Conversely, the diagram to find F, given P, would be exactly the same with the ? and *given* terms interchanged and Equation [2.2] used to calculate F.

It is important to note that the two factors and formulas derived here are *single-payment formulas;* that is, they are used to find the present or future amount when only one payment or receipt is involved. In the next two sections, formulas are developed for calculating the present or future worth when several uniform payments or receipts are involved.

2.2   DERIVATION OF THE UNIFORM-SERIES PRESENT-WORTH FACTOR AND THE CAPITAL-RECOVERY FACTOR (*P/A* AND *A/P*)

The present worth P of the uniform series shown in Figure 2–2 can be determined by considering each A value as a future worth F and using Equation [2.3] with the *P/F* factor and then summing the present-worth values. The general formula is

$$P = A\left[\frac{1}{(1 + i)^1}\right] + A\left[\frac{1}{(1 + i)^2}\right] + A\left[\frac{1}{(1 + i)^3}\right] + \cdots$$

$$+ A\left[\frac{1}{(1 + i)^{n-1}}\right] + A\left[\frac{1}{(1 + i)^n}\right]$$

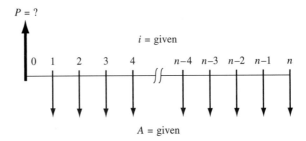

Figure 2-2   Diagram used to determine the present worth of a uniform series.

where the terms in brackets represent the *P/F* factors for years 1 through *n*, respectively. Factoring out *A*,

$$P = A\left[\frac{1}{(1 + i)^1} + \frac{1}{(1 + i)^2} + \frac{1}{(1 + i)^3} + \cdots \right.$$
$$\left. + \frac{1}{(1 + i)^{n-1}} + \frac{1}{(1 + i)^n}\right] \qquad \text{[2.4]}$$

Equation [2.4] may be simplified by multiplying both sides by $1/(1 + i)$ to yield

$$\frac{P}{1 + i} = A\left[\frac{1}{(1 + i)^2} + \frac{1}{(1 + i)^3} + \frac{1}{(1 + i)^4} + \cdots \right.$$
$$\left. + \frac{1}{(1 + i)^n} + \frac{1}{(1 + i)^{n+1}}\right] \qquad \text{[2.5]}$$

Subtracting Equation [2.4] from Equation [2.5], simplifying, and then dividing both sides of the relation by $-i/(1 + i)$ leads to an expression for *P* when $i \neq 0$

$$P = A\left[\frac{(1 + i)^n - 1}{i(1 + i)^n}\right] \qquad i \neq 0 \qquad \text{[2.6]}$$

The term in brackets is called the *uniform-series present-worth factor* (US-PWF), or the *P/A factor*. This equation will give the present worth *P* of an equivalent uniform annual series *A* which begins at the end of year 1 and extends for *n* years at an interest rate *i*.

The *P/A* factor in brackets in Equation [2.6] may also be determined by treating Equation [2.4] as a geometric progression, which has a general form for its closed-end sum *S* of

$$S = \frac{(\text{last term})(\text{common ratio}) - \text{first term}}{\text{common ratio} - 1}$$

The common ratio between terms is $1/(1 + i)$. For simplification purposes, set $y = 1 + i$ and form the S expression above and simplify.

$$S = \frac{1/y^n y - 1/y}{1/y - 1}$$

$$= \frac{y^n - 1}{iy^n}$$

$$= \frac{(1 + i)^n - 1}{i(1 + i)^n}$$

By rearranging Equation [2.6], we can express A in terms of P:

$$A = P\left[\frac{i(1 + i)^n}{(1 + i)^n - 1}\right] \tag{2.7}$$

The term in brackets, called the *capital-recovery factor* (CRF), or *A/P factor,* yields the equivalent uniform annual worth A over n years of a given investment P when the interest rate is i.

It is very important to remember that these formulas are derived with the present worth P and the first uniform annual worth amount A *one year (period) apart.* That is, the present worth P *must always* be located *one period prior* to the first A. The correct use of these factors is illustrated in Section 2.7.

2.3   DERIVATION OF THE SINKING-FUND FACTOR AND THE UNIFORM-SERIES COMPOUND-AMOUNT FACTOR (*A/F* AND *F/A*)

While the *sinking-fund factor* (SFF), or *A/F* factor, and the *uniform-series compound-amount factor* (USCAF), or *F/A* factor, could be derived using the *F/P* factor, the simplest way to derive the formulas is to substitute into those already developed. Thus, if P from Equation [2.3] is substituted into Equation [2.7], the following formula results:

$$A = F\frac{1}{(1 + i)^n}\frac{i(1 + i)^n}{(1 + i)^n - 1} \tag{2.8}$$

$$= F\left[\frac{i}{(1 + i)^n - 1}\right]$$

The expression in brackets in Equation [2.8] is the sinking-fund, or *A/F*, factor. Equation [2.8] is used to determine the uniform annual worth series that would be equivalent to a given future worth F. This is shown graphically in Figure 2–3. Note that the uniform series A begins at the end of period 1 and continues *through the period of the given F.*

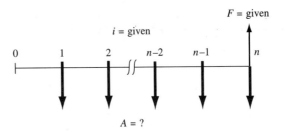

Figure 2–3   Transformation of a given *F* value into an equivalent *A* series.

Equation [2.8] can be rearranged to express *F* in terms of *A*:

$$F = A\left[\frac{(1 + i)^n - 1}{i}\right]$$ **[2.9]**

The term in brackets is called the *uniform-series compound-amount factor* (USCAF), or *F/A* factor, which, when multiplied by the given uniform annual amount *A*, yields the future worth of the uniform series. The cash-flow diagram for this case would appear the same as in Figure 2–3, except that *A* is given and *F* = ?. Again, it is important to remember that the future amount *F* occurs in the same period as the last *A*.

As an exercise, show that the *F/A* factor can be obtained by multiplying the formulas from the *F/P* factor in Equation [2.2] and the *P/A* factor in Equation [2.6] for a given *i* and *n*, that is, *F/A* = (*F/P*)(*P/A*) in factor terms.

Problem 2.1

2.4   STANDARD FACTOR NOTATION AND USE OF INTEREST TABLES

We introduced shortened terms for each factor as it was derived. These are used to avoid the cumbersome task of writing out the formulas each time one of the factors is used. A standard notation has been adopted which includes the interest rate and the number of periods and is always in the general form (*X/Y,i,n*). The first letter *X* inside the parentheses represents what you want to find, while the second letter *Y* represents what is given. For example, *F/P* means *find F when given P*. The *i* is the interest rate in percent, and the *n* represents the number of periods involved. Thus, (*F/P*,6%,20) means obtain the factor which when multiplied by a given *P* allows you to find the future amount of money *F* that will be accumulated in 20 periods if the interest rate is 6% per period.

The standard notation is simpler than factor names for identifying factors and will be used exclusively hereafter. Table 2–1 shows the standard notation

Table 2–1	Standard factor notations
Factor Name	**Standard Notation**
Single-payment present-worth	$(P/F,i,n)$
Single-payment compound-amount	$(F/P,i,n)$
Uniform-series present-worth	$(P/A,i,n)$
Capital-recovery	$(A/P,i,n)$
Sinking-fund	$(A/F,i,n)$
Uniform-series compound-amount	$(F/A,i,n)$

Table 2–2	Computations using standard notation			
To Find	**Given**	**Factor**	**Equation**	**Formula**
P	F	$(P/F,i,n)$	$P = F(P/F,i,n)$	$P = F[1/(1 + i)^n]$
F	P	$(F/P,i,n)$	$F = P(F/P,i,n)$	$F = P(1 + i)^n$
P	A	$(P/A,i,n)$	$P = A(P/A,i,n)$	$P = A\{[(1 + i)^n - 1]/i(1 + i)^n\}$
A	P	$(A/P,i,n)$	$A = P(A/P,i,n)$	$A = P\{i(1 + i)^n/[(1 + i)^n - 1]\}$
A	F	$(A/F,i,n)$	$A = F(A/F,i,n)$	$A = F\{i/[(1 + i)^n - 1]\}$
F	A	$(F/A,i,n)$	$F = A(F/A,i,n)$	$F = A\{[(1 + i)^n - 1]/i\}$

for the formulas derived thus far. For ready reference the formulas used in computations are collected in Table 2–2, and they are shown inside the front cover of the text. It is also easy to use the standard notation to remember how the factors may be derived. For example, the A/F factor may be derived by multiplying the P/F and A/P factor formulas. In equation form, this is,

$$A = F(P/F,i,n)(A/P,i,n)$$

$$= F(A/F,i,n)$$

The equivalent of algebraic cancellation of the P makes this relation easier to remember.

In order to simplify the routine engineering economy calculations involving the factors, tables of factor values have been prepared for interest rates from 0.25 to 50% and time periods from 1 to large n values, depending on the i value. These tables, found at the rear of the book, are arranged with various factors across the top and the number of periods n down the left column. The word *discrete* in the title of each table is printed to emphasize that these tables are for factors which utilize the end-of-period convention (Section 1.9) and that interest is compounded once each interest period. For a given factor, interest rate, and time, the correct factor value is found in the respective interest-rate table at the intersection of the given factor and n. For example, the value of the factor $(P/A,5\%,10)$

Table 2–3	Examples of interest table values			
Standard Notation	i	n	Table	Factor Value
$(F/A,10\%,3)$	10	3	15	3.3100
$(A/P,7\%,20)$	7	20	12	0.09439
$(P/F,25\%,35)$	25	35	25	0.0004

is found in the P/A column of Table 10 at period 10 as 7.7217. The value 7.7217 could, of course, have been computed using the mathematical expression for this factor in Equation [2.6].

$$(P/A,5\%,10) = \frac{(1+i)^n - 1}{i(1+i)^n}$$

$$= \frac{1.05^{10} - 1}{0.05(1.05)^{10}}$$

$$= 7.7217$$

Table 2–3 presents several examples of the use of the interest tables.

Problem 2.2

2.5   DEFINITION AND DERIVATION OF GRADIENT FORMULAS

A *uniform gradient* is a *cash-flow series* which either increases or decreases uniformly. That is, the cash flow, whether income or disbursements, changes by the same arithmetic amount each interest period. The *amount* of the increase or decrease is the *gradient*. For example, if an automobile manufacturer predicts that the cost of maintaining a robot will increase by $500 per year until the machine is retired, a gradient series is involved and the amount of the gradient is $500. Similarly, if the company expects income to decrease by $3000 per year for the next 5 years, the decreasing income represents a negative gradient in the amount of $3000 per year.

The formulas previously developed for uniform-series cash flows were generated on the basis of year-end amounts of equal value. In the case of a gradient, each year-end cash flow is different, so a new formula must be derived. To do so, it is convenient to assume that the cash flow that occurs at the end of year (period) 1 is not part of the gradient series but is rather a *base amount*. This is convenient because in actual applications, the base amount is usually larger or

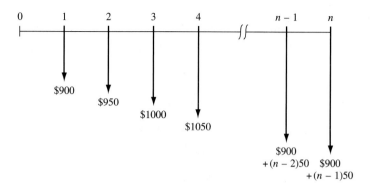

Figure 2–4   Diagram of a uniform-gradient series with a gradient of $50.

smaller than the gradient increase or decrease. For example, if you purchase a used car with a 1-year 12,000-mile warranty, you might reasonably expect to have to pay for only the gasoline during the first year of operation. Let us assume that this cost is $900; that is, $900 is the base amount. After the first year, however, you would have to absorb the cost of repair or replacement yourself, and these costs could reasonably be expected to increase each year that you own the car. So if you estimate that your operating and repair costs will increase by $50 each year, the amount you would pay after the second year would be $950, after the third $1000, and so on to year n, when the total cost would be $900 + (n - 1)50$. The cash-flow diagram for this is shown in Figure 2–4. Note that the gradient ($50) is first observed between year 1 and year 2, and the base amount ($900) is not equal to the gradient. We define the symbol G for gradients as

G = uniform arithmetic change in the magnitude of receipts or
      disbursements from one time period to the next

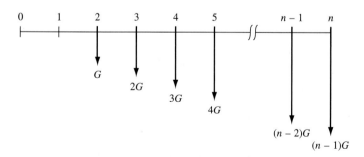

Figure 2–5   Uniform-gradient series ignoring the base amount.

The value of G may be positive or negative. If we ignore the base amount, we can construct a generalized uniformly increasing gradient cash-flow diagram as shown in Figure 2–5. Note that the gradient begins between years 1 and 2. This is called a *conventional gradient*.

Example 2.1

The Free Spirit Company expects to realize a revenue of $47,500 next year from the sale of its soda product. However, sales are expected to increase uniformly with the introduction of a new drink to a level of $100,000 in 8 years. Determine the gradient and construct the cash-flow diagram.

Solution

The base amount is $47,500 and the revenue gain is

$$\text{Revenue gain in 8 years} = 100,000 - 47,500 = \$52,500$$

$$\text{Gradient} = \frac{\text{gain}}{n - 1}$$

$$= \frac{52,500}{8 - 1} = \$7500 \text{ per year}$$

The cash-flow diagram is shown in Figure 2–6.

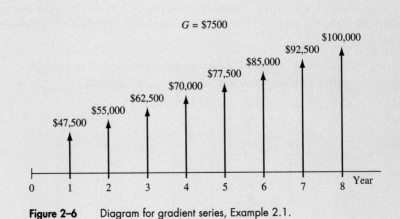

Figure 2–6   Diagram for gradient series, Example 2.1.

There are several ways by which the uniform-gradient factors can be derived. We will use the single-payment present-worth factor ($P/F,i,n$), but the same result can be obtained using the F/P, F/A, or P/A factor.

Referring to Figure 2–5, we find that the present worth at year 0 of the gradient payment is equal to the sum of the present worths of the individual payments.

$$P = G(P/F,i,2) + 2G(P/F,i,3) + 3G(P/F,i,4) + \cdots$$
$$+ [(n - 2)G](P/F,i,n - 1) + [(n - 1)G](P/F,i,n)$$

Factoring out G yields

$$P = G[(P/F,i,2) + 2(P/F,i,3) + 3(P/F,i,4) + \cdots$$
$$+ (n - 2)(P/F,i,n - 1) + (n - 1)(P/F,i,n)]$$

Replacing the symbols with the appropriate P/F factor expression in Equation [2.3] yields

$$P = G\left[\frac{1}{(1 + i)^2} + \frac{2}{(1 + i)^3} + \frac{3}{(1 + i)^4} + \cdots \right. \tag{2.10}$$
$$\left. + \frac{n - 2}{(1 + i)^{n-1}} + \frac{n - 1}{(1 + i)^n}\right]$$

Multiplying both sides of Equation [2.10] by $(1 + i)^1$ yields

$$P(1 + i)^1 = G\left[\frac{1}{(1 + i)^1} + \frac{2}{(1 + i)^2} + \frac{3}{(1 + i)^3} + \cdots \right. \tag{2.11}$$
$$\left. + \frac{n - 2}{(1 + i)^{n-2}} + \frac{n - 1}{(1 + i)^{n-1}}\right]$$

Subtracting Equation [2.10] from Equation [2.11] and then simplifying yields

$$P = \frac{G}{i}\left[\frac{(1 + i)^n - 1}{i(1 + i)^n} - \frac{n}{(1 + i)^n}\right] \tag{2.12}$$

Equation [2.12] is the general relation to convert a uniform gradient G (not including the base amount) for n years into a present worth at year 0; that is, Figure 2–7a is converted into the equivalent cash flow shown in Figure 2–7b. The *uniform-gradient present-worth factor,* or *P/G factor,* may be expressed as follows in two equivalent forms:

$$(P/G,i,n) = \frac{1}{i}\left[\frac{(1 + i)^n - 1}{i(1 + i)^n} - \frac{n}{(1 + i)^n}\right]$$
$$= \frac{(1 + i)^n - in - 1}{i^2(1 + i)^n}$$

Note that the gradient starts in year 2 in Figure 2–7a and P is located in year 0. Equation [2.12] is represented in standard factor notation as

$$P = G(P/G,i,n)$$

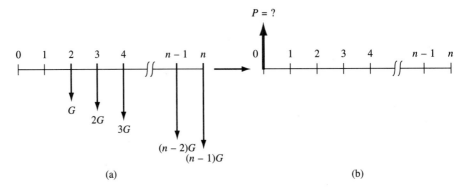

Figure 2–7   Conversion diagram from a uniform gradient to a present worth.

The equivalent uniform annual worth of a uniform gradient G is found by multiplying the present worth in Equation [2.12] by the $(A/P,i,n)$ factor expression in Equation [2.7]. Using standard factor notation,

$$A = G(P/G,i,n)(A/P,i,n)$$
$$= G(A/G,i,n)$$

In the standard form, the equivalent of algebraic cancellation of P can be used to obtain the $(A/G,i,n)$ factor. In equation form,

$$A = \frac{G}{i}\left[\frac{(1+i)^n - 1}{i(1+i)^n} - \frac{n}{(1+i)^n}\right]\left[\frac{i(1+i)^n}{(1+i)^n - 1}\right] \qquad \text{[2.13]}$$
$$= G\left[\frac{1}{i} - \frac{n}{(1+i)^n - 1}\right]$$

The expression in brackets of the simplified form in Equation [2.13] is called the *uniform-gradient annual-worth factor* and is identified by $(A/G,i,n)$. This factor converts Figure 2–8a into Figure 2–8b. Realize that the annual worth is nothing but an A value equivalent to the gradient (not including the base amount). Note from Figure 2–8 that the gradient starts in year 2 and the A values occur from year 1 to year n inclusive.

In standard factor notation, the formulas used to compute P and A from uniform- or arithmetic-gradient cash flows are

$$P = G(P/G,i,n) \qquad \text{[2.14]}$$
$$A = G(A/G,i,n) \qquad \text{[2.15]}$$

The P/G and A/G factors are the two right-most columns in the factor Tables 1 through 29. Table 2–4 lists several examples of gradient factors taken from these tables.

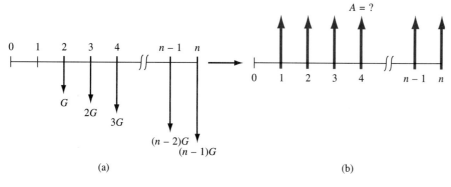

Figure 2–8   Conversion diagram of a uniform gradient to an equivalent uniform annual series.

Throughout this chapter we assume the n value is given in years. In Chapter 3 you will learn how to use the factor tables for interest periods other than years. An F/G factor (*uniform-gradient future-worth factor*) could readily be generated by multiplying the P/G and F/P factors for the same interest rate and n values as follows:

$$(P/G,i,n)(F/P,i,n) = (F/G,i,n)$$

Such a factor would yield an F value in the same year as the last gradient amount. As an exercise, carry out the multiplication suggested above to obtain the following F/G equation:

$$F = \frac{G}{i}\left[\frac{(1 + i)^n - 1}{i} - n\right]$$

Problems 2.3 to 2.7

Table 2–4	Examples of gradient factor values				
Value to be Computed	**Standard Notation**	**i,%**	**n**	**Table**	**Factor**
P	(P/G,5%,10)	5	10	10	31.6520
P	(P/G,30%,24)	30	24	26	10.9433
A	(A/G,6%,19)	6	19	11	7.2867
A	(A/G,35%,8)	35	8	27	2.0597

2.6   DERIVATION OF PRESENT WORTH OF GEOMETRIC (ESCALATING) SERIES

In Section 2.5 uniform-gradient factors were introduced which could be used for calculating the present worth or equivalent uniform annual worth of a series of payments which increase or decrease by a constant arithmetic amount in consecutive payment periods. Oftentimes, cash flows change by a *constant percentage* in consecutive payment periods, e.g., 5% per year. This type of cash flow, called an *escalating* or *geometric series,* is shown in general form in Figure 2–9, where D represents the dollar amount in year 1 and E represents the geometric growth rate in decimal form. The equation for calculating the present worth P_E of an escalating series is found by computing the present worth of the cash flows in Figure 2–9 using the P/F factor, $1/(1 + i)^n$.

$$P_E = \frac{D}{(1 + i)^1} + \frac{D(1 + E)}{(1 + i)^2} + \frac{D(1 + E)^2}{(1 + i)^3} + \cdots + \frac{D(1 + E)^{n-1}}{(1 + i)^n}$$

$$= D\left[\frac{1}{1 + i} + \frac{1 + E}{(1 + i)^2} + \frac{(1 + E)^2}{(1 + i)^3} + \cdots + \frac{(1 + E)^{n-1}}{(1 + i)^n}\right] \qquad [2.16]$$

Multiply both sides by $(1 + E)/(1 + i)$, subtract Equation [2.16] from the result, factor out P_E, and obtain

$$P_E\left(\frac{1 + E}{1 + i} - 1\right) = D\left[\frac{(1 + E)^n}{(1 + i)^{n+1}} - \frac{1}{1 + i}\right] \qquad [2.17]$$

Solve for P_E and simplify to obtain

$$P_E = \frac{D\left[\dfrac{(1 + E)^n}{(1 + i)^n} - 1\right]}{E - i} \qquad E \neq i \qquad [2.18]$$

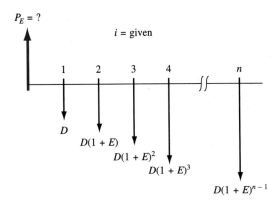

$P_E = ?$

$i = $ given

1   2   3   4 n

D

$D(1 + E)$

$D(1 + E)^2$

$D(1 + E)^3$

$D(1 + E)^{n-1}$

Figure 2–9   Cash-flow diagram of geometric gradient and its present worth P_E.

where P_E is the present worth of an escalating series starting in year 1 at D dollars. For the condition $E = i$, Equation [2.18] becomes*

$$P_E = D \frac{n}{1 + E} \qquad E = i \qquad \text{[2.19]}$$

The equivalent present-worth value P_E occurs in the year prior to the cash flow D, as shown in Figure 2–9. Note that P_E is for the *entire amount of the geometric series*, not only the amount G applicable when using the P/G factor for arithmetic gradients.

Problem 2.8

2.7   INTERPOLATION IN INTEREST TABLES

Sometimes it is necessary to locate a factor value for an interest rate i or number of periods n that is not in the interest tables. When this occurs, the desired factor value can be obtained in one of two ways: (1) by using the formulas derived in Sections 2.1 to 2.3 and 2.5 (and summarized inside the front cover) or (2) by interpolating between the tabulated values. It is generally easier and faster to use the formulas from a calculator or spreadsheet which has them preprogrammed. Furthermore, the value obtained through interpolation is not exactly the correct value, since we are linearly interpreting nonlinear equations. Nevertheless, interpolation is acceptable and is considered sufficient in most cases as long as the values of i or n are not too distant from each other.

The first step in linear interpolation is to set up the known (values 1 and 2) and unknown factors as shown in Table 2–5. A ratio equation is then set up and solved for c, as follows:

$$\frac{a}{b} = \frac{c}{d} \qquad \text{or} \qquad c = \frac{a}{b}d \qquad \text{[2.20]}$$

where a, b, c, and d represent the differences between the numbers shown in the interest tables. The value of c from Equation [2.20] is added to or subtracted from value 1, depending on whether the factor is increasing or decreasing in value, respectively. The following examples illustrate the procedure just described.

---

*Use L'Hospital's rule to modify Equation [2.18].

$$\frac{dP_E}{dE} = \frac{D[n(1 + E)^{n-1}]}{(1 + i)^n}$$

Substitute $E = i$ to obtain Equation [2.19]. Cash-flow calculations involving geometric series are discussed in Section 2.10.

Table 2-5	Linear interpolation setup	
i or n		**Factor**

Example 2.2

Determine the value of the A/P factor for an interest rate of 7.3% and n of 10 years, that is, $(A/P,7.3\%,10)$.

Solution

The values of the A/P factor for interest rates of 7 and 8% are listed in Tables 12 and 13, respectively. We have the following situation:

The unknown X is the desired factor value. From Equation [2.20],

$$c = \left(\frac{7.3 - 7}{8 - 7}\right)(0.14903 - 0.14238)$$

$$= \frac{0.3}{1}0.00665 = 0.00199$$

Since the factor is increasing in value as the interest rate increases from 7 to 8%, the value of c must be *added* to the value of the 7% factor. Thus,

$$X = 0.14238 + 0.00199 = 0.14437$$

Comment

It is good practice to check the reasonableness of your final answer by verifying that X lies *between* the values of the known factors used in the interpolation in approximately the correct proportions. In this case, since 0.14437 is less than 0.5 of the distance between 0.14238 and 0.14903, the answer seems reasonable. Rather than interpolating, it is usually a simpler procedure to use the formula to compute the factor value directly (and it is more accurate). The correct factor value is 0.144358.

Example 2.3

Find the value of the $(P/F,4\%,48)$ factor.

Solution

From interest factor Table 9 for 4% interest, the values of the P/F factor for 45 and 50 years can be found as follows:

From Equation [2.20],

$$c = \frac{a}{b}d = \frac{48 - 45}{50 - 45}(0.1712 - 0.1407) = 0.0183$$

Since the value of the factor decreases as n increases, c is subtracted from the factor value for $n = 45$.

$$X = 0.1712 - 0.0183 = 0.1529$$

Additional Example 2.15
Problems 2.9 and 2.10

2.8   PRESENT-WORTH, FUTURE-WORTH, AND EQUIVALENT-UNIFORM-ANNUAL-WORTH CALCULATIONS

The first and probably most important step in solving engineering economy problems is construction of a cash-flow diagram. In addition to more clearly illustrating the problem situation, the cash-flow diagram helps in determining which formulas should be used and whether the cash flows as presented allow straightforward application of the formulas as derived in the preceding sections. Obviously, the formulas can be used only when the cash flow of the problem conforms exactly to the cash-flow diagram for the formulas. For example, the uniform-series factors could not be used if payments or receipts occurred *every other year* instead of every year. It is very important, therefore, to remember the conditions for which the formulas apply. The correct use of the formulas for finding P, F, or A is illustrated in Examples 2.4 to 2.8. The equations used are shown in Table 2–2 and inside the front cover of this book. See the Additional Examples for cases in which some of these formulas cannot be applied.

Example 2.4

An independent tile contractor conducted an audit of some old records and found that the cost of office supplies varied as shown in the pie chart. If the contractor wanted

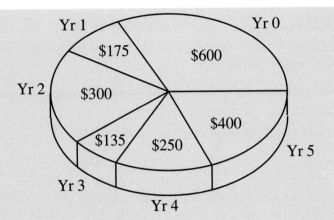

to know the equivalent value in year 10 of only the three largest amounts, what was that total at an interest rate of 5% per year?

Solution

The first step is to draw the cash-flow diagram from the contractor's perspective. Figure 2–10 indicates that an F value is to be computed. Since each value is different and does not take place each year, the future worth F can be determined by summing the equivalent individual single costs at year 10. Thus,

$$F = 600(F/P,5\%,10) + 300(F/P,5\%,8) + 400(F/P,5\%,5)$$

$$= 600(1.6289) + 300(1.4775) + 400(1.2763)$$

$$= \$1931.11$$

Comment

The problem could also be solved by finding the present worth in year 0 of the $300 and $400 costs using the P/F factors and then finding the future worth of the total in year 10.

$$P = 600 + 300(P/F,5\%,2) + 400(P/F,5\%,5)$$

$$= 600 + 300(0.9070) + 400(0.7835)$$

$$= \$1185.50$$

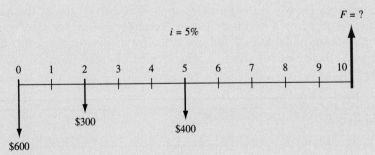

Figure 2–10   Diagram for a future worth in year 10, Example 2.4.

$$F = 1185.50(F/P,5\%,10)$$

$$= 1185.50(1.6289)$$

$$= \$1931.06$$

It should be obvious that there are a number of ways the problem could be worked, since any year could be used to find the equivalent total of the costs before finding the future value in year 10. As an exercise, you should work the problem using year 5 for the equivalent total and then determine the final amount in year 10. All answers should be the same. Any minor differences in amount here and in all future computations of this type are due to round-off error and the different number of significant digits used in the factor and dollar amounts to arrive at the final answer.

Example 2.5

How much money would a man have in his investment account after 8 years if he deposited $1000 per year for 8 years at 14% per year starting 1 year from now?

Solution

The cash-flow diagram is shown in Figure 2–11. Since the payments start at the end of year 1 and end in the year the future worth is desired, the F/A formula can be used. Thus,

$$F = 1000(F/A,14\%,8) = 1000(13.2328) = \$13,232.80$$

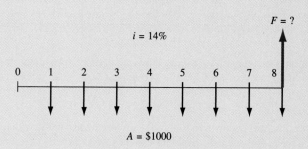

Figure 2–11   Diagram to find F for a uniform series, Example 2.5.

Example 2.6

How much money should you be willing to spend now in order to avoid spending $500 seven years from now if the interest rate is 18% per year?

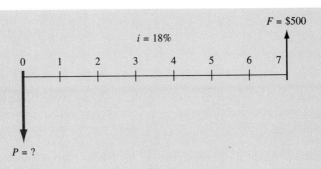

Figure 2–12   Diagram for Example 2.6.

Solution

The cash-flow diagram, which appears in Figure 2–12, allows use of the P/F factor directly. F is given and P is to be computed.

$$P = \$500(P/F,18\%,7) = 500(0.3139) = \$156.95$$

Comment

The same problem may be stated in other ways. What is the present worth of $500 seven years from now if the interest rate is 18% per year? What present amount would be equivalent to $500 seven years hence if the interest rate is 18% per year? What initial investment is equivalent to spending $500 seven years from now at an interest rate of 18% per year? In each case F is given and P is to be computed. Although there are several ways to state the same problem, the cash-flow diagram remains the same in each case.

Example 2.7

How much money should you be willing to pay now for an investment that is guaranteed to yield $600 per year for 9 years starting next year, at an interest rate of 16% per year?

Solution

The cash-flow diagram is shown in Figure 2–13. Since the cash-flow diagram fits the P/A uniform-series formula, the problem can be solved directly.

$$P = 600(P/A,16\%,9) = 600(4.6065) = \$2673.90$$

Comment

You should recognize that P/F factors can be used for each of the nine receipts and the resulting present worths added to get the correct answer. Another way is to find the future worth F of the $600 payments and then find the present worth of the F value. There are many ways to solve an engineering economy problem. Usually only the most direct method is presented here, but you should work the problems in at least one other way to become more familiar with the use of the formulas.

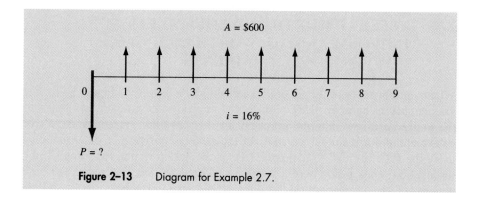

Figure 2–13   Diagram for Example 2.7.

Example 2.8

How much money must Carol deposit every year starting 1 year from now at $5\frac{1}{2}\%$ per year in order to accumulate $6000 seven years from now?

Solution

The cash-flow diagram from Carol's perspective is shown in Figure 2–14. The cash-flow diagram fits the A/F formula as derived.

$$A = \$6000\ (A/F,5.5\%,7) = 6000(0.12096) = \$725.76 \text{ per year}$$

Comment

The A/F factor value of 0.12096 is computed using the factor formula in Equation [2.8].

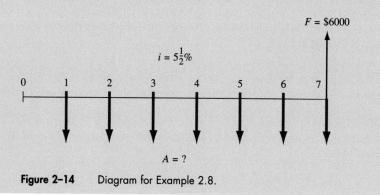

Figure 2–14   Diagram for Example 2.8.

Additional Example 2.16
Problems 2.11 to 2.29

2.9   PRESENT WORTH AND EQUIVALENT UNIFORM ANNUAL WORTH OF UNIFORM CONVENTIONAL GRADIENTS

When a uniform conventional gradient cash flow is involved (Figure 2–5), the gradient begins between years 1 and 2, and year 0 for the gradient and year 0 of the entire cash-flow diagram coincide. In this case the present worth P_G or equivalent uniform annual worth A_G of the *gradient only* can be determined by using the P/G formula, Equation [2.14], or the A/G formula, Equation [2.15], respectively. (See inside book cover also.) The cash flow that forms the base amount of the gradient must be considered separately. Thus, for cash-flow situations involving conventional gradients:

1. The base amount is the uniform-series amount A that begins in year 1 and extends through year n.
2. For an increasing gradient, the gradient amount must be added to the uniform-series amount.
3. For a decreasing gradient, the gradient amount must be subtracted from the uniform-series amount.

The general equations for calculating the total present worth P_T of conventional gradients, therefore, are

$$P_T = P_A + P_G \quad \text{and} \quad P_T = P_A - P_G$$

The present-worth calculation for an increasing gradient is illustrated in the next example.

Example 2.9

A couple plans to start saving money by depositing $500 into their savings account 1 year from now. They estimate that the deposits will increase by $100 each year for 9 years thereafter. What would be the present worth of the investments if the interest rate is 5% per year?

Solution

The cash-flow diagram from the couple's perspective is shown in Figure 2–15. Two computations must be made: the first to compute the present worth of the base amount P_A, and a second to compute the present worth of the gradient P_G. Then the total present worth P_T is equal to P_G plus P_A, since P_A and P_G both occur in year 0. This is clearly illustrated by the partitioned cash-flow diagram in Figure 2–16. The present worth is

$$P_T = P_A + P_G$$

$$= 500(P/A,5\%,10) + 100(P/G,5\%,10)$$

$$= 500(7.7217) + 100(31.652)$$

$$= \$7026.05$$

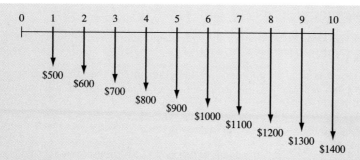

Figure 2–15   Cash flow, Example 2.9.

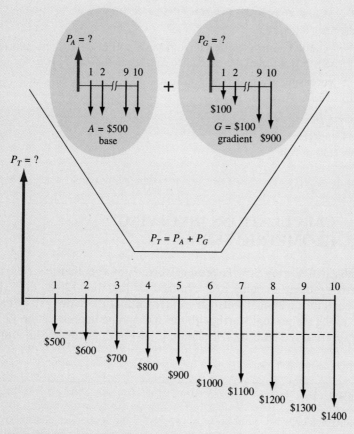

Figure 2–16   Partitioned diagram for Example 2.9.

Comment

It is important to emphasize again that the P/G factor determines the present worth of the *gradient only*. Any other cash flow involved must be considered separately.

Example 2.10

Rework Example 2.9 solving for the equivalent-uniform-annual-worth series.

Solution

Here, too, it is necessary to consider the gradient and the other costs involved in the cash flow separately. Using the cash flows of Figure 2–16, the total annual worth A_T is

$$A_T = A_1 + A_G$$

where A_1 is the equivalent annual worth of the $500 base amount and A_G is the equivalent annual worth of the gradient.

$$A_T = 500 + 100(A/G,5\%,10) = 500 + 100(4.0991)$$

$$= \$909.91 \text{ per year for years 1 through 10}$$

Comment

It is often helpful to remember that if the present worth is already calculated (as in Example 2.9), it can simply be multiplied by the appropriate A/P factor to get A. Here,

$$A_T = P_T(A/P,5\%,10) = 7026.05(0.12950)$$

$$= \$909.87$$

Problems 2.30 to 2.38

2.10   CALCULATIONS INVOLVING GEOMETRIC SERIES

As discussed in Section 2.6, the present worth P_E of a geometric series (base and geometric-gradient amount) is determined by Equation [2.18] or [2.19]. The equivalent uniform annual worth or future worth of the series can be calculated by converting the present worth with the appropriate interest factor, i.e., A/P or F/P, respectively. The use of Equation [2.18] is illustrated in Example 2.11.

Example 2.11

A pickup truck big-wheel modification costs $8000 and is expected to last 6 years with a $1300 salvage value. The maintenance cost is expected to be $1700 the first year, increasing by 11% per year thereafter. Determine the equivalent present worth of the modification and maintenance cost if the interest rate is 8% per year. When determining P for this example, use minus signs for negative cash flows and plus signs to indicate a positive cash flow for the salvage value.

Solution

The cash-flow diagram is shown in Figure 2–17. Since $E \neq i$, Equation [2.18] is used to calculate P_E. The total P_T is

$$P_T = -8000 - P_E + 1300(P/F,8\%,6)$$

$$= -8000 - 1700\frac{[(1 + 0.11)^6/(1 + 0.08)^6] - 1}{0.11 - 0.08} + 1300(P/F,8\%,6) \quad \textbf{[2.21]}$$

$$= -8000 - 1700(5.9559) + 819.26 = \$-17,305.85$$

Comment

The equivalent uniform annual worth of the truck could be determined by multiplying the $\$-17,305.85$ by $(A/P,8\%,6)$.

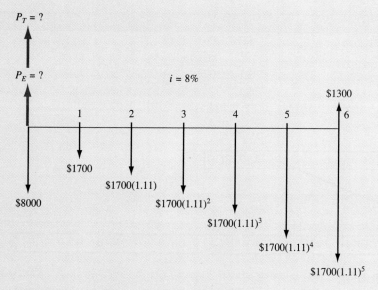

$P_T = ?$

$P_E = ?$

$i = 8\%$

$\$1300$

1   2   3   4   5   6

$\$1700$

$\$1700(1.11)$

$\$8000$

$\$1700(1.11)^2$

$\$1700(1.11)^3$

$\$1700(1.11)^4$

$\$1700(1.11)^5$

Figure 2–17   Cash-flow diagram for Example 2.11.

Example 2.11
(Spreadsheet)

A pickup truck big-wheel modification costs $8000 and is expected to last 6 years with a $1300 salvage value. The maintenance cost is expected to be $1700 the first year, increasing by 11% per year thereafter. Use spreadsheet analysis to determine the equivalent present worth of the modification and maintenance costs if the interest rate is 8% per year.

Solution

Figure 2–18 presents a spreadsheet with the present worth in cell B8. The relation used to compute $P_T = -\$17,306$ is shown below the Excel spreadsheet on the screen capture. It is equivalent to Equation [2.21] in Example 2.11 above.

Comment
You should work this example on your own spreadsheet system to get familiar with its use for engineering economy problems. Also, try to do your homework assignments using a spreadsheet as much as possible.

Problems 2.39 to 2.45

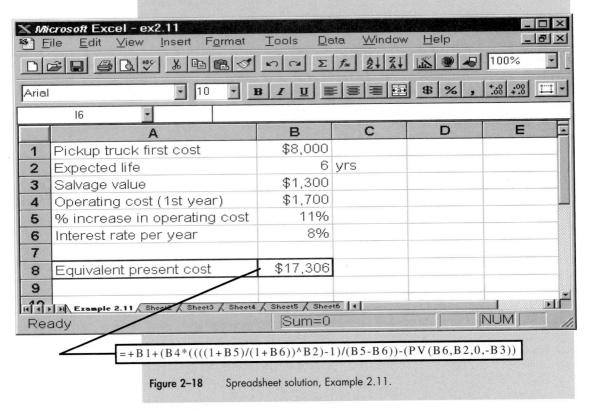

$$=+B1+(B4*((((1+B5)/(1+B6))^B2)-1)/(B5-B6))-(PV(B6,B2,0,-B3))$$

Figure 2–18   Spreadsheet solution, Example 2.11.

2.11   CALCULATION OF UNKNOWN INTEREST RATES

In some cases, the amount of money deposited and the amount of money received after a specified number of years are known, and it is the interest rate or rate of return that is unknown. When a single payment and single receipt, a uniform series of payments or receipts, or a uniform conventional gradient of payments or receipts are involved, the unknown rate can be determined by direct solution of the time value of money equation for *i*. When nonuniform payments or several factors are involved, however, the problem must be solved by a trial-and-error or numerical method. In this section, single-payment, uniform-series, or conventional gradient-series cash-flow problems are considered. The more complicated trial-and-error problems are deferred until Chapter 7, which discusses rate-of-return analysis.

The single-payment formulas can be easily rearranged and expressed in terms of i, but for the uniform series and gradient equations, it is commonly necessary to *solve for the value of the factor* and determine the interest rate from interest factor tables. Both situations are illustrated in the examples that follow.

Example 2.12

(a) If Carol can make a business investment requiring an expenditure of $3000 now in order to receive $5000 five years from now, what would be the rate of return on the investment?

(b) If Carol can receive 7%-per-year interest from a certificate of deposit, which investment should be made?

Solution

(a) The cash-flow diagram is shown in Figure 2–19. Since the single-payment formulas are involved in this problem, the i can be determined directly from the formula:

$$P = F(P/F,i,n) = F\frac{1}{(1 + i)^n}$$

$$3000 = 5000\frac{1}{(1 + i)^5}$$

$$0.600 = \frac{1}{(1 + i)^5}$$

$$i = \left(\frac{1}{0.6}\right)^{0.2} - 1 = 0.1076 \ (10.76\%)$$

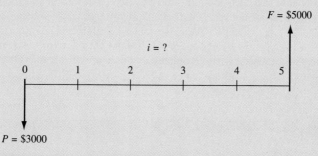

Figure 2–19   Diagram used to determine i, Example 2.12a.

Alternatively, the interest rate can be found by setting up the P/F or F/P equations, solving for the factor value, and interpolating. Using P/F,

$$P = F(P/F,i,n)$$

$$3000 = 5000(P/F,i,5)$$

$$(P/F,i,5) = 3000/5000$$

$$= 0.6000$$

From the interest tables, a P/F factor of 0.6000 for $n = 5$ lies between 10 and 11%. Interpolating between these two values using Equation [2.20], we have

$$c = \frac{0.6209 - 0.6000}{0.6209 - 0.5935}(11 - 10)$$

$$= \frac{0.0209}{0.0274}(1)$$

$$= 0.7628$$

Therefore, add c to the 10% base factor.

$$i = 10 + 0.76 = 10.76\%$$

It is good practice to insert the calculated value back into the equation to verify the correctness of the answer. Thus,

$$3000 = 5000(P/F,10.76\%,5)$$

$$= 5000\frac{1}{(1 + 0.1076)^5}$$

$$= 5000(0.5999)$$

$$= 3000$$

(b) Since 10.76% is greater than the 7% available in certificates of deposit, Carol should make the business investment.

Comment

Since the higher rate of return would be received on the business investment, Carol would probably select this option instead of the certificates of deposit. However, the degree of risk associated with the business investment was not specified. Obviously, the amount of risk associated with a particular investment is an important parameter and oftentimes causes selection of the lower-rate-of-return investment. Unless specified to the contrary, the problems in this text will assume equal risk for all alternatives.

Example 2.13

Parents wishing to save money for their child's education purchased an insurance policy that will yield $10,000 fifteen years from now. The parents must pay $500 per

year for 15 years starting 1 year from now. What will be the rate of return on their investments?

Solution

The cash-flow diagram is shown in Figure 2–20. Either the A/F or F/A factor can be used. Using A/F,

$$A = F(A/F,i,n)$$

$$500 = 10,000(A/F,i,15)$$

$$(A/F,i,15) = 0.0500$$

From the interest tables under the A/F column for 15 years, the value 0.0500 lies between 3 and 4%. By interpolation, $i = 3.98\%$.

Comment

For confirmation, insert the value $i = 3.98\%$ into the A/F formula to determine that 0.0500 is obtained.

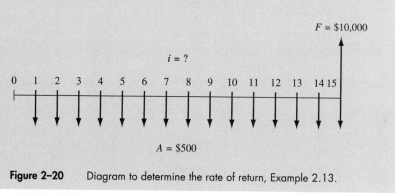

Figure 2–20   Diagram to determine the rate of return, Example 2.13.

Problems 2.46 to 2.51

2.12   CALCULATION OF UNKNOWN YEARS

In breakeven economic analysis, it is sometimes necessary to determine the number of years (periods) required before an investment pays off. Other times it is desirable to determine when given amounts of money will be available from a proposed investment. In these cases, the unknown value is n, and techniques similar to those of the preceding section on unknown interest rates can be used to find n.

Some of these problems can be solved directly for n by proper manipulation of the single-payment and uniform-series formulas. Alternatively, you can solve for the factor value and interpolate in the interest tables, as illustrated below.

Example 2.14

How long will it take for $1000 to double if the interest rate is 5% per year?

Solution

The cash-flow diagram is shown in Figure 2–21. The n value can be determined using either the F/P or P/F factor. Using the P/F factor,

$$P = F(P/F,i,n)$$

$$1000 = 2000(P/F,5\%,n)$$

$$(P/F,5\%,n) = 0.500$$

From the 5% interest table, the value 0.500 under the P/F column lies between 14 and 15 years. By interpolation, $n = 14.2$ years.

Comment

Problems of this type become more complicated when two or more nonuniform payments are involved. See the Additional Examples for an illustration using trial and error.

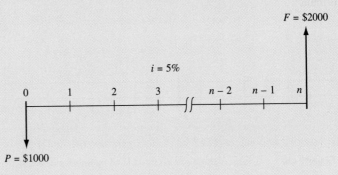

Figure 2–21   Diagram to determine an n value, Example 2.14.

Additional Example 2.17
Problems 2.52 to 2.57

ADDITIONAL EXAMPLES

Example 2.15

INTEREST FACTORS, SECTION 2.7   A new building has been purchased by Waldorf Concession Stands, Inc. The present worth of future maintenance costs is to be calculated with a *P/A* factor. If $i = 13\%$ per year and the life is expected to be 42 years, find the correct factor value.

Solution

The formula for the *P/A* factor is

$$(P/A,13\%,42) = \frac{(1 + i)^n - 1}{i(1 + i)^n}$$

$$= \frac{(1 + 0.13)^{42} - 1}{0.13(1 + 0.13)^{42}}$$

$$= \frac{168.549}{22.0412} = 7.647$$

The *P/A* factor value could also be determined by interpolation in the interest tables. However, since there are no table values here for $i = 13\%$ or $n = 42$, a two-way interpolation would be required. It is obviously easier and more accurate to use the factor formula.

Example 2.16

P, F, AND A CALCULATIONS, SECTION 2.8   Explain why the uniform-series factors cannot be used to compute *P* or *F* directly for any of the cash flows shown in Figure 2–22.

Solution

(a) The *P/A* factor cannot be used to compute *P* since the $100-per-year receipt does not occur each year from year 1 through year 5.

(b) Since there is no $A = \$550$ in year 5, the *F/A* factor cannot be used. The relation $F = 550(F/A,i,4)$ would furnish the future worth in year 4, not year 5 as desired.

(c) The first $A = \$1000$ value occurs in year 2. Use of the relation $P = 1000(P/A,i,4)$ will compute *P* in year 1, not year 0.

(d) The receipt values are unequal; thus the relation $F = A(F/A,i,3)$ cannot be used to compute *F*.

Comment

There are ways to compute *P* or *F* without resorting to only *P/F* and *F/P* factors; these methods are discussed in Chapter 4.

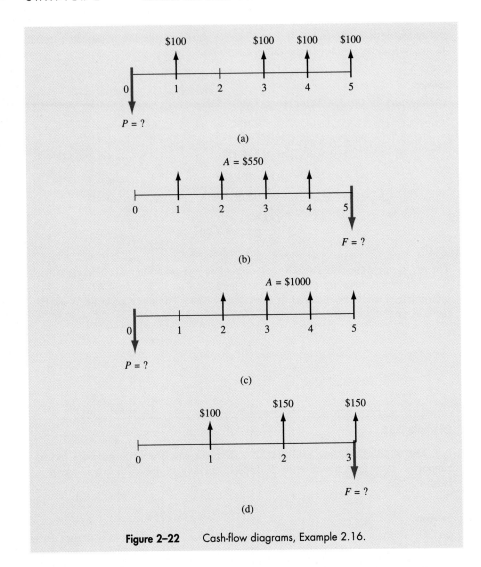

Figure 2–22   Cash-flow diagrams, Example 2.16.

Example 2.17

UNKNOWN n VALUE, SECTION 2.12   If Jeremy deposits $2000 now, $500 three years from now, and $1000 five years from now, how many years from now will it take for his total investment to amount to $10,000 if the interest rate is 6% per year?

Solution

The cash-flow diagram (Figure 2–23) requires that the following equation be correct.

$$F = P_1(F/P,i,n) + P_2(F/P,i,n - 3) + P_3(F/P,i,n - 5)$$

$$10,000 = 2000(F/P,6\%,n) + 500(F/P,6\%,n - 3) + 1000(F/P,6\%,n - 5)$$

Select various values of n and solve the equation. Interpolation for n will be necessary to obtain an exact equality. The procedure shown in Table 2–6 indicates that 20 years is too long and 15 years is too short. Therefore, we interpolate between 15 and 20 years.

$$c = \frac{10,000 - 7590.10}{(10,157.20 - 7590.10)} (20 - 15) = 4.69$$

$$n = 15 + c = 19.69 \approx 20 \text{ years}$$

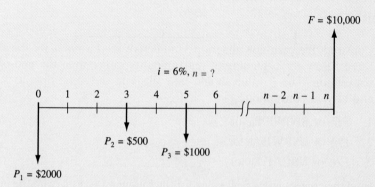

Figure 2–23   Diagram to determine n for a nonuniform series, Example 2.17.

Comment

Since interest is compounded at the end of each year, the money could actually not be withdrawn until year 20, at which time the accrued amount would be $10,157.20.

Table 2–6	Trial-and-error solution for n, Example 2.17				
n	$2,000 \times$ $(F/P,6\%,n)$	$500 \times$ $(F/P,6\%,n - 3)$	$1,000 \times$ $(F/P,6\%,n - 5)$	F	Remark
15	4,793.20	1,006.10	1,790.80	7,590.10	Too small
20	6,414.20	1,346.40	2,396.60	10,157.20	Too large

CHAPTER SUMMARY

Formulas were introduced in this chapter which allow you to make equivalence calculations for present, future, annual, and gradient cash flows. Facility in using these formulas and the standard factor notation that represents them is critical for allowing you to successfully advance through the other chapters in this book. By using these formulas and/or the tables generated from the factor part of these formulas, you can convert single cash flows into uniform cash flows or gradients into present worths and much more. You can also solve these formulas for rate of return (i) or time (n). By understanding how to manipulate these formulas, you have a truly powerful set of tools that will help you negotiate not only the remainder of this book, but also many of the experiences encountered in everyday living.

PROBLEMS

2.1 Construct the cash-flow diagrams and derive the formulas for the factors listed below for beginning-of-year amounts rather than the end-of-year convention. The P value should take place at the same time as for the end-of-year convention.

1. P/F or SPPWF factor
2. P/A or USPWF factor
3. F/A or USCAF factor

2.2 Find the correct numerical value for the following factors from the interest tables:

1. $(F/P,10\%,28)$
2. $(A/F,1\%,1)$
3. $(A/P,30\%,22)$
4. $(P/A,10\%,25)$
5. $(P/F,16\%,35)$

2.3 Construct a cash-flow diagram for the following transactions.

Year, k	0	1	2	3–10
Deposit, $	10,000	200	400	$400 + 300 (k - 3)$

2.4 Construct a cash-flow diagram for the following transactions.

Year, k	0	1	2–8
Transaction	$-6000	1000	$2000 - 100(k - 2)$

2.5 Construct a cash-flow diagram for the following transactions.

Year, k	0	1–4	5–7
Transaction	$-8000	1000	$800 - 100(k + 2)$

2.6 Find the value of $(F/G,10\%,10)$ using the F/A and A/G factors.

2.7 Find the value of the factor to convert a gradient with $n = 10$ into a present value using an interest rate of 16% per year.

2.8 What is the difference between (*a*) a geometric and an escalating series, (*b*) a gradient and an escalating series, and (*c*) a gradient and a geometric series?

2.9 Find the numerical value of the following factors by (*a*) interpolation and (*b*) using the appropriate formula:

1. $(F/P,6\%,23)$
2. $(P/A,16.3\%,15)$
3. $(A/G,12.7\%,20)$
4. $(A/F,28\%,30)$

2.10 Find the numerical value of the following factors by (*a*) interpolation and (*b*) using the appropriate formula:

$\left(\dfrac{P}{F}\, n,i\right)$

1. $(F/A,2\%,92)$
2. $(P/F,15\%,39)$
3. $(P/G,16\%,21)$
4. $(A/G,23\%,20)$

2.11 What is the present worth of a future cost of $7000 in year 20 if the interest rate is 15% per year?

2.12 How much money could you afford to spend now in lieu of spending $40,000 five years from now if the interest rate is 12% per year?

2.13 An advertisement in the newspaper is offering a second mortgage note for sale. The note for $25,000 is due 7 years from now. If you want to make a rate of return of 20% on any investments you make, how much should you pay for the note?

2.14 A married couple is planning to buy a new sport utility vehicle (SUV) 5 years from now. They expect the SUV to cost $32,000 at the time of purchase. If they want to have half of the cost for a down payment, how much must they save each year if they can earn 10% per year on their savings?

2.15 If the couple in the previous problem expects to inherit some money 2 years from now, how much would they have to set aside in a lump sum at that time in order to have their down payment? Assume $i = 10\%$ per year.

2.16 If you purchase a piece of equipment which has a cost of $23,000, what amount of money will you have to make each year to recover your investment in 6 years if you (*a*) borrow the money at an interest rate of 15% per year, or (*b*) pay for the equipment from money you had saved which was earning 10%-per-year interest?

2.17 How much money would you have 12 years from now if you take your Christmas bonus of $2500 each year and (*a*) place it under your mattress, (*b*) put it in an interest-bearing checking account at 3% per year, or (*c*) buy stock in a mutual fund which earns 16% per year?

2.18 How much money can you borrow now if you promise to repay the loan in 10 year-end payments of $3000, starting 1 year from now, if the interest rate is 18% per year?

2.19 To keep up with your increasing number of accounts receivable, you are considering the purchase of a new computer system. If you go the "cheap" way, you can buy a basic system now for $6000 and then upgrade at the end of year 1 for $2000 and again at the end of year 3 for $2800. Alternatively, you can buy a first-class system now that will provide the same level of service as the upgraded cheap system for the same length of time. If you can invest money at 20% per year, how much could you afford to spend now for the first-class system?

2.20 What is the future worth in year 25 of $3000 at $t = 0$, $7500 at $t = 4$ years, and $5200 at $t = 12$ years if the interest rate is 15% per year?

2.21 How much money would be accumulated in year 10 if you deposit $1000 in years 0, 2, 4, 6, 8, and 10 at an interest rate of 12% per year?

2.22 How much money must you deposit in year 6 if you deposit $5000 now and you want to have $12,000 at the end of year 11? Assume your deposits earn interest at 6% per year.

2.23 How much money could a start-up software company borrow now if it promises to repay the loan with three equal payments of $7000 in years 2, 6, and 10 if the interest rate on the loan is 13% per year?

2.24 If you borrow $11,000 now to buy a tricked-out 250-cc motocross bike, how much will you have to pay at the end of year 3 to pay off the loan if you make a $3000 payment at the end of year 1? Assume $i = 10\%$ per year.

2.25 If you are paying off a loan of $10,000 by making equal year-end payments for 5 years, how much principal reduction will you get in (*a*) the second payment and (*b*) the last payment if the interest rate on the loan is 17% per year?

2.26 A discount furniture store in planning an expansion which will cost $250,000 three years from now. If the company plans to set aside money at the end of each of the next 3 years, how much must it set aside in year 1 if each of the next two deposits will be twice as large as that one? Assume the deposits will earn interest at 10% per year.

2.27 How much money will be in your retirement account if you invest $9000 per year for 35 years at an interest rate of 6 ½% per year?

2.28 Because of your company's good credit rating, a distributor will allow you to purchase products costing up to $15,000 with no interest charge as long as you repay the loan within 2 years. If you purchase $15,000 worth of materials now and repay the total amount in one lump sum at the end of year 2, what is the amount of the effective discount you get if the prevailing interest rate is 15 ½% per year?

2.29 What compound interest rate is equivalent to a 15%-per-year simple-interest rate over a 20-year time period?

2.30 A cash-flow sequence starts in year 1 at $1000 and increases by $100 each year through year 7. Do the following: (*a*) Draw the cash-flow diagram, (*b*) determine the amount of the cash flow in year 7, (*c*) locate the gradient present worth on the diagram, (*d*) determine the value of *n* for the gradient.

2.31 A company which manufactures automobile wiring harnesses has budgeted $300,000 now to pay for a certain plastic part over the next 5 years. If the company expects to spend $50,000 in year 1, how much of an increase per year is the company expecting in the cost of this part? Assume the $300,000 is in an account earning 12%-per-year interest.

2.32 For the cash flow shown below, calculate (*a*) the equivalent uniform annual worth in years 1 through 4 and (*b*) the present worth in year 0. Assume $i = 14\%$ per year.

Year	1	2	3	4
Cash flow	$4000	3200	2400	1600

2.33 For a cash-flow sequence described by $(500 + 30k)$, where *k* is in years (*a*) draw the cash-flow diagram for years 1 through 9, (*b*) determine the value of *G*, (*c*) determine the cash-flow amount in year 5, (*d*) determine the present worth of the cash flow in years 1–14 if $i = 12\%$ per year.

2.34 The utility bill of a small paper recycling center has been increasing by about $428 per year. If the utility cost in year 1 was $3000, what is the equivalent uniform annual worth through year 8 if the interest rate is 15% per year?

2.35 The receipts from certain mineral rights have been following a decreasing gradient for the past 4 years. The first receipt was $10,500 and the second one was $9800. (*a*) In how many years from now will the income stream go to zero? (*b*) What is the future worth (in the last year money is received) of the remaining series of receipts at an interest rate of 11% per year?

2.36 For the cash flow shown below, determine the value of *G* that will make the equivalent annual worth equal to $800 at an interest rate of 20% per year.

Year	0	1	2	3	4
Cash flow	0	$200	200 + G	200 + 2G	200 + 3G

2.37 Find the value of G for the cash flow in Problem 2.36 if the future worth (year 4) of the cash flow is $3000 at an interest rate of 18% per year.

2.38 A major drug company anticipates that in future years could be involved in litigation regarding perceived side effects of one of its antidepressant drugs. In order to prepare a "war chest" the company wants to have $20 million available 5 years from now. The company expects to set aside $5 million the first year and uniformly increasing amounts in each of the next four. If the company can earn 11% per year on the money it sets aside, by how much must it increase the amount set aside each year to achieve its goal of $20 million at the end of 5 years?

2.39 Assume you were told to prepare a table of factor values (like those in the back of your book) for calculating the present worth of a geometric series. Determine the first three values (i.e., for $n = 1, 2,$ and 3) for an interest rate of 10% per year and an escalation rate of 6% per year.

2.40 Calculate the present worth of a geometric series of payments wherein the amount in year 1 is 500 and each succeeding amount increases by 10% per year. Use an interest rate of 15% per year and a 7-year time period.

2.41 The present worth of a geometric series of cash flows was found to be $65,000. If the series extended through 15 years and the interest rate was 18% per year, what was the escalation rate if the cash flow in year 1 was $6000?

2.42 In order to have money available for replacing their family vehicle, a couple planned to have $38,000 available in 6 years by investing in a global mutual fund. If they plan to increase their savings by 7% each year, how much must they invest in year 1 if they expect to earn 14% per year on their investment?

2.43 Find the present worth of a series of cash flows which starts at $800 in year 1 and increases by 10% per year for 20 years. Assume the interest rate is 10% per year.

2.44 Assume you want to start saving money for a "rainy day." If you invest $1000 at the end of year 1 and you increase your savings by 8% each year, how much will you have in your account 10 years from now if it earns interest at a rate of 8% per year?

2.45 A company is planning to make deposits such that each one is 6% larger than the preceding one. How large must the second deposit be (at the end of year 2) if the deposits extend through year 15 and the fourth deposit is $1250? Use an interest rate of 10% per year.

2.46 If you invest $5000 now in a franchise which promises that your investment will be worth $10,000 in 3 years, what rate of return would you earn?

2.47 You have just gotten a hot tip that you should buy stock in the GRQ company. The stock is selling for $25 per share. If you buy 500 shares and the stock increases to $30 per share in 2 years, what rate of return would you realize on your investment?

2.48 A certain company pays a bonus to each engineer at the end of each year based on the company's profit for that year. If the company invested $2 million to start up, what rate of return has it made if each engineer's bonus has been $3000 per year for the past 10 years? Assume the company has six engineers and that the bonus money paid represents 4% of the company's profits.

2.49 If you purchased a house 5 years ago which cost $80,000, what rate of return did you make on your investment if you find out that you can sell the house now for $100,000? Assume that closing costs associated with the sale will amount to 10% of the selling price.

2.50 You have just inherited $100,000 from your favorite uncle. His will stipulated that a certain bank will keep the money on deposit for you. His will also stipulated that you can withdraw $10,000 after 1 year, $11,000 after 2 years, and amounts increasing by $1000 per year until the amount is exhausted. If it takes 18 years for the inheritance to go to zero, what interest rate was the money earning while on deposit?

2.51 A small company wants to begin saving money so that it will have enough saved in 3 years to purchase a new computer system which costs $12,000. If the company deposits $3000 at the end of year 1 and then increases its deposit by 15% per year, what rate of return will be required on the investment so that the company can buy the computer 3 years from now?

2.52 How long does it take for money to increase to five times the initial amount at an interest rate of 17% per year?

2.53 How long would it take for you to repay a loan of $30,000 (from your parents) if you pay $2000 per year and the interest rate is (*a*) 0%, (*b*) 5% per year, (*c*) 18% per year?

2.54 If you hit a small lottery for $50,000, how long will you be able to withdraw $10,000 per year if you earn 12% per year on your investments?

2.55 If you want to have $10,000 available for an Australian vacation, when will you be able to go if you deposit $1000 per year into an account which earns 8%-per-year interest?

2.56 A pension fund was started some time ago which now contains $600,000. If the first deposit was $50,000 and each subsequent deposit was decreased by $4000, how long ago was the fund started if it earned 11%-per-year interest?

2.57 How long will it take for a savings fund to accumulate to $15,000 if $1000 is deposited at the end of year 1 and the amount of the deposit is increased by 10% each year? Assume that the interest rate is 10% per year.

Nominal and Effective Interest Rates and Continuous Compounding

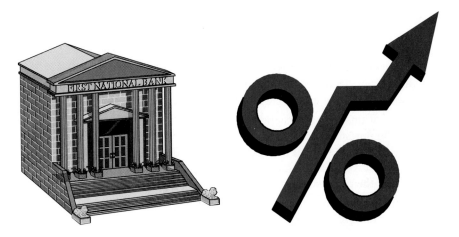

This chapter teaches you how to make engineering economy computations using interest periods and compounding frequencies which are other than 1 year. The material of this chapter is helpful for handling personal financial matters which often involve monthly, daily, or continuous time periods.

LEARNING OBJECTIVES

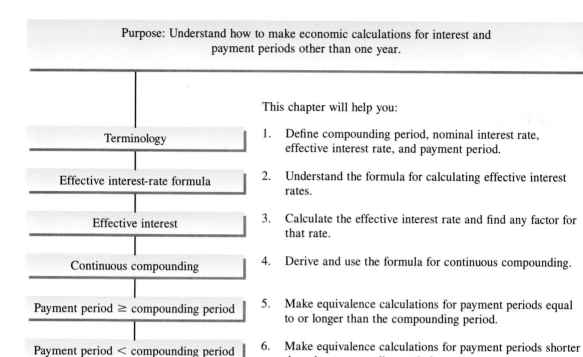

Purpose: Understand how to make economic calculations for interest and payment periods other than one year.

	This chapter will help you:
Terminology	1. Define compounding period, nominal interest rate, effective interest rate, and payment period.
Effective interest-rate formula	2. Understand the formula for calculating effective interest rates.
Effective interest	3. Calculate the effective interest rate and find any factor for that rate.
Continuous compounding	4. Derive and use the formula for continuous compounding.
Payment period $\geq$ compounding period	5. Make equivalence calculations for payment periods equal to or longer than the compounding period.
Payment period $<$ compounding period	6. Make equivalence calculations for payment periods shorter than the compounding period.

3.1   NOMINAL AND EFFECTIVE RATES

In Chapter 1, the concepts of simple- and compound-interest rates were introduced. The basic difference between the two is that compound interest includes interest on the interest earned in the previous period while simple interest does not. In essence, nominal and effective interest rates have the same relationship to each other as do simple and compound interest. The difference is that nominal and effective interest rates are used when the compounding period (or interest period) is less than 1 year. Thus, when an interest rate is expressed over a period of time shorter than a year, such as 1% per month, the terms nominal and effective interest rates must be considered.

Nominal/Effective used < 1 year

A dictionary definition of the word *nominal* is "purported, so-called, ostensible, or professed." These synonyms imply that a nominal interest rate is not a correct, actual, genuine, or effective rate. As discussed below, nominal interest rates must be converted into effective rates in order to accurately reflect time-value considerations. Before discussing effective rates, however, let us define a nominal interest rate, r, as the period interest rate times the number of periods. In equation form,

$$r = \text{interest rate per period} \times \text{number of periods}$$

A nominal interest rate can be found for any time period which is longer than the originally stated period. For example, a period interest rate listed as 1.5% per month could also be expressed as a nominal 4.5% per quarter (that is, 1.5% per month × 3 months per quarter), 9.0% per semiannual period, 18% per year, or 36% per 2 years, etc. The nominal interest rate obviously ignores the time value of money and the frequency with which interest is compounded. When the time value of money is taken into consideration in calculating interest rates from period interest rates, the rate is called an effective interest rate. Just as was true for nominal interest rates, effective rates can be determined for any time period longer than the originally stated period as shown in the next two sections of this chapter. It is important to recognize that all the formulas derived in Chapter 2 were based on compound interest and, therefore, only effective interest rates can be used in the equations.

A discussion of nominal and effective interest rates is not complete without commenting about the various ways that interest rates can be expressed. There are three general ways of expressing interest rates as shown by the three groups of statements in Table 3–1. The three statements in the top part of the table show that an interest rate can be stated over some designated time period without specifying the compounding period. Such interest rates are assumed to be effective rates with the *compounding period (CP)* assumed to be the same as that of the stated interest rate.

For the interest statements presented in the middle of Table 3–1, three conditions prevail: (1) The compounding period is identified, (2) this compounding period is shorter than the time period over which the interest is stated,

Table 3–1	Various interest statements and their interpretations	
(1) **Interest Rate Statement**	**(2)** **Interpretation**	**(3)** **Comment**
i = 12% per year i = 1% per month i = 3½% per quarter	i = effective 12% per year compounded yearly i = effective 1% per month compounded monthly i = effective 3½% per quarter compounded quarterly	When no compounding period is given, interest rate is an effective rate, with compounding period assumed to be equal to stated time period.
i = 8% per year, compounded monthly i = 4% per quarter compounded monthly i = 14% per year compounded semiannually	i = nominal 8% per year compounded monthly i = nominal 4% per quarter compounded monthly i = nominal 14% per year compounded semiannually	When compounding period is given without stating whether the interest rate is nominal or effective, it is assumed to be nominal. Compounding period is as stated.
i = effective 10% per year compounded monthly i = effective 6% per quarter i = effective 1% per month compounded daily	i = effective 10% per year compounded monthly i = effective 6% per quarter compounded quarterly i = effective 1% per month compounded daily	If interest rate is stated as an effective rate, then it is an effective rate. If compounding period is not given, compounding period is assumed to coincide with stated time period.

⇐ Exam

and (3) the interest rate is designated neither as nominal nor as effective. In such cases, the interest rate is assumed to be nominal and the compounding period is equal to that which is stated. (We will show how to get effective interest rates from these in the next section.)

For the third group of interest-rate statements in Table 3–1, the word *effective* precedes or follows the specified interest rate and the compounding period is also given. These interest rates are obviously effective rates over the respective time periods stated. Likewise, the compounding periods are equal to those which are stated. Similarly, if the word *nominal* had preceded any of the interest statements, the interest rate would be a nominal rate.

The importance of being able to recognize whether a given interest rate is nominal or effective cannot be overstated with respect to the reader's understanding of the remainder of the material in this chapter and indeed the rest of the book. Table 3–2 contains a listing of several interest statements (column 1) along with their interpretations (columns 2 and 3).

Table 3–2	Specific examples of interest statements and interpretations	
(1) **Interest** **Statement**	**(2)** **Nominal or** **Effective Interest**	**(3)** **Compounding** **Period**
15% per year compounded monthly	Nominal	Monthly
15% per year	Effective	Yearly
Effective 15% per year compounded monthly	Effective	Monthly
20% per year compounded quarterly	Nominal	Quarterly
Nominal 2% per month compounded weekly	Nominal	Weekly
2% per month	Effective	Monthly
2% per month compounded monthly	Effective	Monthly
Effective 6% per quarter	Effective	Quarterly
Effective 2% per month compounded daily	Effective	Daily
1% per week compounded continuously	Nominal	Continuously
0.1% per day compounded continuously	Nominal	Continuously

Exam ⇗

Now that the concept of nominal and effective interest rates has been introduced, in addition to considering the compounding period (which is also known as the interest period), it will also be necessary to consider the frequency of the payments or receipts within the cash-flow time interval. For simplicity, the frequency of the payments or receipts is known as the *payment period (PP)*. It is important to distinguish between the compounding period and the payment period because in many instances the two do not coincide. For example, if a company deposited money each month into an account that pays a nominal interest rate of 14% per year compounded semiannually, the payment period would be 1 month while the compounding period would be 6 months as shown in Figure 3–1. Similarly, if a person deposits money each year into a savings account which compounds interest quarterly, the payment period is 1 year, while the compounding period is 3 months. Hereafter, for problems which involve either uniform-series or uniform-gradient cash-flow amounts, it will be necessary to determine the relationship between the compounding period and the payment period as a first step in the solution of the problem (Section 3.5).

Problems 3.1 and 3.2

i = nominal 14% per year compounded semiannually

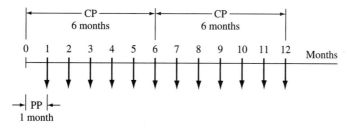

Months

→| PP |←
1 month

Figure 3–1   Cash-flow diagram for a monthly payment period (PP) and semiannual compounding period (CP).

3.2   EFFECTIVE INTEREST-RATE FORMULATION

To illustrate the difference between nominal and effective interest rates, the future worth of $100 after 1 year is determined using both rates. If a bank pays 12% interest compounded annually, the future worth of $100 using an interest rate of 12% per year is

$$F = P(1 + i)^n = 100(1.12)^1 = \$112.00 \qquad \text{[3.1]}$$

On the other hand, if the bank pays interest that is compounded semiannually, the future worth must include the interest on the interest earned in the first period. An interest rate of 12% per year compounded semiannually means that the bank will pay 6% interest after 6 months and another 6% after 12 months (i.e., every 6 months). Figure 3–2 is the cash-flow diagram for semiannual compounding for a nominal interest rate of 12% per year compounded semiannually. The calculation in Equation [3.1] obviously ignores the interest earned in the first period. Taking compounding period 1 interest into consideration, the

$Accum = Pres\left(1 + r_{eff}\right)^t$

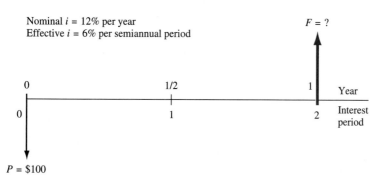

Nominal i = 12% per year
Effective i = 6% per semiannual period

F = ?

Year

Interest period

P = $100

Figure 3–2   Cash-flow diagram for semiannual compounding periods.

future worth values of $100 after 6 months and after 12 months are

$$F_6 = 100(1 + 0.06)^1 = \$106.00$$

$$F_{12} = 106(1 + 0.06)^1 = \$112.36 \tag{3.2}$$

r = nominal i per year

where 6% is the effective semiannual interest rate. In this case, the interest earned in 1 year is $12.36 instead of $12.00. Therefore, the effective annual interest rate is 12.36%.

The equation to determine the effective interest rate from the nominal interest rate may be generalized as follows:

same

$$r_{eff} = \left(1 + \frac{r}{m}\right)^m - 1$$

$$i = \left(1 + \frac{r}{m}\right)^m - 1 \tag{3.3}$$

where i = effective interest rate per period
  r = nominal interest rate per period
  m = number of compounding periods

Equation [3.3] is referred to as the *effective interest-rate equation*. As the number of compounding periods increases, m approaches infinity, in which case the equation represents the interest rate for continuous compounding. A detailed discussion of this subject is presented in Section 3.4.

3.3   CALCULATION OF EFFECTIVE INTEREST RATES

Effective interest rates can be calculated for any time period longer than the actual compounding period through the use of Equation [3.3]. That is, an effective interest rate of 1% per month, for example, can be converted into effective rates per quarter, per semiannual period, per 1 year, per 2 years, or per any other period longer than 1 month (the compounding period). It is important to remember that in Equation [3.3] the time units on i and r must always be the same. Thus, if an effective interest rate, i, per semiannual period is desired, then r must be the nominal rate per semiannual period. The m in Equation [3.3] is always equal to the number of times that interest would be compounded in the period of time over which i is sought. The next example illustrates these relationships.

Example 3.1

A national credit card carries an interest rate of 2% per month on the unpaid balance. (*a*) Calculate the effective rate per semiannual period. (*b*) If the interest rate is stated as 5% per quarter, find the effective rates per semiannual and annual time periods.

Solution

(a) In this part of the example, the compounding period is monthly. Since the effective interest rate per semiannual period is what is desired, the r in Equation [3.3] must be the nominal rate per 6 months, or

r = 2% per month × 6 months per semiannual period

= 12% per semiannual period

The m in Equation [3.3] is equal to 6, since interest would be compounded 6 times in a 6-month time period. Thus, the effective semiannual rate is

$$i \text{ per 6 months} = \left(1 + \frac{0.12}{6}\right)^6 - 1$$

$$= 0.1262 \quad (12.62\%)$$

(b) For an interest rate of 5% per quarter, the compounding period is quarterly. Therefore, in a semiannual period, m = 2 and r = 10%. Thus,

$$i \text{ per 6 months} = \left(1 + \frac{0.10}{2}\right)^2 - 1$$

$$= 0.1025 \quad (10.25\%)$$

The effective interest rate per year can be determined using r = 20% and m = 4, as follows:

$$i \text{ per year} = \left(1 + \frac{0.20}{4}\right)^4 - 1$$

$$= 0.2155 \quad (21.55\%)$$

Comment

Note that the term r/m in Equation [3.3] is always equal to the interest rate (effective) per compounding period. In part (a) this was 2% per month, while in part (b) it was 5% per quarter.

Table 3–3 presents the effective interest rate i for various nominal interest rates r using Equation [3.3] and compounding periods of 6 months, 3 months, 1 month, 1 week, and 1 day. The continuous-compounding column is discussed in the next section. Note that as the interest rate increases, the effect of more frequent compounding becomes more pronounced. When Equation [3.3] is used to find an effective interest rate, the answer is usually an interest rate which is not an integer number, as illustrated in Example 3.1 and Table 3–3. When this occurs, the factor values desired must be obtained either through interpolation in the interest tables or through direct use of the equations developed in Chapter 2. Example 3.2 shows these calculations.

Table 3–3 ₒ	Tabulation of effective annual interest rates for nominal period rates					
Nominal Rate, $r\%$	Semiannually ($m = 2$)	Quarterly ($m = 4$)	Monthly ($m = 12$)	Weekly ($m = 52$)	Daily ($m = 365$)	Continuously ($m = \infty$; $e^r - 1$)
0.25	0.250	0.250	0.250	0.250	0.250	0.250
0.50	0.501	0.501	0.501	0.501	0.501	0.501
0.75	0.751	0.752	0.753	0.753	0.753	0.753
1.00	1.003	1.004	1.005	1.005	1.005	1.005
1.50	1.506	1.508	1.510	1.511	1.511	1.511
2	2.010	2.015	2.018	2.020	2.020	2.020
3	3.023	3.034	3.042	3.044	3.045	3.046
4	4.040	4.060	4.074	4.079	4.081	4.081
5	5.063	5.095	5.116	5.124	5.126	5.127
6	6.090	6.136	6.168	6.180	6.180	6.184
7	7.123	7.186	7.229	7.246	7.247	7.251
8	8.160	8.243	8.300	8.322	8.328	8.329
9	9.203	9.308	9.381	9.409	9.417	9.417
10	10.250	10.381	10.471	10.506	10.516	10.517
11	11.303	11.462	11.572	11.614	11.623	11.628
12	12.360	12.551	12.683	12.734	12.745	12.750
13	13.423	13.648	13.803	13.864	13.878	13.883
14	14.490	14.752	14.934	15.006	15.022	15.027
15	15.563	15.865	16.076	16.158	16.177	16.183
16	16.640	16.986	17.227	17.322	17.345	17.351
17	17.723	18.115	18.389	18.497	18.524	18.530
18	18.810	19.252	19.562	19.684	19.714	19.722
19	19.903	20.397	20.745	20.883	20.917	20.925
20	21.000	21.551	21.939	22.093	22.132	22.140
21	22.103	22.712	23.144	23.315	23.358	23.368
22	23.210	23.883	24.359	24.549	24.598	24.608
23	24.323	25.061	25.586	25.796	25.849	25.860
24	25.440	26.248	26.824	27.054	27.113	27.125
25	26.563	27.443	28.073	28.325	28.390	28.403
26	27.690	28.646	29.333	29.609	29.680	29.693
27	28.823	29.859	30.605	30.905	30.982	30.996
28	29.960	31.079	31.888	32.213	32.298	32.313
29	31.103	32.309	33.183	33.535	33.626	33.643
30	32.250	33.547	34.489	34.869	34.968	34.986
31	33.403	34.794	35.807	36.217	36.327	36.343
32	34.560	36.049	37.137	37.578	37.693	37.713
33	35.723	37.313	38.478	38.952	39.076	39.097
34	36.890	38.586	39.832	40.339	40.472	40.495
35	38.063	39.868	41.198	41.740	41.883	41.907
40	44.000	46.410	48.213	48.954	49.150	49.182
45	50.063	53.179	55.545	56.528	56.788	56.831
50	56.250	60.181	63.209	64.479	64.816	64.872

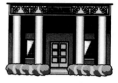

Example 3.2

A university credit union advertises that its interest rate on loans is 1% per month. Calculate the effective annual interest rate and use the interest factor tables to find the corresponding P/F factor for $n = 8$ years.

Solution

Since an annual interest rate is desired, r per year $= (0.01)(12) = 0.12$ and $m = 12$. Substitute $r/m = 0.12/12 = 0.01$ and $m = 12$ into Equation [3.3].

$$i = (1 + 0.01)^{12} - 1$$
$$= 1.1268 - 1$$
$$= 0.1268 \quad (12.68\%)$$

In order to find the P/F factor for $i = 12.68\%$ and $n = 8$, it is necessary to interpolate between $i = 12\%$ and $i = 14\%$ using the interest factor tables.

$$
\begin{array}{c}
b \left[a \left[\begin{array}{l} 12\% \\ 12.68\% \\ 14\% \end{array} \right. \right.
\begin{array}{l} 0.4039 \\ P/F \\ 0.3506 \end{array}
\left. \left. \right] c \right] d
\end{array}
$$

$$c = \frac{0.68}{2}(0.0533) = 0.0181$$

$$(P/F,12.68\%,8) = 0.4039 - 0.0181 = 0.3858$$

An easier, and more accurate, way to find the value of the factor is to substitute $i = 12.68\%$ and $n = 8$ into the P/F factor relation in Equation [2.3].

$$(P/F,12.68\%,8) = \frac{1}{(1 + 0.1268)^8} = 0.3848$$

Additional Example 3.6
Problems 3.3 to 3.16

3.4   EFFECTIVE INTEREST RATES FOR CONTINUOUS COMPOUNDING

As the compounding period becomes shorter and shorter, the value of m, the number of compounding periods per interest period, increases. In the situation where interest is compounded continuously, m approaches infinity and the effective interest-rate formula in Equation [3.3] may be written in a new form. First, recall the definition of the natural logarithm base.

$$\lim_{h \to \infty} \left(1 + \frac{1}{h}\right)^h = e = 2.71828+ \qquad \text{[3.4]}$$

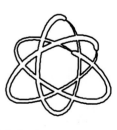

The limit of Equation [3.3] as m approaches infinity is found using $r/m = 1/h$, which makes $m = hr$.

$$\lim_{m \to \infty} i = \lim_{m \to \infty}\left(1 + \frac{r}{m}\right)^m - 1$$

$$= \lim_{h \to \infty}\left(1 + \frac{1}{h}\right)^{hr} - 1 = \lim_{h \to \infty}\left[\left(1 + \frac{1}{h}\right)^h\right]^r - 1 \qquad \textbf{[3.5]}$$

$$i = e^r - 1$$

Equation [3.5] is used to compute the *effective continuous interest rate*. As in Equation [3.3], the time periods on i and r must be the same. As an illustration, for an annual nominal rate of 15% per year ($r = 15\%$ per year), the effective continuous rate per year is

$$i = e^{0.15} - 1 = 0.16183 \qquad (16.183\%)$$

For convenience, Table 3–3 includes the effective continuous rate for many nominal rates as computed by Equation [3.5].

Example 3.3

(*a*) For an interest rate of 18% per year compounded continuously, calculate the effective monthly and annual interest rates.

(*b*) If an investor requires an effective return of at least 15% on his money, what is the minimum annual nominal rate that is acceptable if continuous compounding takes place?

Solution

(*a*) The nominal monthly rate, r, is $^{18}/_{12} = 1\frac{1}{2}\% = 0.015$ per month. From Equation [3.5], the effective monthly rate is

$$i \text{ per month} = e^r - 1 = e^{0.015} - 1$$

$$= 0.01511 \qquad (1.511\%)$$

Similarly, the effective annual rate using $r = 0.18$ per year is

$$i \text{ per year} = e^r - 1 = e^{0.18} - 1 = 0.1972 \qquad (19.72\%)$$

(*b*) In this problem, we must solve Equation [3.5] backward, since we have i and we want r. Thus, for $i = 15\%$ per year, solve for r by taking the natural logarithm.

$$e^r - 1 = 0.15$$

$$e^r = 1.15$$

$$\ln e^r = \ln 1.15$$

$$r = 0.13976 \qquad (13.976\%)$$

Therefore, a rate of 13.976% per year compounded continuously will generate an effective 15%-per-year return.

Comment

The general formula to find the nominal rate when given the effective continuous rate i is $r = \ln(1 + i)$.

Additional Examples 3.7 and 3.8
Problems 3.17 to 3.25

3.5   CALCULATIONS FOR PAYMENT PERIODS EQUAL TO OR LONGER THAN COMPOUNDING PERIODS

When the compounding period of an investment or loan does not coincide with the payment period, it becomes necessary to manipulate the interest rate and/or payment in order to determine the correct amount of money accumulated or paid at various times. Remember that if the payment and compounding periods do not agree, the interest tables cannot be used until appropriate corrections are made. In this section, we consider the situation where the payment period (for example, year) is equal to or longer than the compounding period (for example, month). The two conditions that can occur are as follows:

1. The cash flows require the use of the single-amount factors $(P/F, F/P)$.
2. The cash flows require the use of uniform-series or gradient factors.

3.5.1   Single-Amount Factors

There are essentially an infinite number of correct procedures that can be used when only single factors are involved. This is because there are only two requirements which must be satisfied: (1) An effective rate must be used for i, and (2) the units on n must be the same as those on i. In standard factor notation, then, the single-payment equations can be generalized as follows:

$P = F(P/F$, effective i per period, number of periods)
$F = P(F/P$, effective i per period, number of periods)

Thus, for a nominal interest rate of 12% per year compounded monthly, any of the i and corresponding n values shown in Table 3–4 could be used (as well as many others not shown) in the single-payment formulas. For example, if the equivalent effective rate per month is used for $i(1\%)$, then the n term must be in months (12). If an effective quarterly interest rate is used for i, that is, $(1.03)^3 - 1$ or 3.03%, then the n term must be in quarters (4).

Table 3–4	Various i and n values for single-amount equations using $r = 12\%$ per year, compounded monthly	
Effective Interest Rate, i	**Units for n**
1% per month	Months
3.03% per quarter	Quarters
6.15% per 6 months	Semiannual periods
12.68% per year	Years
26.97% per 2 years	2-year periods

Example 3.4

If a woman deposits $1000 now, $3000 four years from now, and $1500 six years from now at an interest rate of 12% per year compounded semiannually, how much money will she have in her account 10 years from now?

Solution

The cash-flow diagram is shown in Figure 3–3. Let us assume that we have decided to use an *annual* interest rate in solving the problem. Since only effective interest rates can be used in the equations, the first step is to find the effective annual rate. From Table 3–3, for $r = 12\%$ and semiannual compounding, effective $i = 12.36\%$; or by Equation [3.3],

$$i \text{ per year} = \left(1 + \frac{0.12}{2}\right)^2 - 1$$

$$= 0.1236 \qquad (12.36\%)$$

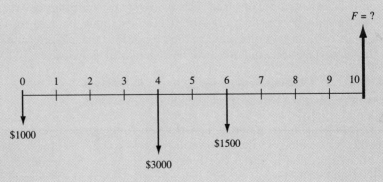

Figure 3–3   Cash-flow diagram, Example 3.4.

Exam draw cash flow

Since i has units of *per year,* n must be expressed in years. Thus,

$$F = 1000(F/P,12.36\%,10) + 3000(F/P,12.36\%,6) + 1500(F/P,12.36\%,4)$$

$$= \$11,634.50$$

Alternatively, we can use the effective rate of 6% per semiannual period and then use semiannual periods for n. In this case, the future worth is

$$F = 1000(F/P,6\%,20) + 3000(F/P,6\%,12) + 1500(F/P,6\%,8)$$

$$= \$11,634.50$$

Comment

The second method is the easier of the two because the interest tables can be used directly without interpolation or without the formulas to calculate the factors.

3.5.2   Uniform-Series and Gradient Factors

When the cash flow of the problem dictates the use of one or more of the uniform-series or gradient factors, the relationship between the compounding period, CP, and payment period, PP, must be determined. The relationship will be one of the following three cases:

Case 1. The payment period is equal to the compounding period, PP = CP.

Case 2. The payment period is longer than the compounding period, PP > CP.

Case 3. The payment period is shorter than the compounding period, PP < CP.

In this section, the procedure for solving problems which belong to one of the first two categories will be presented. Case 3 problems are discussed in the following section. For either case 1 or case 2, where PP = CP or PP > CP, the following procedure always applies:

Step 1. Count the number of payments and use that number as n. For example, if payments are made quarterly for 5 years, n is 20 quarters.

Step 2. Find the *effective* interest rate over the *same time period* as n in step 1. For example, if n is expressed in quarters, then the effective interest rate per quarter must be found.

Step 3. Use these values of n and i (and only these!) in the standard factor notation equations or formulas.

Table 3–5 shows the correct formulation (column 4) for several hypothetical cash-flow sequences (column 1) and interest rates (column 2). Note in column 4 that n is always equal to the number of payments and i is an effective rate expressed over the same time period as n.

Table 3–5	Examples of n and i values where PP = CP or PP > CP		
(1) **Cash-flow** **Sequence**	**(2)** **Interest Rate**	**(3)** **What to Find;** **What is Given**	**(4)** **Standard** **Notation**
$500 semiannually for 5 years	16% per year compounded semiannually	Find P; given A	$P = 500(P/A,8\%,10)$
$75 monthly for 3 years	24% per year compounded monthly	Find F; given A	$F = 75(F/A,2\%,36)$
$180 quarterly for 15 years	5% per quarter	Find F; given A	$F = 180(F/A,5\%,60)$
$25 per month increase for 4 years	1% per month	Find P; given G	$P = 25(P/G,1\%,48)$
$5000 per quarter for 6 years	1% per month	Find A; given P	$A = 5000(A/P,3.03\%,24)$

Exam ⇒

Example 3.5

If a woman deposits $500 every 6 months for 7 years, how much money will she have in her investment portfolio after the last deposit if the interest rate is 20% per year compounded quarterly?

Solution

The cash-flow diagram is shown in Figure 3–4. Since the compounding period (quarterly) is shorter than the payment period (semiannually), this is a case 2 problem. The first step is to determine that n, the number of payments, is 14. The future worth is

$$F = 500(F/A,i,14)$$

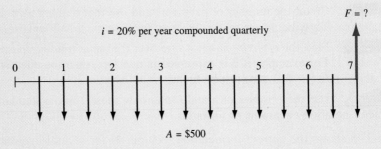

$i = 20\%$ per year compounded quarterly

$F = ?$

$A = \$500$

Figure 3–4   Diagram of semiannual deposits used to determine F, Example 3.5.

Now, since n is expressed in semiannual periods, an effective semiannual interest rate per semiannual period is needed. Use Equation [3.3] with $r = 0.10$ per 6-month period and $m = 2$ quarters per semiannual period.

$$i \text{ per 6 months} = \left(1 + \frac{0.10}{2}\right)^2 - 1$$

$$= 0.1025 \quad (10.25\%)$$

Alternatively, the effective semiannual interest rate could have been obtained from Table 3–3 by using the r value of 10% with semiannual compounding to get $i = 10.25\%$.

The value $i = 10.25\%$ seems reasonable, since we expect the effective rate to be slightly higher than the nominal rate of 10% per 6-month period. The effective rate can now be used to find the future worth of the semiannual deposits, where $n = 2(7) = 14$ periods.

$$F = A(F/A,10.25\%,14)$$

$$= 500(28.4891)$$

$$= \$14,244.50$$

Comment

It is important to note that the effective interest rate per payment period (6 months) is used for i and that the total number of payment periods is used for n.

Additional Example 3.9
Problems 3.26 to 3.32

3.6   CALCULATIONS FOR PAYMENT PERIODS SHORTER THAN COMPOUNDING PERIODS

This is the case 3 situation previously described in Section 3.5.2. When the payment period is shorter than the compounding period (PP < CP), the procedure to calculate the future worth or present worth depends on the conditions specified (or assumed) regarding interperiod compounding. *Interperiod compounding,* as used here, refers to the handling of the payments made *between* compounding periods. There are three possible scenarios:

1. There is no interest paid on the money deposited (or withdrawn) between compounding periods.

2. The money deposited (or withdrawn) between compounding periods earns simple interest.

3. All interperiod transactions earn compound interest.

Only scenario number 1 (no interest on interperiod transactions) will be considered here, since most real-world transactions fall into this category. If no interest

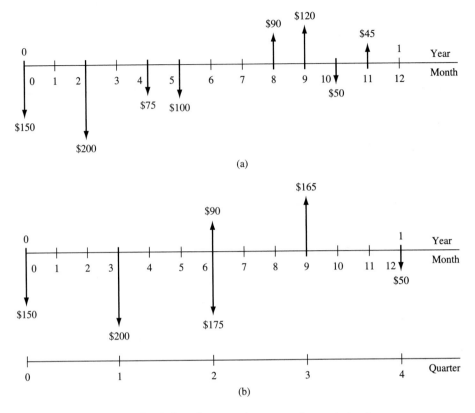

Figure 3–5   Diagram of cash flows for quarterly compounding periods using no interperiod interest.

is to be paid on interperiod transactions, then any amount of money that is deposited or withdrawn between compounding periods is regarded as having been *deposited at the end of the compounding period* or *withdrawn at the beginning of the compounding period*. This is the usual mode of operation of banks and other lending institutions. Thus, if the compounding period were a *quarter,* the actual transactions shown in Figure 3–5a would be treated as shown in Figure 3–5b. To find the present worth of the cash flow represented by Figure 3–5b, the nominal yearly interest rate is divided by 4 (since interest is compounded quarterly) and the appropriate *n* value is used in the *P/F* or *F/P* factor. If, for example, the interest rate is 12% compounded quarterly for the Figure 3–5 cash flows,

$$P = -150 - 200(P/F,3\%,1) - 85(P/F,3\%,2) + 165(P/F,3\%,3)$$

$$- 50(P/F,3\%,4)$$

$$= -\$317.73$$

Problems 3.33 to 3.37

ADDITIONAL EXAMPLES

Example 3.6

EFFECTIVE INTEREST RATES, SECTION 3.3   Ms. Jones plans to place money in a JUMBO certificate of deposit that pays 18% per year compounded daily. What effective rate will she receive (*a*) yearly and (*b*) semiannually?

Solution

(*a*) Using Equation [3.3], with $r = 0.18$ and $m = 365$,

$$i \text{ per year} = \left(1 + \frac{0.18}{365}\right)^{365} - 1$$

$$= 0.19716 \quad (19.716\%)$$

That is, Ms. Jones will get an effective 19.716% per year on her deposit.

(*b*) Here $r = 0.09$ per 6 months and $m = 182$ days:

$$i \text{ per 6 months} = \left(1 + \frac{0.09}{182}\right)^{182} - 1$$

$$= 0.09415 \quad (9.415\%)$$

Example 3.7

EFFECTIVE INTEREST RATES, SECTIONS 3.1 AND 3.4   Mr. Adams and Ms. James both plan to invest $5000 for 10 years at 10% per year. Compute the future worth for both individuals if Mr. Adams gets interest compounded annually and Ms. James gets continuous compounding.

Solution

Mr. Adams. For annual compounding the future worth is

$$F = P(F/P,10\%,10) = 5000(2.5937) = \$12,969$$

Ms. James. Using the continuous-compounding relation, Equation [3.5], first find the effective *i* per year.

$$i = e^{0.10} - 1 = 0.10517 \quad (10.517\%)$$

The future worth is

$$F = P(F/P,10.517\%,10) = 5000(2.7183) = \$13,591$$

Comment

Continuous compounding represents a $622, or 4.8%, increase in earnings. Just for comparison, note that a savings and loan association might compound daily, which yields an effective rate of 10.516% ($F = \$13,590$), whereas 10% continuous compounding offers only a very slight increase to 10.517%.

Example 3.8

CONTINUOUS COMPOUNDING, SECTION 3.4   If $2000 is deposited each year for 10 years at an interest rate of 10% per year, compare the present worth for (*a*) annual and (*b*) continuous compounding.

Solution

(*a*) For annual compounding,

$$P = 2000(P/A,10\%,10) = 2000(6.1446) = \$12,289$$

(*b*) For continuous compounding,

$$i \text{ per year} = e^{0.10} - 1 = 0.10517 \quad (10.517\%)$$

$$P = 2000(P/A,10.517\%,10)$$

$$= 2000(6.0104)$$

$$= \$12,021$$

As expected, the present worth for continuous compounding is less than that for annual compounding because higher interest rates require greater discounts of future cash flows.

Example 3.9

PAYMENT AND COMPOUNDING PERIODS, SECTION 3.5   Ms. Warren wants to purchase a previously owned compact car for $8500. She plans to borrow the money from her credit union and to repay it monthly over a period of 4 years. If the nominal interest rate is 12% per year compounded monthly, what will her monthly installments be?

Solution

The compounding period equals the payment period (case 1), with an effective monthly rate of $i = 1\%$ per month and $n = 12(4) = 48$ payments. Therefore, the monthly payments are

$$A = 8500(A/P,1\%,48) = 8500(0.02633) = \$223.84$$

Example 3.9
(Spreadsheet)

PAYMENT AND COMPOUNDING PERIODS, SECTION 3.5   Ms. Warren wants to purchase a previously owned compact car for $8500. She plans to borrow the money from her credit union and to repay it monthly over a period of 4 years. If the nominal interest rate is 12% per year compounded monthly, determine her monthly installments using a spreadsheet.

Solution

Set up the i and n values for months; $i = 0.12/12$ and $n = 48$ payments. Then apply the PMT function on the Excel spreadsheet to determine that $A = \$223.84$. See Figure 3–6 for a screen capture of one way to determine the monthly payment amount.

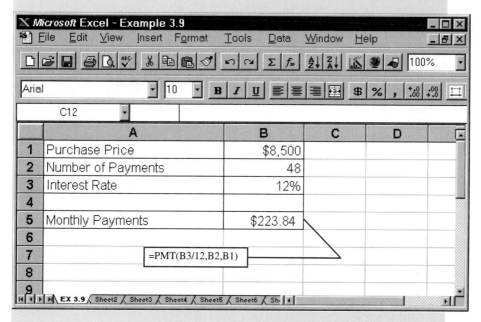

Figure 3–6   Spreadsheet for monthly compounding and payments, Example 3.9.

CHAPTER SUMMARY

Since many real-world problems involve payment and compounding periods which are not equal to 1 year, it is necessary to understand nominal and effective interest rates (Equation [3.3]). If the compounding period is made infinitely small, interest is compounded continuously (Equation [3.5]). All the engineering economy formulas require the use of effective interest rates only. In some instances, there are several effective rates that can be used to solve the problem (such as when only single-payment amounts are involved). In other cases, only one effective rate can be used (uniform-series and gradient problems).

For uniform-series and gradient problems, the interest and payment periods must agree. If the compounding period is shorter than the payment period, then the interest rate is manipulated to obtain an effective rate over the payment period. When the compounding period is longer than the payment period, the payments are manipulated so that the cash flows coincide with compounding periods (deposits are moved to the end and withdrawals are moved to the beginning of the period).

PROBLEMS

✓ **3.1** What is the difference between a nominal and a simple interest rate?

✓ **3.2** What is meant by (a) interest period, and (b) payment period?

3.3 What is the nominal interest rate per month for an interest rate of (a) 0.50% every 2 days and (b) 0.1% per day? Assume 30 days per month.

3.4 Identify the following interest rates as either nominal or effective. (a) $i = 1\frac{1}{2}\%$ per month, (b) $i = 3\%$ per quarter compounded quarterly, (c) $i = 0.5\%$ per week compounded daily, (d) $i = $ effective 16% per year compounded monthly.

3.5 For the interest statements in Problem 3.4, identify the compounding period.

3.6 Complete the following statement: An interest rate of 4% per month is a (an) _____ rate when compounded monthly, a (an) _____ rate when compounded weekly, and a (an) _____ rate when compounded daily.

3.7 If interest is compounded daily, what is the value of m in Equation [3.3] if you want to find an effective interest rate (a) per year, (b) per week, (c) per day?

3.8 What are the nominal and effective interest rates per year for an interest rate of 0.015% per day?

3.9 What effective interest rate per month is equivalent to a weekly interest rate of 0.3%? Assume 4 weeks per month.

3.10 What nominal rate per month is equivalent to a nominal 20% per year compounded daily? Assume 30 days per month, 365 days per year.

3.11 What effective interest rate per quarter is equivalent to a nominal 12% per year compounded monthly?

3.12 What nominal interest rate per month is equivalent to an effective 14% per year compounded (a) monthly, (b) daily? Assume 30 days per month, 365 days per year.

3.13 What quarterly interest rate is equivalent to an effective annual rate of 6% per year compounded quarterly?

3.14 What nominal rate per 3 years is equivalent to a monthly rate of $1\frac{1}{2}\%$?

3.15 Which interest rate is better: 20% per year compounded yearly or 18% per year compounded hourly? Assume 8760 hours per year.

3.16 Determine the value of the F/P factor for 5 years if the interest rate is 1% per month compounded daily. Assume 30 days per month.

3.17 What nominal interest rate per year compounded continuously would be equal to 25% per year compounded semiannually?

3.18 At what compounding frequency would a nominal 10.2%-per-year rate be equal to a nominal rate of 10% per year compounded continuously?

3.19 What nominal and effective rates per month are equivalent to a nominal 12%-per-year compounded continuously?

3.20 What is the difference in the present worth of $50,000 eight years from now if the interest rate is 13% per year compounded semiannually or continuously?

3.21 As an inducement to attract depositors, a bank has offered customers interest rates which increase with the size of the deposit. For example, from $1000 to $9999 the interest rate offered is 8% per year. Above $10,000, the rate is 9% per year compounded continuously. Since you only have $9000 to deposit, you are thinking about borrowing $1000 from a credit union so that you will have $10,000 and be able to take advantage of the higher interest rate. What is the maximum effective interest rate per year you could pay on the borrowed $1000 that would make your plan to borrow at least as attractive as not doing so?

3.22 You are contemplating borrowing $200,000 to start a mail-order radio-controlled airplane business. Credit union A has offered to loan you the money at an interest rate of 14% per year compounded continuously. Credit union B has offered the money with the stipulation that you repay it by making monthly payments of $5000 for 5 years. From which credit union should you borrow the money?

3.23 A small sign company borrowed $175,000 with an agreement to repay the loan in monthly installments. The first payment 1 month from now will be $1000 with each succeeding payment increasing by $100. If the interest rate is 10% per year compounded continuously, how long will it be before the company repays the loan completely?

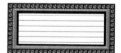

3.24 How much money could you borrow if you promise to make quarterly payments of $650 for 7 years, if the interest rate is 16% per year compounded continuously?

3.25 A woman deposited $125 each month for 10 years. If she received interest at a rate of 8% per year compounded continuously, how much did she have immediately after her last deposit?

3.26 A jeans washing company is buying an ozone system for its washing machines and for treatment of its dye wastewater. The initial cost of the ozone system is $750,000. How much money must the company save each quarter (in chemical costs and fines) in order to justify the investment if the system will last 5 years and the interest rate is (*a*) 16% per year compounded quarterly and (*b*) 12% per year compounded monthly?

3.27 Improvements to the runways at a regional airport are expected to cost $1.7 million. To pay for the improvements, the airport authority will increase the landing fees of commercial planes. If the airport wants to recover its investment in 4 years, how much money must be generated from the increased fees each year at an interest rate of 8% per year compounded semiannually?

3.28 In order to improve the elapsed times of nitro-burning dragsters and funny cars, a local drag strip is planning to smooth the track surface by laser planing it for $750,000. The new surface is expected to last for 3 years. If the track gets 216,000 spectators per year, by how much will the average ticket price have to increase if the interest rate is 1% per month? Assume uniform distribution of spectators.

3.29 If you wanted to accumulate $800,000 for your retirement 30 years from now, by how much would you have to increase your monthly deposit (uniformly) each month if your first deposit is $100 and the interest rate is 7% per year compounded weekly? Assume 4 weeks per month.

3.30 A young couple planning for their future is considering purchasing a franchise for a pizza business. The franchise will cost $35,000, and the couple will be required to purchase certain supplies from the franchisor. If the average profit on a pizza is $1, how many pizzas will they have to sell each week for 5 years just to pay for the cost of the franchise if the interest rate is 3½% per quarter compounded daily? Assume 30 days per month, 52 weeks per year.

3.31 If the building and equipment in Problem 3.30 costs $220,000, how many pizzas must be sold per day to pay off the franchise, building, and equipment in 4 years at an interest rate of 12% per year compounded continuously? Assume 365 days per year.

3.32 A sharp investor buys 2000 shares of stock for $23 per share. If she sells the stock after 7 months for $26 per share, what nominal and effective rates of return per year did she make?

3.33 How much money would be in the account of a person who deposited $1000 now and $100 every month and withdrew $100 every 2 months for 3 years? Use an interest rate of 6% per year with no interperiod interest paid.

3.34 How much money would be in a savings account in which a person had deposited $100 every month for 5 years at an interest rate of 5% per year compounded quarterly? Assume no interperiod interest.

3.35 A tool-and-die company expects to have to replace one of its lathes in 5 years at a cost of $18,000. How much would the company have to deposit every month in order to accumulate $18,000 in 5 years if the interest rate is 6% per year compounded semiannually? Assume no inter-period interest.

3.36 What monthly deposit would be equivalent to a deposit of $600 every 3 months for 2 years if the interest rate is 6% per year compounded semiannually? Assume no interperiod interest on all deposits.

3.37 How many monthly deposits of $75 would a person have to make in order to accumulate $15,000 if the interest rate is 6% per year compounded semiannually? Assume no interperiod interest is paid.

4

Use of Multiple Factors

Because many of the cash-flow situations encountered in real-world engineering problems do not fit exactly the cash-flow sequences for which the equations in Chapter 2 were developed, it is common to combine the equations. For a given sequence of cash flows, there are usually many ways to determine the equivalent present-worth, future-worth, or annual-worth cash flows. In this chapter, you will learn how to combine several of the engineering economy factors in order to address these more complex situations.

Purpose: Understand how to combine several factors to evaluate PW, FW, and AW of complex cash-flow sequences.

Locate PW or FW

Shifted series

Shifted series and single amounts

Shifted series and single amounts

Shifted gradients

Decreasing gradients

This chapter will help you:

1. Determine the location of present worth (PW) or future worth (FW) for randomly placed uniform series.

2. Determine PW, FW, or annual worth (AW) of a series starting at a time other than year 1.

3. Calculate PW or FW of randomly placed single amounts and uniform-series amounts.

4. Calculate AW of randomly placed single amounts and uniform-series amounts.

5. Make equivalence calculations for cash flows involving shifted uniform or geometric gradients.

6. Make equivalence calculations for cash flows involving decreasing gradients.

4.1   LOCATING THE PRESENT WORTH AND FUTURE WORTH

When a uniform series of payments begins at a time other than at the end of interest period 1, several methods can be used to find the present worth. For example, the present worth of the uniform series of disbursements shown in Figure 4–1 could be determined by any of the following methods:

- Use the P/F factor to find the present worth of each disbursement at year 0 and add them.

- Use the F/P factor to find the future worth of each disbursement in year 13, add them, and then find the present worth of the total using $P = F(P/F,i,13)$.

- Use the F/A factor to find the future amount by $F = A(F/A,i,10)$, and then compute the present worth using $P = F(P/F,i,13)$.

- Use the P/A factor to compute the "present worth" (which will be located in year 3, not year 0), and then find the present worth in year 0 by using the $(P/F,i,3)$ factor. (Present worth is enclosed in quotation marks to represent the present worth as determined by the P/A factor and to differentiate it from the present worth in year 0.)

This (and the next section) illustrate the last method for calculating the present worth of a uniform series that does not begin at the end of period 1. For Figure 4–1, the "present worth" obtained using the P/A factor would be located in year 3, not year 4. This is shown in Figure 4–2. Note that P is located *1 year*

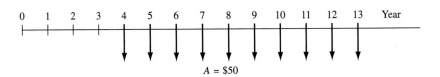

Figure 4–1   A randomly placed uniform series.

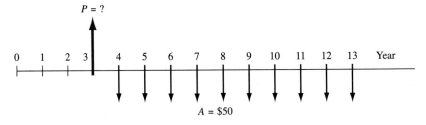

Figure 4–2   Location of P for the randomly placed uniform series in Figure 4–1.

prior to the beginning of the first annual amount. Why? Because the *P/A* factor was derived with the *P* in time period 0 and the *A* beginning at the end of interest period 1; that is, the *P* is always one interest period ahead of the first *A* value (Section 2.2). The most common mistake made in working problems of this type is improper placement of *P*. Therefore, it is extremely important that you remember the following rule. *The present worth is always located one interest period prior to the first uniform-series amount when using the P/A factor.*

On the other hand, the *F/A* factor was derived in Section 2.3 with the future worth *F* located in the *same* period as the last payment. Figure 4–3 shows the location of the future worth when *F/A* is used for the Figure 4–1 cash flow. Thus, *the future worth is always located in the same period as the last uniform payment when using the F/A factor.*

It is also important to remember that the number of periods *n* that should be used with the *P/A* or *F/A* factors is equal to the number of payments. It is generally helpful to *renumber* the cash-flow diagram to avoid errors in counting. Figure 4–4 shows the cash-flow diagram of Figure 4–1 renumbered for determination of *n*. Note that in this example $n = 10$.

Additional Example 4.13
Problem 4.1

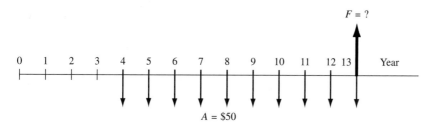

Figure 4–3   Placement of *F* for the uniform series of Figure 4–1.

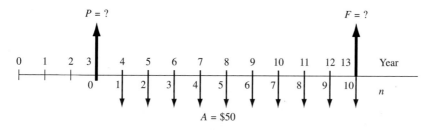

Figure 4–4   Renumbering of payments in Figure 4–1 to show that $n = 10$ for the *P/A* or *F/A* factors.

4.2   CALCULATIONS FOR A UNIFORM SERIES THAT BEGINS AFTER PERIOD 1

As stated in Section 4.1, there are many methods that can be used to solve problems having a uniform series that is shifted, that is, it begins at a time other than the end of period 1. However, it is generally more convenient to use the uniform-series factors than the single-amount factors. There are specific steps which should be followed in order to avoid errors:

1. Draw a cash-flow diagram of the receipts and disbursements.
2. Locate the present worth or future worth of each series on the cash-flow diagram.
3. Determine n by renumbering the cash-flow diagram.
4. Draw the cash-flow diagram representing the desired equivalent cash flow.
5. Set up and solve the equations.

These steps are illustrated in Examples 4.1 and 4.2.

Example 4.1

A person buys a small piece of land for $5000 down and deferred annual payments of $500 a year for 6 years starting 3 years from now. What is the present worth of the investment if the interest rate is 8% per year?

Solution

The cash-flow diagram is shown in Figure 4–5. The symbol P_A is used throughout this chapter to represent the present worth of a uniform annual series A, and P_A' represents the present worth at a time other than period 0. Similarly, P_T represents the total present worth at time 0. The correct placement of P_A' and diagram renumbering to obtain n are also indicated in Figure 4–5. Note that P_A' is located in year 2,

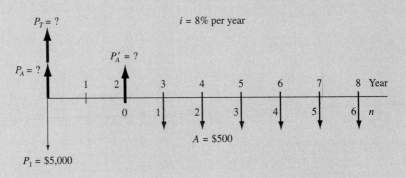

Figure 4–5   Placement of present-worth values, Example 4.1.

not year 3, and $n = 6$, not 8, for the P/A factor. First find the value of P'_A of the shifted series.

$$P'_A = \$500(P/A,8\%,6)$$

Since P'_A is located in year 2, it is necessary to find P_A in year 0:

$$P_A = P'_A(P/F,8\%,2)$$

The total present worth is determined by adding P_A and the initial investment P_1.

$$
\begin{aligned}
P_T &= P_1 + P_A \\
&= 5000 + 500(P/A,8\%,6)(P/F,8\%,2) \\
&= 5000 + 500(4.6229)(0.8573) \\
&= \$6981.60
\end{aligned}
$$

Example 4.2

Calculate the 8-year equivalent uniform annual worth amount at 16% per year interest for the uniform disbursements shown in Figure 4–6.

Solution

Figure 4–7 shows the original cash-flow diagram from Figure 4–6 and the desired equivalent diagram. In order to convert uniform cash flows that begin sometime after period 1 into an equivalent uniform worth over all periods, first convert the uniform series into a present-worth or future-worth amount. Then either the A/P factor or the A/F factor determines the equivalent annual worth. Both methods are illustrated here.

1. **Present-worth method.**   (Refer to Figure 4–7.) Calculate the P'_A for the shifted series.

$$P'_A = 800(P/A,16\%,6)$$

$$P_T = P'_A(P/F,16\%,2) = 800(P/A,16\%,6)(P/F,16\%,2)$$

$$= 800(3.6847)(0.7432) = \$2190.78$$

where P_T is the total present worth of the cash flow. The equivalent series A' *for 8 years* can now be determined via the A/P factor (Figure 4–7b).

$$A' = P_T(A/P,16\%,8) = \$504.36$$

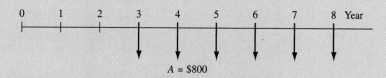

$A = \$800$

Figure 4–6   Series of uniform disbursements, Example 4.2.

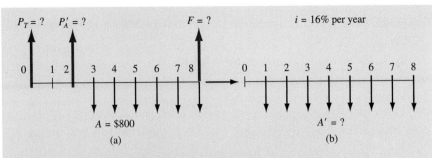

Figure 4-7   Desired equivalent diagram for the uniform series of Figure 4-6.

2. **Future-worth method.**   (Refer to Figure 4–7.) The first step is to calculate the future worth F in year 8.

$$F = 800(F/A,16\%,6) = \$7184$$

The A/F factor is now used to obtain A'.

$$A' = F(A/F,16\%,8) = \$504.46$$

Comment

In the present-worth method, note that P_A' is located in year 2, not year 3. After the present worth is determined, the equivalent series is calculated using $n = 8$. In the future-worth method, $n = 6$ is used to find F, and $n = 8$ is used to find the equivalent 8-year series.

Additional Example 4.14
Problems 4.2 to 4.16

4.3   CALCULATIONS INVOLVING UNIFORM-SERIES AND RANDOMLY PLACED AMOUNTS

When a cash flow includes both a uniform series and randomly placed single amounts, the procedures of Section 4.2 are applied to the uniform series and the single-amount formulas are applied to the one-time cash flows. This type of problem, illustrated in Examples 4.3 through 4.5, is merely a combination of previous types.

Example 4.3

A couple owning 50 hectares of valuable land have decided to sell the mineral rights on their property to a mining company. Their primary objective is to obtain long-term investment income and sufficient money to finance the college education of their two children. Since the children are currently 12 and 2 years of age, the couple estimates

that the children will start college in 6 and 16 years, respectively, from the present time. They therefore make a proposal to the mining company that it pay $20,000 per year for 20 years beginning 1 year hence, plus $10,000 six years from now and $15,000 sixteen years from now. If the company wants to pay off its lease immediately, how much should it pay now if the investment could make 16% per year?

Solution

The cash-flow diagram is shown in Figure 4–8 from the couple's perspective. This problem is solved by finding the present worth of the 20-year uniform series and adding it to the present worth of the two one-time amounts.

$$P = 20{,}000(P/A{,}16\%{,}20) + 10{,}000(P/F{,}16\%{,}6) + 15{,}000(P/F{,}16\%{,}16)$$

$$= \$124{,}075$$

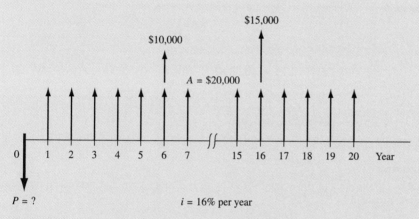

Figure 4–8   Diagram including a uniform series and single amounts, Example 4.3.

Comment

Note that the uniform series started at the end of year 1 so that the present worth obtained with the P/A factor represents the present worth at year 0. So, it is not necessary to use the P/F factor on the uniform series.

Example 4.4

If the uniform series described in Example 4.3 begins in year 3 after the contract is signed, determine the present worth at $t = 0$.

Solution

The cash-flow diagram is shown in Figure 4–9, with the n scale shown above the time axis. The n value for the uniform series is still 20.

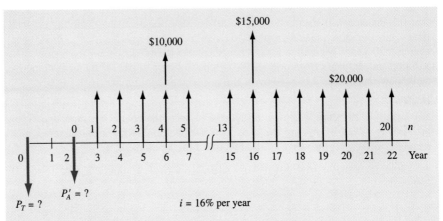

Figure 4–9   Diagram from Figure 4–8 with the A series shifted 2 years.

$$P'_A = 20,000(P/A,16\%,20)$$

$$P_T = P'_A(P/F,16\%,2) + 10,000(P/F,16\%,6) + 15,000(P/F,16\%,16)$$

$$= 20,000(P/A,16\%,20)(P/F,16\%,2) + 10,000(P/F,16\%,6)$$

$$+ 15,000(P/F,16\%,16)$$

$$= \$93,625$$

Comment

Delaying the start of the annual series by only 2 more years has decreased the total present worth by $30,450.

Example 4.5

The couple in Example 4.3 plan to sell their mineral rights. Determine for the couple the future worth of all receipts if the cash flows actually occur as presented in Figure 4–9 and earnings are at 16% per year.

Solution

The future worth of the uniform series and the single-amount receipts can be calculated as follows in year 22.

$$F = 20,000(F/A,16\%,20) + 10,000(F/P,16\%,16) + 15,000(F/P,16\%,6)$$

$$= \$2,451,626$$

Comment

Although the determination of n is straightforward, make sure that you fully understand how the values 20, 16, and 6 are obtained. Also, in order to get a feel for the effect that interest rates have on the time value of money, rework Examples 4.3

through 4.5 using $i = 6\%$ per year instead of 16%. You should get the results shown below. Obviously, a 10%-per-year difference in i has a significant effect on P and F.

i	Example 4.3	Example 4.4	Example 4.5
6%	$P = \$242,352$	$P_T = \$217,118$	$F = \$782,381$
16%	$P = \$124,075$	$P_T = \$93,625$	$F = \$2,451,626$

Additional Example 4.15
Problems 4.17 to 4.23

4.4 EQUIVALENT UNIFORM ANNUAL WORTH OF BOTH UNIFORM-SERIES AND RANDOMLY PLACED AMOUNTS

To calculate the equivalent uniform annual series A of cash flows which include randomly placed single amounts and uniform-series amounts, the most important fact to remember is to *first convert everything to a present worth or a future worth*. Then the equivalent uniform series is obtained with the appropriate A/P or A/F factor.

Example 4.6

Calculate the 20-year equivalent uniform annual worth for the cash flows described in Example 4.3 (Figure 4–8).

Solution

Figure 4–10 is the equivalent cash-flow diagram. From Figure 4–8 it is evident that the uniform series of $20,000 per year is already distributed through all 20 years. It is therefore necessary to convert only the single amounts to an equivalent uniform annual series and add it to the $20,000. This can be done by either (*a*) the present-worth or (*b*) the future-worth method.

(*a*) $A = 20,000 + 10,000(P/F,16\%,6)(A/P,16\%,20)$

$\qquad + \ 15,000(P/F,16\%,16)(A/P,16\%,20)$

$\quad = 20,000 + [10,000(P/F,16\%,6) + 15,000(P/F,16\%,16)](A/P,16\%,20)$

$\quad = \$20,928$ per year

(*b*) $A = 20,000 + 10,000(F/P,16\%,14)(A/F,16\%,20)$

$\qquad + \ 15,000(F/P,16\%,4)(A/F,16\%,20)$

$\quad = 20,000 + [10,000(F/P,16\%,14) + 15,000(F/P,16\%,4)](A/F,16\%,20)$

$\quad = \$20,928$ per year

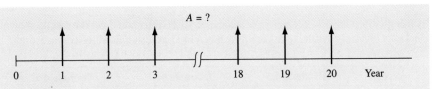

Figure 4–10   Equivalent 20-year series of Figure 4–8, Example 4.6.

Comment
Note that it is necessary to take the single payments to either end of the time scale before annualizing. Not doing so will result in unequal amounts in some years.

Example 4.7

Convert the cash flows of Figure 4–9 to an equivalent uniform annual series over 22 years. Use $i = 16\%$ per year.

Solution
Since the uniform series is not distributed through all 22 years, first find the present worth or future worth of the series. This was done in Examples 4.4 and 4.5, respectively. The equivalent uniform annual worth is then determined by multiplying the values previously obtained by the $(A/P,16\%,22)$ factor or the $(A/F,16\%,22)$ factor as follows:

Using A/P:

$$A = P_T(A/P,16\%,22) = 93,625(0.166353) = \$15,575$$

Using A/F:

$$A = F(A/F,16\%,22) = 2,451,626(0.006353) = \$15,575$$

Comment
When a uniform series begins at a time other than period 1, or when the series occurs over a different time period than the desired equivalent series, remember that the equivalent present or future worth must be obtained before the equivalent uniform series is determined.

Problems 4.24 to 4.28

4.5   PRESENT WORTH AND EQUIVALENT UNIFORM SERIES OF SHIFTED GRADIENTS

In Section 2.5, Equation [2.12] was derived to calculate the present worth of a uniform gradient. You will recall that the equation was derived for a present worth in year 0 with the gradient starting between periods 1 and 2 (see Figure 2–5). Therefore, the present worth of a uniform gradient will always be located 2 periods before the gradient starts. Examples 4.8 and 4.9 illustrate where the present worth of the gradient P_G is located.

Example 4.8

For the cash flows of Figure 4–11, locate the gradient present worth.

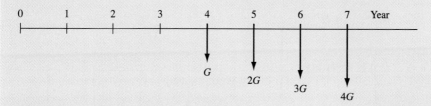

Figure 4–11   Diagram of a gradient, Example 4.8.

Solution

The present worth of the gradient, P_G, is shown in Figure 4–12. Since the present worth of a gradient series is located 2 periods before the start of the gradient, P_G is placed at the end of year 2. It is usually advantageous to renumber the diagram so that the gradient year 0 and the number of years n of the gradient are determined. To accomplish this, determine where the gradient begins, label that time as year 2, and then work backward and forward. In this example, since the gradient started between actual years 3 and 4, gradient year 2 is placed below year 4. Gradient year 0 is then located by moving back 2 years.

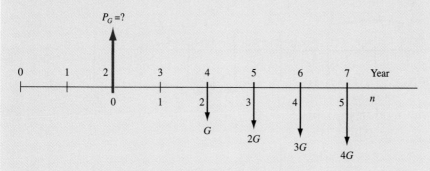

Figure 4–12   Diagram locating the present worth of the gradient in Figure 4–11.

Example 4.9

For the cash flows in Figure 4–13, explain why the present worth of the gradient is located in year 3.

Solution

The gradient of $50 begins between years 4 and 5 of the original cash-flow diagram. Therefore, actual year 5 represents gradient year 2; the present worth of the gradient is then located in year 3. If Figure 4–13 is divided into two cash-flow diagrams, the location of the gradient becomes quite clear, as presented in Figure 4–14.

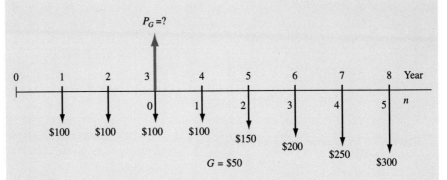

Figure 4–13   Location of the present worth of a gradient.

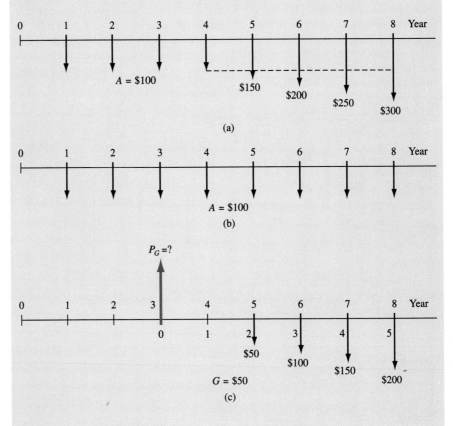

Figure 4–14   Partitioned cash flow of Figure 4–13: (a) = (b) + (c).

When the gradient of a cash-flow sequence starts between periods 1 and 2, it is called a conventional gradient, as discussed in Section 2.5. When a gradient begins at a time before or after period 2, it is called a *shifted gradient*. To determine the n value for a gradient factor, use the same renumbering procedure of the two previous examples. The cash flows in Figure 4–15*a* to *c* are illustrations. The gradients G, number of years n, and gradient factors (P/G and A/G) used to calculate the present worth and annual series of the gradients are shown on each diagram, assuming an interest rate of 6% per year.

It is important to note that the A/G factor *cannot* be used to find an equivalent A value in periods 1 through n for cash flows involving a shifted gradient. Consider the cash-flow diagram of Figure 4–16. To find the equivalent annual worth disbursement in years 1 through 7, first find the present worth of the

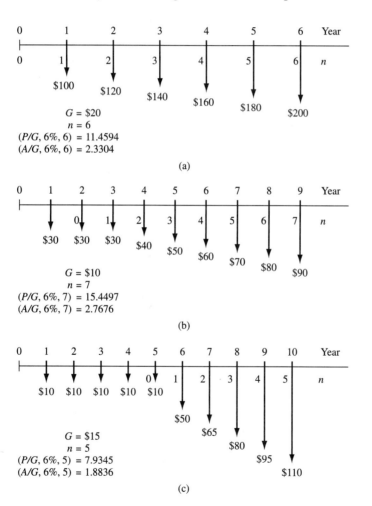

Figure 4–15   Determination of G and n values used in gradient factors.

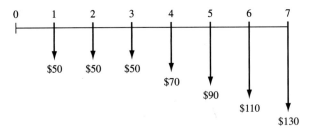

Figure 4-16   Diagram illustrating a shifted gradient.

gradient, take this present worth back to year 0, and then annualize the present worth from year 0 with the A/P factor. Or, move all the cash flows to year 7 and use the A/F factor. If you apply the annual-series gradient factor $(A/G,i,n)$ directly, the gradient is converted into an equivalent annual worth over years 3 through 7 only. For this reason, if a uniform series is desired through all the periods, the first step is always to find the present worth of the gradient at actual year 0. The steps involved are illustrated in Example 4.10.

Example 4.10

Set up the relations to compute the equivalent annual worth for the cash-flow amounts in Figure 4–16.

Solution

The solution steps to compute the equivalent amount A are

1. Consider the $50 base amount as an annual amount for all 7 years (Figure 4–17).

2. Find the present worth P_G of the $20 gradient that starts in actual year 4 as shown on the gradient-year time scale where $n = 5$.

$$P_G = 20(P/G,i,5)$$

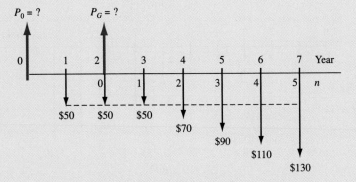

Figure 4-17   Diagram used to determine A for a shifted gradient, Example 4.10.

3. Bring the gradient present worth back to actual year 0.

$$P_0 = P_G(P/F,i,2) = 20(P/G,i,5)(P/F,i,2)$$

4. Annualize the gradient present worth from year 0 through year t to obtain A_G.

$$A_G = P_0(A/P,i,7)$$

5. Finally, add the base amount to the gradient annual worth to determine A.

$$A = 20(P/G,i,5)(P/F,i,2)(A/P,i,7) + 50$$

If the cash-flow sequence involves a geometric (escalating) gradient at the rate E and the gradient starts at a time other than between interest periods 1 and 2, the same procedure applies. In this case, P_E is located on the diagram in a manner similar to that for P_G above. Example 4.11 shows these calculations.

Example 4.11

Calculate the equivalent present worth of $35,000 now and an annual series of $7000 per year for 5 years beginning 1 year from now, which starts to increase annually at 12% thereafter for the next 8 years. Use an interest rate of 15% per year.

Solution

Figure 4–18 presents the cash flows. The total present worth P is found using $E = 0.12$ and $i = 0.15$. Equation [2.18] determines the present worth P_E for the

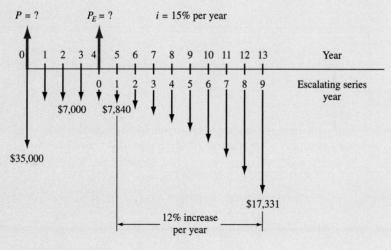

Figure 4-18   Cash-flow diagram, Example 4.11.

entire geometric series at actual year 4.

$$P = 35{,}000 + 7000(P/A,15\%,4) + \left[7000\frac{(1.12/1.15)^9 - 1}{0.12 - 0.15}\right](P/F,15\%,4)$$

$$= 35{,}000 + 19{,}985 + 28{,}247$$

$$= \$83{,}232$$

Note that $n = 4$ in the P/A factor because the $7000 in year 5 is the amount D in Equation [2.18], shown in brackets. The P_E in year 4 (Figure 4–18) is then moved to year 0 with the $(P/F,15\%,4)$ factor.

Additional Examples 4.16 and 4.17
Problems 4.29 to 4.45

4.6   DECREASING GRADIENTS

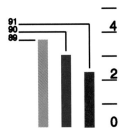

The use of the gradient factors is the same for increasing and decreasing gradients, except that in the case of decreasing gradients the following are true:

1. The base amount is equal to the *largest* amount attained in the gradient series.

2. The gradient term is *subtracted* from the base amount instead of added; thus, the term $-G(A/G,i,n)$ or $-G(P/G,i,n)$ is used in the computations.

The present worth of the gradient will still take place 2 periods before the gradient starts and the A value will start at period 1 and continue through period n.

Example 4.12

Find the (*a*) present worth and (*b*) annual worth of the receipts shown in Figure 4–19 for $i = 7\%$ per year.

Solution

(*a*) The cash flows of Figure 4–19 are partitioned in Figure 4–20. The dashed line in Figure 4.20*a* indicates that the $100 conventional gradient is subtracted from an annual receipt of $900 starting in year 2. The total present worth is

$$P_T = P_A - P_G = 900(P/A,7\%,6) - 100(P/G,7\%,6)$$

$$= 900(4.7665) - 100(10.9784)$$

$$= \$3192.01$$

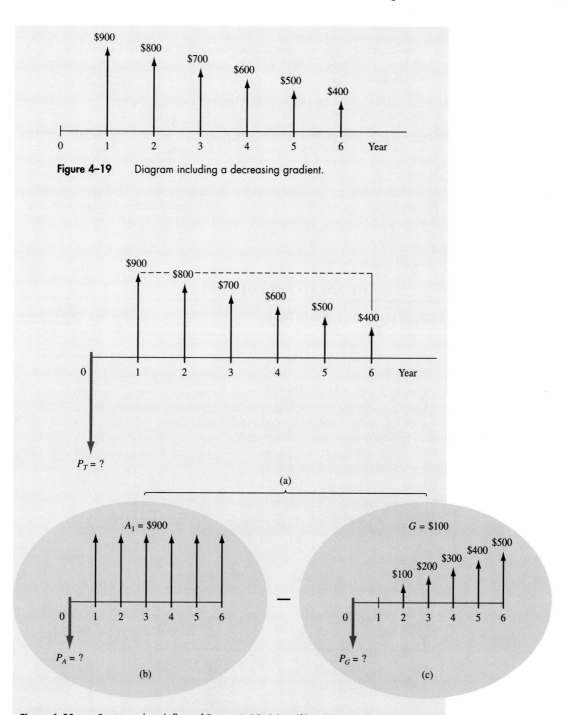

Figure 4–19   Diagram including a decreasing gradient.

Figure 4–20   Partitioned cash flow of Figure 4–19: (a) = (b) − (c).

(b) The annual-worth series is comprised of two components: the base amount and the gradient's equivalent uniform amount. The annual-receipt series ($A_1 = \$900$) is the base amount, and the gradient uniform series A_G is subtracted from A_1.

$$A = A_1 - A_G = 900 - 100(A/G,7\%,6)$$

$$= 900 - 100(2.3032)$$

$$= \$669.68 \text{ per year for years 1 to 6}$$

Comment

Shifted decreasing gradients are treated in a manner similar to shifted increasing gradients. For an example that combines conventional increasing and shifted decreasing gradients, see Additional Example 4.18.

Additional Example 4.18
Problems 4.46 to 4.50

ADDITIONAL EXAMPLES

Example 4.13

LOCATING P AND F, SECTION 4.1   A family decides to buy a new refrigerator on credit. The payment plan calls for a $100 down payment now (the month is March) and $55 a month from June to November with interest at 1½% per month compounded monthly. Construct the cash-flow diagram and indicate P in the month in which you can compute an equivalent value using one P/A and one F/P factor. Give the n values for all computations.

Solution

Since the payment period (months) equals the compounding period, the 1.5% interest tables can be used. Figure 4–21 places the P in May. The relation using only the

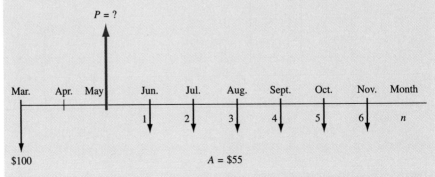

Figure 4–21   Placement of an equivalent amount using only P/A and F/P factors, Example 4.13.

two factors is

$$P = 100((F/P,1.5\%,2) + 55(P/A,1.5\%,6)$$

where $n = 2$ for the F/P factor and $n = 6$ for the P/A factor.

Comment

The placement of P is controlled by the uniform series A, since the P/A factor is inflexible in this regard.

Example 4.14

SHIFTED UNIFORM SERIES, SECTION 4.2   Consider the two uniform series in Figure 4–22. Compute the present worth at 15% per year using three different methods.

Solution

There are numerous ways to find the present worth. The two simplest are probably the future-worth and present-worth methods. For a third method, we will use the intermediate year 7 as an anchor point. This is called the *intermediate-year method*.

(a) **Present-worth method.**   (See Figure 4–23a.) The use of P/A factors for the uniform series, followed by the use of P/F factors to obtain the present worth in year 0, allows us to find P_T.

$$P_T = P_{A1} + P_{A2}$$

$$P_{A1} = P'_{A1}(P/F,15\%,2) = A_1(P/A,15\%,3)(P/F,15\%,2)$$

$$= 1000(2.2832)(0.7561)$$

$$= \$1726$$

$$P_{A2} = P'_{A2}(P/F,15\%,8) = A_2(P/A,15\%,5)(P/F,15\%,8)$$

$$= 1500(3.3522)(0.3269)$$

$$= \$1644$$

$$P_T = 1726 + 1644 = \$3370$$

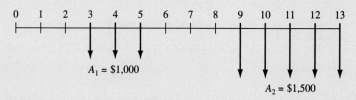

$i = 15\%$ per year

$A_1 = \$1{,}000$

$A_2 = \$1{,}500$

Figure 4–22   Uniform series to compute a present worth by several factors, Example 4.14.

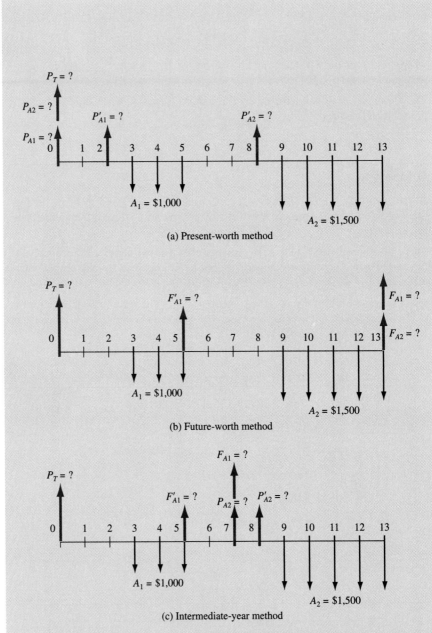

(a) Present-worth method

(b) Future-worth method

(c) Intermediate-year method

Figure 4–23   Computation of the present worth of Figure 4–22 by three methods.

(*b*) **Future-worth method.**   (See Figure 4–23*b*.) Using the *F/A*, *F/P*, and *P/F* factors, we have

$$P_T = (F_{A1} + F_{A2})(P/F,15\%,13)$$

$$F_{A1} = F'_{A1}(F/P,15\%,8) = A_1(F/A,15\%,3)(F/P,15\%,8)$$

$$= 1000(3.4725)(3.0590)$$

$$= \$10{,}622$$

$$F_{A2} = A_2(F/A,15\%,5) = 1500(6.7424)$$

$$= \$10{,}113$$

$$P_T = (F_{A1} + F_{A2})(P/F,15\%,13) = 20{,}735(0.1625)$$

$$= \$3369$$

(*c*) **Intermediate-year method.**   (See Figure 4–23*c*.) Find the present worth of both series at year 7 and then use the *P/F* factor.

$$P_T = (F_{A1} + P_{A2})(P/F,15\%,7)$$

The P_{A2} value is computed as a present worth; but to find the total value P_T at year 0, it must be treated as an *F* value. Thus,

$$F_{A1} = F'_{A1}(F/P,15\%,2) = A_1(F/A,15\%,3)(F/P,15\%,2)$$

$$= 1000(3.4725)(1.3225)$$

$$= \$4592$$

$$P_{A2} = P'_{A2}(P/F,15\%,1) = A_2(P/A,15\%,5)(P/F,15\%,1)$$

$$= 1500(3.3522)(0.8696)$$

$$= \$4373$$

Now,

$$P_T = (F_{A1} + P_{A2})(P/F,15\%,7)$$

$$= 8965(0.3759)$$

$$= \$3370$$

Example 4.14
(Spreadsheet)

SHIFTED UNIFORM SERIES, SECTION 4.2   Consider the two uniform series in Figure 4–22. Compute the *P* value at 15% per year using a spreadsheet and the present-worth method.

Solution

The spreadsheet solution in Figure 4–24 uses the PV function to calculate the value for each amount in year 0, and the Excel SUM operator adds the P values in cell C3, where $P = \$3370$.

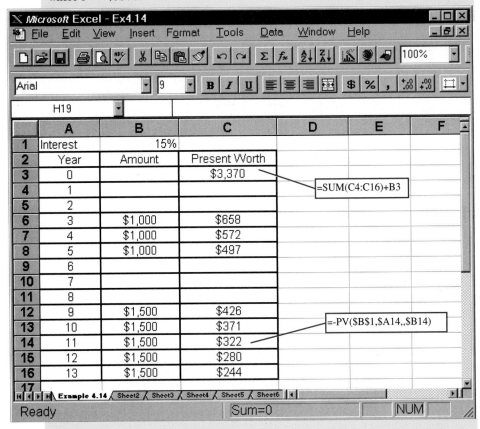

Figure 4–24   Spreadsheet solution for Example 4.14 using the present-worth method.

Example 4.15

PRESENT WORTH WITH MULTIPLE FACTORS, SECTION 4.3   Calculate the total present worth of the following series of cash flows at $i = 18\%$ per year.

Year	0	1	2	3	4	5	6	7
Cash flow, $	+460	+460	+460	+460	+460	+460	+460	−5000

Solution

The cash-flow diagram is shown in Figure 4–25. Since the receipt in year 0 is equal to the receipts of the A series, the P/A factor can be used for either 6 or 7 years. The problem is worked both ways below.

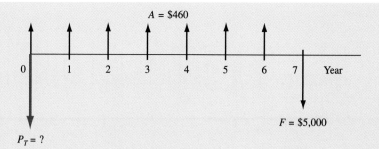

Figure 4–25   Cash-flow diagram, Example 4.15.

1. **Using P/A and $n = 6$.**   For this case, the receipt P_1 in year 0 is added to the present worth of the remaining amounts, since the P/A factor for $n = 6$ will place P_A in year 0. Thus,

$$P_T = P_1 + P_A - P_F$$

$$= 460 + 460(P/A,18\%,6) - 5000(P/F,18\%,7)$$

$$= \$499.40$$

Note that the present worth of the \$5000 cash flow is negative, since it is a negative cash flow.

2. **Using P/A and $n = 7$.**   By using the P/A factor for $n = 7$, the "present worth" is located in year -1, not year 0, because the P is 1 period prior to the first A. It is therefore necessary to move the P_A value 1 year forward with the F/P factor. Thus,

$$P = 460(P/A,18\%,7)(F/P,18\%,1) - 5000(P/F,18\%,7)$$

$$= \$499.38$$

Comment

Rework the problem by first finding the future worth of the series and then solving for P. Within round-off error, you should obtain the same answer as above.

Example 4.16

SHIFTED GRADIENTS, SECTION 4.5   Determine the amount of the gradient, the location of the gradient present worth, and the gradient factor n values for the cash flows of Figure 4–26.

Solution

You should construct your own cash-flow diagram to locate the gradient present worth and determine the n values. If we label the series from year 1 to year 4 as G_1 with a base amount of \$25, $G_1 = \$15$, n_1 equals 4 years, and P_{G1} occurs in year 0. For the second series use G_2. The base amount is also \$25, but $G_2 = \$5$ starting in actual year 7, and n_2 equals 7 years. Here P_{G2} is located in year 5.

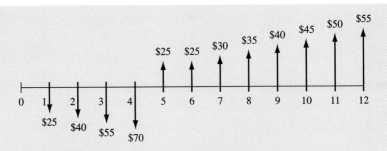

Figure 4–26   Cash flow of two gradients, Example 4.16.

Comment
Even though Figure 4–26 shows a series of disbursements and receipts, both gradients are increasing. Decreasing gradients are discussed in Section 4.6.

Example 4.17

SHIFTED GRADIENTS, SECTION 4.5   For $i = 8\%$ and the cash flows of Figure 4–27, compute the equivalent (*a*) annual worth, and (*b*) present worth.

Solution

(*a*) The dashed lines of Figure 4–27 should help in the solution for present worth and equivalent annual worth. For the annual series, use the steps outlined in Example 4.10.

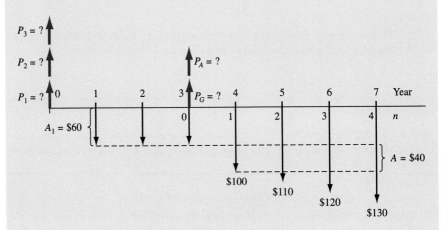

Figure 4–27   Shifted gradient, Example 4.17.

1. $A_1 = \$60$ for 7 years

    $A = \$40$ base amount of gradient for 4 years

    $A_3 =$ equivalent series of base amount for 7 years

    $= P_2(A/P,8\%,7)$

where

$P_2 =$ present worth of $A = \$40$ series

$= 40(P/A,8\%,4)(P/F,8\%,3)$

$= \$105.17$

$A_3 = 105.17(A/P,8\%,7)$

$= \$20.20$

2. $P_G =$ present worth in year 3 of the \$10 uniform gradient

    $= G(P/G,8\%,4) = 10(4.6501)$

    $= \$46.50$

3. $P_1 =$ present worth of P_G in actual year 0

    $= P_G(P/F,8\%,3) = 46.50(0.7938)$

    $= \$36.91$

4. $A_2 =$ equivalent 7-year A value of gradient

    $= P_1(A/P,8\%,7) = 36.91(0.19207)$

    $= \$7.09$

5. The equivalent annual worth is

    $A = A_1 + A_2 + A_3 = 60.00 + 7.09 + 20.20$

    $= \$87.29$

(b) To find P for Figure 4–27, note that the calculations in steps 2 and 3 yield $P_1 = \$36.91$. The \$40 uniform series has a present worth of P_2.

$$P_2 = 40(P/A,8\%,4)(P/F,8\%,3) = 40(3.3121)(0.7938)$$

$$= \$105.17$$

The \$60 uniform annual series has a present worth of P_3.

$$P_3 = 60(P/A,8\%,7) = 60(5.2064)$$

$$= \$312.38$$

The total present worth is the sum.

$$P_T = P_1 + P_2 + P_3 = 36.91 + 105.17 + 312.38$$

$$= \$454.46$$

This is equivalent to finding P_T using the A determined in part (a).

$$P_T = A(P/A,8\%,7) = 87.29(5.2064) = \$454.46$$

Example 4.18

CONVENTIONAL AND SHIFTED GRADIENTS, SECTIONS 4.5 AND 4.6   Assume that you are planning to invest money at 7% per year as shown by the increasing gradient of Figure 4–28. In addition, you expect to withdraw according to the decreasing gradient shown. Find the net present worth and equivalent annual worth for the entire cash-flow sequence.

Solution

For the investment sequence, G is $500, the base amount is $2000, and n equals 5. For the withdrawal sequence through year 10, G is $-1000, the base amount is $5000, and n equals 5. There is a 2-year annual series with A equal to $1000 in years 11 and 12. For the investment sequence,

$$P_I = \text{present worth of investment deposits}$$

$$= 2000(P/A,7\%,5) + 500(P/G,7\%,5)$$

$$= 2000(4.1002) + 500(7.6467)$$

$$= \$12,023.75$$

For the withdrawal sequence, let P_W represent the present worth of the withdrawal base amount and gradient series in years 6 through 10 (P_2), plus the present worth of withdrawals in years 11 and 12 (P_3). Then

$$P_W = P_2 + P_3$$

$$= P_G(P/F,7\%,5) + P_3$$

$$= [5,000(P/A,7\%,5) - 1000(P/G,7\%,5)](P/F,7\%,5)$$

$$\quad + 1000(P/A,7\%,2)(P/F,7\%,10)$$

$$= [5000(4.1002) - 1000(7.6467)](0.7130) + 1000(1.8080)(0.5083)$$

$$= \$9165.12 + 919.00 = \$10,084.12$$

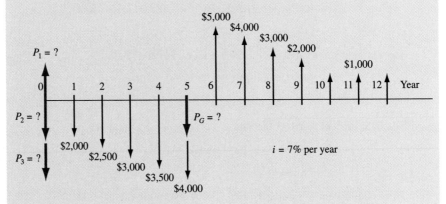

Figure 4–28   Investment and withdrawal cash-flow sequences, Example 4.18.

Since P_I is actually a negative cash flow and P_W is positive, the total present worth is

$$P_T = P_W - P_I = 10,084.12 - 12,023.75$$

$$= \$-1939.63$$

The A value may be computed using

$$A = P(A/P,7\%,12)$$

$$= \$-244.20$$

The interpretation of these results is as follows: In present-worth equivalence, you will invest \$1939.63 more than you expect to withdraw. This is equivalent to an annual savings of \$244.20 per year for the 12-year period.

CHAPTER SUMMARY

In Chapter 2, we derived equations which are used to calculate the present, future, or annual worth of specific cash-flow situations. In Chapter 3, we showed that these equations are useful for time periods shorter or longer than 1 year. In this chapter, we have shown that these equations also apply to cash-flow situations different from those for which the basic relations are derived. For example, when a uniform series does not begin in period 1, we still use the P/A factor to find the "present worth" of the series, except the P value is located not at time 0, but one interest period ahead of the first A value. For arithmetic gradients, the P value is 2 periods ahead of where the gradient starts. With this type of information, we are now able to solve for any symbol value—P, A, or F—for virtually any conceivable cash-flow sequence.

CASE STUDY #2, CHAPTER 4:   FINANCING A HOME PURCHASE

Introduction

Purchasing a house is probably the largest single financial commitment the average person will ever make. Undoubtedly, the most important factor enabling such transactions to occur is the financing. There are many methods of financing the purchase of residential property, each having advantages which make it the method of choice under a given set of circumstances. The selection of one method from several for a given set of conditions is the topic of this case study. Four financing methods, identified as plans A, B, C, and D, are evaluated.

Plan	Description
A	30-year fixed rate of 10%-per-year interest, 5% down payment
B	30-year Adjustable-Rate Mortgage (ARM), 9% first 3 years, 9½% in year 4, 10¼% in years 5 through 10 (assumed), 5% down payment
C	15-year fixed rate of 9½%-per-year interest, 5% down payment
D	Owner financing at 8½% per year, $20,000 down payment, with a balloon payment in year 10.

Other assumptions include

- Price of house is $150,000.
- House will be sold in 10 years for $170,000 (net proceeds after selling expenses).
- Taxes and insurance (T&I) are $300 per month.
- Amount available: maximum of $40,000 for down payment, $1600 per month, including T&I.
- New loan expenses: origination fee of 1%, appraisal fee $300, survey fee $200, attorney's fee $200, processing fee $350, escrow fees $150, other costs $300.
- Any money not spent on the down payment or monthly payments will earn tax-free interest at ¼% per month.

The criterion used here: Select the financing plan which has the largest amount of money remaining at the end of the 10-year ownership period. Therefore, calculate the future worth of each plan and select the one with the largest future-worth value.

Analysis of Financing Plans

Plan A: 30-Year Fixed Rate   The amount of money required up front is

(a) Down payment (5% of $150,000)	$7,500
(b) Origination fee (1% of $142,500)	1,425
(c) Appraisal	300
(d) Survey	200
(e) Attorney's fee	200
(f) Processing	350
(g) Escrow	150
(h) Other (recording, credit report, etc.)	300
Total	$10,425

The amount of the loan is $142,500. The equivalent annual principal and interest (P&I) payment at 10% for a 30-year period is

$$A = 142,500(A/P,10/12\%,360)$$

$$= \$1250.56$$

When T&I are added to P&I, the total monthly payment PMT_A is

$$PMT_A = 1250.56 + 300$$

$$= \$1550.56$$

We can now determine the future worth of plan A by summing the future worths of the remaining funds not used for the down payment and up-front fees (F_{1A}) and for monthly payments (F_{2A}), along with the increase in the value of the house (F_{3A}). Since non-expended money earns interest at ¼% per month, in 10 years the money not spent on the down payment would accumulate to

$$F_{1A} = (40,000 - 10,425)(F/P,0.25\%,120)$$

$$= \$39,907.13$$

The available money not spent on monthly payments is $49.44 (i.e., $1600 − 1550.56). Its future worth is

$$F_{2A} = 49.44(F/A,0.25\%,120)$$

$$= \$6908.81$$

Net money available from the sale of the house is the difference between the net selling price (after expenses) and the balance of the loan. The balance of the loan is

$$Loan\ balance = 142,500(F/P,10/12\%,120) - 1250.56(F/A,10/12\%,120)$$

$$= 385,753.40 - 256,170.92$$

$$= \$129,582.48$$

Since the net proceeds from the sale of the house are $170,000,

$$F_{3A} = 170,000 - 129,582.48 = \$40,417.52$$

The total future worth of plan A is

$$F_A = F_{1A} + F_{2A} + F_{3A}$$

$$= 39,907.13 + 6908.81 + 40,417.52$$

$$= \$87,233.46$$

Plan B: 30-Year Adjustable-Rate Mortgage (ARM)   Adjustable-Rate Mortgages are tied to some index such as U.S. Treasury bonds. For this example, we have assumed that the rate is 9% for the first 3 years, 9½% in year 4, and 10¼% in years 5 through 10. Since this option also requires 5% down, the up-front money required will be the same as for plan A, that is, $10,425.

The monthly P&I amount for the first 3 years is
$$A = 142,500(A/P,9/12\%,360)$$
$$= \$1146.58$$

The total payment for the first 3 years is
$$\text{PMT}_B = \$1146.58 + 300$$
$$= \$1446.58$$

At the end of year 3, the interest rate changes to $9\frac{1}{2}\%$ per year. This new rate applies to the balance of the loan at that time:
$$\text{Loan balance at end of year 3} = 142,500(F/P,0.75\%,36)$$
$$- 1146.58(F/A,0.75\%,36)$$
$$= \$139,297.08$$

The P&I payment for year 4 is now
$$A = 139,297.08(A/P,9.5/12\%,324)$$
$$= \$1195.67$$

The total payment for year 4 is
$$\text{PMT}_B = 1195.67 + 300$$
$$= \$1495.67$$

At the end of year 4, the interest rate changes again, this time to $10\frac{1}{4}\%$ per year and stays at this rate for the remainder of the 10-year period. The loan balance at the end of year 4 is
$$\text{Loan balance at end of year 4} = 139,297.08(F/P,9.5/12\%,12)$$
$$- 1195.67(F/A,9.5/12\%,12)$$
$$= \$138,132.42$$

The new P&I amount is
$$A = 138,132.42(A/P,10.25/12\%,312)$$
$$= \$1269.22$$

The new total payment for years 5 through 10 is
$$\text{PMT}_B = 1269.22 + 300$$
$$= \$1569.22$$

The loan balance at the end of 10 years is
$$\text{Loan balance after 10 years} = 138,132.42(F/P,10.25/12\%,72)$$
$$- 1269.22(F/A,10.25/12\%,72)$$
$$= \$129,296.16$$

The future worth of plan B can now be determined. The future worth of the money not spent on a down payment is the same as for plan A.

$$F_{2B} = (40,000 - 10,425)(F/P,0.25\%,120)$$
$$= \$39,907.13$$

The future worth of the money not spent on monthly payments is more complex than in plan A.

$$F_{2B} = (1600 - 1446.58)(F/A,0.25\%,36)(F/P,0.25\%,84)$$
$$+ (1600 - 1495.67)(F/A,0.25\%,12)(F/P,0.25\%,72)$$
$$+ (1600 - 1569.22)(F/A,0.25\%,72)$$
$$= 7118.61 + 1519.31 + 2424.83$$
$$= \$11,062.75$$

The amount of money left from the sale of the house is

$$F_{3B} = 170,000 - 129,296.16$$
$$= \$40,703.84$$

The total future worth of plan B is

$$F_B = F_{1B} + F_{2B} + F_{3B}$$
$$= \$91,673.72$$

Plan C: 15-Year Fixed Rate   The calculations for this plan are the same as those for plan A except that $i = 9\frac{1}{2}\%$ per year and $n = 180$ periods instead of 360. However, for a 5% down payment, the P&I is now $1488.04 which will yield a total payment of $1788.04. This is greater than the $1600 maximum payment available. Therefore, the down payment will have to be increased to $25,500, making the loan amount $124,500. This will make the P&I amount $1300.06 for a total monthly payment of $1600.06.

The amount of money required up front is now $28,245 (the origination fee has also changed). The plan C values for F_{1C}, F_{2C}, and F_{3C} are shown below.

$$F_{1C} = (40,000 - 28,245)(F/P,0.25\%,120)$$
$$= \$15,861.65$$
$$F_{2C} = 0$$
$$F_{3C} = 170,000 - [124,500(F/P,9.5/12\%,120)$$
$$- 1300.06(F/A,9.5/12\%,120)]$$
$$= \$108.097.93$$
$$F_C = F_{1C} + F_{2C} + F_{3C}$$
$$= \$123,959.58$$

Plan D: Owner-Financed With an owner-financed loan, many of the up-front costs are not incurred—origination fee, survey fee, processing fee, etc. For this example, we will assume that the up-front costs are limited to the $20,000 down payment and other costs of $500.

The P&I payment is

$$A = 130,000(A/P,8.5/12\%,360)$$
$$= \$999.59$$

The total monthly payment is

$$PMT_D = 999.59 + 300$$
$$= \$1299.59$$

The future worth is found using calculations similar to plan A.

$$F_{1D} = (40,000 - 20,500)(F/P,0.25\%,120)$$
$$= \$26,312.39$$
$$F_{2D} = (1600 - 1299.59)(F/A,0.25\%,120)$$
$$= \$41,979.72$$
$$F_{3D} = 170,000$$
$$- [130,000(F/P,8.5/12\%,120) - 999.59(F/A,8.5/12\%,120)]$$
$$= \$54,817.16$$

The total future worth of plan D is

$$F_D = 26,312.39 + 41,979.72 + 54,817.16$$
$$= \$123,109.27$$

Conclusion

The analysis reveals that plan C is the preferred financing option. It yields slightly more future-worth money than plan D and considerably more than plan A or B once the house sale 10 years after purchase is complete.

Questions to Consider

1. What is the total amount of interest paid in plan A through the 10-year period?
2. What is the total amount of interest paid in plan B through year 4?
3. What is the maximum amount of money available for a down payment under plan A, if $40,000 is the total amount available?
4. By how much does the payment increase in plan A for each 1% increase in interest rate?
5. If you wanted to "buy down" the interest rate from 10% to 9% in plan A, how much extra down payment would you have to make?

PROBLEMS

4.1 On May 1, 1953, your father started a savings account by depositing $50 per month at the local bank. If the interest rate on the account was 0.25% per month, in what month and year would (*a*) the *P* be located if you used the *P/A* factor with $i = 0.25\%$ and (*b*) the *F* be located if you used $i = 0.25\%$ and $n = 30$ months?

4.2 Determine the amount of money a person must deposit 3 years from now in order to be able to withdraw $10,000 per year for 10 years beginning 15 years from now if the interest rate is 11% per year?

4.3 How much money would you have to deposit for 5 consecutive months starting 2 years from now if you wanted to be able to withdraw $50,000 twelve years from now? Assume the interest rate is a nominal 6% per year compounded monthly.

4.4 How much money would a man have to deposit each year for 6 years starting 4 years from now if he wanted to have $12,000 eighteen years from now? Assume the interest rate is a nominal 8% per year compounded yearly.

4.5 Calculate the present worth of the following series of incomes and expenses if the interest rate is 8% per year compounded yearly.

Year	Income, $	Expense, $
0	12,000	1,000
1–6	800	100
7–11	900	200

4.6 What monthly deposits must be made in order to accumulate $4000 five years from now, if the first deposit will be made 6 months from now and the interest is a nominal 9% per year compounded monthly?

4.7 If your father deposits $40,000 now into an account which will earn interest at a rate of 7% per year compounded quarterly, how much money can he withdraw every 6 months, if he makes his first withdrawal 15 years from now and he wants to make a total of 10 withdrawals?

4.8 What is the present worth of the following series of income and disbursements if the interest rate is a nominal 8% per year compounded semiannually?

Year	Income, $	Expense, $
0	0	9000
1–5	6000	2000
6–8	6000	3000
9–14	8000	5000

4.9 It is desired to make a lump-sum investment on a girl's sixth birthday to provide her with $1500 on each birthday from the seventeenth to the twenty-second, both inclusive. If the interest rate is 8% per year, what lump sum must be invested?

4.10 For the cash-flow diagram shown below, calculate the amount of money in year 4 that would be equivalent to all the cash flow shown if the interest rate is 10% per year.

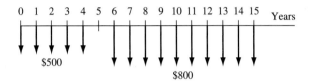

4.11 Rework Problem 4.10 using an interest rate of 1% per month.

4.12 If a man deposits $100 per month for 5 years into a savings account, with the first deposit made 1 month from now, how much will he have in his account after he has made the last deposit, assuming the interest rate is 0.5% per month for the first 3 years and 0.75% per month thereafter?

4.13 An individual borrows $8000 at an interest rate of a nominal 12% per year compounded semiannually, and he desires to repay the money with five equal semiannual payments, the first payment to be made 3 years after receiving the money. What should be the size of the payments?

4.14 You have just graduated from college and plan to begin a retirement fund. It is your desire to withdraw money every year for 30 years starting 25 years from now. The retirement fund earns 9% interest. What uniform annual amount will you be able to withdraw when you retire in 25 years?

4.15 How much money would you have to deposit each month starting 5 months from now if you wanted to have $5000 three years from now, assuming the interest rate is a nominal 8% per year compounded monthly?

4.16 How much must be deposited January 1, 1999, and every 6 months thereafter until July 1, 2004, in order to withdraw $1000 every 6 months for 5 years starting January 1, 2005? Interest is a nominal 12% per year compounded semiannually.

4.17 A couple purchases an insurance policy which they plan to use to partially finance their daughter's college education. If the policy provides $35,000 ten years from now, what additional lump-sum deposit must the couple make 12 years from now in order for their daughter to be able to withdraw $20,000 per year for 5 years beginning 18 years from now? Assume the interest rate is 10% per year.

4.18 A businessman purchased a building and insulated the ceiling with 6 inches of foam. This cut the heating bill by $25 per month and the air conditioning cost by $20 per month. Assuming that the winter season is the first 6 months of the year and the summer season is the next 6 months, what was the equivalent amount of his savings after the first 3 years at an interest rate of 1% per month?

4.19 In what year would you have to make a single deposit of $10,000 if you had already deposited $1000 each year in years 1 through 4 and you wanted to have $17,000 eighteen years from now? Use an interest rate of 7% per year compounded annually.

4.20 Calculate the value of x in the cash flow shown below such that the equivalent total value in month 4 is $9000 using an interest rate of 1.5% per month.

Month	Cash flow, $	Month	Cash flow, $
0	200	6	x
1	200	7	x
2	600	8	x
3	200	9	900
4	200	10	500
5	x	11	500

4.21 Find the value of x below such that the positive cash flows will be exactly equivalent to the negative cash flows if the interest rate is 14% per year, compounded semiannually.

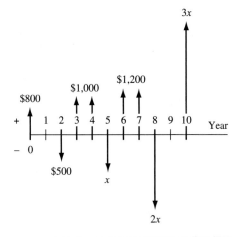

4.22 Find the value of x in the diagram below that would make the equivalent present worth of the cash flow equal to $22,000, if the interest rate is 13% per year.

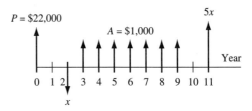

4.23 Calculate the amount of money in year 7 that would be equivalent to the following cash flows if the interest rate is a nominal 16% per year compounded quarterly.

Year	0	1	2	3	4	5	6	7	8	9
Amount, $	900	900	900	900	1300	1300	−1300	500	900	900

4.24 Determine the uniform annual payments which would be equivalent to the cash flow shown below. Use an interest rate of 12% per year.

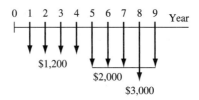

4.25 Calculate the annual worth (years 1 through 10) of the following series of disbursements. Assume that $i = 10\%$ per year compounded semiannually.

Year	Disbursement, $	Year	Disbursement, $
0	3,500	6	5,000
1	3,500	7	5,000
2	3,500	8	5,000
3	3,500	9	5,000
4	5,000	10	15,000
5	5,000		

4.26 A petroleum company is planning to sell a number of existing oil wells. The wells are expected to produce 100,000 barrels of oil per year for 14 more years. If the selling price per barrel of oil is currently $35, how much would you be willing to pay for the wells if the price of oil is expected to decrease by $2 per barrel every 3 years, with the first decrease to occur immediately after the start of year 2? Assume that the interest rate is 12% per year and that oil sales are made at the end of each year.

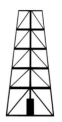

4.27 A large manufacturing company purchased a semiautomatic machine for $18,000. Its annual maintenance and operation cost was $2700. After 4 years from the initial purchase, the company decided to purchase an additional unit for the machine which would make it fully automatic. The additional unit had a first cost of $9100. The cost for operating the machine in fully automatic condition was $1200 per year. If the company used the machine for a total of 13 years, after which time it was worthless, what was the equivalent uniform annual worth of the machine at an interest rate of 9% per year compounded semiannually?

4.28 Calculate the annual worth (years 1 through 8) of the trucking company's receipts and disbursements shown below at an interest rate of 12% per year compounded monthly.

Year	0	1	2	3	4	5	6	7	8
Cash flow, $	−10,000	4,000	−2,000	4,000	4,000	4,000	−1,000	5,000	5,000

4.29 The by-product department of a meat-packing plant has a cooker which has the cost stream shown below. If the interest rate is 15% per year, determine the present worth of the costs.

Year	Cost, $	Year	Cost, $
0	5000	6	8000
1	5000	7	9000
2	5000	8	9100
3	5000	9	9200
4	6000	10	9300
5	7000	11	9400

4.30 A person borrows $10,000 at an interest rate of 8% compounded annually and wishes to repay the loan over a 4-year period with annual payments such that the second payment is $500 greater than the first payment, the third payment is $1000 greater than the second payment, and the fourth payment is $2000 greater than the third payment. Determine the size of the first payment.

4.31 For the cash flow shown below, calculate the equivalent uniform annual worth in periods 1 through 12, if the interest rate is a nominal 8% per year compounded semiannually.

Semiannual Period	Amount, $
0	100
1	100
2	100
3	100
4	100
5	150
6	200
7	250
8	300
9	350
10	400
11	450
12	500

4.32 For the cash flow shown below, find the value of x that will make the equivalent annual worth in years 1 through 9 equal to $2000 at an interest rate of 12% per year compounded quarterly.

Year	0	1	2	3	4	5	6	7	8	9
Cash flow, $	200	300	400	500	x	600	700	800	900	1000

4.33 A person borrows $8000 at a nominal interest rate of 7% per year compounded semiannually. It is desired to repay the loan in 12 semiannual payments, with the first payment to start 1 year from now. If the payments are to increase by $50 each time, determine the size of the first payment.

4.34 Find the value of G in the diagram below that would make the income stream equivalent to the disbursement stream, using an interest rate of 12% per year.

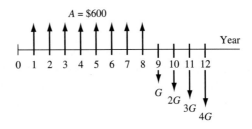

4.35 For the diagram shown below, find the first year and the value of the last receipt in the income stream that would make the receipts at least as large as the $500 investments at years 0, 1, and 2. Use an interest rate of 13% per year.

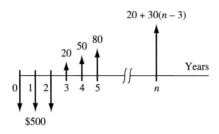

4.36 Calculate the value of x for the cash-flow series shown below such that the equivalent annual worth in months 1 through 14 is $5000, using an interest rate of a nominal 12% per year compounded monthly.

Month	Cash Flow, $
0	100
1	$100 + x$
2	$100 + 2x$
3	$100 + 3x$
4	$100 + 4x$
5	$100 + 5x$
6	$100 + 6x$
7	$100 + 7x$
8	$100 + 8x$
9	$100 + 9x$
10	$100 + 10x$
11	$100 + 11x$
12	$100 + 12x$
13	$100 + 13x$
14	$100 + 14x$

4.37 Assuming the cash flow in Problem 4.36 represents deposits, find the value of x that will make the deposits equal to $9000 in month 9 if the interest rate is 12% per year compounded quarterly. Assume no interperiod interest.

4.38 Solve for the value of G such that the left cash-flow diagram below is equivalent to the one on the right. Use an interest rate of 13% per year.

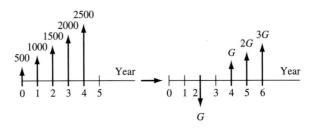

4.39 For the diagram below, solve for the value of x, using an interest rate of 12% per year.

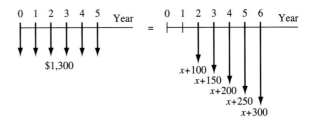

4.40 Mr. Alum Nye is planning to make a contribution to the junior college from which he graduated. He would like to donate an amount of money now so that the college can support students. Specifically, he would like to provide financial support for tuition for five students per year for a total of 20 years (i.e., 21 grants), with the first tuition grant to be made immediately and continuing at 1-year intervals. The cost of tuition at the school is $3800 per year and is expected to stay at that amount for 4 more years. After that time, however, the tuition cost will increase by 10% per year. If the school can deposit the donation and earn interest at a rate of a nominal 8% per year compounded semiannually, how much must Mr. Alum Nye donate?

4.41 Calculate the present worth (at time 0) of a lease which requires a payment now of $20,000 and amounts increasing by 6% per year. Assume the lease is for a total of 10 years. Use an interest rate of 14% per year.

4.42 Calculate the annual worth of a machine which has an initial cost of $29,000, a life of 10 years, and an annual operating cost of $13,000 for the first 4 years, increasing by 10% per year thereafter. Use an interest rate of 15% per year.

4.43 Calculate the end-of-period annual worth to the A-1 Box Company of leasing a computer if the yearly cost is $15,000 for year 1 and $16,500 for year 2, with costs increasing by 10% each year thereafter. Assume that the lease payments must be made at the beginning of the year and that a 7-year study period is to be used. The company's minimum attractive rate return is 16% per year.

4.44 Calculate the present worth of a machine which costs $55,000 and has an 8-year life with a $10,000 salvage value. The operating cost of the machine is expected to be $10,000 in year 1 and $11,000 in year 2, with amounts increasing by 10% per year thereafter. Use an interest rate of 15% per year.

4.45 Calculate the equivalent annual worth of a machine which costs $73,000 initially and will have a $10,000 salvage value after 9 years. The operating cost is $21,000 in year 1, $22,050 in year 2, with amounts increasing by 5% each year thereafter. The minimum attractive rate of return is 19% per year.

4.46 Find the present worth (at time 0) of the cash flows shown in the diagram below. Assume $i = 12\%$ per year compounded semiannually.

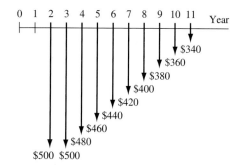

4.47 If you start a bank account by depositing $2000 six months from now, how long will it take for you to deplete the account if you start withdrawing money 1 year from now according to the following plan: withdraw $500 the first month, $450 the second month, $400 the next month, and amounts decreasing by $50 per month until the account is depleted? Assume that the account earns interest at a rate of a nominal 12% per year compounded monthly.

4.48 Compute the annual worth of the following cash flows at $i = 12\%$ per year.

Year	0	1–4	5	6	7	8	9	10
Amount, $	5000	1000	950	800	700	600	500	400

4.49 For the diagram shown below, calculate the amount of money in year 15 that would be equivalent to the amounts shown, if the interest rate is 1% per month.

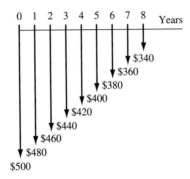

4.50 Compute the present worth and equivalent uniform annual worth at $i = 10\%$ per year for the cash flows below.

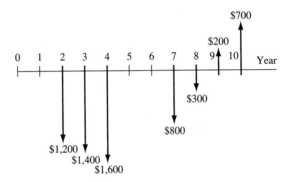

TWO

*N*ow that you have learned to correctly account for the time value of money and the effective interest rate in a project, you are ready to evaluate complete alternatives. In this level you will learn how to perform an economic-based analysis of one or more alternatives, and you will learn how to make a choice of the economically best method from two or more alternatives.

There are four basic methods used to perform an economic analysis: present worth (PW), equivalent uniform annual worth (AW), rate of return (ROR), and benefit/cost ratio (B/C). All four methods will give identical decisions for alternative selection when applied to the same set of cost and revenue estimates and when the comparisons are properly conducted.

CHAPTER

---

chapter

5

Present-Worth and Capitalized-Cost Evaluation

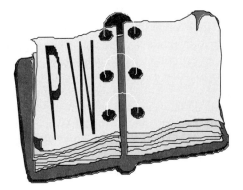

A future amount of money converted into its equivalent present value has a magnitude of the present-worth amount that is always less than that of the actual cash flow, because for any interest rate greater than zero, all P/F factors have a value less than 1.0. For this reason, present-worth calculations are often referred to as *discounted cash-flow* (DCF) methods. Similarly, the interest rate used in making the calculations is referred to as the *discount rate*. Other terms frequently used in reference to present-worth calculations are present worth (PW), present value (PV), and net present value (NPV). Regardless of what they are called, present-worth calculations are routinely used to make economic-related decisions. Up to this point, present-worth computations have been made from cash flows associated with only a single project or alternative. In this chapter, techniques for comparing alternatives by the present-worth method are treated. While the illustrations may be based on comparing two alternatives, the same procedures are followed in a present-worth evaluation of three or more alternatives.

LEARNING OBJECTIVES

Purpose: Understand how to compare alternatives on a present-worth or capitalized-cost basis.

PW of equal-life alternatives

PW of different-life alternatives

Life-Cycle Cost (LCC)

Capitalized cost (CC)

Alternative selection using CC

This chapter will help you:

1. Select the best of equal-life alternatives using present-worth calculations.

2. Select the best of different-life alternatives using present-worth calculations.

3. Describe a life-cycle cost analysis for the major cost categories of an alternative.

4. Make capitalized-cost calculations.

5. Select the best alternative using capitalized-cost calculations.

Instructor Mannual

5.1   PRESENT-WORTH COMPARISON OF EQUAL-LIFE ALTERNATIVES

The *present-worth (PW)* method of alternative evaluation is very popular because future expenditures or receipts are transformed into *equivalent dollars now*. That is, all the future cash flows associated with an alternative are converted into present dollars. In this form, it is very easy, even for a person unfamiliar with economic analysis, to see the economic advantage of one alternative over another.

The comparison of alternatives having equal lives by the present-worth method is straightforward. If both alternatives are used in identical capacities for the same time period, they are termed *equal-service* alternatives. Frequently, an alternative's cash flows represent disbursements only; that is, no receipts are estimated. For example, we might be interested in identifying the process which has the lowest equivalent initial cost, operating cost, and maintenance. Other times, the cash flows will include both receipts and disbursements. Receipts, for example, could come from product sales, equipment salvage values, or realizable savings associated with a particular aspect of the alternative. Since a majority of the problems we will consider have both receipts and disbursements, we represent disbursements as negative cash flows and receipts as positive. (Deviation from this sign convention occurs only when there could be no mistake in interpreting the final results, such as for personal account transactions).

Thus, whether alternatives involve disbursements only, or receipts and disbursements, the following guidelines are applied to select an alternative using the present worth measure of worth:

One alternative.   If PW $\geq$ 0, the requested rate of return is met or exceeded and the alternative is financially viable.

Two or more alternatives.   When only one can be selected (i.e., alternatives are mutually exclusive), *select the alternative with the PW value that is numerically larger,* that is, less negative or more positive, indicating a lower PW of costs or larger PW of net cash flow of receipts and disbursements.

Hereafter we use the symbol PW, rather than *P*, to indicate an alternative's present-worth amount. Example 5.1 illustrates a present-worth comparison.

Example 5.1

Make a present-worth comparison of the equal-service machines for which the costs are shown below, if $i = 10\%$ per year.

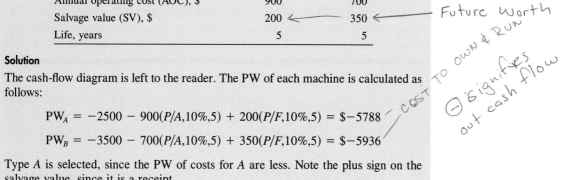

	Type A	**Type B**
First cost (P), $	2500	3500
Annual operating cost (AOC), $	900	700
Salvage value (SV), $	200	350
Life, years	5	5

Future worth

cost to own & run

⊖ signifies

out cash flow

Solution

The cash-flow diagram is left to the reader. The PW of each machine is calculated as follows:

$$PW_A = -2500 - 900(P/A,10\%,5) + 200(P/F,10\%,5) = \$-5788$$

$$PW_B = -3500 - 700(P/A,10\%,5) + 350(P/F,10\%,5) = \$-5936$$

Type A is selected, since the PW of costs for A are less. Note the plus sign on the salvage value, since it is a receipt.

Additional Example 5.5
Problems 5.1 to 5.7

5.2   PRESENT-WORTH COMPARISON OF DIFFERENT-LIFE ALTERNATIVES

When the present-worth method is used for comparing mutually exclusive alternatives that have different lives, the procedure of the previous section is followed with one exception: *The alternatives must be compared over the same number of years.* This is necessary, since, by definition, a present-worth comparison involves calculating the equivalent present value of all future cash flows for each alternative. A fair comparison can be made only when the present worths represent costs and receipts associated with equal service, as described in the preceding section. Failure to compare equal service will always favor the shorter-lived alternative (for costs), even if it were not the most economical one, because fewer periods of costs are involved. The equal-service requirement can be satisfied by either of two approaches:

MUST COMPARE EQUAL SERVICE

1. Compare the alternatives over a period of time equal to the *least common multiple (LCM)* for their lives.

2. Compare the alternatives using a *study period of length* n *years,* which does not necessarily take into consideration the lives of the alternatives. This is also called the *planning horizon* approach.

For the LCM approach, equal service is achieved by making the comparison over the least common multiple of lives between the alternatives, which automatically makes their cash flows extend to the same time period. That is, the cash flow for one "cycle" of an alternative is assumed to be duplicated for the least common multiple of years in terms of constant-value dollars (discussed in Chapter 12). Then service is compared over the same total life for each alternative. For example, if it is desired to compare alternatives which have lives of 3 years and 2 years, respectively, the alternatives are evaluated over a period of 6 years. It is important to remember that when an alternative has a positive (negative) terminal salvage value, this must also be included and shown as an income (a cost) on the cash-flow diagram in each life cycle. Such a procedure obviously requires that some assumptions be made about the alternatives in their subsequent life cycles. Specifically, these assumptions are

• The alternatives under consideration will be needed for the least common multiple of years or more.

• The respective costs of the alternatives will be the same in all subsequent life cycles as they were in the first one.

As will be shown in Chapter 12, this second assumption is valid when the cash flows are expected to change by exactly the inflation or deflation rate that is applicable through the LCM time period. If the cash flows are expected to change by any other rate, then a study-period–based PW analysis must be conducted using constant-value dollars as described in Chapter 12. This also holds true when the assumption about the length of time the alternatives are needed cannot be made.

For the second or study-period approach, a time horizon is chosen over which the economic analysis is to be conducted, and only those cash flows which occur during that time period are considered relevant to the analysis. Any cash flows occurring beyond the stated horizon, whether incoming or outgoing, are ignored. An estimated realistic salvage value (or residual value) at the end of the study period for both alternatives must be made and used. The time horizon chosen might be relatively short, especially when short-term business goals are very important, or vice versa. In any case, once the horizon has been selected and the cash flows are estimated for each alternative, the PW values are determined and the most economical one is chosen. The study period, or planning horizon, concept is especially useful in replacement analysis as discussed in Chapter 10.

Although the planning-horizon analysis may be relatively straightforward and more realistic for many real-world situations, we also use the LCM method in our examples and problems to reinforce your understanding of equal service. Example 5.2 shows evaluations based on the LCM and planning-horizon techniques.

Example 5.2

A plant superintendent is trying to decide between two excavating machines with the estimates presented below.

	Machine A	Machine B
First cost, $	11,000	18,000
Annual operating cost, $	3,500	3,100
Salvage value, $	1,000	2,000
Life, years	6	9

(*a*) Determine which one should be selected on the basis of a present-worth comparison using an interest rate of 15% per year.

(*b*) If a study period of 5 years is specified and the salvage values are not expected to change, which alternative should be selected?

(*c*) Which machine should be selected over a 6-year horizon if the salvage value of machine *B* is estimated to be $6000 after 6 years?

Solution

(*a*) Since the machines have different lives, they must be compared over their LCM, which is 18 years. For life cycles after the first, the first cost is repeated in year 0 of the new cycle, which is the last year of the previous cycle. These are years 6 and 12 for machine *A*, and year 9 for *B*. The cash-flow diagram in Figure 5–1

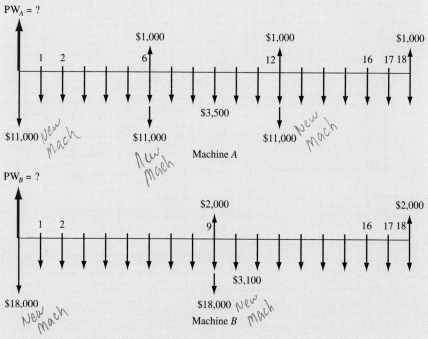

Figure 5–1   Cash-flow diagram for different-life alternatives, Example 5.2*a*.

assists in performing the PW analysis.

$$PW_A = -11,000 - 11,000(P/F,15\%,6) + 1000(P/F,15\%,6)$$

$$- 11,000(P/F,15\%,12) + 1000(P/F,15\%,12) + 1000(P/F,15\%,18)$$

$$- 3500(P/A,15\%,18)$$

$$= \$-38,559$$

$$PW_B = -18,000 - 18,000(P/F,15\%,9) + 2000(P/F,15\%,9)$$

$$+ 2000(P/F,15\%,18) - 3100(P/A,15\%,18)$$

$$= \$-41,384$$

Machine A is selected, since machine A costs less in PW terms than machine B.

(b) For a 5-year planning horizon no cycle repeats are necessary, and $SV_A = \$1000$ and $SV_B = \$2000$ in year 5. The PW analysis is

$$PW_A = -11,000 - 3500(P/A,15\%,5) + 1000(P/F,15\%,5)$$

$$= \$-22,236$$

$$PW_B = -18,000 - 3100(P/A,15\%,5) + 2000(P/F,15\%,5)$$

$$= \$-27,397$$

Machine A is still the better choice.

(c) For a 6-year planning horizon, $SV_B = \$6000$ in year 6, and the PW equations are

$$PW_A = -11,000 - 3500(P/A,15\%,6) + 1000(P/F,15\%,6) = \$-23,813$$

$$PW_B = -18,000 - 3100(P/A,15\%,6) + 6000(P/F,15\%,6) = \$-27,138$$

Machine A is still favored.

Comments

In part (a) and Figure 5–1, the salvage value of each machine is recovered *after each life cycle,* that is, in years 6, 12, and 18 for machine A, and in years 9 and 18 for machine B. In part (b), it is assumed that the salvage values will not change when the horizon is shortened. This clearly is not usually the case. In part (c), machine A is still favored even though the salvage value of machine B is increased from $2000 to $6000. As an exercise, determine the minimum salvage value of machine B that would make it more attractive than machine A.

Additional Examples 5.6 and 5.7
Problems 5.8 to 5.19

5.3   LIFE-CYCLE COST

The term *life-cycle cost (LCC)* is interpreted to mean the total of every cost estimate considered possible for a system with a long life, from the design phase, through the manufacturing and the field-use phases, onto the scrap phase followed by replacement with a new, more advanced system. The LCC includes all estimated service, retrofit, upgrade, scrap, and anticipated recycle costs. Application is usually made to projects which will require research and development time to design and test a product or system intended to perform a specific task.

The technique of LCC analysis is applied by large contractor corporations to government-sponsored systems, especially defense-related projects. For some systems, the total cost over the system life is many multiples of the initial cost. The LCC concept is just as meaningful for smaller systems, for example, an automobile where the manufacturer and a series of owners experience many costs in addition to the initial design, manufacture and purchase costs as the car is maintained, repaired, and finally disposed of.

The total anticipated costs of an alternative are usually estimated using major cost categories such as

Research and development costs.   All expenditures for design, prototype fabrication, testing, manufacturing planning, engineering services, software engineering, software development, and the like for a product or service.

Production costs.   The investment necessary to produce or acquire the product, including expenses to employ and train personnel, transport subassemblies and the final product, build new facilities, and acquire equipment.

Operating and support costs.   All costs incurred to operate, maintain, inventory, and manage the product for its entire anticipated life. This may include periodic rework costs and average costs if the system requires recall or major in-service repairs, based upon cost experiences for other systems already developed.

Present-worth computations, using the *P/F* factor to discount the costs in each category to the time that the analysis is performed, are applied to complete the LCC analysis. The major difference between LCC analysis and the analyses we have performed thus far is the scope of the effort to include all types of costs over the long-term future of the system. Also, LCC analysis is most useful when performed for systems with relatively long lives, say 15 to 30 years. Examples are radar, aircraft, and weapon systems and advanced manufacturing systems. Public sector projects can be evaluated using the LCC approach, but because of the difficulty in estimating benefits, revenues, taxpayer costs, MARR, and other factors which have human lives and welfare at stake, public sector projects are more commonly evaluated using benefit/cost analysis (Chapter 9).

The approach of the LCC evaluation is to determine the cost of each alternative for its entire life and select the one with the minimum LCC. Actually, a PW analysis and comparison with all definable costs estimated for the life of each alternative is the same as the LCC analysis. For a more complete description of the cost-estimation procedures and analyses for LCC refer to the books by Seldon and by Ostwald that are listed in the bibliography.

Problems 5.37 and 5.38

5.4   CAPITALIZED-COST CALCULATIONS

Capitalized cost (CC) refers to the present worth of a project that is assumed to last forever. Some public works projects such as dams, irrigation systems, and railroads fall into this category. In addition, permanent university or charitable-organization endowments are evaluated using capitalized-cost methods. In general, the procedure followed in calculating the capitalized cost of an infinite sequence of cash flows is as follows:

1. Draw a cash-flow diagram showing all nonrecurring (one-time) costs (and/or incomes) and at least two cycles of all recurring (periodic) costs and receipts.
2. Find the present worth of all nonrecurring amounts.
3. Find the equivalent uniform annual worth (*A* value) through one life cycle of all recurring amounts and add this to all other uniform amounts occurring in years 1 through infinity. This results in a total equivalent uniform annual worth (AW).
4. Divide the AW obtained in step 3 by the interest rate *i* to get the capitalized cost.
5. Add the value obtained in step 2 to the value obtained in step 4.

The purpose for beginning the solution by drawing a cash-flow diagram should be evident from previous chapters. However, the cash-flow diagram is probably more important in CC calculations than elsewhere, because it facilitates the differentiation between nonrecurring and recurring (periodic) amounts.

Since the capitalized cost is another term for the present worth of a perpetual cash-flow sequence, the present worth of all nonrecurring amounts is determined (step 2). In step 3 the AW (which has been called *A* previously) of all recurring and uniform annual amounts is calculated. Then, step 4, which is effectively A/i, determines the present worth (capitalized cost) of the perpetual annual series using the equation:

$$\text{Capitalized cost} = \frac{\text{AW}}{i} \qquad \text{or} \qquad \text{PW} = \frac{\text{AW}}{i} \qquad \text{[5.1]}$$

Equation [5.1] is derived from the $(P/A,i,n)$ factor when $n = \infty$. The equation for P using the P/A factor is

$$P = A\frac{(1 + i)^n - 1}{i(1 + i)^n}$$

If the numerator and denominator are divided by $(1 + i)^n$, the equation becomes

$$P = A\frac{1 - \dfrac{1}{(1 + i)^n}}{i}$$

Now, as n approaches ∞, the numerator term becomes 1, yielding $P = A/i$.

The validity of Equation [5.1] can be illustrated by considering the time value of money. If $10,000 is deposited into a savings account at 20%-per-year interest compounded annually, the maximum amount of money that can be withdrawn at the end of every year for *eternity* is $2000, or the amount equal to the interest accumulated each year. This leaves the original $10,000 deposit to earn interest so that another $2000 will be accumulated the next year. Mathematically, the amount of money that can be accumulated and withdrawn in each consecutive interest period for an infinite number of periods is

$$A = Pi \qquad\qquad [5.2]$$

Thus, for the example,

$$A = 10,000(0.20) = \$2000 \text{ per year}$$

The capitalized-cost calculation in Equation [5.1] is Equation [5.2] solved for P.

$$P = \frac{A}{i} \qquad\qquad [5.3]$$

After the present worths of all cash flows have been obtained, the total capitalized cost is simply the sum of these present worths. Capitalized-cost calculations are illustrated in Example 5.3.

Example 5.3

Calculate the capitalized cost of a project that has an initial cost of $150,000 and an additional investment cost of $50,000 after 10 years. The annual operating cost will be $5000 for the first 4 years and $8000 thereafter. In addition, there is expected to be a recurring major rework cost of $15,000 every 13 years. Assume that $i = 15\%$ per year.

Solution

The 5-step procedure outlined above is followed.

1. Draw a cash-flow diagram for two cycles (Figure 5–2).

2. Find the present worth P_1 of the nonrecurring costs of $150,000 now and $50,000 in year 10:

$$P_1 = -150,000 - 50,000(P/F,15\%,10) = \$-162,360$$

3. Convert the recurring cost of $15,000 every 13 years into an annual worth A_1 for the first 13 years.

$$A_1 = -15,000(A/F,15\%,13) = \$-437$$

Note that the same value, $A = \$-437$, applies to all the other 13-year periods as well.

4. The capitalized cost for the $-5000 annual-cost series may be computed by either of two ways: Consider a series of $-5000 from now to infinity and find the present worth of $-\$8000 - (\$-5000) = \$-3000$ from year 5 on, or find the present worth of $-5000 for 4 years and the present worth of $-8000 from year 5 to infinity. Using the first method, we find that the annual cost (A_2) is $-5000, and the present worth (P_2) of $-3000 from year 5 to infinity, using Equation [5.3] and the P/F factor, is

$$P_2 = \frac{-3000}{0.15}(P/F,15\%,4) = \$-11,436$$

The two annual costs are converted into a capitalized cost (P_3):

$$P_3 = \frac{A_1 + A_2}{i} = \frac{-437 + (-5000)}{0.15} = \$-36,247$$

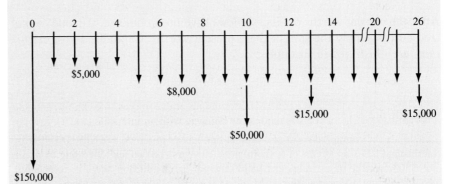

i = 15% per year

Figure 5–2   Diagram used to compute capitalized cost, Example 5.3.

5. The total capitalized cost PW_T is obtained by adding the three PW values.

$$PW_T = P_1 + P_2 + P_3 = \$-210{,}043$$

Comment

The P_2 value is calculated using $n = 4$ in step 4 in the P/F factor because the present worth of the annual \$3000 cost is computed in year 4, since P is always 1 period ahead of the first A. You should rework the problem using the second method suggested for calculating P_2.

Problems 5.20 to 5.27

5.5 CAPITALIZED-COST COMPARISON OF TWO ALTERNATIVES

When two or more alternatives are compared on the basis of their capitalized cost, the procedure of Section 5.4 is followed for each alternative. Since the capitalized cost represents the present total cost of financing and maintaining a given alternative forever, the alternatives will automatically be compared for the same number of years (i.e., infinity). The alternative with the smaller capitalized cost will represent the most economical one. As in present-worth and all other alternative evaluation methods, it is only the differences in cash flow between the alternatives which must be considered for comparative purposes. Therefore, whenever possible, the calculations should be simplified by eliminating the elements of cash flow which are common to both alternatives. On the other hand, if true capitalized-cost values are needed instead of just comparative ones, the actual cash flows rather than the differences should be used. True capitalized-cost values would be needed, for example, if one wanted to know the actual or true financial obligations associated with a given alternative. Example 5.4 shows the procedure for comparing two alternatives on the basis of their capitalized cost.

Example 5.4

Two sites are currently under consideration for a bridge to cross the Ohio River. The north site, which connects a major state highway with an interstate loop around the city, would alleviate much of the local through traffic. The disadvantages of this site are that the bridge would do little to ease local traffic congestion during rush hours, and the bridge would have to stretch from one hill to another to span the widest part of the river, railroad tracks, and local highways below. This bridge would therefore be a suspension bridge. The south site would require a much shorter span allowing for construction of a truss bridge but would require new road construction.

The Brooklyn Bridge

The suspension bridge would have a first cost of $30 million with annual inspection and maintenance costs of $15,000. In addition, the concrete deck would have to be resurfaced every 10 years at a cost of $50,000. The truss bridge and approach roads are expected to cost $12 million and have annual maintenance costs of $8000. The bridge would have to be painted every 3 years at a cost of $10,000. In addition, the bridge would have to be sandblasted every 10 years at a cost of $45,000. The cost of purchasing right-of-way is expected to be $800,000 for the suspension bridge and $10.3 million for the truss bridge. Compare the alternatives on the basis of their capitalized cost if the interest rate is 6% per year.

Solution

Construct the two cash-flow diagrams first.
Capitalized cost of suspension bridge

$$P_1 = \text{present worth of initial cost} = -30.0 - 0.8 = \$-30.8 \text{ million}$$

The recurring operating cost is $A_1 = \$-15,000$, while the annual equivalent of the resurface cost is

$$A_2 = -50,000(A/F,6\%,10) = \$-3794$$

$$P_2 = \text{capitalized cost of recurring costs} = \frac{A_1 + A_2}{i}$$

$$= \frac{-15,000 + (-3794)}{0.06}$$

$$= \$-313,233$$

Finally, the total capitalized cost (PW_S) is

$$\text{PW}_S = P_1 + P_2 = \$-31,113,233 \text{ (about } \$-31.1 \text{ million)}$$

Capitalized cost of truss bridge

$$P_1 = -12.0 + (-10.3) = -\$22.3 \text{ million}$$

$$A_1 = \$-8000$$

$$A_2 = \text{annual cost of painting} = -10000(A/F,6\%,3)$$

$$= \$-3141$$

$$A_3 = \text{annual cost of sandblasting} = -45,000(A/F,6\%,10)$$

$$= \$-3414$$

$$P_2 = \frac{A_1 + A_2 + A_3}{i} = \$-242,583$$

The total capitalized cost PW_T of the truss bridge alternative is

$$\text{PW}_T = P_1 + P_2 = \$-22,542,583 \text{ (about } \$-22.5 \text{ million)}$$

Build the truss bridge, since its capitalized cost is lower.

To determine a capitalized-cost for an alternative which has a finite life, simply calculate the AW for one life cycle and divide the resulting *A* value by the interest rate. This is illustrated in Additional Example 5.8.

Additional Example 5.8
Problems 5.28 to 5.36

ADDITIONAL EXAMPLES

Example 5.5

EQUAL-LIFE PW ANALYSIS, SECTION 5.1   A traveling saleswoman expects to purchase a used car this year. She has collected or estimated the following information: first cost is $10,000; trade-in value will be $500 after 4 years; annual maintenance and insurance costs are $1500; and additional annual income due to ability to travel is $5000. Will the saleswoman be able to make a rate of return of 20% per year on her purchase?

Solution

Compute the PW value of the investment at $i = 20\%$.

$$PW = -10,000 - 1500(P/A,20\%,4) + 5000(P/A,20\%,4)$$
$$+ 500(P/F,20\%,4)$$
$$= \$-698$$

No, she will not make 20%, since the PW is less than zero.

Comment

If the PW value had been greater than zero, the rate of return would have exceeded 20%. In Chapter 7, calculations similar to those above will be made to determine the actual rate of return of estimated cash flows.

[handwritten annotation] Solution is usually negative ⊖ because it indicates a cash outflow.

Example 5.6

DIFFERENT-LIFE PW ANALYSIS, SECTION 5.2   AAA Cement plans to open a new rock pit. Two plans have been devised for movement of raw material from the quarry to the plant. Plan *A* requires the purchase of two earth movers and construction of an unloading pad at the plant. Plan *B* calls for construction of a conveyor system from the quarry to the plant. The costs for each plan are detailed in Table 5–1. Using PW analysis, (*a*) determine which plan should be selected if money is presently worth 15% per year, and (*b*) select the better plan for a 12-year period, assuming the salvage

Table 5–1	Details of plans to move rock from quarry to cement plant		
	Plan *A*		Plan *B*
	Mover	**Pad**	**Conveyor**
Initial cost, $	45,000	28,000	175,000
Annual operating cost, $	6,000	300	2,500
Salvage value, $	5,000	2,000	10,000
Life, years	8	12	24

value of the mover after 4 years will be $20,000 and the salvage value of the conveyor after 12 years will be $25,000.

Solution

(*a*) Evaluation must take place over the LCM of 8 and 12, that is, 24 years. Reinvestment in the two movers will occur in years 8 and 16, and the unloading pad must be repurchased in year 12. No reinvestment is necessary for plan *B*. You should construct your own cash-flow diagram for each plan to follow the PW analysis.

To simplify computations, we can use the fact that plan *A* will have an extra AOC over plan *B* in the amount of $2(6000) + 300 - 2500 = \$9800$ per year.

PW of plan A

$$PW_A = PW_{movers} + PW_{pad} + PW_{AOC}$$

$$PW_{movers} = -2(45,000)[1 + (P/F,15\%,8) + (P/F,15\%,16)]$$
$$+ 2(5000)[(P/F,15\%,8) + (P/F,15\%,16) + (P/F,15\%,24)]$$
$$= \$-124,355$$

$$PW_{pad} = -28,000[1 + (P/F,15\%,12)] + 2000[(P/F,15\%,12)$$
$$+ (P/F,15\%,24)]$$
$$= \$-32,790$$

$$PW_{AOC} = -9800(P/A,15\%,24) = -\$63,051$$

$$PW_A = \$-220,196$$

PW of plan B

$$PW_B = PW_{conveyer} = -175,000 + 10,000(P/F,15\%,24)$$
$$= \$-174,651$$

Since the PW of cost of *B* is less than that of *A*, the conveyor should be constructed.

(b) For plan A, repurchase the movers in year 8 and salvage them after only 4 years at $20,000 each. The PW$_\text{pad}$ is the same as in part (a).

$$PW_\text{movers} = -2(45,000)[1 + (P/F,15\%,8)] + 2(5000)(P/F,15\%,8)$$
$$+ 2(20,000)(P/F,15\%,12)$$
$$= \$-108,676$$

$$PW_\text{pad} = \$-32,790$$

$$PW_\text{AOC} = -9800(P/A,15\%,12) = \$-53,122$$

$$PW_A = PW_\text{movers} + PW_\text{pad} + PW_\text{AOC} = \$-194,588$$

For plan B, only 12 of the expected 24 years are considered.

$$PW_B = -175,000 + 25,000(P/F,15\%,12) = \$-170,328$$

Still select plan B; however, the difference in PW values is now much less due to the shortened evaluation period of 12 years. Why do you think this is the case?

Example 5.7

DIFFERENT-LIFE PW ANALYSIS, SECTION 5.2   A restaurant owner is trying to decide between two different garbage disposals. A regular steel (RS) disposal has an initial cost of $65 and a life of 4 years. The alternative is a corrosion-resistant disposal constructed primarily of stainless steel (SS). The initial cost of the SS disposal is $110, but it is expected to last 10 years. Because the SS disposal has a slightly larger motor, it is expected to cost about $5 per year more to operate than the RS disposal. If the interest rate is 16% per year, (a) which disposal should be selected, assuming both have a negligible salvage value, and (b) which disposal should be selected if a 4-year planning horizon is used and it is assumed that the 4-year-old SS disposal can be resold for $50?

Solution

(a) The cash-flow diagram (Figure 5–3) uses a comparison period of 20 years with reinvestment in year 10 for the SS disposal and in years 4, 8, 12, and 16 for the RS disposal. The present-worth calculations are as follows:

$$PW_\text{RS} = -65 - 65(P/F,16\%,4) - 65(P/F,16\%,8) - 65(P/F,16\%,12)$$
$$- 65(P/F,16\%,16)$$
$$= \$-137.72$$

$$PW_\text{SS} = -110 - 110(P/F,16\%,10) - 5(P/A,16\%,20)$$
$$= \$-164.58$$

The RS disposal should be purchased, since its PW of costs is less.

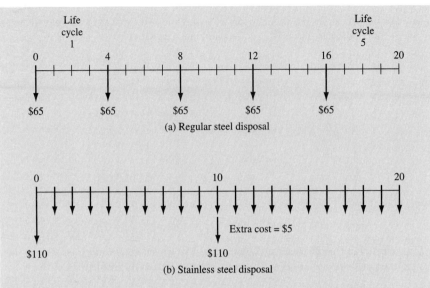

Figure 5–3   Present-worth comparison of two different-life assets, Example 5.7.

(*b*)   For a 4-year planning horizon,

$$PW_{RS} = -65$$

$$PW_{SS} = -110 - 5(P/A,16\%,4) + 50(P/F,16\%,4)$$

$$= \$-96.37$$

The RS disposal should be purchased. Note that the relatively high resale value of \$50 after 4 years for the SS disposal is not able to reduce the PW of costs enough to change the decision.

Comment

In the solution presented, the extra operating cost of \$5 per year is regarded as an expense for the SS disposal. However, the same decision is reached if the \$5 per year is shown as an income for the RS disposal, but the present worths of both will be lower by $5(P/A,16\%,20)$. This illustrates that unless the absolute money values are sought, it is only important to consider *differences* in cash flow for alternative evaluation.

Example 5.8

CAPITALIZED COST OF TWO ALTERNATIVES, SECTION 5.5   A city engineer is considering two alternatives for the local water supply. The first alternative involves construction of an earthen dam on a nearby river, which has a highly variable flow. The dam will form a reservoir so the city may have a dependable source of water. The initial cost of the dam is expected to be \$8 million and will require annual upkeep costs of \$25,000. The dam is expected to last indefinitely.

Alternatively, the city can drill wells as needed and construct pipelines for transporting the water to the city. The engineer estimates that an average of 10 wells will be required initially at a cost of $45,000 per well, including the pipeline. The average life of a well is expected to be 5 years with an annual operating cost of $12,000 per well. If the city uses an interest rate of 15% per year, determine which alternative should be selected on the basis of their capitalized costs.

Solution

The cash-flow diagram is left to the reader. The capitalized cost of the dam is calculated using Equation [5.1] for the annual upkeep cost component.

$$PW_{dam} = PW \text{ of dam} + PW \text{ of annual costs}$$

$$= -8,000,000 + \frac{-25,000}{0.15} = \$-8,166,667$$

To calculate the capitalized cost of the wells, first convert the recurring costs and annual operating costs to an AW and then divide by the interest rate of 0.15.

$$AW_{wells} = AW \text{ of investment} + \text{annual operating costs}$$

$$= -45,000(10)(A/P,15\%,5) - 12,000(10)$$

$$= \$-254,242$$

The capitalized cost of the wells, using Equation [5.1], is

$$PW_{wells} = \frac{-254,242}{0.15} = \$-1,694,947$$

The wells are significantly cheaper than the dam.

Comment

The capitalized cost of the wells could also have been obtained by using the A/F factor for calculating the AW of future wells. The value obtained should then be divided by i and added to the initial investment cost $P_1 = \$-450,000$ for the 10 wells.

$$PW_A = \frac{AW}{i} = \frac{-450,000(A/F,15\%,5) - 120,000}{0.15}$$

$$= -\$1,244,947$$

$$PW_{wells} = PW_A + P_1$$

$$= -1,244,947 + (-450,000) = -\$1,694,947$$

Example 5.8
(Spreadsheet)

CAPITALIZED COST OF TWO ALTERNATIVES, SECTION 5.5   Use spreadsheet analysis to work Example 5.8.

Solution

The relations used above to determine PW_{dam} and PW_{wells} are included in Figure 5–4 as a spreadsheet solution. The PMT function in cell B12 calculates AW_{wells} = $\$-254{,}242$. Select the wells alternative.

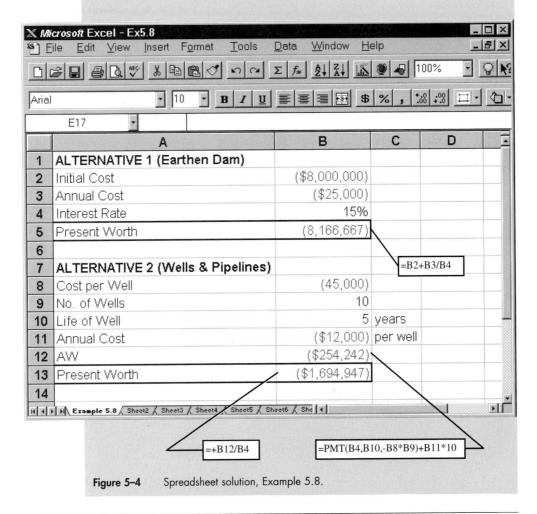

	A	B	C	D
1	ALTERNATIVE 1 (Earthen Dam)			
2	Initial Cost	($8,000,000)		
3	Annual Cost	($25,000)		
4	Interest Rate	15%		
5	Present Worth	(8,166,667)		
6				
7	ALTERNATIVE 2 (Wells & Pipelines)			
8	Cost per Well	(45,000)		
9	No. of Wells	10		
10	Life of Well	5 years		
11	Annual Cost	($12,000) per well		
12	AW	($254,242)		
13	Present Worth	($1,694,947)		
14				

=B2+B3/B4

=+B12/B4

=PMT(B4,B10,-B8*B9)+B11*10

Figure 5–4   Spreadsheet solution, Example 5.8.

CHAPTER SUMMARY

The present-worth method of comparing alternatives involves converting all cash-flow values into present dollars. When the alternatives have different lives, the comparison must be made for equal-service periods. This is done by either making the comparison over the least common multiple of their lives or by selecting a planning horizon and making the present-worth calculation over that time period for both alternatives, regardless of their lives. It is important to remember that with either method, you are comparing the alternatives for equal service. Any remaining life of an alternative is recognized through its residual or salvage value.

Life-cycle cost (LCC) analysis is a present-worth analysis performed for alternatives which have relatively long lives and which include estimates for all phases of a system—design, manufacturing, field use, expected upgrade and re-work, etc. Defense systems and some public works systems are evaluated using the LCC approach.

The present worth of an alternative with an infinite life is called its capitalized cost. This value is rather easily calculated because the P/A factor reduces to $1/i$ in the limit when $n = \infty$.

PROBLEMS

Ans

5.1 A homeowner who is redoing her bathrooms is trying to decide between low-flush toilets (13 liters per flush) and ultralow-flush toilets (6 liters per flush). For the color of toilet she wants, the store carries only one model of each. The low-flush model will cost $90, and the ultralow-flush model will cost $150. If the cost of water is $1.50 per 400 liters, determine which toilet should be purchased on the basis of a present-worth analysis using an interest rate of 10% per year. Assume the toilets will be flushed an average of 10 times per day and will be replaced in 10 years.

$P_{W_{low}}$ -1999.31

$P_{W_{ultra}}$ -200.45

5.2 The manager of a canned-food processing plant is trying to decide between two different labeling machines. Machine *A* will have a first cost of $15,000, an annual operating cost of $2800, and a service life of 4 years. Machine *B* will cost $21,000 to buy and have an annual operating cost of $900 during its 4-year life. At an interest rate of 9% per year, which should be selected on the basis of a PW analysis?

5.3 A house remodeling contractor is trying to decide between buying and renting a steamer for removing wallpaper. The heating unit and the necessary appurtenances (pan, hose, etc.) will cost $190 to purchase. It will have a 10-year useful life if the heating element is replaced in 5 years at a cost of $75. Alternatively, the contractor can rent an identical unit for $20 per day. If he expects to need the steamer an average of 3 days per year, should he purchase a unit or rent one at an interest rate of 8% per year? Use the PW method.

5.4 An investor is trying to decide whether he should invest the $30,000 he received from the sale of his boat in the stock market or in a small fast-food restaurant with three other partners. If he buys the stock, he will receive 3500 shares which pay dividends of $1 per share each quarter. He expects the stock to appreciate to $40,000 six years from now. If he invests in the restaurant, he will have to put up another $10,000 one year from now, but

$P_{W_{stock}}$ $48,950$

$P_{W_{rest}}$ $-6,772$

starting 2 years from now, his share of the profits will be $9000 per year for 5 years, after which time he will receive $35,000 from the sale of the business. Using a PW analysis and an interest rate of 12% per year compounded quarterly, which investment should he make?

5.5 Compare the alternatives below on the basis of their present worths using an interest rate of 14% per year compounded monthly. Maintenance and operations (M&O) costs are given.

	Alternative X	Alternative Y
Initial cost, $	40,000	60,000
Monthly M&O cost, $	5,000	—
Semiannual M&O cost, $	—	13,000
Salvage value, $	10,000	8,000
Life, years	5	5

5.6 A company which manufactures glass-door fireplace screens makes two different types of mounting brackets for the frame. An L-shaped bracket is used for relatively small fireplace openings, and a U-shaped bracket is used for all others. The company currently includes both types of brackets in one box, and the purchaser discards the one not needed. The cost of these two brackets with screws and other parts is $3.25. If the frame of the fireplace screen is redesigned, a single universal bracket can be used which will cost $1.10 to make. However, retooling for manufacturing the new bracket will cost $5000. In addition, inventory write-downs, retraining, and repackaging costs will amount to another $7000. If the company sells 900 of the fireplace screen units per year, should the company keep the old brackets or go with the new ones, assuming the company's interest rate is 18% per year and it wants to recover its investment within 5 years? Use the present-worth method.

5.7 The supervisor of a country club swimming pool is trying to decide between two methods for adding chlorine. If gaseous chloride is added, a chlorinator, which has an initial cost of $8000 and a useful life of 5 years, will be required. The chlorine will cost $200 per year and the labor cost will be $400 per year. Alternatively, dry chlorine can be added manually at a cost of $500 per year for chlorine and $1500 per year for labor. If the interest rate is 8% per year, which method should be used on the basis of a present-worth analysis?

5.8 Two types of miniblinds can be purchased for a certain window. The cheaper one is made of vinyl, costs $9.50, and is expected to last 4 years. The aluminum blind will cost $24, but it will have a useful life of 8 years. At an interest rate of 12% per year, which blind should be purchased on the basis of their present worths?

5.9 Two types of materials can be used for roofing a commercial building which has 1500 square meters of roof. Asphalt shingles will cost $14 per square meter installed and are guaranteed for 10 years. Fiberglass shingles will cost $17 per square meter installed, but they are guaranteed for 20 years. If the fiberglass shingles are selected, the owner will be able to sell the building for $2500 more than if the asphalt shingles are used. Which shingles should be used if the minimum attractive rate of return is 17% per year and the owner plans to sell the building in (*a*) 12 years? (*b*) 8 years?

P_w Asphalt $ - 25,368

$P_{w\,FG} = $ $ - 25,120

5.10 Compare the following machines on the basis of their present worths. Use $i = 12\%$ per year.

	New Machine	Used Machine
First cost, $	44,000	23,000
Annual operating cost, $	7,000	9,000
Annual repair cost, $	210	350
Overhaul every 2 years, $	—	1,900
Overhaul every 5 years, $	2,500	—
Salvage value, $	4,000	3,000
Life, years	14	7

5.11 Two methods can be used for producing a certain machine part. Method 1 costs $20,000 initially and will have a $5000 salvage value after 3 years. The operating cost with this method is $8500 per year. Method 4 has an initial cost of $15,000, but it will last only 2 years. Its salvage value is $3000. The operating cost for method 4 is $7000 per year. If the minimum attractive rate of return is 12% per year, which method should be used on the basis of a present-worth analysis?

5.12 Compare the two plans below using the present-worth method at $i = 13\%$ per year.

	Plan A	Plan B Machine 1	Machine 2
First cost, $	10,000	30,000	5,000
Annual operating cost, $	500	100	200
Salvage value, $	1,000	5,000	−200
Life, years	40	40	20

5.13 Rework Problem 5.12 using a study period of 10 years. Assume the salvage values will be $3000, $10,000, and $1000 for plan A, machine 1, and machine 2, respectively.

5.14 The ABC Company is considering two types of siding for its proposed new building. Anodized metal siding will require very little maintenance, and minor repairs will cost only $500 every 3 years. The initial cost of the siding will be $250,000. If a concrete facing is used, the building will have to be painted now and every 5 years at a cost of $80,000. The building is expected to have a useful life of 15 years, and the "salvage value" will be $25,000 greater if the metal siding is used. Compare the present worths of the two methods at an interest rate of 15% per year.

5.15 Compare the alternatives shown below on the basis of a present-worth analysis. Use an interest rate of 1% per month.

	Alternative Y	Alternative Z
First cost, $	70,000	90,000
Monthly operating cost, $	1,200	1,400
Salvage value, $	7,000	10,000
Life, years	3	6

5.16 A food processing company is evaluating various methods for disposing of the sludge from the wastewater treatment plant. Under consideration is disposal of the sludge by spraying or incorporation into the soil. If the spraying alternative is selected, an underground distribution system will be constructed at a cost of $60,000. The salvage value after 20 years is expected to be $10,000. Operation and maintenance of the system is expected to cost $26,000 per year.

Alternatively, the company can use Big Foot trucks to transport and dispose of the sludge by incorporation below the soil surface. Three trucks will be required at a cost of $120,000 per truck. The operating cost of the trucks, including driver, routine maintenance, and overhauls, is expected to be $42,000 per year. The used trucks can be sold after 10 years for $30,000 each. If the trucks are used, field corn can be planted and sold for $20,000 per year. For spraying, grass must be planted and harvested, and because of the presence of the contaminated sludge on the cuttings, the grass will have to be disposed of at a landfill for a cost of $14,000 per year. If the company's minimum attractive rate of return is 18% per year, which method should be selected based on present-worth analysis?

5.17 Compare the alternatives shown below on the basis of a present-worth comparison. The interest rate is 16% per year.

	Alternative R1	Alternative R2
First cost, $	147,000	56,000
Annual cost, $	11,000 in year 1: increasing by $500 per year	30,000 in year 1: increasing by $1,000 per year
Salvage value, $	5,000	2,000
Life, years	6	3

5.18 Rework Problem 5.17 above, using a study period of 4 years. Assume the salvage value for R1 after 4 years will be $18,000 and that R2 will have a residual (salvage) value of $14,000 when it is only 1 year old.

5.19 Compare the alternatives below on the basis of a present-worth analysis, using an interest rate of 14% per year. The index k goes from 1 to 10 years.

	Plan A	Plan B
First cost, $	28,000	36,000
Installation cost, $	3,000	4,000
Annual maintenance cost, $	1,000	2,000
Annual operating cost, $	$2,200 + 75k$	$800 + 50k$
Life, years	5	10

5.20 A city that is attempting to attract a professional football team is planning to build a new football stadium costing $220 million. Annual upkeep is expected to amount to $625,000 per year, but this will be paid for by the professional team using the stadium. The artificial turf will have to be replaced every 10 years at a cost of $950,000. Painting every 5 years will cost $65,000. If the city expects to maintain the facility indefinitely, what will be its capitalized cost? Assume that $i = 8\%$ per year.

5.21 What is the capitalized cost of $100,000 in year 0, $150,000 in year 2, and a uniform annual amount of $50,000 per year forever, beginning in year 5? Use an interest rate of 10% per year.

5.22 Calculate the capitalized cost of $60,000 in year 0 and uniform beginning-of-year rent payments of $25,000 for an infinite time using an interest rate of (a) 12% per year, and (b) 16% per year compounded monthly.

5.23 Determine the capitalized cost of a national park which will have an initial cost of $5 million, an annual operating cost of $18,000, and periodic improvement costs of $120,000 every 10 years at an interest rate of 9% per year.

$Pw_{6P} = \$ -5,287,760$

5.24 As a wealthy university grad, you plan to set up an account which can be drawn upon by a person teaching engineering economy. You plan to deposit $1 million now with the stipulation that nothing can be withdrawn for the first 10 years. Between years 11 and 20, only the interest can be withdrawn. In addition, one-half of the balance can be withdrawn in year 20, with the remainder to be withdrawn via an annuity in years 21 through 25. If the account earns interest at a rate of 10% per year, what is the amount that can be withdrawn in (a) year 15? (b) year 20? (c) year 25?

5.25 An alumnus of a private university desires to establish a permanent scholarship in his name. He plans to donate $20,000 per year for 10 years starting 1 year from now and leave $100,000 when he dies. If the alumnus

$F = \$ -567,516$

dies 15 years from now, how much money will be in the account immediately after the $100,000 deposit, assuming the account earned interest at a rate of 9% per year.

5.26 The first cost of a small dam is expected to be $3 million. The annual maintenance cost is estimated to be $10,000 per year; a $35,000 outlay will be required every 5 years. In addition, an expenditure of $5000 in year 10 will be required, increasing by $1000 per year through year 20, after which it will remain constant. If the dam is expected to last forever, what will be its capitalized cost at an interest rate of 10% per year?

5.27 A grateful alumna of a public university wants to establish a permanent scholarship endowment in her name. She wants the endowment to provide $20,000 per year for an infinite time, with the first scholarship to be given 5 years from now. She plans to make her first deposit now and increase each succeeding annual deposit by $5000 through year 4. If the fund earns interest at a rate of 0.5% per month, (a) what is the amount of her first deposit, and (b) what is the capitalized cost (in year 0) of the endowment?

5.28 Compare the machines shown below on the basis of their capitalized costs using an interest rate of 16% per year.

	Machine M	Machine N
First cost, $	31,000	43,000
Annual operating cost, $	18,000	19,000
Salvage value, $	5,000	7,000
Life, years	3	5

5.29 Data for machines X and Y are shown below. If the interest rate is 12% per year, which machine should be selected on the basis of their capitalized costs?

	Machine X	Machine Y
First cost, $	25,000	55,000
Annual operating cost, $	8,000	6,000
Annual increase in operating cost	5%	3%
Salvage value, $	12,000	9,000
Life, years	5	10

5.30 A moving and storage company is considering two possibilities for warehouse operations. Proposal 1 requires the purchase of a forklift for $15,000 and of 500 pallets that cost $5 each. The average life of a pallet is assumed to be 2 years. If the forklift is purchased, the company must hire an operator for $23,000 annually and spend $600 per year in maintenance and operation. The life of the forklift is expected to be 12 years, with a $4000 salvage value.

Alternatively, proposal 2 requires that the company hire two people to operate power-driven hand trucks at a cost of $19,000 per person. One hand truck will be required at a cost of $2000. The hand truck will have a life of 6 years with no salvage value. If the company's minimum attractive rate of return is 14% per year, which alternative should be selected on the basis of capitalized costs?

5.31 Compare the alternatives shown below on the basis of their capitalized costs. Use i = 14% per year.

	Alternative U	Alternative W
First cost, $	8,500,000	50,000,000
Annual operating cost, $	8,000	7,000
Salvage value, $	5,000	2,000
Life, years	5	∞

5.32 Compare the alternatives shown below on the basis of their capitalized costs, using an interest rate of 15% per year.

	Alternative C	Alternative D
First cost, $	160,000	25,000
Annual operating cost, $	15,000	3,000
Overhaul every 4 years, $	12,000	—
Salvage value, $	1,000,000	4,000
Life, years	∞	7

5.33 Compare the alternatives shown below on the basis of their capitalized costs, using an interest rate of an effective 11% per year compounded semiannually.

	Alternative MAX	Alternative MIN
First cost, $	150,000	900,000
Annual operating cost, $	50,000	10,000
Salvage value, $	8,000	1,000,000
Life, years	5	∞

Cap costmax = $793,062
Cap cost min = $ -988,496
Select alt MAX

5.34 Compare the alternatives shown below on the basis of their capitalized costs. Use an interest rate of 1% per month.

	Project C	Project D
First cost, $	8,000	99,000
Monthly operating cost, $	1,500	4,000
Increase per month in operating cost, %	10	. . .
Salvage value, $	1,000	47,000
Life, years	10	∞

5.35 Compare the alternatives shown below using a capitalized-cost comparison and an interest rate of 14% per year compounded semiannually. Assume no interperiod interest.

	Alternative PL	Alternative KN
First cost, $	160,000	25,000
Monthly operating cost, $	15,000	3,000
Overhaul every 4 years, $	12,000	. . .
Salvage value, $	1,000,000	4,000
Life, years	∞	7

5.36 Compare the alternatives shown below on the basis of their capitalized costs. Use $i = 14\%$ per year compounded quarterly.

	Alternative U	Alternative R
First cost, $	8,500,000	20,000,000
Semiannual operating cost, $	80,000	70,000
Salvage value, $	50,000	2,000
Life, years	5	∞

5.37 A medium-sized municipality wants to develop an intelligent software system to assist in project selection during the next 10 years. A life-cycle cost approach has been used to categorize costs into development, programming, operating, and support costs for each alternative. There are three alternatives under consideration identified as *A* (tailored system),

B (adapted system), and *C* (current system). The costs are summarized below. Use a present-worth analysis and an interest rate of 10% per year to identify the best alternative.

Alternative	Cost Component	Cost
A	Development	$100,000 now, $150,000 year 1
	Programming	$45,000 now, $35,000 year 1
	Operation	$50,000 years 1 through 10
	Support	$30,000 years 1 through 10
B	Development	$10,000 now
	Programming	$45,000 year 0, $30,000 year 1
	Operation	$80,000 years 1 through 10
	Support	$40,000 years 1 through 10
C	Operation	$150,000 years 1 through 10

5.38 A window manufacturing company is considering the development of a decision support system to reduce the amount of scrap glass. The current cost for scrap glass loss is $12,000 per year for glass and for labor is 0.75 work-year at a cost of $45,000 per work-year. If the current system is minimally upgraded now for $25,000, it will serve the company's needs for 10 more years. A new system which will also serve the company for 10 years would have the projected costs shown below.

$100,000 per year equipment costs (years 0 and 1)

$120,000 per year for 2 years to develop (years 1 and 2)

$20,000 per year to maintain old system (years 1, 2, and 3)

$10,000 per year to maintain hardware and software (years 3 through 10)

$40,000 per year personnel costs (years 3 through 10)

$30,000 per year scrap glass cost (years 3 through 10)

Sales of the system to other companies are expected to net revenues of $20,000 beginning in year 5, increasing by $5000 per year through year 10. If a 20%-per-year return is required, which alternative has the lower LCC for the 10-year period?

chapter

6

Equivalent-Uniform-Annual-Worth Evaluation

The objectives of this chapter are to explain and demonstrate the primary methods of calculating the equivalent uniform annual worth (AW) of an asset and how to select the better of two alternatives on the basis of an annual-worth comparison. Although the word *annual* is included in the name of the method, the procedures developed in this chapter can be used to find an equivalent uniform series over any interest period desired, as per Chapter 3. Additionally, the word *cost* is often used interchangeably with *worth* in describing a series so that annual cost (AC) and AW mean the same thing. However, AW more properly describes the cash flow because oftentimes the uniform series developed represents an income rather than a cost. Regardless of which term is used to describe the resulting uniform cash flow, the alternative selected as better will be the same as that chosen by the present-worth method or any other evaluation method when the comparisons are properly conducted.

Purpose: Make annual-worth calculations and compare alternatives using the annual-worth basis.

This chapter will help you:

One life cycle

1. Understand why AW needs to be calculated over only one life cycle.

AW by salvage sinking-fund method

2. Calculate AW using the salvage sinking-fund method.

AW by salvage present-worth method

3. Calculate AW using the salvage present-worth method.

AW by capital-recovery-plus-interest method

4. Calculate AW using the capital-recovery-plus-interest method.

Alternative selection

5. Select the best alternative on the basis of an AW analysis.

Perpetual series

6. Calculate the AW of a permanent investment involving a perpetual uniform series.

6.1   ANNUAL-WORTH VALUES FOR ONE OR MORE LIFE CYCLES

The AW method is commonly used for comparing alternatives. As illustrated in Chapter 4, the AW means that all incomes and disbursements (irregular and uniform) are converted into an equivalent uniform annual (end-of-period) amount, which is the *same each period*. The major advantage of this method over all the other methods is that it does not require making the comparison over the least common multiple (LCM) of years when the alternatives have different lives. That is, the AW value of the alternative is calculated for *one life cycle only*. Why? Because, as its name implies, the AW is an equivalent annual worth over the life of the project. If the project is continued for more than one cycle, the equivalent annual worth for the next cycle and all succeeding cycles is assumed to be exactly the same as for the first, provided all actual cash flows are the same for each cycle in constant-value dollars (discussed in Chapter 12).

The repeatability of the uniform annual series through various life cycles can be demonstrated by considering the cash-flow diagram in Figure 6–1, which represents two life cycles of an asset with a first cost of $20,000, an annual operating cost of $8000, and a 3-year life.

The AW for one life cycle (i.e., 3 years) would be calculated as follows:

$$AW = -20{,}000(A/P,22\%,3) - 8000$$

$$= \$-17{,}793$$

The AW for two life cycles is calculated as

$$AW = -20{,}000(A/P,22\%,6) - (20{,}000)(P/F,22\%,3)(A/P,22\%,6) - 8000$$

$$= \$-17{,}793$$

The AW value for the first life cycle is exactly the same as that for two life cycles. This same AW will be obtained when three, four, or any other number of life cycles are evaluated. Thus, the AW for one life cycle of an alternative represents the equivalent uniform annual worth of that alternative *every time the cycle is repeated*.

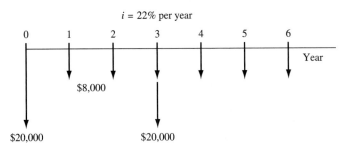

Figure 6–1   Cash-flow diagram for two life cycles of an asset.

When information is available indicating that the estimated cash flows will not be the same in succeeding life cycles (or more specifically, that they will change by an amount other than the expected inflation or deflation rate), then a study period or planning horizon is chosen and used as discussed later in Section 6.5. In this text, unless otherwise specified, it is assumed that all future costs will change exactly in accordance with the inflation or deflation rate for that time.

Problem 6.1

6.2   AW BY THE SALVAGE SINKING-FUND METHOD

When an asset has a terminal salvage value (SV), there are several ways by which the AW can be calculated. This section presents the salvage sinking-fund method, probably the simplest of the three methods discussed in this chapter. This is the method that will usually be used in this text. In the salvage sinking-fund method, the initial cost P is first converted into an equivalent uniform annual amount using the A/P factor. Since the salvage value is (normally) a positive cash flow, after conversion to an equivalent uniform amount via the A/F factor, it is added to the annual equivalent of the first cost. These calculations can be represented by the general equation:

$$AW = -P(A/P,i,n) + SV(A/F,i,n) \qquad [6.1]$$

Naturally, if the alternative has any other cash flows, they must be included in the complete AW computation. This is illustrated in Example 6.1.

Example 6.1

Calculate the AW of a tractor attachment that has an initial cost of $8000 and a salvage value of $500 after 8 years. Annual operating costs for the machine are estimated to be $900, and an interest rate of 20% per year is applicable.

Solution

The cash-flow diagram (Figure 6–2) indicates that

$$AW = A_1 + A_2$$

where A_1 = annual cost of initial investment with salvage value considered, Equation [6.1]

$$= -8000(A/P,20\%,8) + 500(A/F,20\%,8) = \$-2055$$

$$A_2 = \text{annual operating cost} = \$-900$$

The annual worth for the attachment is

$$AW = -2055 - 900 = \$-2955$$

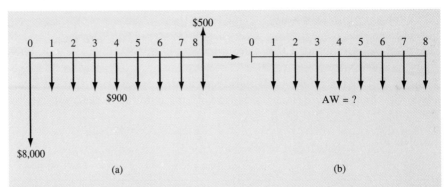

Figure 6–2   (a) Diagram for machine costs, and (b) conversion to an AW.

Comment

Since the operating cost is already expressed as an annual cost for the asset's life, no conversion is necessary.

The simplicity of the salvage sinking-fund method should be obvious from the straightforward calculations shown in Example 6.1. The steps can be summarized as follows *using the correct cash-flow signs:*

1. Annualize the initial investment cost over the life of the asset using the A/P factor.
2. Annualize the salvage value using the A/F factor.
3. Combine the annualized salvage value with the annualized investment cost.
4. Combine any uniform annual amounts with the value from step 3.
5. Convert any other cash flows into equivalent uniform annual worths and combine them with the value obtained in step 4.

Steps 1 through 3 are accomplished by Equation [6.1]. The uniform annual amount in step 3 may be the annual operating cost (AOC), as in Example 6.1.

Additional Example 6.7
Problems 6.2 to 6.4

6.3   AW BY THE SALVAGE PRESENT-WORTH METHOD

The salvage present-worth method also converts investments and salvage values into an AW. The present worth of the salvage is removed from the initial investment cost, and the resulting difference is annualized with the A/P factor for the life of the asset. The general equation is

$$AW = [-P + SV(P/F,i,n)](A/P,i,n)$$ [6.2]

The steps to determine the complete asset AW are

1. Calculate the present worth of the salvage value via the P/F factor.
2. Combine the value obtained in step 1 with the investment cost P.
3. Annualize the resulting difference over the life of the asset using the A/P factor.
4. Combine any uniform annual worths with the value from step 3.
5. Convert any other cash flows into an equivalent uniform annual worth and combine with the value obtained in step 4.

Steps 1 through 3 are accomplished by using Equation [6.2].

Example 6.2

Compute the AW of the attachment detailed in Example 6.1 using the salvage present-worth method.

Solution

Using the steps outlined above and Equation [6.2],

$$AW = [-8000 + 500(P/F,20\%,8)](A/P,20\%,8) - 900 = \$-2955$$

Problem 6.5

6.4 AW BY THE CAPITAL-RECOVERY-PLUS-INTEREST METHOD

The final procedure presented for calculating the AW of an asset having a salvage value is the capital-recovery-plus-interest method. The general equation for this method is

$$AW = -(P - SV)(A/P,i,n) - SV(i) \qquad [6.3]$$

Subtracting the salvage value from the investment cost, that is, $P - SV$, *before* multiplying by the A/P factor, recognizes that the salvage value will be recovered. However, the fact that the salvage value will not be recovered for n years is taken into account by charging the interest lost, $SV(i)$, during the asset's life. Failure to include this term assumes that the salvage value was obtained in year 0 instead of year n. The steps to be followed for this method are as follows:

1. Reduce the initial cost by the amount of the salvage value.
2. Annualize the value in step 1 using the A/P factor.
3. Multiply the salvage value by the interest rate.
4. Combine the values obtained in steps 2 and 3.
5. Combine any uniform annual amounts.
6. Convert all other cash flows into equivalent uniform amounts and combine them with the value from step 5.

Steps 1 through 4 are accomplished by applying Equation [6.3].

[handwritten margin note: Ans: should all be the same regardless of method.]

Example 6.3

Use the values of Example 6.1 to compute the AW using the capital-recovery-plus-interest method.

Solution

From Equation [6.3] and the steps above,

$$AW = -(8000 - 500)(A/P,20\%,8) - 500(0.20) - 900 = \$-2955$$

While it makes no difference which method is used to compute the AW, it is good practice to consistently use one method, thus avoiding errors caused by mixing techniques. We will generally use the salvage sinking-fund method (Section 6.2).

Problem 6.6

6.5   COMPARING ALTERNATIVES BY ANNUAL WORTH

The annual-worth method for comparing alternatives is probably the simplest to perform of the evaluation techniques presented in this book. The alternative selected has the lowest equivalent cost or largest equivalent income, if revenues are included. As discussed in other chapters, nonquantifiable data are always considered in real-world selection decisions, but, in general, the alternative having the highest net worth is selected.

Perhaps the most important rule to remember when making AW comparisons is that *only one life cycle* of each alternative must be considered. This is because the AW will be the same for any number of life cycles as it is for one (Section 6.1). This procedure is, of course, subject to the assumptions underlying this method. These assumptions are similar to those applicable to a present-worth analysis with the LCM for lives; namely, (1) the alternatives will be needed for their LCM of years, or, if not, the annual worth will be the same for any fraction of the asset's life cycle as it is for the entire cycle, and (2) the cash flows in succeeding life cycles will change by exactly the inflation or deflation rate. When information is available indicating that one or the other of these assumptions may not be valid, the planning-horizon approach is followed. That is, the disbursements and incomes actually expected to occur over some specified study period—the planning horizon—must be identified and converted into annual worths. Examples 6.4 and 6.5 illustrate these procedures.

Example 6.4

The following costs are estimated for two equal-service tomato-peeling machines to be evaluated by a canning plant manager.

	Machine A	Machine B
First cost, $	26,000	36,000
Annual maintenance cost, $	800	300
Annual labor cost, $	11,000	7,000
Extra annual income taxes, $	—	2,600
Salvage value, $	2,000	3,000
Life, years	6	10

If the minimum required rate of return is 15% per year, help the manager decide which machine to select.

Solution

The cash-flow diagram for each alternative is shown in Figure 6–3. The AW of each machine using the salvage sinking-fund method, Equation [6.1], is calculated as follows:

$$AW_A = -26,000(A/P,15\%,6) + 2000(A/F,15\%,6) - 11,800 = \$-18,442$$

$$AW_B = -36,000(A/P,15\%,10) + 3000(A/F,15\%,10) - 9900 = \$-16,925$$

Select machine B, since the AW of costs is smaller.

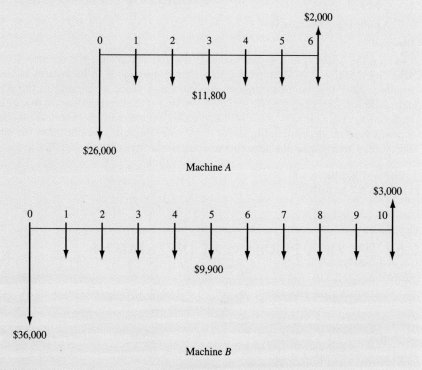

Machine A

Machine B

Figure 6–3   Cash flows for two tomato-peeling machines, Example 6.4.

Example 6.5

(a) Assume the company in Example 6.4 is planning to exit the tomato-canning business in 4 years. At that time, the company expects to sell machine A for $12,000 or machine B for $15,000. All other costs are expected to remain the same. Which machine should the company purchase under these conditions?

(b) If all costs, including the salvage values, will be as originally estimated, which machine should the company purchase using the 4-year horizon?

Solution

(a) The planning horizon is now 4 years, not the 6- or 10-year lives. Only 4 years of the estimated costs are relevant to the analysis. Recompute the AW amounts.

$$AW_A = -26,000(A/P,15\%,4) + 12,000(A/F,15\%,4) - 11,800 = \$-18,504$$

$$AW_B = -36,000(A/P,15\%,4) + 15,000(A/F,15\%,4) - 9900 = \$-19,506$$

Now, select machine A, since its cost is smaller. This is a decision reversal from Example 6.4.

(b) The only change from part (a) is in the salvage values of the machines. The AW values are now

$$AW_A = -26,000(A/P,15\%,4) + 2000(A/F,15\%,4) - 11,800 = \$-20,506$$

$$AW_B = -36,000(A/P,15\%,4) + 3000(A/F,15\%,4) - 9900 = \$-21,909$$

Again, select machine A.

Comment

Limiting the analysis to a 4-year planning horizon did change the decision from selecting machine B in Example 6.4 to selecting machine A in both cases in this example. Note that no cost estimates beyond the 4 years are present in the AW calculations of this example. We will discover later (Chapter 16) that this oversight can, under certain circumstances of future cash-flow patterns, contribute to making a poor economic decision in the longer term. Also, this situation illustrates the importance of recognizing the assumptions inherent to the evaluation methods.

Additional Example 6.8
Problems 6.7 to 6.18

6.6   AW OF A PERMANENT INVESTMENT

This section discusses the annual-worth equivalent of the capitalized cost introduced in Sections 5.4 and 5.5. Evaluation of flood-control projects, irrigation canals, bridges, or other large-scale projects requires the comparison of alternatives which have such long lives that they may be considered infinite in economic analysis terms. For this type of analysis, it is important to recognize that the annual worth of the initial investment is simply equal to the annual interest earned on the lump-sum investment, as expressed by Equation [5.2], that is, $A = Pi$.

Cash flows recurring at regular or irregular intervals are handled exactly as in conventional AW computations; that is, they are converted into equivalent uniform annual amounts for one cycle. This automatically annualizes them for each succeeding life cycle, as discussed in Section 6.1. Example 6.6 illustrates AW calculations for a project with a very long life.

Example 6.6

The U.S. Bureau of Reclamation is considering two proposals for increasing the capacity of the main canal in their Lower Valley irrigation system. Proposal A would involve dredging the canal in order to remove sediment and weeds which have accumulated during previous years' operation. Since the capacity of the canal will have to be maintained in the future near its design peak flow because of increased water demand, the bureau is planning to purchase the dredging equipment and accessories for $65,000. The equipment is expected to have a 10-year life with a $7000 salvage value. The annual labor and operating costs for the dredging operation are estimated to total $22,000. In order to control weeds in the canal itself and along the banks, herbicides will be sprayed during the irrigation season. The yearly cost of the weed-control program, including labor, is expected to be $12,000.

Proposal B would involve lining the canal with concrete at an initial cost of $650,000. The lining is assumed to be permanent, but minor maintenance will be required every year at a cost of $1000. In addition, lining repairs will have to be made every 5 years at a cost of $10,000. Compare the two alternatives on the basis of equivalent uniform annual worth using an interest rate of 5% per year.

Solution

Since this is an investment for a permanent project, compute the AW for one cycle. We use the salvage sinking-fund method in the evaluation of proposal A. The cash-flow diagrams are left to the learner.

Proposal A	
AW of dredging equipment:	
$-65,000(A/P,5\%,10) + 7000(A/F,5\%,10)$	$ -7,861
Annual cost of dredging	$-22,000$
Annual cost of weed control	$-12,000$
	$-41,861
Proposal B	
AW of initial investment: $-650,000(0.05)$	$-32,500
Annual maintenance cost	$-1,000$
Lining repair cost: $-10,000(A/F,5\%,5)$	$-1,800$
	$-35,310

Proposal B should be selected.

Comment

For proposal A, no calculations are necessary for the dredging and weed-control costs since they are already expressed as annual costs. For proposal B, the AW of the initial investment is obtained by multiplying by the interest rate, which is nothing more than Equation [5.2], that is, $AW = A = Pi$. Note the use of the A/F (sinking-fund) factor for the lining repair cost in proposal B. The A/F factor is used instead of A/P, because the lining repair cost begins in year 5, not year 0, and continues indefinitely at 5-year intervals.

If there are *nonrecurring* single or series amounts involved, they are converted to a present worth and then multiplied by the interest rate. See Additional Examples 6.9 and 6.10 for illustrations.

Additional Examples 6.9 and 6.10
Problems 6.19 to 6.25

ADDITIONAL EXAMPLES

Example 6.7

DETERMINATION OF AW, SECTIONS 6.2 TO 6.4   A local pizza shop has just purchased a fleet of five electric-powered mini-vehicles for delivery in an urban area. The initial cost was $4600 per vehicle, and their expected life and salvage values are 5 years and $300, respectively. The combined insurance, maintenance, recharge, and lubrication costs are expected to be $650 the first year and to increase by $50 per year thereafter. Delivery service will generate an estimated extra $1200 per year. If a return of 10% per year is required, use the AW method to determine if the purchase should have been made.

Solution

The cash-flow diagram is shown in Figure 6–4. If we apply Equation [6.1], the first three steps of the salvage sinking-fund method are complete.

$$A_1 = \text{annual cost of fleet purchase}$$
$$= -5(4600)(A/P,10\%,5) + 5(300)(A/F,10\%,5)$$
$$= \$-5822$$

For step 4, the annual disbursement and income series can be combined into an annual net-income series that conveniently follows a decreasing gradient with a base amount of $550 ($1200 − 650). The equivalent annual income A_2 is

$$A_2 = 550 - 50(A/G,10\%,5) = \$460$$

The total AW equals the algebraic sum of the vehicle cost and income AW values.

$$AW = -5822 + 460 = \$-5362$$

Since $AW < 0$, a return of less than 10% per year is expected; the purchase is not justified.

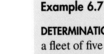

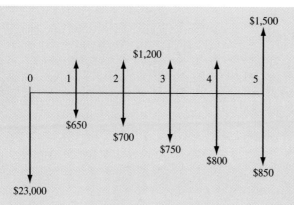

Figure 6–4   Diagram used to compute AW, Example 6.7.

Comment

Try one of the other AW methods of Sections 6.3 and 6.4 to solve the problem. You should get the same AW value.

Example 6.8

ALTERNATIVE EVALUATION, SECTION 6.5   Compare the two plans proposed in Example 5.6 (in Chapter 5) using the AW method.

Solution

Even though the two components of plan A, movers and pads, have different lives, the AW analysis is conducted for only one life cycle of each component. Applying the salvage sinking-fund method, Equation [6.1],

$$AW_A = AW_{movers} + AW_{pad} + AW_{AOC}$$

where $AW_{movers} = -90,000(A/P,15\%,8) + 10,000(A/F,15\%,8) = \$-19,328$

$$Aw_{pad} = -28,000(A/P,15\%,12) + 2000(A/F,15\%,12) = -\$5096$$

$$AW_{AOC} = -\$9800$$

Then,

$$AW_A = -19,328 - 5096 - 9800$$

$$= \$-34,224$$

$$AW_B = AW_{conveyer}$$

$$= -175,000(A/P,15\%,24) + 10,000(A/F,15\%,24)$$

$$= \$-27,146$$

As concluded in the PW analysis of Example 5.6, select plan B for its lower equivalent cost.

Comment

You should recognize a fundamental relation between the PW and AW values for the two examples discussed here. If you have the PW of a given plan, you can get the AW by calculating AW = PW($A/P,i,n$), or if you have the AW, then PW = AW($P/A,i,n$). To obtain the correct PW value, a fundamental question is, What value does n assume? The answer is the LCM of the alternative lives. This is correct since the PW method of evaluation must take place over an equal time period for each alternative to ensure an equal-service comparison. Therefore, the PW values, with round-off considered, are the same as determined in Example 5.6.

$$PW_A = AW_A(P/A,15\%,24) = \$-220,190$$

$$PW_B = AW_B(P/A,15\%,24) = \$-174,652$$

Example 6.9

PERMANENT INVESTMENT, SECTION 6.6   An aggressive stockbroker claims he can make 18% per year on an investor's money. If he invests $10,000 now, $30,000 three years from now, and $6000 per year for 5 years starting 4 years from now, how much money can be withdrawn every year forever beginning 12 years from now, if taxes are disregarded?

Solution

The cash-flow diagram (Figure 6–5) indicates that the uniform amount A that can be withdrawn every year forever starting in year 12 is equal to the amount of interest that accumulates each year on the principal amount available at the time of the first withdrawal. To truly be a perpetual withdrawal series, this amount must remain constant in size.

To solve the problem, therefore, determine the principal amount F_{11} accumulated in year 11 (not year 12) and multiply by the interest rate i to obtain A. The future

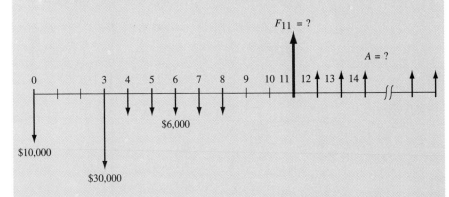

Figure 6–5   Diagram to determine a perpetual annual withdrawal, Example 6.9.

amount in year 11 is

$$F_{11} = 10,000(F/P,18\%,11) + 30,000(F/P,18\%,8)$$
$$+ 6000(F/A,18\%,5) \ (F/P,18\%,3)$$
$$= \$245,050$$

The perpetual withdrawal (or annual worth of the perpetual series) can now be found by multiplying F_{11}, which is relabeled to be a PW value with respect to the perpetual withdrawal series, by $i = 0.18$.

$$AW = PW(i) = 245,050(0.18) = \$44,109$$

Comment

The perpetual withdrawal series generated by the principal amount, sometimes called the *corpus,* has been *endowed.* That is, here the deposits worth \$245,050 in year 11 are an endowment which will produce \$44,109 forever, provided the 18% return is experienced each and every year forever.

Example 6.10

PERMANENT INVESTMENT, SECTION 6.6   If Ms. Kaw deposits \$10,000 now at an interest rate of 7% per year, how many years must the money accumulate before she can withdraw \$1400 per year forever?

Solution

Figure 6–6 presents the cash-flow diagram. The first step is to find the total amount of money, call it PW_n, that must be accumulated in some future year n, just 1 year prior to the first withdrawal of the perpetual AW = \$1400-per-year series. That is,

$$PW_n = \frac{AW}{i} = \frac{1400}{0.7} = \$20,000$$

So, when \$20,000 is accumulated, Ms. Kaw can withdraw \$1400 per year forever. Determine when the initial \$10,000 deposit will accumulate to \$20,000. This is easily

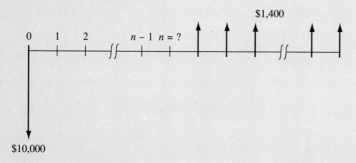

Figure 6–6   Diagram to determine n for a perpetual withdrawal, Example 6.10.

done using the F/P factor and solving for n using logarithms.

$$20{,}000 = 10{,}000(F/P,7\%,n)$$

$$(1.07)^n = 2.00$$

$$n \log (1.07) = \log (2.00)$$

$$n = \frac{\log (2.00)}{\log (1.07)} = 10.24 \text{ years}$$

Or by interpolation in the interest tables, $(F/P,7\%,n) = 2.00$ when $n = 10.24$ years. This withdrawal can begin in year 11.

Example 6.10
(Spreadsheet)

If Ms. Kaw deposits $10,000 now at an interest rate of 7% per year, how many years must the money accumulate before she can withdraw $1400 per year forever?

Solution

As calculated above, it is necessary for the $10,000 to grow to $20,000 in order to take out $1400 forever. Figure 6–7 presents a spreadsheet solution for 10.24 years using the NPER function with interest at 7% per year.

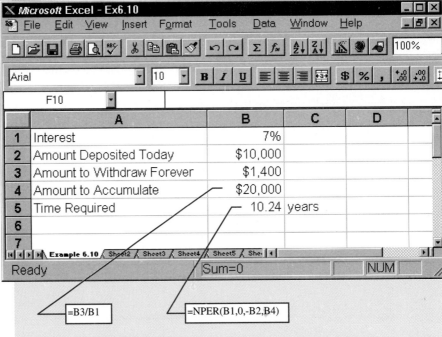

Figure 6–7   Spreadsheet solution to find the number of years to double an amount at 7% per year, Example 6.10.

CHAPTER SUMMARY

The annual-worth method of comparing alternatives is usually computationally simpler than the present-worth method because the AW comparison is performed for only one life cycle, since, by definition, the annual worth through the first life cycle will be the same as the annual worth for the second, third, and all succeeding life cycles. When a planning horizon or study period is specified, the AW calculation is determined for that time period, regardless of the lives of the alternatives. As in the present-worth method, any remaining life of an alternative at the end of its service period or the planning horizon is recognized by adjusting its estimated salvage value.

For infinite-life alternatives, their initial cost can be annualized simply by multiplying P or PW times i, that is, AW $=$ PW(i). For finite-life alternatives, the AW through one life cycle is equal to the perpetual equivalent annual worth.

PROBLEMS

6.1 Machine A has an initial cost of $5000, a life of 3 years, and zero salvage. Machine B has an initial cost of $12,000, a life of 6 years, and a $3000 salvage value. What time period must be used to find (*a*) the present worth of machine A? (*b*) the annual worth of machine A? (*c*) the annual worth of machine B?

6.2 A contractor purchased a used crane for $9000. His operating cost will be $2300 per year, and he expects to sell it for $6000 five years from now. What is the equivalent annual worth of the crane at an interest rate of 12% according to the salvage sinking-fund method?

6.3 Find the annual-worth amount (per month) of a truck which had a first cost of $38,000, an operating cost of $2000 per month, and a salvage value of $11,000 after 4 years at an interest rate of 9% per year compounded monthly. Use the salvage sinking-fund method.

6.4 If the operating cost of the truck in Problem 6.3 was $2000 in month 1, $2010 in month 2, and amounts increasing by $10 each month thereafter, what would be its annual worth at 12%-per-year interest compounded monthly? Use the salvage sinking-fund method.

6.5 Work the following problems by the salvage present-worth method: (*a*) Problem 6.2, (*b*) Problem 6.3.

6.6 Work the following problems by the capital-recovery-plus-interest method: (*a*) Problem 6.2, (*b*) Problem 6.3.

6.7 Two machines are under consideration for purchase by a metal-fabricating company. Machine X will have a first cost of $15,000, an annual maintenance and operation cost of $3000, and a $3000 salvage value. Machine

Y will have a first cost of $22,000, an annual cost of $1500, and a $5000 salvage value. If both machines are expected to last for 10 years, determine which machine should be selected on the basis of annual-worth values using an interest rate of 10% per year.

6.8 A public utility is trying to decide between two different sizes of pipe for a new water main. A 250-millimeter line will have an initial cost of $35,000, whereas a 300-millimeter line will cost $55,000. Since there is more head loss through the 250-millimeter pipe, the pumping cost for the smaller line is expected to be $3000 per year more than for the 300-millimeter line. If the pipes are expected to last for 20 years, which size should be selected if the interest rate is 13% per year? Use an annual-worth analysis.

6.9 A commercial developer is trying to determine if it would be economically feasible to install rainwater drains in a large shopping center currently under construction. In the 3 years required for construction, 12 heavy thundershowers are expected. If no drains are installed, the cost of refilling the washed-out area is expected to be $1000 per thunderstorm. Alternatively, a corrugated steel drainpipe could be installed which will prevent the soil erosion. The installation cost of the pipe would be $6.50 per meter, with a total length of 2000 meters. After the 3-year construction period, some of the pipe could be recovered with an estimated value of $3000. Assuming that the thunderstorms occur at 3-month intervals, determine which alternative should be selected, if the interest rate is a nominal 16% per year compounded quarterly. Use the annual-worth method.

6.10 Machines that have the following costs are under consideration for a continuous production process. Using an interest rate of 12% per year, determine which alternative should be selected on the basis of an annual-worth analysis.

	Machine G	Machine H
First cost, $	62,000	77,000
Annual operating cost, $	15,000	21,000
Salvage value, $	8,000	10,000
Life, years	4	6

6.11 Compare the machines below on the basis of their annual worths, using an interest rate of 14% per year.

	Machine P	Machine Q
First cost, $	29,000	37,000
Annual operating cost, $	4,000	5,000
Life, years	3	5
Annual maintenance cost, $	3,000	3,500
Overhaul every 2 years, $	3,700	2,000

6.12 Two fabricating machines are presently under consideration for purchase by the L Tech Metal Fabricating Company. The manual model will cost $25,000 to buy with an 8-year life and a $5000 salvage value. Its annual operating cost will be $15,000 for labor and $1000 for maintenance. A computer-controlled model will cost $95,000 to buy, and it will have a 12-year life if upgraded at the end of year 6 for $15,000. Its terminal salvage value will be $23,000. The annual costs for the computer-controlled model will be $7500 for labor and $2500 for maintenance. If the company's minimum attractive rate of return is 20%, which machine would be preferred on the basis of the equivalent annual cost of each?

Aw manual
$-22,212

Aw computer
$-31,950

Aw_m < Au comp
Select Aw comp

6.13 If an 8-year planning horizon is used in Problem 6.12, which machine should be selected, assuming the salvage values remain the same as estimated?

6.14 Compare the alternatives below on the basis of their annual worths at 10%-per-year interest.

	Alternative P	Alternative Q
First cost, $	30,000	42,000
Annual operating cost years 1 through 4, $	15,000	6,000
Annual cost decrease years 5 through n, $	500	—
Salvage value, $	7,000	11,000
Life, years	10	12

6.15 A consulting engineering firm is trying to decide between purchasing and leasing cars. It estimates that medium-sized cars will cost $12,000 and will have a probable trade-in value in 4 years of $2800. The annual cost of such items as fuel and repairs is expected to be $950 the first year and to increase by $50 per year. Alternatively, the company can lease the same cars for $4500 per year payable at the beginning of each year. Since some maintenance is included in the lease price, the annual maintenance and operation expenses are expected to be $100 per year lower if the cars are leased. If the company's minimum attractive rate of return is 10% per year, which alternative should be selected on the basis of its annual worth?

6.16 The Mining Company is considering purchasing a machine which costs $30,000 and is expected to last 11 years, with a $3000 salvage value. The annual operating expenses are expected to be $8000 for the first 3 years, but owing to increased use, the operating costs will increase by $200 per year for the next 8 years. Alternatively, the company can purchase a highly automated machine at a cost of $58,000. This machine will last only 6 years because of its higher technology and delicate design, and its salvage value will be $15,000. Because it is so automated, its operating cost will be only $4000 per year. If the company's minimum attractive rate of return is 18% per year, which machine should be selected on the basis of an annual-worth analysis?

6.17 A company is considering the implementation of one of two processes identified as Q and R. Process Q will have a first cost of $43,000, a quarterly operating cost of $10,000, and a $5000 salvage value at the end of its 6-year life. Process R will have a first cost of $31,000 with a quarterly operating cost of $39,000. It will have an 8-year life with a $2000 value at that time. If the interest rate is a nominal 12% per year compounded monthly, which alternative would be preferred using annual-worth analysis?

6.18 Compare the alternatives shown below on the basis of an annual-worth analysis. Use an interest rate of 18% per year compounded monthly.

	Alternative Y	Alternative Z
First cost, $	20,000	31,000
Quarterly operating cost, $	4,000	5,000
Monthly income, $	600	900
Salvage value, $	3,000	6,000
Life, years	5	4

[handwritten left margin: $Aw_y = \$- 1{,}190$
$Aw\,z = \$- 1{,}566$
select machine Y *]*

6.19 What is the perpetual annual worth of $50,000 now and another $50,000 three years from now at an interest rate of 10% per year?

6.20 Determine the perpetual uniform annual worth of an initial cost of $600,000, annual costs of $25,000, and periodic costs every 4 years of $70,000 for an infinite time. Use an interest rate of 12% per year.

6.21 An alumna of a small college decided to set up a permanent scholarship endowment in her name. Her initial donation was $2 million. The fund earned interest at a rate of 9% year, but the fund trustee took 3% of the earnings as a management fee. (*a*) How much scholarship money was available each year, assuming all earnings (after the management fee) were awarded for scholarships? (*b*) If the donor added another $1 million 5 years after the fund was established, how much was available each year for scholarships thereafter?

6.22 Determine the annual worth of the following cash flows at an interest rate of 12% per year.

[handwritten left margin: $Pw = \$- 46{,}990$
$Aw = \$- 5{,}639$ *]*

Year	0	1	2–6	7–12	13 on
Cash flow, $	−50,000	−6,000	−2,000	+3,000	+4,000

6.23 It is desired to determine the equivalent annual worth of establishing, upgrading, and maintaining a permanent national park. The park service expects to purchase the land for $10 million. They estimate that improvements to be made every 3 years through year 15 will cost $500,000 each time. In addition, annual costs of $40,000 will be required for the first 10 years, after which time the costs will be $55,000 per year. If the interest rate is 10% per year, what is the equivalent annual worth if the park is kept forever?

6.24 Find the equivalent perpetual annual worth of an initial investment of $250,000; an annual cost of $8000 for the first 2 years, increasing by $500 per year through year 15; and a cost of $13,000 per year in years 16 through infinity. Use an interest rate of (a) 10% per year and (b) 12% per year compounded monthly.

6.25 Compare the alternatives shown below on the basis of their annual worths. Use an interest rate of (a) 8% per year and (b) 10% per year compounded quarterly. The index k ranges from 1 to 10 years.

	Alternative G	Alternative H
First cost, $	40,000	300,000
Annual cost, $	$5,000 - 100(k - 2)$	1,000
Salvage value, $	8,000	50,000
Life, years	10	∞

chapter

Rate-of-Return Computations for a Single Project

In this chapter, the procedure to correctly compute the rate of return (ROR) for one project based on a present-worth or annual-worth equation is discussed. Since ROR calculations frequently require a manual trial-and-error method, or a spreadsheet system for quicker resolution, procedures for estimating the interest rate that will satisfy the ROR equation are presented.

One of the dilemmas of ROR analysis is that, in some cases, multiple values of i will satisfy the ROR equation. We explain how to recognize this possibility and the correct approach to obtain the ROR value by using a reinvestment rate for the project's positive net cash flows. Only one project is considered in this chapter; however, the next chapter applies the principles discussed here for selection from two and more alternatives.

Purpose: Understand rate-of-return (ROR) analysis and perform ROR calculations for one project.

	This chapter will help you:
Definition	1. Understand the basis of rate-of-return calculations.
ROR using PW	2. Calculate the rate of return using a present-worth equation.
ROR using AW	3. Calculate the rate of return using an annual-worth equation.
Multiple RORs	4. Determine the maximum number of possible ROR values for a nonconventional (nonsimple) cash-flow sequence.
Composite ROR	5. Calculate the composite rate of return using a stated reinvestment rate.

7.1 OVERVIEW OF RATE OF RETURN AND ITS COMPUTATION

If money is borrowed, the interest rate is applied to the *unpaid balance* so that the total loan amount and interest are paid in full exactly with the last loan payment. From the perspective of the lender or investor, when money is lent or invested, there is an *unrecovered balance* at each time period. The interest rate is the return on this unrecovered balance so that the total amount and interest are recovered exactly with the last payment or receipt. Rate of return defines both of these situations.

> *Rate of return (ROR)* is the rate of interest paid on the unpaid balance of borrowed money, or the rate of interest earned on the unrecovered balance of an investment, so that the final payment or receipt brings the balance to exactly zero with interest considered.

The rate of return is expressed as a percent per period, for example, $i = 10\%$ per year. It is stated as a positive percentage; that is, the fact that interest paid on a loan is actually a negative rate of return from the borrower's perspective is not considered. The numerical value of i can range from -100% to infinity, that is, $-100\% < i < \infty$. In terms of an investment, a return of $i = -100\%$ means the entire amount is lost.

The definition above does not state that the rate of return is on the initial amount of the investment, but it is rather on the *unrecovered* balance, which varies with time. The example below illustrates the difference between these two concepts.

Example 7.1

At $i = 10\%$ per year, a $1000 investment is expected to produce a net cash flow of $315.47 for each of 4 years.

$$A = \$1000(A/P,10\%,4) = \$315.47$$

This represents a 10%-per-year rate of return on the unrecovered balance. Compute the amount of the unrecovered investment for each of the 4 years using (*a*) the rate of return on the unrecovered balance and (*b*) the rate of return on the initial $1000 investment. (*c*) Explain why all the initial $1000 investment is not recovered by the approach in part (*b*).

Solution

(*a*) Table 7–1 presents the unrecovered balance figures for each year using the 10% rate on the unrecovered balance at the beginning of the year. After 4 years the total $1000 investment is recovered and the balance in column 6 is exactly zero.

(*b*) Table 7–2 shows the unrecovered balance figures if the 10% return is always figured on the initial investment of $1000. Column 6 in year 4 shows a remaining

Table 7-1 Unrecovered balances using a rate of return of 10%

(1) Year	(2) Beginning Unrecovered Balance	(3) = 0.10(2) Interest on Unrecovered Balance	(4) Cash Flow	(5) = (4) − (3) Recovered Amount	(6) = (2) + (5) Ending Unrecovered Balance
0	—	—	$−1,000.00	—	$−1,000.00
1	$−1,000.00	$100.00	+315.47	$ 215.47	−784.53
2	−784.53	78.45	+315.47	237.02	−547.51
3	−547.51	54.75	+315.47	260.72	−286.79
4	−286.79	28.68	+315.47	286.79	0
		$261.88		$1,000.00	

Table 7-2 Unrecovered balances using a 10% return on the initial investment

(1) Year	(2) Beginning Unrecovered Balance	(3) = 0.10(2) Interest on Initial Investment	(4) Cash Flow	(5) = (4) − (3) Recovered Amount	(6) = (2) + (5) Ending Unrecovered Balance
0	—	—	$−1,000.00	—	$−1,000.00
1	$−1,000.00	$100	+315.47	$215.47	−784.53
2	−784.53	100	+315.47	215.47	−569.06
3	−569.06	100	+315.47	215.47	−353.59
4	−353.59	100	+315.47	215.47	−138.12
		$400		$861.88	

unrecovered amount of $138.12, because only $861.88 is recovered in the 4 years (column 5).

(c) A total of $400 in interest must be earned if the 10% return each year is figured on the initial amount of $1000. However, only $261.88 in interest must be earned if a 10% return on the unrecovered balance is used. There is more of the annual cash flow available to reduce the remaining investment when the rate is applied to the unrecovered balance as in part (a) and Table 7–1.

As defined above, rate of return is the interest rate on the unrecovered balance; therefore, the computations in *Table 7–1 for part (a) present a correct interpretation of a 10% rate of return*. Clearly, an interest rate applied only to the principal represents a higher rate than is stated. In practice, a so-called add-on interest rate is frequently based on principal only, as in part (b).

To determine the rate of return i of a project's cash flows, set up the ROR relation. The present worth of investments or disbursements, PW_D, is equated to the present worth of incomes or receipts, PW_R. Equivalently, the two can be subtracted and set equal to zero. That is,

$$PW_D = PW_R$$

$$0 = -PW_D + PW_R \qquad \text{[7.1]}$$

The annual-worth approach utilizes the AW values in the same fashion to solve for i.

$$AW_D = AW_R$$

$$0 = -AW_D + AW_R \qquad \text{[7.2]}$$

The i value which makes these equations numerically correct is the root of the ROR relation. This i value is referred to by several terms in addition to rate of return. Some are internal rate of return (IRR), breakeven rate of return, profitability index, and return on investment (ROI). They are represented by the notation i^* (i star).

Problems 7.1 and 7.2

7.2   RATE-OF-RETURN CALCULATIONS USING A PRESENT-WORTH EQUATION

In Section 2.11 the method for calculating the rate of return on an investment was illustrated when only one engineering economy factor was involved. In this section, a present-worth equation is the basis for calculating the rate of return on an investment when several factors are involved. To understand rate-of-return calculations more clearly, remember that the basis for engineering economy calculations is *equivalence,* or time value of money. In previous chapters, we have shown that a present sum of money is equivalent to a larger sum at a future date, provided the interest rate is greater than zero. In rate-of-return calculations, the objective is to find the interest rate $i*$ at which the present sum and future sum are equivalent. The calculations made here are the reverse of calculations made in previous chapters, where the interest rate i was known.

The backbone of the rate-of-return method is a ROR relation. For example, if you deposit $1000 now and are promised payments of $500 three years from now and $1500 five years from now, the rate-of-return relation using PW is

$$1000 = 500(P/F,i*,3) + 1500(P/F,i*,5) \qquad \text{[7.3]}$$

where the value of $i*$ to make the equality correct is to be computed (see Figure 7–1). If the $1000 is moved to the right side of Equation [7.3], we have

$$0 = -1000 + 500(P/F,i*,3) + 1500(P/F,i*,5) \qquad \text{[7.4]}$$

Equation [7.4] applies the general form of Equation [7.1], which will be used in setting up all rate-of-return calculations based upon present worth. The equation is solved for i to obtain $i* = 16.9\%$. Since receipts and disbursements are usually involved in a given project, a value of $i*$ can be found; moreover, the rate of return will always be greater than zero if the total amount of receipts is greater than the total amount of disbursements, when the time value of money is considered.

It should be evident that rate-of-return relations are merely a rearrangement of a present-worth equation. That is, if the above interest rate were known to be 16.9%, and it were desired to find the present worth of $500 three years from now and $1500 five years from now, the equation would be

$$P = 500(P/F,16.9\%,3) + 1500(P/F,16.9\%,5) = \$1000$$

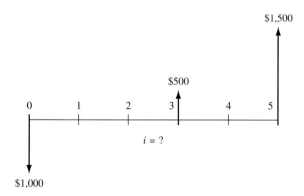

Figure 7–1   Cash flow for which a value of i is to be determined.

which is easily rearranged to the form of Equation [7.4]. This illustrates that rate-of-return and present-worth equations are set up in exactly the same fashion. The only differences are what is given and what is sought.

There are two common ways to determine $i*$ once the PW relation is established: manual solution via trial and error and computer solution via spreadsheet. The second is faster; the first helps you better understand how ROR computations work. We summarize both methods here and in Example 7.2.

$i*$ Using Manual Trial and Error   The general procedure used to make a rate-of-return calculation using a present-worth equation and manual trial-and-error computations is

1.   Draw a cash-flow diagram.
2.   Set up the rate-of-return equation in the form of Equation [7.1].
3.   Select values of i by trial and error until the equation is balanced.

When using the trial-and-error method to determine $i*$, it is advantageous to get fairly close to the correct answer on the first trial. If the cash flows are combined in such a manner that the income and disbursements can be represented by a *single factor* such as P/F or P/A, it is possible to look up the interest rate (in the tables) corresponding to the value of that factor for n years as discussed in Chapter 2. The problem, then, is to combine the cash flows into the format of only one of the standard factors. This may be done through the following procedure:

1.   Convert all *disbursements* into either single amounts (P or F) or uniform amounts (A) by neglecting the time value of money. For example, if it is desired to convert an A into an F value, simply multiply the A by the number of years n. The scheme selected for movement of cash flows should be the one which minimizes the error caused by neglecting the time value of money. That is, if most of the cash flow is an A and a small amount is an F, convert the F into an A rather than the other way around, and vice versa.
2.   Convert all *receipts* to either single or uniform values.
3.   Having combined the disbursements and receipts so that either a P/F, P/A, or A/F format applies, use the interest tables to find the approximate interest rate at which the P/F, P/A, or A/F value, respectively, is satisfied for the proper n value. The rate obtained is a good ballpark figure to use in the first trial.

It is important to recognize that the rate of return obtained in this manner is only an *estimate* of the actual rate of return, because the time value of money is neglected. The procedure is illustrated in the next example.

Example 7.2

If $5000 is invested now in common stock that is expected to yield $100 per year for 10 years and $7000 at the end of 10 years, what is the rate of return?

Solution

Use the manual trial-and-error procedure based on a PW equation.

1. Figure 7–2 shows the cash-flow diagram.
2. Use Equation [7.1] format.

$$0 = -5000 + 100(P/A,i^*,10) + 7000(P/F,i^*,10)$$

3. Use the interest-rate estimation procedure to determine i for the first trial. All income will be regarded as a single F in year 10 so that the P/F factor can be used.

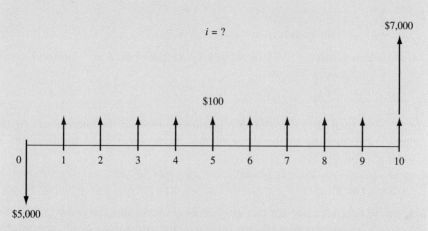

Figure 7–2   Cash flow for a stock investment, Example 7.2.

The P/F factor is selected because most of the cash flow (i.e., $7000) already fits this factor and errors created by neglecting the time value of the remaining money will be minimized. Only for the first estimate of i, define $P = \$5000$, $n = 10$, and $F = 10(100) + 7000 = 8000$. Now we can state that

$$5000 = 8000(P/F,i,10)$$

$$(P/F,i,10) = 0.625$$

The approximate i is between 4% and 5%. Therefore, use $i = 5\%$ to estimate the actual rate of return.

$$0 = -5000 + 100(P/A,5\%,10) + 7000((P/F,5\%,10)$$

$$0 < \$69.46$$

We are too large on the positive side, indicating that the return is more than 5%. Try $i = 6\%$.

$$0 = -5000 + 100(P/A,6\%,10) + 7000(P/F,6\%,10)$$

$$0 > -\$355.19$$

Since the interest rate of 6% is too high, interpolate between 5% and 6% (Section 2.7) to obtain

$$i* = 5.00 + \frac{69.46 - 0}{69.46 - (-355.19)}(1.0)$$

$$= 5.00 + 0.16 = 5.16\%$$

Comment

Note that 5% rather than 4% was used for the first trial. The higher value was used because, by assuming that the ten $100 amounts were equivalent to a single $1000 in year 10, the approximate rate estimated from the P/F factor was lower than the true value. This is due to the neglect of the time value of money.

i* Using a Spreadsheet   The general procedure based on a present-worth equation and spreadsheet solution is

1. Draw a cash-flow diagram.
2. Set up the rate-of-return relation in the form of Equation [7.1].
3. Enter onto the spreadsheet the cash-flow values in exactly the same order in which they occur.
4. Set up the spreadsheet system's internal rate-of-return (IRR) function to obtain the correct $i*$ (preferably to an accuracy of two decimal places).

Since most spreadsheets use the entry order to determine the order and size of the cash flows in the PW relations, it is vital that you enter all values carefully. Be sure to enter '0' in periods where there is no cash flow. Example 7.2 (Spreadsheet) illustrates a simple application of this procedure. Refer to Appendix A for additional information on using Excel (and other) spreadsheet systems.

**Example 7.2
(Spreadsheet)**

If $5000 is invested now in common stock that is expected to yield $100 per year for 10 years and $7000 at the end of 10 years, what is the rate of return?

Solution

Use the spreadsheet procedure steps. Figure 7–2 gives the cash flows. The PW relation is

$$0 = -5000 + 100(P/A,i*,10) + 7000(P/F,i*,10)$$

The spreadsheet entries are presented in Figure 7–3, column B with the year values shown for information only. The IRR function provides the $i*$ value in cell B13 as $i* = 5.16\%$, which is the same as the trial-and-error value. Note that the year values 1 through 10 are not used in the spreadsheet function, so it is very important that the cash-flow values are correctly entered into the cells.

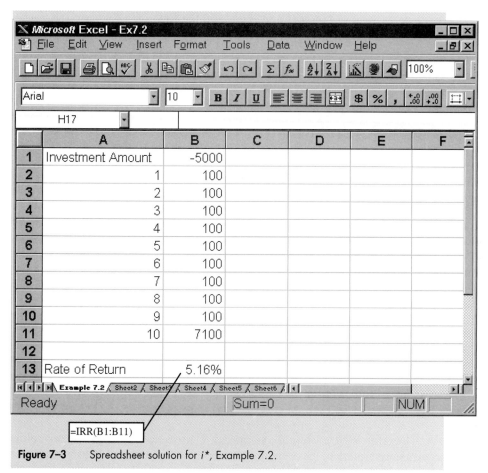

Figure 7–3   Spreadsheet solution for *i**, Example 7.2.

Additional Example 7.6
Problems 7.3 to 7.22

7.3   RATE-OF-RETURN CALCULATIONS USING AN ANNUAL-WORTH EQUATION

Just as $i*$ can be found using a PW equation, it may also be determined using the AW form of Equation [7.2]. This method is preferred, for example, when uniform annual cash flows are involved. The procedure is as follows:

1. Draw a cash-flow diagram.
2. Set up the relations for the AW of disbursements, AW_D, and receipts, AW_R, with $i*$ as an unknown.
3. Set up the rate-of-return relation in the form of Equation [7.2], $0 = -AW_D + AW_R$.

4.   Select values of i by trial and error until the equation is balanced. If necessary, interpolate to determine i^*.

The estimation procedure in Section 7.2 for the first i value may be used here also.

Example 7.3

Use AW computations to find the rate of return for the cash flows in Example 7.2.

Solution

1.   Figure 7–2 shows the cash-flow diagram.
2.   The AW relations for disbursements and receipts are

$$AW_D = -5000(A/P,i,10)$$

$$AW_R = 100 + 7000(A/F,i,10)$$

3.   The AW formulation using Equation [7.2] is

$$0 = -5000(A/P,i^*,10) + 100 + 7000(A/F,i^*,10)$$

4.   Trial-and-error solution yields the results:

At $i = 5\%$, $0 < \$+9.02$,
At $i = 6\%$, $0 > \$-48.26$

Interpolation yields $i^* = 5.16\%$, as before.

Thus, for ROR calculations you can choose the PW, AW, or any other equivalence equation. It is generally better to get accustomed to using only one of the methods in order to avoid errors. If i^* is determined using a spreadsheet, it is most likely approximated with PW-based computations, not AW-based ones.

Problems 7.3 to 7.22

7.4   MULTIPLE VALUES AS POSSIBLE RATES OF RETURN

In the two previous sections a unique rate-of-return value i^* was determined for the given cash-flow sequences. In the cases thus far, the algebraic signs on the *net cash flows* change only once, usually from minus in year 0 to plus for the rest of the periods. This is called a *conventional (or simple) cash-flow sequence*. If there is more than one sign change, the series is called *nonconventional* or *nonsimple*. As shown in the examples in Table 7–3, the series of positive or negative net cash-flow signs may be one or more in length.

| Table 7–3 | Possible conventional (simple) and nonconventional (nonsimple) net cash-flow sequences for a 6-year project |

Type of Sequence	Sign on Net Cash Flow							Number of Sign Changes
	0	**1**	**2**	**3**	**4**	**5**	**6**	
Conventional	−	+	+	+	+	+	+	1
Conventional	−	−	−	+	+	+	+	1
Conventional	+	+	+	+	+	−	−	1
Nonconventional	−	+	+	+	−	−	−	2
Nonconventional	+	+	−	−	−	+	+	2
Nonconventional	−	+	−	−	+	+	+	3

When there is more than one sign change, that is, when the net cash flow is nonconventional, it is possible that there will be multiple $i*$ values in the minus 100% to plus infinity range which will balance the rate-of-return equation. The total number of real-number $i*$ values is always less than or equal to the number of sign changes in the sequence. This is called *Descartes' rule* and is derived from the fact that the equation set up by Equation [7.1] or [7.2] to find $i*$ is an nth-order polynomial. (It is possible to determine that imaginary values or infinity may also satisfy the equation.) Example 7.4 presents the determination and graphical interpretation of multiple rate-of-return possibilities.

Example 7.4

A European-based company has marketed a synthetic oil lubricant for 3 years, with the following net cash flows in thousands of U.S. dollars.

Year	0	1	2	3
Cash flow ($1000)	$+2000	−500	−8100	+6800

(a) Plot the value of the present worth versus interest rates of 5, 10, 20, 30, 40, and 50%.

(b) Determine whether the cash-flow series is conventional or nonconventional and estimate the rates of return from the graph in part (a).

Solution

(a) The PW relation is:

$$PW = 2000 - 500(P/F,i,1) - 8100(P/F,i,2) + 6800(P/F,i,3)$$

The present-worth amounts for each i value are

i, %	5	10	20	30	40	50
PW, $	+51.44	−39.55	−106.13	−82.01	−11.85	+81.85

Figure 7–4 shows that the present worth has a parabolic shape and crosses the *i* axis two times. Both the linear segments and a smooth (approximating) parabolic curve are drawn.

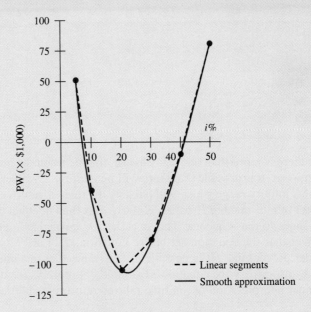

Figure 7–4   Present worth of cash flows at several interest rates, Example 7.4.

(*b*)   The sequence is nonconventional or nonsimple, because of two sign changes for the cash flows (plus to minus in years 0 to 1, and minus to plus in years 2 to 3). The two values (i_1^* and i_2^*) may be graphically determined from Figure 7–4 to be approximately

$$i_1^* = 8\% \quad \text{and} \quad i_2^* = 41\%$$

Comment

If two *i* values are solved for mathematically, the more exact values are found to be 7.47% and 41.35%. If there were three sign reversals in the cash-flow sequence, it is possible for there to be three possible *i** values, but another criterion must be checked first as discussed after Example 7.4 (Spreadsheet).

**Example 7.4
(Spreadsheet)**

For the nonconventional net cash-flow sequence in Example 7.4, use a spreadsheet to determine the multiple *i** values.

Solution

The sequence can have up to two values at which PW = 0, as we have already determined. The spreadsheet solution is shown in Figure 7–5. Row 8 cells indicate the i^* value "guesses" entered into the Excel IRR function. By guessing different values to initiate the solution, the multiple values of i^* can be determined when they exist, as shown in row 10 cells. Two different values are indicated: 7.47% and 41.35%. Note in cell G8 that negative guesses may be entered.

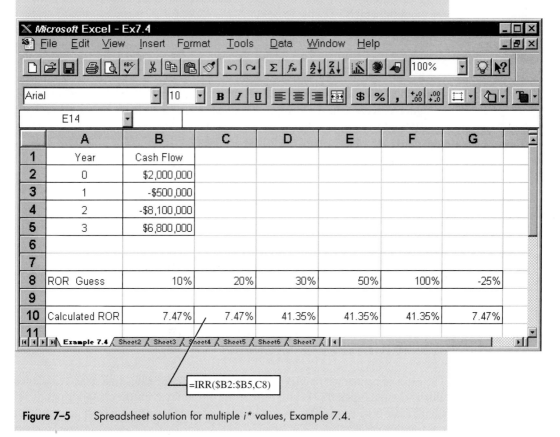

	A	B	C	D	E	F	G
1	Year	Cash Flow					
2	0	$2,000,000					
3	1	-$500,000					
4	2	-$8,100,000					
5	3	$6,800,000					
6							
7							
8	ROR Guess	10%	20%	30%	50%	100%	-25%
9							
10	Calculated ROR	7.47%	7.47%	41.35%	41.35%	41.35%	7.47%

=IRR($B2:$B5,C8)

Figure 7–5   Spreadsheet solution for multiple i^* values, Example 7.4.

There is an additional test that should be applied to a cash-flow sequence to determine if there is only one, real-number, positive i^* value. *Norstrom's criterion* states that only one sign change in the series of cumulative cash flows (no time value considered), which starts negatively, indicates that there is one positive root to the polynomial relation. First, determine the series

$$S_t = \text{cumulative cash flows through period } t$$

Observe the sign of S_0 and count the sign changes in the series $S_0, S_1, \ldots, S_n$. For Example 7.4,

$$S_0 = \$+2000 \qquad S_1 = +1500 \qquad S_2 = -6600 \qquad S_3 = +200$$

The first sign is + and there are two sign changes. These facts indicate there is not a single positive $i*$. As we determined, there are two roots to the PW equation: 7.47% and 41.35%.

In many cases some of the multiple $i*$ values will seem ridiculous because they are too large or too small (negative). For example, values of 10, 150, and 750% for a sequence with three sign changes are difficult to use in practical decision making. (Obviously, one advantage of the PW and AW methods for alternative analysis is that unrealistic rates do not enter into the analysis.) In determining which $i*$ value to select as *the* ROR value, it is common to neglect negative and large values or to simply never compute them. *Actually, the correct approach is to determine the unique composite rate of return,* as described in the next section.

If a standard spreadsheet system, such as Excel, is used, it will normally determine only one real-number root, unless different 'ROR guess' amounts are entered sequentially. This one $i*$ value determined from Excel is usually a realistically valued root, because the $i*$ which solves the PW relation is determined by the spreadsheet's built-in trial-and-error method, starting with a default value, commonly 10%, or with the user-supplied guess as illustrated in the previous example.

For the interested reader, further background materials on multiple $i*$ and rate-of-return computations introduced here and in the next section are available in references after Section 7.5.

Additional Example 7.7
Problems 7.23 to 7.30

7.5   COMPOSITE RATE OF RETURN: REMOVING MULTIPLE $i*$ VALUES

The rates of return we have computed thus far are the rates that exactly balance plus and minus cash flows with time value of money considered and for conventional cash-flow sequences. Any measure-of-worth method which accounts for the time value of money can be used in calculating this balancing rate, such as PW, AW, or FW. Regardless of which method is used, the interest rate obtained from these calculations is known as the *internal rate of return, IRR.* Simply stated, the internal rate of return is the rate of return on the unrecovered balance of an investment, as defined in Section 7.1. The funds that remain unrecovered are still inside the investment and hence the name, internal rate of return. The general terms rate of return and interest rate usually imply internal rate of return. The interest rates quoted or calculated in previous chapters were all internal rates.

The concept of unrecovered balance becomes important when positive net cash flows are generated (thrown off) before the end of a project. A positive net cash flow, once generated, becomes *released* or *external funds to the project* and is not considered further in an internal rate-of-return calculation. These positive net cash flows may cause a nonconventional cash-flow sequence and multiple $i*$ values, as we learned in the previous section. However, there is a method to explicitly consider these funds, as discussed in this section. Additionally, the dilemma of multiple $i*$ roots is eliminated.

First, let's further examine the internal rate-of-return calculations for the following cash flows: $10,000 is invested at $t = 0$, $8000 is received in year 2, and $9000 is received in year 5. The PW equation to determine $i*$ for the rate of return is

$$0 = -10,000 + 8000(P/F,i,2) + 9000(P/F,i,5)$$

$$i = 16.815\%$$

If this rate is used for the unrecovered balances, the investment will be recovered exactly at the end of year 5 as the following computations verify.

Unrecovered balance at end of year 2 immediately before $8000 receipt:

$$-10,000(F/P,16.815\%,2) = -10,000(1 + 0.16815)^2 = \$-13,645.74$$

Unrecovered balance at end of year 2 immediately after $8000 receipt:

$$-13,645.74 + 8000 = \$-5645.74$$

Unrecovered balance at end of year 5 immediately before $9000 receipt:

$$-5645.74(F/P,16.815\%,3) = \$-8999.47$$

Unrecovered balance at end of year 5 immediately after $9000 receipt:

$$-\$8999.47 + 9000 = \$0.53 \text{ (considering round-off error)}$$

In this calculation, no consideration is given to the $8000 available after year 2. Therefore, a good question is: What happens if funds released from a project *are* considered in calculating the overall rate of return of a project? After all, something must be done with the released funds. One possibility is to assume the money is reinvested at some stated rate. A second is to simply assume the reinvestments occur at the MARR, if it is a believable rate for this purpose. In addition to accounting for all the money released during the project period and reinvested at a realistic rate, the approach discussed below has the advantage of converting a nonconventional cash-flow sequence (with multiple $i*$ values) into a conventional sequence with one root, which can be considered *the* rate of return for making a decision about a project.

The rate of earnings used for the released funds is called the *reinvestment rate,* symbolized by c. This rate, established outside (external to) the particular project being evaluated, depends upon the market rate available for investments. If a company is making, say, 8% on its daily investments, then $c = 8\%$. It may be practical to set c equal to the MARR already established. Now, the one interest rate that satisfies the rate-of-return equation is called the *composite rate of return (CRR)* and will be symbolized by i'. By definition

> The *composite rate of return*, i', is the unique rate of return for a project which assumes that net positive cash flows, which represent money not immediately needed by the project, are reinvested at the reinvestment rate c.

The term composite is used here to describe this rate of return because it is derived using another interest rate, namely c, the reinvestment rate. If c happens to equal any one of the i^* values, then the composite rate i' will equal that i^* value. The CRR is known by other terms, two of which are return on invested capital (RIC) and external rate of return (ERR).

The correct procedure to determine i', called the *project-net-investment procedure*, is summarized now. The technique involves finding the future worth of the net investment amount one period (year) in the future. Find the project's net-investment value F_t in year t from F_{t-1} by using the F/P factor for 1 year at the reinvestment rate c if the previous net investment F_{t-1} is positive (extra money generated by project), or at the CRR rate i' if F_{t-1} is negative (project used all available funds). To do this mathematically, for each year t set up the relation:

$$F_t = F_{t-1}(1+i) + C_t \qquad \text{[7.5]}$$

where $t = 1, 2, \ldots, n$

n = total years in project

C_t = net cash flow in year t

$$i = \begin{cases} c & \text{if } F_{t-1} > 0 \quad \text{(net positive investment)} \\ i' & \text{if } F_{t-1} < 0 \quad \text{(net negative investment)} \end{cases}$$

The equation for F_n obtained using this procedure is set equal to zero and solved for i' by computer or by manual trial and error. The i' value obtained is unique for a stated reinvestment rate c.

The development of F_1 through F_3 for the cash-flow sequence below, which is graphed in Figure 7–6a, is illustrated for a reinvestment rate c of 15%.

Year	Cash flow, $
0	50
1	−200
2	50
3	100

The net investment for year $t = 0$ is

$$F_0 = \$50$$

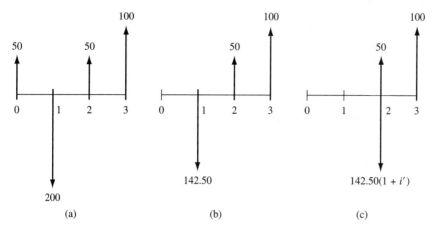

Figure 7–6   Cash-flow sequence for which the composite rate of return i' is computed: (a) original form, (b) equivalent form in year 1, and (c) equivalent form in year 2.

which is positive, so it returns $c = 15\%$ during the first year. By Equation [7.5], F_1 is

$$F_1 = 50(1 + 0.15) - 200 = \$-142.50$$

This result is shown in Figure 7–6b. Since the project net investment is now negative, the value F_1 earns interest at the i' rate for year 2. Therefore, for year 2,

$$F_2 = F_1(1+i') + C_2 = -142.50(1+i') + 50$$

The i' value is to be determined. Since F_2 will be negative for all $i' > 0$, use i' to set up F_3 as shown in Figure 7–6c.

$$F_3 = F_2(1+i') + C_3 = [-142.50(1+i') + 50](1+i') + 100 \qquad [7.6]$$

Setting Equation [7.6] equal to zero and solving for i' will result in the unique composite rate of return i'. If there had been more F_t expressions, i' would be used in all subsequent equations (see Example 7.8). The project-net-investment procedure to find i' may be summarized as follows:

1. Draw a cash-flow diagram of the original net cash-flow sequence.
2. Develop the series of project net investments using Equation [7.5] and the stated c value. The result is the F_n expression in terms of i'.
3. Set the F_n expression equal to zero and find the i' value to balance the equation.

Several comments are in order. If the reinvestment rate c is equal to the internal rate-of-return value $i*$ (or only one of the $i*$ values when there are multiple ones), the i' that is calculated will be exactly the same as $i*$, that is, $c = i* = i'$. The closer the c value is to i', the smaller the difference between the composite and internal rates. It is reasonable to assume that c = MARR, if all extra funds from the project can realistically earn at the MARR rate.

A summary of the relations between c, i', and i^* follows, and the relations are demonstrated in Example 7.5.

Relation between Reinvestment Rate c and IRR i^*	Relation between CRR i' and IRR i^*
$c = i^*$	$i' = i^*$
$c < i^*$	$i' < i^*$
$c > i^*$	$i' > i^*$

Remember: This entire procedure is necessary only in the event that multiple i^* values are indicated, as discussed in Section 7.4. Multiple i^* values are present when a nonconventional cash-flow sequence does not have one positive root, as determined by Norstrom's criterion. Additionally, none of the steps in this procedure apply if the present-worth or annual-worth method is used correctly to evaluate a project, as presented in Chapters 5 and 6, respectively.

Example 7.5

Compute the composite rate of return for the synthetic oil lubricant investment in Example 7.4 if the reinvestment rate is (*a*) 7.47% and (*b*) 20%.

Solution

(*a*) Use the procedure stated above to determine i' for c = 7.47%.

1. Figure 7–7 shows the original cash flow.
2. The first project-net-investment expression is F_0 = \$+2000. Since $F_0 > 0$, use c = 7.47% to write F_1 by Equation [7.5].

$$F_1 = 2000(1.0747) - 500 = \$1649.40$$

Again $F_1 > 0$, so use c = 7.47% to determine F_2.

$$F_2 = 1649.40(1.0747) - 8100 = -\$6327.39$$

Figure 7–8 shows the equivalent cash flow at this time. Since $F_2 < 0$, use i' to express F_3.

$$F_3 = -6327.39(1+i') + 6800$$

3. Set F_3 = 0 and solve for i' directly.

$$-6327.39(1+i') + 6800 = 0$$

$$1 + i' = \frac{6800}{6327.39} = 1.0747$$

$$i' = 0.0747 \quad (7.47\%)$$

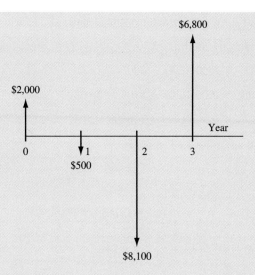

Figure 7–7   Original cash flow (in thousands), Example 7.5.

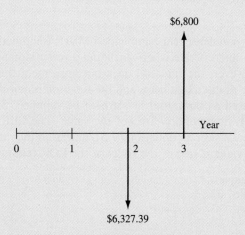

Figure 7–8   Equivalent cash flow (in thousands) of Figure 7–7 with reinvestment at $c = 7.47\%$.

The CRR is 7.47%, which is the same as c, the reinvestment rate, and the i_t^* value determined in Example 7.4 (Spreadsheet). Note that 41.35%, which is the second i^* value, no longer balances the rate-of-return equation. The equivalent future-worth result for the cash flow in Figure 7–8, if i' were 41.35%, is

$$6327.39(F/P,41.35\%,1) = 8943.77 \ne 6800$$

(b) For $c = 20\%$, the project-net-investment series is

$$F_0 = +2000 \qquad\qquad (F_0 > 0, \text{ use } c)$$

$$F_1 = 2000(1.20) - 500 = \$1900 \qquad (F_1 > 0, \text{ use } c)$$

$$F_2 = 1900(1.20) - 8100 = \$-5820 \qquad (F_2 < 0, \text{ use } i')$$

$$F_3 = -5820(1+i') + 6800$$

Set $F_3 = 0$ and solve for i' directly.

$$1 + i' = \frac{6800}{5820} = 1.1684$$

$$i' = 0.1684 \qquad (16.84\%)$$

The CRR is $i' = 16.84\%$ at a reinvestment rate of 20%, which is a marked increase from $i' = 7.47\%$ at $c = 7.47\%$.

Comment

If the value $c = 41.35\%$ is used, the equation will be balanced with $i' = 41.35\%$, which is the second i^* value. This is possible only when receipts are reinvested at exactly i^*, because only then does $c = i^* = i'$.

There is a spreadsheet function called MIRR (modified IRR) which determines a unique interest rate when you input a reinvestment rate c for positive cash flows. However, the function does not implement the project-net-investment procedure as discussed here, and the function requires that a finance rate for the funds used as the initial investment be supplied. So, the formulas for MIRR and CRR computation are not the same. The MIRR will not produce exactly the same answer as Equation [7.5], unless all the rates happen to be the same, and this value is one of the roots of the ROR relation.

Additional Example 7.8
Problems 7.31 to 7.32

REFERENCES FOR SECTION 7.4

Bussey, L. E., and T. G. Eschenbach. *The Economic Analysis of Industrial Projects,* 2d ed. Englewood Cliffs, NJ: Prentice-Hall, 1992, pp. 188–203.

Fleischer, G. A. *Introduction to Engineering Economy.* Boston: PWS, 1994, pp. 124–30.

McLean, J. G. "How to Evaluate New Capital Investments." *Harvard Business Review,* Nov.–Dec. 1958, pp. 58–69.

Newnan, D. G. *Engineering Economic Analysis,* 6th ed. San Jose, CA: Engineering Press, 1996, pp. 221–33.

Park, C. S. *Contemporary Engineering Economics,* 2d ed. Menlo Park, CA: Addison-Wesley, 1997, pp. 302–9, 347–62.

ADDITIONAL EXAMPLES

Example 7.6
(Spreadsheet)

ROR USING A PW RELATION, SECTION 7.2   Assume that a couple invests $10,000 now and $500 three years from now and will receive $500 one year from now, $600 two years from now, and amounts increasing by $100 per year for a total of 10 years. They will also receive lump-sum payments of $5000 in 5 years and $2000 in 10 years. Calculate the rate of return on their investment using the trial-and-error and spreadsheet methods.

Solution

The conventional (simple) cash-flow diagram in Figure 7–9 is used to set up a rate-of-return relation using present worth.

$$0 = -10,000 - 500(P/F,i^*,3) + 500(P/A,i^*,10) + 100(P/G,i^*,10)$$

$$+ 5000(P/F,i^*,5) + 2000(P/F,i^*,10)$$

$$i = ?$$

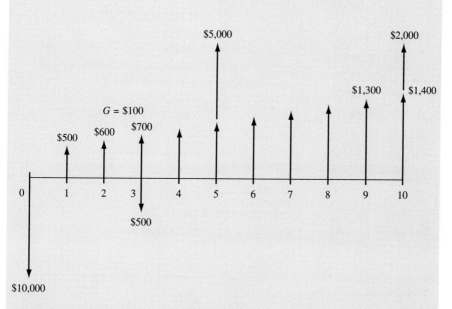

Figure 7–9   Cash-flow diagram for Example 7.6.

Solving by manual trial and error and interpolating between $i = 7\%$ and $i = 8\%$, we find $i^* = 7.8\%$. Solving via spreadsheet, the results are presented in Figure 7–10, column B, where $i^* = 7.79\%$.

Comment

Note that in the manual solution, the single values at years 3, 5, and 10 were handled separately so that the P/A and P/G factors could be used on the remaining cash flows.

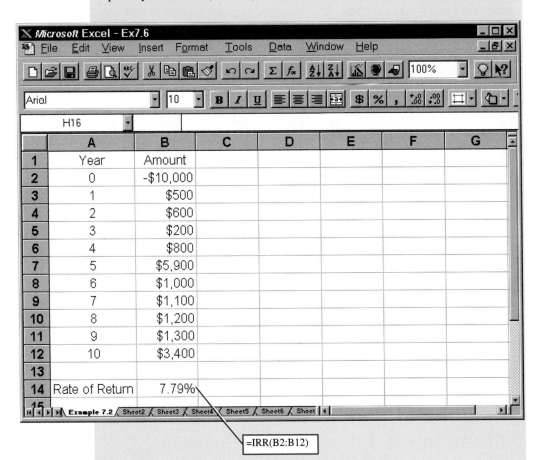

	A	B
1	Year	Amount
2	0	-$10,000
3	1	$500
4	2	$600
5	3	$200
6	4	$800
7	5	$5,900
8	6	$1,000
9	7	$1,100
10	8	$1,200
11	9	$1,300
12	10	$3,400
13		
14	Rate of Return	7.79%

=IRR(B2:B12)

Figure 7–10   Spreadsheet solution for $i^* = 7.79\%$, Example 7.6.

Example 7.7

MULTIPLE POSSIBLE RATES OF RETURN, SECTION 7.4.   Assume net cash flows for 10 years (Table 7–4) are estimated for an in-use asset. The negative net cash flow in year 4 is the result of an upgrade to the asset. Determine the number of i^* roots and estimate their values graphically.

Table 7–4		Net cash-flow series and cumulative cash-flow series, Example 7.7				
	Cash Flow, $				**Cash Flow, $**	
Year	**Net**	**Cumulative**	**Year**	**Net**	**Cumulative**	
1	200	+200	6	500	−350	
2	100	+300	7	400	+50	
3	50	+350	8	300	+350	
4	−1800	−1450	9	200	+550	
5	600	−850	10	100	+650	

Solution

Descartes' rule indicates a nonconventional (nonsimple) net cash-flow series with up to two roots, and Norstrom's criterion for the cumulative net cash-flow series in Table 7–4 starts positively and has more than one sign change, thus indicating that no unique positive root will be found. A PW-based ROR relation is used to solve for two i^* values.

$$0 = 200(P/F,i,1) + 100(P/F,i,2) + \cdots + 100(P/F,i,10) \qquad \textbf{[7.7]}$$

The results of the right side of Equation [7.7] are calculated for different values of i and plotted in Figure 7–11 with a sketch of the smooth curve superimposed.

i (%)	10	20	25	30	37	40	50	60
PW, $	+196	+42	+12	−2	−8	−7	+2	+14

Two values which appear to satisfy Equation [7.7] are (approximately) $i_1^* = 29\%$ and $i_2^* = 49\%$.

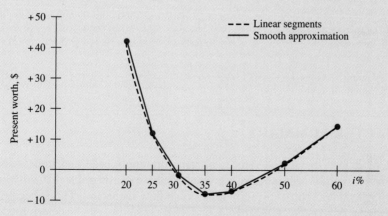

Figure 7–11   Plot of present worth versus i, Example 7.7.

Example 7.8

COMPOSITE RATE OF RETURN, SECTION 7.5   Determine the composite rate of return for Example 7.7 if the reinvestment rate is 15% per year.

Solution

Use the steps summarized in Section 7.5 to write the project-net-investment series for $t = 1$ to $t = 10$.

$$F_0 = 0$$

$$F_1 = \$200 \qquad (F_1 > 0, \text{ use } c)$$

$$F_2 = 200(1.15) + 100 = \$330 \qquad (F_2 > 0, \text{ use } c)$$

$$F_3 = 330(1.15) + 50 = \$429.50 \qquad (F_3 > 0, \text{ use } c)$$

$$F_4 = 429.50(1.15) - 1800 = -\$1306.08 \qquad (F_4 < 0, \text{ use } i')$$

$$F_5 = -1306.08(1+i') + 600$$

Since we do not know if F_5 is greater than zero or less than zero, all remaining expressions use i'.

$$F_6 = F_5(1 + i') + 500 = [-1306.08(1+i') + 600](1+i') + 500$$

$$F_7 = F_6(1 + i') + 400$$

$$F_8 = F_7(1 + i') + 300$$

$$F_9 = F_8(1 + i') + 200$$

$$F_{10} = F_9(1 + i') + 100$$

To find i', the expression $F_{10} = 0$ is solved. Computerized or manual solution determines that $i' = 21.25\%$.

Comment

At this time you might want to rework this problem with a reinvestment rate of 29% or 49%, as found in Example 7.7, to see that the i' value will be the same as these reinvestment rates; that is, if $c = 29\%$, then $i' = 29\%$. In the example above, $c = 15\%$ is less than $i* = 29\%$, so $i' = 21.25\% < i*$, as discussed in Section 7.5.

CHAPTER SUMMARY

Rate of return, or interest rate, is a term commonly used and understood by almost everybody. Most people, however, can have considerable difficulty in calculating a rate of return correctly for all cash-flow sequences. This is because for some types of sequences, more than one ROR possibility can be determined. The maximum number of values possible is equal to the number of changes in the signs of the net cash-flow series (Descartes' rule). Also, a single positive rate can be found if the cumulative net cash-flow series starts negatively and has only one sign change (Norstrom's criterion).

For all cash-flow sequences where there is an indication of multiple roots to the ROR equation, a decision must be made about whether to calculate the internal rates or the one composite rate of return. While the internal rate is usually easier to calculate, the composite rate is the correct approach and it has two advantages: the possibility of multiple rates of return is eliminated, and released cash flows can be treated using realistic reinvestment rates.

PROBLEMS

7.1 Rate of return is defined as the interest paid or received on what?

7.2 Rate of return is also commonly known by other names. List three of them.

7.3 If a company spends $12,000 now and $5000 per year for 10 years, with the first $5000 expenditure made 4 years from now, what rate of return would it make if its income were $4000 per year starting in year 8 and continuing through year 25?

7.4 An entrepreneur purchased a dump truck for the purpose of offering a short-haul earth-moving service. He paid $14,000 for the truck and sold it 5 years later for $3000. His operation and maintenance expense while he owned the truck was $3500 per year. In addition, he had the truck engine overhauled for $900 at the end of the third year. If his revenue was $15,000 each year that he owned the truck, what was his rate of return?

7.5 What rate of return would the entrepreneur in Problem 7.4 have made if his revenue was $20,000 in year 1, decreasing by $3000 per year through year 5?

7.6 A real-estate investor purchased a piece of property for $90,000 cash. He rented the house for $1000 per month. At the end of year 2, the renter moved out and the investor spent $8000 for remodeling. It took him 6 months to sell the house for $105,000. He had to pay a real-estate agency 6% of the sales price and provide a title policy for $1200. What rate of return did he make on his investment?

7.7 An alumna wants to establish a permanent endowment in her name at a small private college. She plans to donate $100,000 and have the income used to purchase lab supplies, but she will make the donation only if a permanent plaque with her name on it is placed on the door to the lab. If the college agrees to her request and the lab supplies cost $12,000 per year, what rate of return is required on the endowment?

7.8 A private landfill owner was required to install a plastic liner to prevent leachate from contaminating the groundwater. The fill area was 40,000 m^2 and the liner cost was $5 per square meter. The installation cost was $4000. In order to recover the investment, the owner charged $8 for pickup truck loads, $25 for dump truck loads, and $65 for compactor truck loads. If the monthly distribution has been 200 pickup loads, 40 dump truck loads, and 100 compactor truck loads, what rate of return did the landfill owner make on her investment (*a*) per month and (*b*) effectively per year?

7.9 A frugal businessman who objects to the high cost of funeral services has designed a "reusable casket." It consists of a plastic capsule that fits inside a traditional decorative casket shell. After the services are over, the capsule can be buried and the shell reused. He expects to be able to sell the capsule for $225. If a low-end traditional casket costs $2100, what rate of return is required for a family that disdains a traditional casket to have earned enough in interest on the savings to be able to buy another capsule in 10 years?

7.10 A plaintiff in a successful lawsuit was awarded a judgment of $3000 per month for 3 years. The plaintiff needs a fairly large sum of money for an investment and has offered the defendant the opportunity to pay off the award in a lump-sum amount of $75,000 now. If the defendant accepts the offer and is paid the $75,000 now, what rate of return is made on the "investment"?

7.11 An investor purchased 900 shares of a stock index mutual fund at $18.25 per share. The stock price moves in accordance with the S&P 500 index. If the S&P 500 index rose from 801.32 to 871.66 in 5 months, (*a*) what rate of return per month did the investor make, and (*b*) what was the value of the 900 shares of stock after the index rise?

7.12 A major university is considering a plan to build a 7-megawatt cogeneration plant to provide its power needs. The cost of the plant is expected to be $31 million. The university consumes 36,000 megawatt-hours per year at a cost of $110 per megawatt-hour. The university can produce its power at half the cost for which it now buys it. (*a*) What rate of return will it make on its investment if the power plant lasts 30 years? (*b*) If the university can sell an average of 10,000 megawatt-hours per year back to the utility at $85 per megawatt-hour, what rate of return will it make?

7.13 A middle-age couple is considering the purchase of a first-to-die life insurance policy. This type of policy pays off when the first person dies. For this couple, whose ages are 55 and 53, the annual cost is $5000 for $250,000 worth of coverage (beginning-of-year payments). The insurance agent stated that after 10 years (i.e., 11 payments) the earnings on the accumulated cash value should be equal to the annual premium. If the first mate passes away immediately after the tenth year, (*a*) what rate of return will the insurance company require to just break even, that is, have $250,000 and (*b*) what rate of return would the couple have made?

7.14 If the insurance agent is correct in Problem 7.13 and no more payments are made after year 10, what rate of return has the couple made if the first one passes away (*a*) 20 years after the policy was purchased? (*b*) What is the rate of return if the first person dies immediately after the fifth premium payment?

7.15 A major credit card issuer advertises "interest-free cash" in bold letters. The ad states, "Just pay off your balance in full each month and for a small transaction fee,* your cash advance is interest free." For the *, there is a footnote statement that reads, "The cash advance transaction fee is 2.5% of each cash advance with a minimum of $2 and no maximum." If a person obtains an advance of $100 and then pays the total amount due 1 month later, what effective interest rate is paid (*a*) per month? (*b*) per year?

7.16 If the person in Problem 7.15 above obtains a cash advance of $50, what effective rates per month and per year are paid, if the balance is paid in full at the end of the month?

7.17 A well-known credit card company offers no transaction fees for cash advances and no interest charges during the month the cash advance is made. The annual percentage rate of 19.8% applies only if the balance is not paid in full each month. However, to obtain the credit card, an annual fee of $25 is required. (*a*) If a person obtains such a credit card and borrows $100 every other month (starting in month 1 and ending in month 11) at the beginning of the month and repays it at the end of that month (and, therefore, incurs no extra charges), determine the nominal and effective interest rates per year for a person paying only the annual fee. (*b*) Consider these same questions, if $2000 is withdrawn each time.

7.18 A 55-year-old male wants to leave money to his heirs. He investigated purchasing life insurance. A so-called estate-protected whole life plan would cost him $297 per month or $3374 per year. This policy provides $100,000 of life insurance and different cash values depending upon how long the insured survives. For example, the policy has guaranteed cash values of $52,690 and $86, 574 in 15 and 20 years, respectively. If the insured decides to cash in the policy after 15 years, what rate of return is made per year, if he made premium payments (*a*) monthly? (*b*) annually? Assume the premium payments were made at the beginning of each period.

7.19 A "10 paid life" insurance policy can be purchased for $557 per month or $6330 per year, which will provide $100,000 in death benefits. The policy will also accumulate cash value which can be withdrawn by the policy holder, if desired. For example, the policy holder could withdraw $27,506 in 5 years or $78,160 in 10 years. The death benefits at these times would be $103,000 and $117, 566, respectively. What rate of return would the policy holder make if he lives 10 years and withdraws the accumulated cash value at the end of year 10, assuming he made 10 annual beginning-of-year premium payments?

7.20 If the policy holder in Problem 7.19 above dies immediately after 5 years, what effective rate of return per month was made on the money he paid in premiums, assuming he made 61 beginning-of-month premium payments?

7.21 The manager of a pension fund purchased 10,000 shares of common stock in a company which contracts for stonewashing of jeans. The stock was purchased 3½ years ago when it was selling at $14⅜ per share. If the stock dividends were $0.63 per quarter, what rate of return per quarter did the manager make if he sold the stock for $16⅞ per share? Assume the stock was purchased and sold immediately after the dividend date, that is, on the ex-dividend date.

7.22 An investor who understands the principle of leverage purchased a fixer-upper house for $103,000. The investor made a down payment of $11,000, paid closing costs of $3200, and remodeled the house at a cost of $12,000. After making six monthly house payments of $935 each, the investor sold the house for $138,000, from which selling expenses of $7800 were deducted as well as the principal balance of $90,000. What nominal rate of return per year did she make, assuming the remodeling costs were uniformly distributed over the 6-month ownership period?

7.23 What is meant by conventional cash flow?

7.24 What is the range of values possible when solving a rate-of-return equation? What is the significance of a 100% rate of return?

7.25 Application of Descartes' rule will provide you with what information?

7.26 What is Norstrom's criterion and what is it used for?

7.27 A project has just been implemented which will involve the cash flows shown below. (*a*) How many possible rate-of-return values are there for this cash flow? (*b*) Find all the values between 0 and 100%.

Year	0	1	2	3
Cash flow, $	−20,000	15,000	15,000	−2,000

7.28 Find all rate-of-return values between −10% and 200% for the following cash flows:

Year	0	1	2	3	4
Cash flow, $	−7000	−2000	4000	5000	6000

7.29 For the cash-flow sequences shown at the top of the next page, determine (*a*) the number of possible *i* values that may balance the present-worth relation and (*b*) all real values between 0 and 100%.

Year	Expense, $	Income, $
0	50,000	0
1	22,000	19,000
2	30,000	27,000
3	40,000	25,000
4	20,000	29,000
5	13,000	21,000

7.30 In Problem 7.29, if the net cash flow in year 6 is estimated at $+50,000, (a) how many possible rate-of-return values are there? (b) Determine all values between -50% and $+150\%$.

7.31 Use a reinvestment rate of (a) 15% and (b) 50% to find the composite rate of return for Problem 7.27.

7.32 Use a reinvestment rate of (a) 7% and (b) 25% to find the composite rate of return in Problem 7.28.

chapter

8

Rate-of-Return Evaluation for Multiple Alternatives

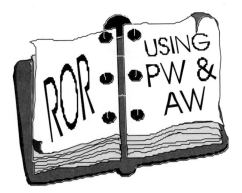

This chapter presents the methods by which two or more alternatives can be evaluated using a rate-of-return (ROR) comparison. The ROR evaluation correctly performed will result in the same selection as the PW and AW analyses, but the computational procedure is considerably different for ROR evaluations.

LEARNING OBJECTIVES

Purpose: Understand how to select the best alternative on the basis of rate-of-return analysis on incremental cash flows.

This chapter will help you:

Incremental analysis	1. State why an incremental investment analysis is necessary for comparing alternatives by the ROR method.
Tabulation	2. Prepare a tabulation of incremental cash flow for two alternatives.
ROR on incremental investment	3. Interpret the meaning of ROR on the incremental investment.
Alternative selection	4. Select the better of two alternatives using an ROR relation based on present worth.
Alternative selection	5. Select the better of two alternatives using an ROR relation based on annual worth.
Alternative selection	6. Select the best of multiple alternatives using the ROR method.

8.1   UNDERSTANDING INCREMENTAL ANALYSIS

When two or more alternatives are under consideration and only one is to be selected, engineering economy can identify the one alternative which is considered the best economically. As we have learned, the PW and AW techniques discussed in Chapters 5 and 6 can obviously be used to do so. Now the procedure for using rate-of-return techniques to identify the best is presented.

Let's assume that a company uses a MARR of 16% per year, that the company has $90,000 available for investment, and that two alternatives (A and B) are being evaluated. Alternative A requires an investment of $50,000 and will yield an internal rate of return (IRR) of 35% per year. Alternative B requires $85,000 and will yield an IRR of 29% per year. Intuitively we may conclude that the better alternative is the one which yields the higher IRR, A in this case. However, this is not necessarily so because, while A has the higher projected return, it also requires an initial investment which is much less than the total money available ($90,000). In a case such as this, a logical question is: What happens to the capital that is left over? It is generally assumed that excess funds will be invested at the company's MARR, as we learned in the previous chapter. Using this assumption, it is possible to determine the consequences of the alternative investments. If alternative A is selected, $50,000 will be invested at a rate of 35% per year. The $40,000 left over will be invested at the MARR of 16% per year. The rate of return on the total capital available, then, will be the weighted average of these values. Thus, if alternative A is selected,

$$\text{Overall ROR}_A = \frac{50,000(0.35) + 40,000(0.16)}{90,000} = 26.6\%$$

If alternative B is selected, $85,000 will be invested at a yield of 29% per year and the remaining $5000 earns 16% per year. Now, the weighted average is

$$\text{Overall ROR}_B = \frac{85,000(0.29) + 5000(0.16)}{90,000} = 28.3\%$$

Assuming the MARR is realized, these calculations show that, even though the IRR for alternative A is higher, alternative B presents the better overall ROR for the total of $90,000. If either a PW or AW comparison were conducted using $i = 16\%$ per year, alternative B would be chosen.

This simple example illustrates a major fact about the rate-of-return method for comparing alternatives; that is, under some circumstances, overall project ROR values do not provide the same ranking of alternatives as do PW and AW analyses. This situation does not occur if we conduct an *incremental-investment rate-of-return* analysis as described in the following sections.

The multiple alternative evaluation discussions in this chapter and the previous chapters on PW and AW analyses refer to selection between alternatives which are termed *mutually exclusive*. This means that only one (the best)

alternative is selected from the entire group of alternatives available. An example of mutually exclusive alternative selection is a contractor who wants to purchase one forklift and has several models from different companies as choices. He or she selects only one, because the alternatives are mutually exclusive. All the PW and AW analyses discussed thus far are for mutually exclusive alternatives, as are the ROR techniques presented in this chapter.

When more than one alternative can be selected from those available, such as when an investor wants to purchase all stocks which are expected to yield a rate of return of at least 25% per year, the alternatives are said to be independent. *Independent alternatives* need only be compared to the do-nothing alternative and are evaluated against a predetermined standard (such as a stated MARR). Therefore, they are not compared against each other. The evaluation techniques for independent projects are discussed further in Chapters 9 and 17.

Problems 8.1 to 8.3

8.2   TABULATION OF INCREMENTAL CASH FLOW FOR TWO ALTERNATIVES

The concept of cash flow was discussed in Chapter 7 with respect to ROR calculations for a single alternative. Now, it is necessary to prepare an *incremental cash-flow tabulation* between two alternatives so that an incremental ROR analysis can be conducted in order to select one. Therefore, a standardized format for the tabulation will simplify interpretation of the final results. The column headings for an incremental cash-flow tabulation are shown in Table 8–1. If the alternatives have *equal lives,* the year column will go from 0 to n, the life of the alternatives. If the alternatives have *unequal lives,* the year column will go from 0 to the LCM of the two lives when a PW or AW equation

	(1) Cash Flow	(2) Cash Flow	(3) = (2) − (1) Incremental Cash Flow
Year	**Alternative *A***	**Alternative *B***	
0			
1			
2			
.			
.			
.			

Table 8–1   Format for incremental cash-flow tabulation

MUST COMPARE EQUAL SERVICE

is used. The use of the LCM is necessary because ROR analysis on the incremental cash flow must be done over the same number of periods for each alternative (as is the case with PW comparisons). If the LCM of lives is necessary, reinvestment in each alternative is shown at appropriate times, in the same manner as for PW analysis. The alternative with the *larger initial investment* will always be regarded as *alternative B*. That is,

$$\text{Incremental cash flow} = \text{cash flow}_B - \text{cash flow}_A$$

The next two examples demonstrate cash-flow tabulation for equal-life and different-life alternatives.

If the incremental cash-flow column has more than one change in sign, there may be multiple rate-of-return values, in which case the procedures discussed in Chapter 7 are applied.

Example 8.1

A tool and die company is considering the purchase of an additional drill press. The company has the opportunity to buy a slightly used machine for $15,000 or a new one for $21,000. Because the new machine is a more sophisticated model with automatic features, its operating cost is expected to be $7000 per year, while the used machine is expected to cost $8200 per year. Each machine is expected to have a 25-year life with a 5% salvage value. Tabulate the incremental cash flow of the two alternatives.

Solution

Incremental cash flow is tabulated in Table 8–2. The subtraction performed is (new − used) since the new machine will cost more. The salvage values in year 25 are separated from ordinary cash flow for clarity. When disbursements are the same for a number of consecutive years, it saves time to make a single cash-flow listing, as is done for years 1 to 25 of the example. However, remember that several years were combined when adding to get the column totals.

Comment

When the cash-flow columns are subtracted, the difference between the totals of the two cash-flow series should equal the total of the incremental cash-flow

Table 8–2	Cash-flow tabulation for Example 8.1		
	Cash Flow		**Incremental Cash Flow**
Year	**Used Press**	**New Press**	**(New − Used)**
0	$ −15,000	$ −21,000	$ −6,000
1–25	−8,200	−7,000	+1,200
25	+750	+1,050	+300
Total	$−219,250	$−194,950	$+24,300

column. This will provide a check of your addition and subtraction in preparing the tabulation.

Example 8.2

The Fresh-Pak Seafood Company has under consideration two different types of conveyors. Type A has an initial cost of $7000 and a life of 8 years. Type B has an initial cost of $9500 and a life expectancy of 12 years. The annual operating cost for type A is expected to be $900, while the cost for type B is expected to be $700. If the salvage values are $500 and $1000 for type A and type B conveyors, respectively, tabulate the incremental cash flow using their LCM.

Solution

The LCM between 8 and 12 is 24 years. The incremental cash-flow tabulation for 24 years is given in Table 8–3. Note that the reinvestment and salvage values are shown in years 8 and 16 for type A and in year 12 for type B.

Table 8–3	Cash flow for 24 years for different-life assets, Example 8.2		

Year	Cash Flow Type A	Cash Flow Type B	Incremental Cash Flow (B − A)
0	$ −7,000	$ −9,500	$−2,500
1–7	−900	−700	+200
8	−7,000 / −900 / +500	−700	+6,700
9–11	−900	−700	+200
12	−900	−9,500 / −700 / +1,000	−8,300
13–15	−900	−700	+200
16	−7,000 / −900 / +500	−700	+6,700
17–23	−900	−700	+200
24	−900 / +500	−700 / +1,000	+700
	$−41,100	$−33,800	$+7,300

Problems 8.4 to 8.7

8.3   INTERPRETATION OF RATE OF RETURN ON EXTRA INVESTMENT

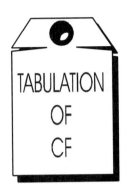

As we have learned, the first step in calculating the ROR on the extra investment is the preparation of a table including incremental cash-flow values. The value in this column reflects the *extra investment* required to be budgeted if the alternative with the larger first cost were selected. This is important in an ROR analysis in order to determine an IRR of the extra funds expended for the larger-investment alternative. If the incremental cash flows of the larger investment don't justify it, we should select the cheaper alternative. But what about the investment amount common to both alternatives? Is it justified automatically? Basically, yes, since *one of the mutually exclusive alternatives must be selected*. If not, the *do-nothing* alternative must be considered one of the selectable alternatives, and then the evaluation is between three alternatives. Example 8.4, later in this chapter, addresses the question above using a PW equation to find the IRR between two alternatives.

In Example 8.1 the new drill press requires an extra investment of $6000 (Table 8–2). Since one of the two machines must be selected, if the new machine is purchased, there will be a "savings" of $1200 per year for 25 years, plus $300 in year 25 as a result of the difference in salvage values. The decision to buy the used or new machine can be made on the basis of the profitability of investing the extra $6000 in the new machine. If the equivalent worth of the savings is greater than the equivalent worth of the extra investment using the company's MARR, the extra investment should be made (i.e., the larger first-cost proposal should be accepted). On the other hand, if the extra investment is not justified by the savings, the lower first-cost proposal should be accepted.

If the new machine is selected, there will be a total net savings of $24,300 (Table 8–2). Keep in mind that this figure does not take into account the time value of money, since this total was obtained by adding the incremental cash-flow values without using the interest factors. It cannot be used as a basis for the alternative decision. The totals at the bottom of the table serve only as a check against the additions and subtractions for the individual years. In fact, the $24,300 is the PW of the incremental cash flow at $i = 0\%$.

It is important to recognize that the rationale for making the selection decision is the same as if only *one alternative* were under consideration, that alternative being the one represented by the cash-flow difference column. When viewed in this manner, it is obvious that unless this investment yields a rate of return equal to or greater than the MARR, the investment should not be made. However, if the rate of return on the extra investment equals or exceeds the MARR, the investment should be made (meaning that the higher-priced alternative should be selected).

As further clarification of this extra-investment rationale, consider the following: the rate of return attainable through the incremental cash flow is an alternate to investing at the company's MARR. In Section 8.1, we stated that any excess funds not invested in the project are assumed to be invested at the

MARR. Clearly, if the rate of return available through the incremental cash flow equals or exceeds the MARR, the alternative associated with the extra investment should be selected.

Problem 8.8

8.4   RATE-OF-RETURN EVALUATION USING A PW EQUATION

In this section we discuss the primary approach to making alternative selections by the ROR method. The ROR technique described in Chapter 7 and the incremental cash-flow tabulation of Section 8.2 are combined to evaluate two mutually exclusive alternatives by the ROR method based on a PW equation. We start here with alternatives which have only negative cash flows (except for any salvage value). Since all cash flows are costs, it is not possible to calculate a rate of return for the individual alternatives. So, the incremental cash-flow series is analyzed using a PW equation. The method involving alternatives with positive cash flows is illustrated in Section 8.6.

   The LCM of lives must be used in the PW equation formulation. Because of the reinvestment requirement for PW analysis for different-life assets, the incremental cash-flow series may well contain the dilemma of several sign changes, indicating multiple $i*$ values. Though incorrect, this dilemma may be neglected in actual practice. The correct approach is to establish the reinvestment rate c and follow the approach of Section 7.5. This means that the unique composite rate of return (CRR) for the incremental cash-flow series is determined. These three required elements—incremental cash-flow series, LCM, and multiple roots—are the primary reasons that the ROR method is often applied incorrectly in engineering economy analyses of multiple alternatives. As stated before, it is always possible to use a PW or AW analysis *at an established MARR* in lieu of the ROR method.

   The complete procedure (manual or spreadsheet) for an ROR analysis for two alternatives involving only negative cash flows is

1. Order the alternatives by initial investment starting with the smaller one. The one with the larger initial investment is in the column labeled B in Table 8–1.

2. Develop the cash-flow and incremental cash-flow series using the LCM of years, assuming reinvestment in alternatives, as necessary.

3. Draw an incremental cash-flow diagram (if needed for your use).

4. Count the number of sign changes in the incremental cash-flow series to determine if multiple rates of return are present (Section 7.4). If necessary, use Norstrom's criterion on the cumulative (incremental) cash-flow series to determine if a single positive root exists. (Refer to Section 7.4 after Example 7.4.)

5. Set up the PW equation for the incremental cash flows in the form of Equation [7.1] and determine the return i^*_{B-A} using manual trial and error, or enter the incremental cash-flow values from step 2 into a spreadsheet system to determine i^*_{B-A}.

6. If $i^*_{B-A} <$ MARR, select alternative A. If $i^*_{B-A} >$ MARR, the extra investment is justified; select alternative B.

In step 4, an indication of multiple roots requires that the project-net-investment technique, Equation [7.5], be applied in step 5 to make $i' = i^*$.

In step 5, if manual trial and error is used to calculate the rate of return, time may be saved if the i^*_{B-A} value is bracketed, rather than approximated by a point value using linear interpolation, provided that a single ROR value is not needed. For example, if the MARR is 15% per year and you have established that i^*_{B-A} is in the 15-to-20% range, an exact value is not necessary to accept B, since you already know that $i^*_{B-A} \geq$ MARR.

The IRR function on a spreadsheet will normally determine one i^* value, and multiple guess values are input to find the multiple roots for a nonconventional sequence, as discussed in Example 7.4 (Spreadsheet). If one of these multiple roots is the same as the expected reinvestment rate, this root can be used as the ROR value, and the project-net-investment technique is not necessary. In this case only, $i' = i^*$, as concluded at the end of Section 7.5.

Example 8.3

A leather clothes manufacturer is considering the purchase of one new industrial sewing machine, which is either semiautomatic or fully automatic. The estimates are

	Semiautomatic	Fully Automatic
First cost, $	8,000	13,000
Annual disbursements, $	3,500	1,600
Salvage value, $	0	2,000
Life, years	10	5

Determine which machine should be selected if the MARR is 15% per year.

Solution

Use the procedure described above to estimate i^*_{f-s}.

1. Alternative A is the semiautomatic (s) and alternative B is the fully automatic (f) machine.

2. The cash flows for the LCM of 10 years are tabulated in Table 8–4.

3. The incremental cash-flow diagram is shown in Figure 8–1.

Table 8–4	Cash-flow tabulation for Example 8.3		
Year	(1) Cash Flow, Semiautomatic	(2) Cash Flow, Fully Automatic	(3) = (2) − (1) Incremental Cash Flow
0	$-8,000	$-13,000	$ -5,000
1–5	-3,500	-1,600	+1,900
5	—	$\begin{cases} +2,000 \\ -13,000 \end{cases}$	-11,000
6–10	-3,500	-1,600	+1,900
10	—	+2,000	+2,000
	$-43,000	$-38,000	$ +5,000

4. There are three sign changes in the incremental cash-flow series (Figure 8–1 or Table 8–4), indicating as many as three roots. There are also three sign changes in the cumulative incremental series, which starts negatively at $S_0 = \$-5000$ and continues to $S_{10} = \$+5000$, indicating no single positive root exists. (Develop the cumulative series and apply Norstrom's criterion yourself.)

5. The rate-of-return equation based on PW of incremental cash flows is

$$0 = -5000 + 1900(P/A,i,10) - 11,000(P/F,i,5) + 2000(P/F,i,10) \quad \text{[8.1]}$$

If it is reasonable to assume that the reinvestment rate to equal to a resulting $i*$ value, Equation [7.5] will result in a CRR of $i' = i*$. Solution of Equation [8.1] for the first root discovered results in an i^*_{f-s} between 12 and 15%. By interpolation $i^*_{f-s} = 12.65\%$. (See the comment below about the other roots.)

6. Since the rate of return of 12.65% on the extra investment is less than the 15% MARR, the lower-cost semiautomatic machine should be purchased.

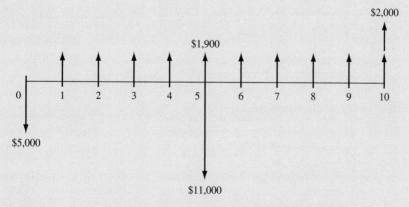

Figure 8–1   Diagram of incremental cash flow, Example 8.3.

Comment

In step 4, the presence of up to three $i*$ values is indicated. The analysis above finds one of the roots at 12.65%. When we use ROR as $i* = 12.65\%$, we assume that any positive project-net-investments are reinvested at $c = 12.65\%$. If this is not a reasonable assumption for the engineering economist to make, the project-net-investment procedure must be applied and a better rate c must be used to find a different value of i' to compare with an MARR of 15%.

The other two roots are very large positive and negative numbers, as the IRR and NPV functions of Excel can help you discover. So, they are not useful to the analysis.

Example 8.3
(Spreadsheet)

Use spreadsheet analysis to find the ROR value to select the better sewing machine in Example 8.3.

Solution

Use the six-step procedure described above to estimate i^*_{f-s} using a spreadsheet system.

1. Alternative A is the semiautomatic (s) and alternative B is the fully automatic (f) machine.

2. The cash flows for the LCM of 10 years are tabulated in Table 8–4.

3. The incremental cash-flow diagram is shown in Figure 8–1.

4. There are three sign changes in the incremental cash-flow series indicating as many as three roots.

5. Figure 8–2 includes the incremental cash flows from Table 8–4. These are the same as in Equation [8.1], but the PW equation does not need to be developed for spreadsheet solution. Cell B12 shows the $i*$ value of 12.65% using the IRR function.

6. Since the rate of return on the extra investment is less than the 15% MARR, the lower-cost semiautomatic machine should be purchased.

Comment

Once the spreadsheet is established, there is a wide variety of analyses that can be performed using resident functions. For example, cell B13 uses the PW function to verify that the present worth is zero at the calculated $i*$. Cell B14 is the PW at the MARR = 15%, which is negative, thus indicating in yet another way that the extra investment does not return the MARR and is not justified. Finally, cell B16 shows that $i*$ drops from 12.65% to only 4.42% when the year 10 incremental cash flow of $+3900 is omitted. This is the equivalent of changing the planning horizon from 10 to 9 years and determining the effect on the ROR.

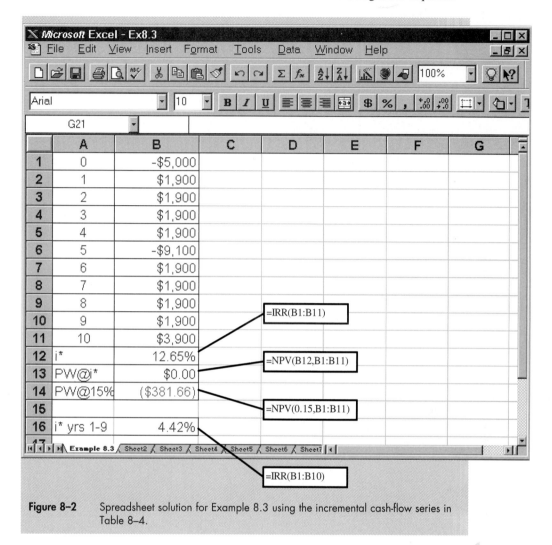

Figure 8–2   Spreadsheet solution for Example 8.3 using the incremental cash-flow series in Table 8–4.

The rate of returns determined thus far can actually be interpreted as *breakeven values*, that is, a rate at which either alternative can be selected. If the incremental cash flow $i*$ is greater than the MARR, the larger-investment alternative is selected. As an illustration, the breakeven rate for the incremental investment in Example 8.3 is 12.65%. Figure 8–3 is a general (not to scale) plot of PW values of the actual cash flows (not incremental cash flows) for different rates of return for each alternative. At values of $i < 12.65\%$, the present worth for the fully automatic machine is less than that of the semiautomatic. For $i > 12.65\%$, the fully automatic PW is larger. Thus at MARR $= 10\%$, select the fully automatic machine because its PW of costs is lower; whereas if the MARR $= 15\%$, as in Example 8.3, select the semiautomatic machine.

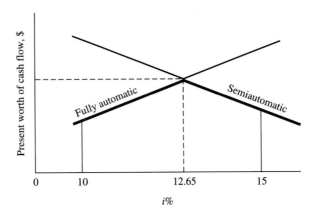

Figure 8–3   Breakeven graph of present worth of cash flows versus rate of return, Example 8.3.

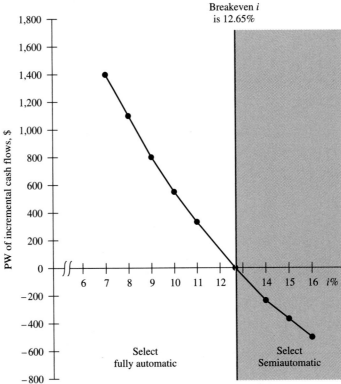

Figure 8–4   Plot of present worth of incremental cash flows, Example 8.3.

Similarly, if the PW of the incremental cash flow (rather than each alternative's actual estimated cash flow) is plotted for various interest rates, the graph shown in Figure 8–4 is obtained, which again shows the $i*$ breakeven at 12.65%. The conclusions are:

1. At a MARR = 12.65%, the alternatives are equally attractive.
2. For any MARR < 12.65%, the extra investment for fully automatic is justified, because the PW of the incremental savings exceeds the PW of the incremental disbursements.
3. For any MARR > 12.65%, the opposite is true; the extra investment for the fully automatic machine should not be made.

Example 8.4

Determine which drill press should be purchased in Example 8.1 using MARR = 15%.

Solution

The alternatives have equal lives and the incremental cash flows of Table 8–2 are used in the six-step procedure described above to set up the PW form of a rate-of-return equation.

$$0 = -6000 + 1200(P/A,i*,25) + 300(P/F,i*,25)$$

There is only one sign change in the series, so $i*$ can be determined uniquely. This breakeven rate is $i* = 19.79\%$. Since 19.79% > 15%, the purchase of the new drill press is justified.

Comment

You can always set up the PW equations and compare the alternatives using the actual (rather than the incremental) cash flows in the format of the general relation

$$0 = PW_{new} - PW_{used}$$

where the *new* alternative has the larger first cost. In this analysis, the PW relations are

$$PW_{new} = -21,000 - 7000(P/A,i,25) + 1050(P/F,i,25)$$

$$PW_{used} = -15,000 - 8200(P/A,i,25) + 750(P/F,i,25)$$

Then

$$0 = PW_{new} - PW_{used}$$

$$= (-21,000 + 15,000) + (-7000 + 8200)(P/A,i,25)$$

$$+ (1050 - 750)(P/F,i,25)$$

$$= -6000 + 1200(P/A,i,25) + 300(P/F,i,25)$$

The reduced form is identical to that used in the solution to this example.

Problems 8.9 to 8.13

8.5 RATE-OF-RETURN EVALUATION USING AN AW EQUATION

As stated earlier, comparing alternatives by the ROR method, correctly performed, leads to the same decision as a PW or AW analysis. This is also correct regardless of whether the incremental cash-flow series ROR is determined via a PW-based or an AW-based equation. However, for an AW-based ROR determination on incremental cash flows, the AW equation on the incremental cash flow *must be written over the least common multiple of lives,* just as for a PW equation. So, unlike the discussion in Chapter 6. there may be no real computational advantage in using an AW-based equation compared to a PW-based one to find thc ROR. Also, as we discussed earlier, spreadsheets usually find i^* values using their own version of trial and error on the PW equation. All in all, this section will help you understand how the AW method can be used to select an alternative using an ROR basis, but you will commonly calculate i^* via a PW relation. Two obvious exceptions are when most of the incremental cash flow is in a uniform series and when alternative lives are equal.

For equal lives, the AW-based ROR equation for incremental cash flow takes the convenient, general form:

$$0 = \pm \Delta P(A/P,i,n) \pm \Delta SV(A/F,i,n) \pm \Delta A \qquad \text{[8.2]}$$

where the Δ (delta) symbol represents the differences in P, SV, and A in the incremental cash-flow tabulation. The same six-step procedure to find the ROR as that in Section 8.4 for present worth is followed, except in step 5 the AW equation is developed. Manual interpolation in the interest tables or a spreadsheet is used to determine i^*_{B-A}. For spreadsheet solution, enter the incremental cash flows over the LCM and use the IRR function.

We emphasize once again that the incremental cash flow can be used for the AW method, but to do so correctly, the LCM of lives must be used, just as in the PW method. If the lives are *unequal,* we can tabulate the incremental cash flows for the LCM and set up the AW-based relation (which may involve determining a PW or FW first).

Also, we always have the equivalent option to perform the analysis using *actual cash flows.* The AW for one cycle of each alternative is developed and i^*_{B-A} is determined using

$$0 = AW_B - AW_A \qquad \text{[8.3]}$$

The incremental cash flow is not needed at all in this analysis, which is the same as you learned in Chapter 6. However, the ROR also represents the i^* for the incremental cash flow between two alternatives.

Examples 8.5 and 8.6 utilize AW equations to find the ROR; the first for equal lives and the second for unequal lives.

Example 8.5

Compare the two drill presses in Example 8.1 using an AW relation to compute the ROR. Assume that MARR = 15% per year.

Solution

Since the lives are equal (25 years), Equation [8.2] may be used to solve for the ROR of the incremental cash flows in Table 8–2.

$$0 = -6000(A/P,i,25) + 300(A/F,i,25) + 1200$$

Equality occurs at $i^*_{n-o} = 19.79\%$, so the extra investment for the new machine is justified, as it was in Example 8.4 using a PW-based equation to determine the same ROR.

Comment

The value 19.79% is the breakeven rate of return. A graph similar to Figure 8–4 for this cash flow series may be constructed in which AW replaces PW. For values of MARR $\geq$ 19.79%, the used drill press should be purchased; for values of MARR < 19.79%, as is the case here, select the new machine.

Example 8.6

Compare the sewing machines of Example 8.3 using an AW-based ROR method and an MARR of 15% per year.

Solution

For reference purposes, the PW-based ROR equation for the incremental cash flow in Example 8.3 shows that the semiautomatic machine should be purchased.

   For the AW relation, there are two equivalent solution approaches. Write Equation [8.2] based on the *incremental* cash-flow series over the LCM of 10 years, or write Equation [8.3] for the *two actual* cash-flow series over one life cycle. Choose the latter method and the ROR is found by Equation [8.3] using the respective lives of 5 years for fully automatic (f) and 10 years for semiautomatic (s),

$$AW_f = -13,000(A/P,i,5) + 2000(A/F,i,5) - 1600$$

$$AW_s = -8000(A/P,i,10) - 3500$$

From the form of Equation [8.3], $0 = AW_f - AW_s$,

$$0 = -13,000(A/P,i,5) + 2000(A/F,i,5) + 8000(A/P,i,10) + 1900$$

The results of manual solution yields an interpolated value of $i^*_{f-s} = 12.65\%$ (same as for the PW relation). The semiautomatic machine should be purchased, since 12.65% is less than a MARR of 15%, meaning that the incremental investment is not justified.

Comment

If a spreadsheet is used, the incremental cash flows over 10 years are determined as in Table 8–4 and entered on the spreadsheet. The IRR value $i^* = 12.65\%$ will result.

It is very important to remember that when an ROR analysis using an AW-based equation is made on the *incremental cash flow,* the least common multiple of lives (10 years in this example) must be used.

Problems 8.14 to 8.17

8.6   SELECTION FROM MULTIPLE ALTERNATIVES USING ROR ANALYSIS

The discussion in this section treats selection from multiple alternatives, that is, more than two, which are mutually exclusive, using the ROR method. Acceptance of one alternative automatically precludes acceptance of any others.

As in any selection problem in engineering economics, there are several correct solution techniques. The PW and AW methods discussed in Chapters 5 and 6 are the most straightforward techniques. They use a specified MARR to compute the total PW or AW for each alternative. The alternative that has the most favorable measure of worth is selected. However, many managers want to know the ROR for each alternative when the results are presented. Primarily because of the wide appeal of knowing ROR values, this method is very popular but often incorrectly applied, as discussed in Chapter 7. It is essential to understand how to correctly perform an ROR analysis based upon PW and AW relations. This is important since the ROR value is based on the incremental cash flows between alternatives to ensure a correct alternative selection. (In essence, this is why the PW and AW methods of previous chapters are more straightforward.)

When the ROR method is applied, the entire investment must return at least the minimum attractive rate of return. When the returns on several alternatives equal or exceed the MARR, at least one of them will be justified in that its ROR > MARR. This is the one requiring the smallest investment. (If not even one investment is justified, the do-nothing alternative is selected.) For all others, the incremental investment must be separately justified. If the return on the extra investment equals or exceeds the MARR, then the extra investment should be made in order to maximize the total return on the money available, as discussed in Section 8.1.

Thus, for ROR analysis of multiple alternatives, the following criteria are used. Select the one alternative that

1.   Requires the *largest investment,* and

2.   Indicates that the *extra investment over another acceptable alternative is justified.*

An important rule to remember when evaluating multiple alternatives by the ROR method is that *an alternative should never be compared with one for which the incremental investment is not justified.* The ROR analysis procedure is

1. Order the alternatives by increasing initial investment (smallest to largest).

2. Determine the nature of the cash-flow series: some positive or all negatives.

    (a) **Some positive cash flows, i.e., incomes.**   Consider the do-nothing alternative as *the defender* and compute the incremental *cash flows* between the do-nothing alternative and the lowest initial-investment alternative (*the challenger*). Go to step 3.

    (b) **All negative cash flows, i.e., costs only.**   Consider the lowest initial-investment alternative as the defender and the next-higher investment as the challenger. Skip to step 4.

3. Set up the ROR relation and determine $i*$ for the defender. (When comparing against the do-nothing alternative, the ROR is actually the overall return for the challenger.) If $i* <$ MARR, remove the lowest-investment alternative from further consideration and compute the overall ROR for the next-higher-investment alternative. Repeat this step until $i* \geq$ MARR for one of the alternatives; then this alternative becomes the defender and the next higher-investment alternative is labeled the challenger.

4. Determine the annual incremental cash flow between the challenger and defender using the relation:

   Incremental cash flow = challenger cash flow − defender cash flow

5. Calculate the $i*$ for the incremental cash-flow series using a PW-based or AW-based equation. (PW is most commonly used.)

6. If $i* \geq$ MARR, the challenger becomes the defender and the previous defender is removed from further consideration. Conversely, if the $i* <$ MARR, the challenger is removed from further consideration and the defender remains the defender against the next challenger.

7. Repeat steps 4 to 6 until only one alternative remains. It is the selected one.

Note that in steps 4 to 6 only two alternatives are compared at any one time. It is very important, therefore, that the correct alternatives be compared. Unless this procedure is followed exactly, the wrong alternative may be selected from the analysis. The procedure is illustrated in Examples 8.7 and 8.8.

Example 8.7

Four different prefab-building locations have been suggested, of which only one will be selected. Cost and annual net cash-flow information are detailed in Table 8–5. The annual net cash-flow series vary due to differences in maintenance, labor costs, transportation charges, etc. If the MARR is 10%, use ROR analysis to select the one economically-best location.

Solution

All alternatives have a 30-year life and the annual cash flows include incomes and

Table 8–5	Estimates for four possible building locations, Example 8.7			
Location	*A*	*B*	*C*	*D*
Building cost, $	−200,000	−275,000	−190,000	−350,000
Annual cash flow, $	+22,000	+35,000	+19,500	+42,000
Life, years	30	30	30	30

Table 8–6	Computation of the rate of return for four alternatives, Example 8.7			
Location	**(1)** *C*	**(2)** *A*	**(3)** *B*	**(4)** *D*
Building cost, $	−190,000	−200,000	−275,000	−350,000
Cash flow, $	+19,500	+22,000	+35,000	+42,000
Projects compared	*C* to do-no	*A* to do-no	*B* to *A*	*D* to *B*
Incremental cost, $	−190,000	−200,000	−75,000	−75,000
Incremental cash flow, $	+19,500	+22,000	+13,000	+7,000
$(P/A,i^*,30)$	9.7436	9.0909	5.7692	10.7143
i^* (%)	9.63	10.49	17.28	8.55
Increment justified?	No	Yes	Yes	No
Project selected	Do-no	*A*	*B*	*B*

disbursements. The procedure outlined above results in the following analysis:

1. The alternatives are ordered by increasing building cost in Table 8–6, first line.
2. Some positive cash flows are present; use step 2, part (*a*) to compare location *C* with the do-nothing (identified as do-no) alternative.
3. The ROR relation is

$$0 = -190,000 + 19,500(P/A,i^*,30)$$

    Table 8–6, column 1 presents the calculated $(P/A,i^*,30)$ factor value of 9.7436 and $i_c^* = 9.63\%$. Since $9.63\% < 10\%$, location *C* is eliminated. Now the comparison is *A* to do-nothing, and column 2 shows that $i_A^* = 10.49\%$. This eliminates the do-nothing alternative; the defender is now *A* and the challenger is *B*.
4. The incremental cash-flow series, column 3, and i^* for *B-to-A comparison* is determined from

$$0 = -275,000 - (-200,000) + (35,000 - 22,000)(P/A,i^*,30)$$
$$= -75,000 + 13,000(P/A,i^*,30) \qquad \text{[8.4]}$$

5. From the interest tables, look up the P/A factor at the MARR, which is $(P/A,10\%,30) = 9.4269$. Now, any P/A value from Equation [8.4] greater than 9.4269 indicates that the i^* will be less than 10% and is therefore unacceptable.

The P/A factor from Equation [8.4] is 5.7692. For reference purposes, $i^* = 17.28\%$.

6. Alternative B is justified incrementally (new defender), thereby eliminating A.

7. Comparing D to B (steps 4 and 5) results in the PW relation $0 = -75,000 + 7000(P/A,i^*,30)$ and a P/A value of $10.7143 (i^*_{D-B} = 8.55\%)$. Location D is thereby eliminated and only alternative B remains; it is selected.

Comment

Remember that an alternative must *always* be compared with an acceptable alternative, and the do-nothing alternative can end up being the only acceptable one. Since C was not justified in this example, location A was not compared with C. Thus, *if* the B-to-A comparison had not indicated that B was incrementally justified, then the D-to-A comparison would be correct instead of D-to-B.

To demonstrate how important it is to properly apply the ROR method in order to not select the wrong alternative consider the following. If the overall rate of return of each alternative in this example is computed (i.e., each alternative is effectively compared with the do-nothing alternative), the results are as follows:

Location	A	B	C	D
Overall i^* (%)	10.49	12.35	9.63	11.56

If we now apply *only* the first criterion stated earlier, that is, make the largest investment that has a MARR of 10% or more, we choose location D. But, as shown above, this is the wrong selection, because the extra investment of $75,000 over location B will not earn the MARR. In fact, it will earn only 8.55% (Table 8–6). Remember, therefore, that incremental analysis is necessary for selection of one alternative from several when the ROR evaluation method is used. This is not necessary for the PW or AW methods to select an alternative, but these methods require that a MARR value be stated to find the PW or AW values.

When the alternatives consist of costs only [step 2, part (*b*) in the procedure], the incremental cash flow is the 'income' or difference between costs for two alternatives, as in Section 8.4. There is no do-nothing alternative. Therefore, the lowest initial-investment alternative is the defender against the next-lowest investment (challenger). This procedure is illustrated in Example 8.8.

Example 8.8

Only one of four different machines is to be purchased for use in seawater cleanup operations. Select a machine using the president's MARR of 13.5% per year and the cost estimates in Table 8–7.

Table 8–7	Costs for four alternative machines, Example 8.8			
Machine	**1**	**2**	**3**	**4**
First cost, $	−5,000	−6,500	−10,000	−15,000
Annual operating cost, $	−3,500	−3,200	−3,000	−1,400
Salvage value, $	+500	+900	+700	+1,000
Life, years	8	8	8	8

Table 8–8	Comparison using rate of return, Example 8.8			
Machine	**1**	**2**	**3**	**4**
Initial cost, $	−5,000	−6,500	−10,000	−15,000
Annual operating cost, $	−3,500	−3,200	−3,000	−1,400
Salvage value, $	+500	+900	+700	+1,000
Plans compared	—	2 to 1	3 to 2	4 to 2
Incremental investment, $	—	−1,500	−3,500	−8,500
Incremental savings, $	—	+300	+200	+1,800
Incremental salvage, $	—	+400	−200	+100
i^*, %	—	14.6	<0	13.6
Increment justified?	—	Yes	No	Yes
Alternative selected	—	2	2	4

Solution

The machines are already ordered according to increasing initial cost, no incomes are involved, and life estimates are all equal. Incremental comparisons are made directly for the two alternatives. This concludes steps 1 and 2 (all negative cash flows) of the procedure. Per step 4 compare machine 2 (challenger) with machine 1 (defender) on an incremental basis. The PW relation is

$$0 = -1500 + 300(P/A,i^*,8) + 400(P/F,i^*,8)$$

The solution yields $i^* = 14.6\%$, which exceeds the MARR of 13.5%. We eliminate machine 1 from further consideration. (If you had trouble obtaining the rate-of-return equation above, prepare a tabulation of cash flow for machines 1 and 2, as in Section 8.2 and enter the values into a spreadsheet.) The remaining calculations are summarized in Table 8–8. When machine 3 is compared with machine 2, $i^* < 0\%$; therefore, machine 3 is eliminated.

The comparison of machine 4 with machine 2 shows that the rate of return on the increment is slightly greater than the MARR, favoring machine 4. Since no additional alternatives are available, machine 4 is selected.

Comment

Remember that when only negative cash flows are estimated, it is implied that one of the machines must be selected; that is, do-nothing is not a feasible alternative. This situation may arise, for example, when the alternatives under consideration are planned as part of a larger project, which is already judged as economical regardless of which alternative is selected in the current evaluation.

**Example 8.8
(Spreadsheet)**

Only one of four different machines is to be purchased for use in seawater cleanup operations. Select a machine using the president's MARR of 13.5% per year and the cost estimates in Table 8–7.

Solution

The spreadsheet in Figure 8–5 on page 252 presents the incremental analysis beginning with row 5 cells using the same logic as for manual selection. Cell E15 indicates an $i^*_{4-2} = 13.6\%$ obtained from the IRR function. Machine 4 is selected, since it justifies the largest investment.

Comment

The spreadsheet in Figure 8–5 does not include logic to select the better alternative at each stage of solution. This feature could be added by constructing the IF-operator logic and developing correct arithmetic operations to obtain the incremental cash flows and i^* values for any selected alternative. In other words, program the remainder of the seven-step procedure used in the manual solution.

Selection from multiple alternatives with unequal lives using i^* values determined from a PW relation requires that the incremental cash flows over the LCM years of the two alternatives' lives be evaluated. Example 8.9 illustrates the calculations for alternatives having different lives. This is another application of the same principle of equal-service comparison we have used previously.

Of course, it is always possible to not use the ROR method and rely on PW or AW analysis of the incremental cash flows at the MARR to make the selection. However, as we learned in the two previous sections, it is still necessary to make the comparison over the LCM number of years for an incremental analysis to be performed correctly. If conventional AW analysis is used, as we applied in Chapter 6, it is not necessary to use the LCM, since no incremental cash-flow analysis is involved. This is illustrated in Example 8.9.

Often the lives of all the alternatives are so long that they can be considered to be infinite. In this situation the capitalized-cost method may be used. See

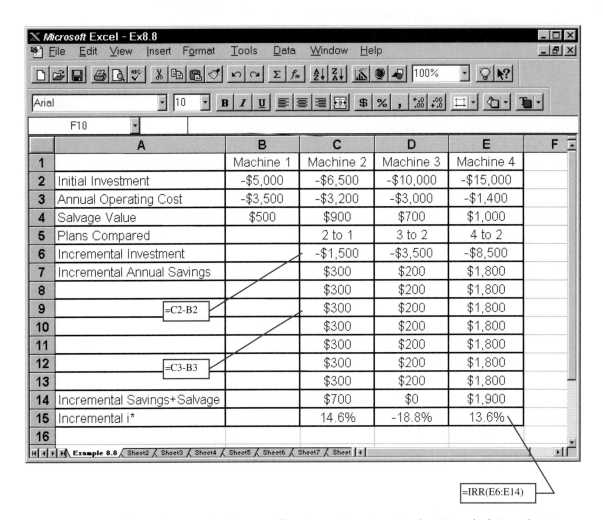

Figure 8–5   Spreadsheet solution to select from mutually exclusive alternatives using the ROR method, Example 8.8.

Example 8.9

For the three mutually exclusive alternatives in Table 8–9, use a MARR of 15% per year to select the best alternative using (*a*) rate-of-return analysis and (*b*) AW analysis on the actual cash-flow estimates with MARR = 15%.

Table 8–9   Three different-life alternatives, Example 8.9

	A	*B*	*C*
Initial cost, $	−6000	−7000	−9000
Salvage value, $	0	+200	+300
Annual cash flow, $	+2000	+3000	+3000
Life, years	3	4	6

Solution

(*a*) Use the procedure at the beginning of this section to compute the rates of return and make the selection.
  1. Ordering by increasing cost is already present.
  2. Since there are positive cash flows involved, compare *A* to do-nothing. The result is $i_A^* = 0\%$ from the PW equation.

$$0 = -6000 + 2000(P/A,i^*,3)$$

  3. Since 0% < 15%, delete *A*, make *B* the challenger, and compute $i_B^* = 26.4\%$ from

$$0 = -7000 + 3000(P/A,i^*, 4) + 200(P/F,i^*,4)$$

  Since 26.4% > 15%, *B* is now the defender and *C* is the challenger.
  4. The incremental cash flow between *C* and *B* (Table 8–10) is determined for 12 years, the LCM of 4 and 6.
  5. Calculate $i_B^* = 19.4\%$ for the (*C* − *B*) incremental cash flow in Table 8–10.
  6. Since 19.4% > 15%, choose alternative *C* over *B*. No alternative remains, so *C* is selected.

(*b*) For the AW analysis of alternative actual cash flows, use *i* = 15% per year and the respective lives as we learned in Chapter 6. (Use of the LCM is not necessary for different-life projects when the AW method is applied to the actual cash-flow estimates.)

$$\text{AW}_A = -6000(A/P,15\%,3) + 2000 = \$-628$$

$$\text{AW}_B = -7000(A/P,15\%,4) + 3000 + 200(A/F,15\%,4) = \$+588$$

$$\text{AW}_C = -9000(A/P,15\%,6) + 3000 + 300(A/F,15\%,6) = \$+656$$

	Table 8–10	Incremental cash-flow tabulation for alternatives *C* and *B*, Example 8.9		
		Cash Flow		**Incremental Cash Flow**
Year	***B***	***C***		***(C − B)***
0	$-7,000	$-9,000		$-2,000
1	+3,000	+3,000		0
2	+3,000	+3,000		0
3	+3,000	+3,000		0
4	-3,800	+3,000		+6,800
5	+3,000	+3,000		0
6	+3,000	-5,700		-8,700
7	+3,000	+3,000		0
8	-3,800	+3,000		+6,800
9	+3,000	+3,000		0
10	+3,000	+3,000		0
11	+3,000	+3,000		0
12	+3,200	+3,300		+100
	$+15,600	$+18,600		$+3,000

Alternative *C* is selected, as expected, because it offers the largest positive AW, indicating a return in excess of 15%.

Comment

In part (*a*), when comparing different-life alternatives by ROR analysis, you can use the LCM of years between only the *two alternatives being compared, not* the LCM of all alternative lives, since actual cash flows are used.

Example 8.10

The Corps of Engineers intends to construct an earthen dam on the Sacochsi River. Six different sites are suggested, and the environmental-impact statements have been approved. The construction costs and average annual benefits (income) are tabulated below. If a MARR of 6% per year for public projects is used and dam life is long enough to be considered infinite for analysis purposes, select the best location from an economic perspective.

Site	Construction Cost P, $ (Millions)	Annual Income A, $
A	6	350,000
B	8	420,000
C	3	125,000
D	10	400,000
E	5	350,000
F	11	700,000

Table 8–11   Capitalized-cost comparison of dam sites, Example 8.10

	C	E	A	B	D	F
P, $ (millions)	3	5	6	8	10	11
A, $ (thousands)	125	350	350	420	400	700
Comparison	C to do-no	E to do-no	A to E	B to E	D to E	F to E
ΔP, $ (millions)	3	5	1	3	5	6
ΔA, $ (thousands)	125	350	0	70	50	350
$\Delta A/i - \Delta P$, $ (millions)	−0.92	0.83	−1.0	−1.83	−4.17	−0.17
Site selected	do-no	E	E	E	E	E

Solution

After ordering the sites by increasing first cost, we can use Equation [5.3], $P = A/i$, for capitalized costs and set it equal to zero to determine if the incremental investment is justified.

$$0 = \frac{\Delta A}{i} - \Delta P \qquad\qquad [8.5]$$

Here ΔP is the incremental investment and ΔA is the incremental income or cash flow forever. Equation [8.5] can be solved for i for each increment of investment and the i compared with the MARR (per step 6 of the analysis procedure).

Alternatively, and more simply, we can use $i = $ MARR and determine if the right side is greater than zero. If so, the incremental investment is justified. Using this alternate approach, Table 8–11 indicates that only site E is justified at $i = 6\%$, so this is the most economical dam site.

Problems 8.18 to 8.30

CHAPTER SUMMARY

Just as present-worth and annual-worth methods will yield the best alternatives from among several, rate-of-return calculations can be used for the same purpose. In using the ROR technique, however, it is necessary to consider the incremental cash flows when selecting between mutually exclusive alternatives. This was not necessary for the PW or AW methods. The incremental-investment evaluation is conducted between only two alternatives at a time beginning with the lowest initial investment alternative. Once an alternative has been eliminated, it is not considered further.

Rate-of-return values have a natural appeal to management, but the ROR analysis is often more difficult to set up and complete than the PW or AW analysis using the stated MARR. Care must be taken to perform an ROR analysis correctly on the incremental cash flows, as discussed in this chapter; otherwise it may give incorrect results.

PROBLEMS

8.1 What is the primary difference between independent and mutually exclusive alternatives?

8.2 Determine the overall rate of return on the following investments:

(*a*) $50,000 at 25% per year

(*b*) $10,000 at 40% per year

(*c*) $40,000 at 2% per year

(*d*) $20,000 at −10% per year

8.3 Why is it not necessary to compare independent alternatives against each other?

Note: For Problems 8.4 to 8.7 prepare only a tabulation of net cash flows over the LCM of years.

8.4 The manager of a canned-food processing plant is trying to decide between two labeling machines. Their respective costs are shown below.

	Machine *A*	Machine *B*
First cost, $	15,000	25,000
Annual operating cost, $	1,600	400
Salvage value, $	3,000	6,000
Life, years	5	10

8.5 A small solid-waste recycling plant is considering two types of storage bins. Their respective costs are shown below.

	Alternative *P*	Alternative *Q*
First cost, $	18,000	35,000
Annual operating cost, $	4,000	3,600
Salvage value, $	3,000	2,500
Life, years	3	9

8.6 A company is considering the adoption of one of two processes identified as *E* and *Z*. Process *E* will have a first cost of $43,000, a monthly operating cost of $10,000, and a $5000 salvage value at the end of its 4-year life. Process *Z* will have a first cost of $31,000 and a quarterly operating cost of $39,000. It will have an 8-year life with a $2000 value at the end of that time. Assume interest will be compounded monthly.

8.7 A manufacturing company is in need of 1000 square meters of office space for 3 years. The company is considering the purchase of land for $20,000 and erecting a temporary metal structure on it at a cost of $70 per square meter. At the end of the 3-year use period, the company expects to be able to sell the land for $25,000 and the building for $5000. Alternatively, the company can lease storage space for $5 per square meter per month payable at the beginning of each year.

8.8 In a tabulation of cash flows for two alternatives, if the rate of return on the net cash flow column exceeds the MARR, which alternative should be selected?

8.9 A couple is trying to decide between purchasing a house and renting one. They can purchase a new house with a down payment of $15,000 and a monthly payment of $750, beginning 1 month from now. Taxes and insurance are expected to be $200 per month. In addition, they expect to paint the house every 4 years at a cost of $600. Alternatively, they can rent a house for $700 per month payable at the beginning of each month with a $900 deposit, which will be returned when they vacate the house. The utilities are expected to average $135 per month whether they purchase or rent. If they expect to sell the house in 6 years for $20,000 more than they paid, should they buy a house or rent one? Assume the MARR is a nominal 12% per year compounded monthly. Use a rate-of-return analysis and assume the principal reduction was $8000 during the time they owned the house.

8.10 The warehouse for a large furniture manufacturing company currently requires too much energy for heating and cooling because of poor insulation. The company is trying to decide between urethane foam and fiberglass insulation. The initial cost of the foam insulation will be $35,000 with no salvage value. The foam will have to be painted every 3 years at a cost of $2500. The energy savings is expected to be $6000 per year.

Alternatively, fiberglass batts can be installed for $12,000. The fiberglass batts would not be salvageable either, but there would be no maintenance costs. If the fiberglass batts would save $1500 per year in energy costs, which method of insulation should the company use at a MARR of 15% per year? Use a 24-year study period and a rate-of-return analysis.

8.11 A meat-packing plant manager is trying to decide between two different methods for cooling cooked hams. The spray method involves spraying water over the hams until the ham temperature is reduced to 30 degrees Celsius. With this method, approximately 80 liters of water is required for each ham. Alternatively, an immersion method can be used in which only 16 liters of water is required per ham. However, this method will require an initial extra investment of $10,000 and extra expenses of $1000 per year, with the equipment expected to last 10 years. The company cooks 60,000 hams per year and pays $0.50 per 1000 liters for water. The company must also pay $0.30 per 1000 liters for wastewater discharged. If the company's minimum attractive rate of return is 18% per year, which method of cooling should be used?

8.12 The owner of the ABC Drive-In Theater is considering two proposals for upgrading the parking ramps. The first proposal involves asphalt paving of the entire parking area. The initial cost of this proposal would be $35,000, and it would require annual maintenance of $250 beginning 3 years after installation. The owner expects to have to resurface the theater in 10 years. Resurfacing will cost only $8000, since grading and surface preparation are not necessary, but the $250 annual maintenance cost will continue.

Alternatively, gravel can be purchased and spread in the drive areas and grass planted in the parking areas. The owner estimates that 29 metric tons of gravel will be needed per year starting 1 year from now at a cost of $90 per metric ton. In addition, a riding lawn mower, which will cost $2100 and have a life of 10 years, will be needed now. The cost of labor for spreading gravel, cutting grass, etc., is expected to be $900 the first year, $950 the second year and will increase by $50 per year thereafter. If the interest rate is 12% per year, which alternative should be selected? Use a rate-of-return analysis and a 20-year study period.

8.13 A state highway department is trying to decide between hot-patching an existing road and resurfacing it. If the hot-patch method is used, approximately 300 cubic meters of material will be required at a cost of $35 per cubic meter (in place). Additionally, the shoulders will have to be improved at the same time at a cost of $5000. The annual cost of routine maintenance on the patched-up road would be $4000. These improvements will last 2 years, at which time they will have to be redone. Alternatively, the state could resurface the road at a cost of $95,000. This surface will last 10 years if the road is maintained at a cost of $1500 per year beginning 4 years from now. No matter which alternative is selected now, the road will be completely rebuilt in 10 years. If the interest rate is 13% per year, which alternative should the state select on the basis of a rate-of-return comparison?

8.14 An environmental engineer is trying to decide between two operating pressures for a wastewater irrigation system. If a high-pressure system is used, fewer sprinklers and less pipe will be required, but the pumping cost will be higher. The alternative is to use lower pressure with more sprinklers. The pumping cost is estimated to be $4 per 1000 cubic meters of wastewater pumped at the high pressure. Twenty-five sprinklers will be required at a cost of $40 per unit. In addition, 1000 meters of aluminum pipe will be required at a cost of $9 per meter. If the lower-pressure system is used, the pumping will cost $2 per 1000 cubic meters of wastewater. Also required will be 85 sprinklers and 3000 meters of pipe. The aluminum pipe is expected to last 10 years and the sprinklers, 5 years. If the volume of wastewater is expected to be 500,000 cubic meters per year, which pressure should be selected if the company's minimum attractive rate of return is 20% per year? The aluminum pipe will have a 10% salvage value. Use the rate-of-return method.

8.15 A production plant manager has been presented with two proposals for automating an assembly process. Proposal A involves an initial cost of $15,000 and an annual operating cost of $2000 per year for the next 4 years. Thereafter, the operating cost is expected to be $2700 per year. This equipment is expected to have a 20-year life with no salvage value. Proposal B requires an initial investment of $28,000 and an annual operating cost of $1200 per year for the first 3 years. Thereafter, the operating cost is expected to increase by $120 per year. This equipment is expected to last 20 years and have a $2000 salvage value. If the company's minimum attractive rate of return is 10%, which proposal should be accepted using an AW-based rate-of-return equation?

8.16 Compare the two processes below on the basis of a rate-of-return analysis at a MARR of 12% per year.

	Process M	Process R
First cost, $	50,000	120,000
Salvage value, $	10,000	18,000
Life, years	5	10
Annual operating cost, $	15,000	13,000
Annual revenues, $	39,000	55,000 for year 1, decreasing by $2500 per year

8.17 A city planning commission is considering two proposals for a new civic center. The do-nothing alternative is not a choice. Proposal *F* requires an initial investment of $10 million now and an expansion cost of $4 million 10 years from now. The annual operating cost is expected to be $250,000 per year. Income from conventions, shows, etc., is expected to be $190,000 the first year and to increase by $20,000 per year for 4 more years and then remain constant until year 10. In year 11 and thereafter income is expected to be $350,000 per year. Proposal *G* requires an initial investment of $13 million now and an annual operating cost of $300,000 per year. However, income is expected to be $260,000, the first year and increase by $30,000 per year through year 7. Thereafter, income will remain at $440,000 per year. Determine which proposal should be selected on the basis of a rate-of-return analysis using a MARR of 7% per year and a 40-year study period.

8.18 In evaluating multiple alternatives, what course of action should you take if the overall rate of return for all the projects is above the MARR and the projects are (*a*) independent? (*b*) mutually exclusive?

8.19 If you are evaluating nine projects, one of which must be selected, which one would you pick if all the rates of return on incremental cash flows were below the MARR?

8.20 (*a*) In evaluating independent alternatives, which alternative are all of them compared against? (*b*) Which alternative is selected if none have a rate of return equal to or greater than the MARR?

8.21 Five different methods can be used for recovering by-product heavy metals from a manufacturing site's liquid waste. The investment costs and incomes associated with each method are shown below. Assuming all methods have a 10-year life with zero salvage value and the company's MARR is 10% per year, determine which one method should be selected using ROR evaluation.

	Method				
	1	**2**	**3**	**4**	**5**
First cost, $	−15,000	−18,000	−25,000	−35,000	−52,000
Salvage value, $	+1,000	+2,000	−500	−700	+4,000
Annual income, $	+4,000	+5,000	+7,000	+9,000	+12,000

8.22 If method 2 in Problem 8.21 has a life of 5 years and method 3 has a life of 15 years, which alternative should be selected?

8.23 Any one of five machines can be used in a certain phase of a canning operation. The costs of the machines are shown below and all are expected to have a 10-year life. If the company's minimum attractive rate of return is 16% per year, determine which machine should be selected on the basis of rate of return.

| | \multicolumn{5}{c}{Machine} |
	1	2	3	4	5
First cost, $	−28,000	−33,000	−32,000	−51,000	−46,000
Annual operating cost, $	−20,000	−18,000	−19,000	−13,000	−14,000

8.24 An independent dirt contractor is trying to determine which size dump truck to buy. The contractor knows that as the bed size increases, the net income increases, but it is uncertain whether the incremental expenditure required for the larger trucks is justified. The cash flows associated with each size truck are shown below. If the contractor's MARR is 18% per year and all trucks are expected to have a useful life of 8 years, determine which size truck should be purchased.

| | \multicolumn{5}{c}{Truck Bed Size, Cubic Meters} |
	8	10	15	20	25
Initial investment, $	−10,000	−12,000	−18,000	−24,000	−33,000
Annual operating cost, $	−5,000	−5,500	−7,000	−11,000	−16,000
Salvage value, $	+2,000	+2,500	+3,000	+3,500	+4,500
Annual income, $	+6,000	+10,000	+14,000	+12,500	+14,500

8.25 Which trucks would the contractor in Problem 8.24 buy if he wanted to have two trucks of different sizes?

8.26 Compare the alternatives below on the basis of a rate-of-return analysis, assuming the MARR is 15% per year.

	Project A	Project B
First cost, $	−40,000	−90,000
Annual operating cost, $	−15,000	−8,000
Annual repair cost, $	−5,000	−2,000
Annual increase in repair cost, $	−1,000	−1,500
Salvage value, $	+8,000	+12,000
Life, years	10	10

8.27 A company is considering the projects shown below, all of which can be considered to last indefinitely. If the company's MARR is 18% per year, determine which should be selected (*a*) if they are independent and (*b*) if they are mutually exclusive.

	A	B	C	D	E
First cost, $	−10,000	−20,000	−15,000	−70,000	−50,000
Annual income, $	+2,000	+4,000	+2,900	+10,000	+6,000
Overall rate of return, %	20	20	19.3	14.3	12

8.28 Alternative *I* requires an initial investment of $20,000 and will yield a rate of return of 25% per year. Alternative *C*, which requires a $30,000 investment, will yield 35% per year. Which of the following statements is true about the rate of return on the $10,000 increment?

(*a*) There is no such thing as a rate of return on an increment of investment.

(*b*) It is greater than 35% per year.

(*c*) It is exactly 35% per year.

(*d*) It is between 25% and 35% per year.

(*e*) It is exactly 25% per year.

(*f*) It is less than 25% per year.

8.29 The four alternatives described below are being evaluated.

Alternative	Initial Investment, $	Alternative Overall Rate of Return, %	Incremental Rate of Return in % When Compared with Alternative		
			A	B	C
A	−40,000	29			
B	−75,000	15	1		
C	−100,000	16	7	20	
D	−200,000	14	10	13	12

(*a*) If the proposals are independent, which should be selected if the MARR is 16% per year?

(*b*) If the proposals are mutually exclusive, which one should be selected if the MARR is 9% per year?

(*c*) If the proposals are mutually exclusive, which one should be selected if the MARR is 7% per year?

(*d*) If the proposals are mutually exclusive, which one is best at a MARR of 14%?

8.30 A rate-of-return analysis was begun for the alternatives shown below.

  (a) Fill in the blanks in the rate of return on the incremental cash-flows portion of the table. (All alternatives have infinite lives.)

  (b) Which should be selected if they are mutually exclusive and the MARR is 16%?

  (c) Which should be selected if you wanted to pick the two best alternatives at a MARR of 18%?

Alternative	Initial Investment, $	Alternative Overall Rate of Return %	Rate of Return on Incremental Cash Flows When Compared with Alternative (%)		
			E	F	G
E	20,000	20		27	
F	35,000	23	27		
G	50,000	16			
H	90,000	19			

9

Benefit/Cost Ratio Evaluation

This chapter explains the evaluation and comparison of alternatives on the basis of the benefit/cost (B/C) ratio. Though this method is sometimes regarded as supplementary, since it is commonly used in conjunction with a present-worth or annual-worth analysis, it is an analytical technique we must all understand. Besides its use in business and industry, the B/C method is utilized on many government and public works projects to determine if the expected benefits provide an acceptable return on the estimated investment and costs.

LEARNING OBJECTIVES

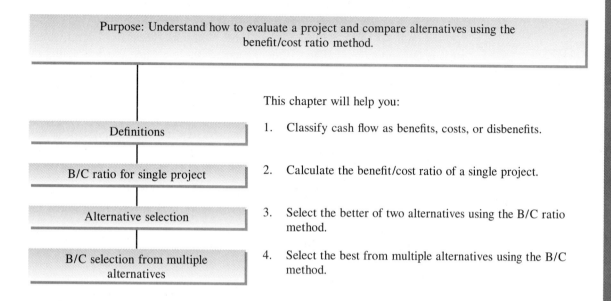

Purpose: Understand how to evaluate a project and compare alternatives using the benefit/cost ratio method.

Definitions

B/C ratio for single project

Alternative selection

B/C selection from multiple alternatives

This chapter will help you:

1. Classify cash flow as benefits, costs, or disbenefits.

2. Calculate the benefit/cost ratio of a single project.

3. Select the better of two alternatives using the B/C ratio method.

4. Select the best from multiple alternatives using the B/C method.

9.1 CLASSIFICATION OF BENEFITS, COSTS, AND DISBENEFITS

The method for selecting alternatives most commonly used by federal, state, provincial, and municipal government agencies for analyzing the desirability of public works projects is the *benefit/cost (B/C) ratio*. As its name suggests, the B/C method of analysis is based on the ratio of the benefits to costs associated with a particular project. A project is considered to be attractive when the benefits derived from its implementation, as reduced by expected disbenefits, exceed its associated costs. Therefore, the first step in a B/C analysis is to determine which of the elements are benefits, disbenefits, and costs. We can use the following descriptions where each must be expressed in monetary terms.

Benefits (B).   Advantages experienced by the owner.

Disbenefits (D).   Disadvantages to the owner when the project under consideration is implemented.

Costs (C).   Anticipated expenditures for construction, operation, maintenance, etc., less any salvage values.

Since B/C analysis is used in economy studies by federal, state, or city agencies, think of the *public as the owner* experiencing the benefits and disbenefits and the *government* as incurring the costs. The determination of whether an item is to be considered a benefit, disbenefit, or cost, therefore, depends on *who is affected* by the consequences. Some examples of each are illustrated in Table 9–1 from the decision makers' viewpoint.

While the examples presented in this chapter are straightforward with regard to identification of benefits, disbenefits, or costs, it should be pointed out that in actual situations, judgments must usually be made which are subject to interpretation, particularly in determining which elements of cash flow should be included in the analysis. For example, improvements in the condition of street pavement might result in fewer traffic accidents, an obvious benefit to the public. However, fewer accidents and injuries mean less work and money for repair

Table 9–1	Examples of benefits, disbenefits, and costs from the decision makers' viewpoint
Item	**Classification**
Expenditure of $11 million for new interstate highway segment	Cost
$50,000 annual income to local residents from tourists because of new water reservoir and recreation area	Benefit
$350,000 per year upkeep cost for irrigation canals	Cost
$25,000 per year loss by farmers because of highway right-of-way purchase	Disbenefit

shops, towing companies, car dealerships, and hospitals—also part of the tax-paying "public." Thus, if the broadest viewpoint is taken, the benefits will almost always be exactly offset by an equal amount of disbenefits. In other instances, it is not easy to place a dollar value on each benefit, disbenefit, or cost involved. In general, however, dollar values are available, or obtainable, but they may take some effort to determine with some precision. The result of a properly completed B/C analysis will agree with that of all the methods studied in preceding chapters (such as PW, AW, and ROR on the incremental investment).

Problems 9.1 and 9.2

9.2 BENEFIT, DISBENEFIT, AND COST CALCULATIONS FOR A SINGLE PROJECT

Before a B/C ratio can be computed, all the identified benefits, disbenefits, and costs must be converted to common dollar units. The unit can be an equivalent present-worth, annual-worth, or future-worth value, but they must all be in the same units. Any method of calculation—PW, AW, or FW—may be used provided the procedures we have learned thus far are followed. Once the numerator (benefits, disbenefits) and denominator (costs) are both expressed in the same units, any of the following versions of the B/C ratio may be applied.

The *conventional B/C ratio,* probably the most widely used, is applied in this text unless specified otherwise. The conventional B/C ratio is calculated as follows:

$$\text{B/C} = \frac{\text{benefits} - \text{disbenefits}}{\text{costs}} = \frac{B - D}{C} \qquad [9.1]$$

A B/C ratio greater than or equal to 1.0 indicates that the project evaluated is economically advantageous. *In B/C analyses, costs are not preceded by a minus sign.*

In Equation [9.1] disbenefits are subtracted from benefits, not added to costs. The B/C value could change considerably if disbenefits are regarded as costs. For example, if the numbers 10, 8, and 8 are used to represent benefits, disbenefits, and costs, respectively, the correct procedure results in B/C = $(10 - 8)/8 = 0.25$, while the incorrect inclusion of disbenefits as costs results in B/C = $10/(8 + 8) = 0.625$, which is over twice the correct B/C value of 0.25. Clearly, then, the method by which disbenefits are handled affects the magnitude of the B/C ratio. However, no matter whether disbenefits are (correctly) subtracted from the numerator or (incorrectly) added to costs in the denominator, a B/C ratio of less than 1.0 by the first method, which is consistent with Equation [9.1], will always yield a B/C ratio less than 1.0 by the latter method, and vice versa.

The *modified B/C ratio,* which is gaining support, includes maintenance and operation (M&O) costs in the numerator and treats them in a manner similar to disbenefits. The denominator, then, includes only the initial investment cost. Once all amounts are expressed in PW, AW, or FW terms, the modified B/C ratio is calculated as

$$\text{Modified B/C} = \frac{\text{benefits} - \text{disbenefits} - \text{M\&O costs}}{\text{initial investment}} \qquad [9.2]$$

Any salvage value is included in the denominator as a negative cost, as previously treated. The modified B/C ratio will obviously yield a different value than the conventional B/C method. However, as with disbenefits, *the modified procedure can change the magnitude of the ratio but not the decision to accept or reject.*

The *benefit and cost difference* measure of worth, which does not involve a ratio, is based on the difference between the PW, AW, or FW of benefits and costs, that is, B − C. If (B − C) ≥ 0, the project is acceptable. This method has the obvious advantage of eliminating the discrepancies noted above when disbenefits are regarded as costs, since B represents *net benefits.* Thus, for the numbers 10, 8, and 8 the same result is obtained regardless of how disbenefits are treated.

Subtracting disbenefits: B − C = (10 − 8) − 8 = −6

Adding disbenefits to costs: B − C = 10 − (8 + 8) = −6

Before calculating the B/C ratio, check if the alternative with the larger AW or PW of costs also yields a larger AW or PW of benefits than less-expensive alternatives, after all the benefits and costs have been expressed in common units. It is possible for one alternative with a larger AW or PW of costs to generate a lower AW or PW of benefits than other alternatives, thus making it unnecessary to consider the more-expensive alternative any further. Example 9.2 on the next page illustrates this point.

Example 9.1

The Wartol Foundation, a nonprofit educational research organization, is contemplating an investment of $1.5 million in grants to develop new ways to teach people the rudiments of a profession. The grants will extend over a 10-year period and will create an estimated savings of $500,000 per year in faculty salaries, student tuition and fees, and other expenses. The foundation uses a rate of return of 6% per year on all grant awards.

Since the new program will be in addition to ongoing activities, an estimated $200,000 per year will be removed from other program funding to support this educational research. To make this program successful, a $50,000-per-year operating expense will be incurred by the foundation from its regular M&O budget. Use the following analysis methods to determine if the program is justified over a 10-year period: (*a*) conventional B/C, (*b*) modified B/C, and (*c*) (B − C) analysis.

Solution

Use annual worth as the common-unit basis.

Benefit. $500,000 per year

Disbenefit. $200,000 per year

O&M cost. $50,000 per year

Investment cost. $1,500,000(A/P,6%,10) = \$203,805$ per year

(*a*)   Use Equation [9.1] for conventional B/C analysis, where O&M is placed in the denominator.

$$B/C = \frac{500,000 - 200,000}{203,805 + 50,000} = 1.18$$

The project is justified, since B/C > 1.0.

(*b*)   By Equation [9.2] the modified B/C treats the O&M cost like a disbenefit.

$$\text{Modified B/C} = \frac{500,000 - 200,000 - 50,000}{203,805} = 1.23$$

The project is also justified by the modified B/C method.

(*c*)   Now, B is the net benefit, and the annual O&M cost is included with C.

$$B - C = (500,000 - 200,000) - (203,805 + 50,000) = \$46,195$$

Since (B − C) > 0, the investment is again justified.

Comment

In part (*a*), if the disbenefits were added to costs, the incorrect B/C value would be

$$B/C = \frac{500,000}{203,805 + 50,000 + 200,000} = 1.10$$

which still justifies the project. However, the disbenefit D = \$200,000 is not a direct cost to this program and should be subtracted from B.

Example 9.2

Alternative routes are being considered by the local highway district for a new bypass. Route *A*, costing \$4,000,000 to build, will provide estimated annual benefits of \$125,000 to local businesses. Route *B*, costing \$6,000,000, can provide \$100,000 in annual benefits. The annual cost of maintenance is \$200,000 for *A* and \$120,000 for *B*. If the life of each road is 20 years and an interest rate of 8% per year is used, which alternative should be selected on the basis of a conventional B/C analysis?

CONSTRUCTION ZONE BEGINS

Solution

The benefits in this example are $125,000 for route A and $100,000 for route B. The AW of costs for each alternative is as follows:

$$AW_A = -4,000,000(A/P,8\%,20) - 200,000 = \$-607,400$$

$$AW_B = -6,000,000(A/P,8\%,20) - 120,000 = \$-731,100$$

Route B has a larger AW of costs than route A by $123,700 per year but offers less benefits by $25,000. Therefore, there is no need to calculate the B/C ratio for route B, since this alternative is obviously inferior to route A. Furthermore, if the decision has been made that either route A or B must be accepted (which would be the case if there are no other alternatives), then no other calculations are necessary; route A is selected.

**Example 9.2
(Spreadsheet)**

Develop a spreadsheet to select the better route in Example 9.2 above using B/C analysis.

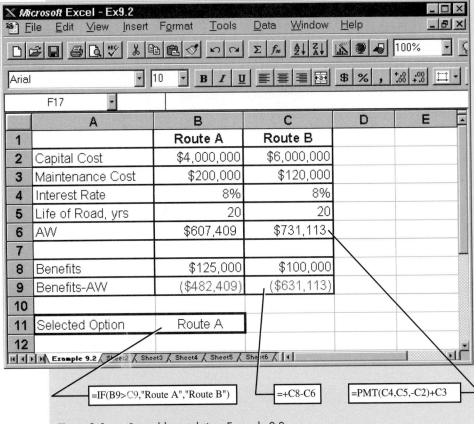

Figure 9-1   Spreadsheet solution, Example 9.2.

Solution

Figure 9–1 is a spreadsheet which shows the costs and benefits for each route and the resulting B − C values in cells (B6) and (C6) using the PMT function. Since B − C is negative for both routes (row 9 cells), neither is economically justified. However, if one must be selected, route *A* is more favorable since its B − C value is closer to zero. In other words, *A* offers a lower annual cost or *loss*.

Problems 9.3 to 9.9

9.3   ALTERNATIVE SELECTION BY BENEFIT/COST ANALYSIS

In computing the B/C ratio by Equation [9.1] for a given alternative, it is important to recognize that the benefits and costs used in the calculation represent the *increments* or *differences* between *two* alternatives. This will always be the case, since sometimes the do-nothing alternative is acceptable. Thus, when it seems as though only one proposal is involved in the calculation, such as whether or not a flood-control dam should be built to reduce flood damage, it should be recognized that the construction proposal is being compared against another alternative—the do-nothing alternative. Although this is also true for the other alternative evaluation techniques previously presented, it is emphasized here because of the difficulty often present in determining the benefits and costs between two alternatives when only costs are involved (as in Example 9.3 below).

Once the B/C ratio for the differences is computed, a B/C ≥ 1.0 means that the extra benefits justify the larger-cost alternative. If B/C < 1.0, the extra investment or cost is not justified and the lower-cost alternative is selected. The lower-cost project may or may not be the do-nothing alternative.

In the previous section it was stated that in a B/C analysis costs are not preceded by a minus sign. The procedure for properly setting up and interpreting B/C ratios to compare alternatives when using a positive-sign-for-costs convention follows. Care must be exercised in setting up the B/C relation to ensure that the interpretation of the result is correct. In this regard, the alternative with the larger total cost (amount used in the denominator) should always be the alternative to be justified; that is, accept the larger-cost alternative only if the incremental benefits justify the added investment cost. This is correct even if the larger-cost alternative is the existing (do-nothing) alternative.

For example, let's assume that the county's annual cost of maintaining an in-place flood-control structure is $10,000 and that the structure provides benefits (flood protection) estimated to be worth $20,000 per year. By leasing a certain machine for $3000 per year, the annual maintenance cost will decrease to $5000 and the flood protection benefits will increase to $24,000 per year. How is the B/C ratio set up? The benefits and costs are summarized below.

	Present Condition (PC)	Improved Condition (IC)
Benefits per year, $	20,000	24,000
Costs per year, $	10,000	8,000

In this case, it is obvious by inspection that the improved condition (IC) should be accepted, since there are increased benefits ($4000) at a lower cost ($2000). Since PC is the do-nothing alternative, an incorrect analysis using IC as the comparative base with PC (do-nothing) shows benefits of $4000 per year and costs of $-$2000 per year. For this incorrect analysis, the B/C ratio is

$$\text{B/C} = \frac{\$4000}{-\$2000} = -2.0$$

Since B/C $<$ 1, the improved condition would be rejected, which is obviously wrong. This error is not made if the alternative with the larger cost is always the one being evaluated for justification. *This means the denominator is always a positive value indicating the incremental cost.* The numerator can be positive, negative, or zero, depending on whether the incremental cost (denominator) yields incremental benefits or not. In this case, the alternative with the larger cost is PC, so it is the base for comparison—the one to accept or reject. There are actually $4000 less benefits which result from the $2000 incremental cost for the present condition. Therefore, the sign on the incremental benefit is negative. The B/C ratio is

$$\text{B/C} = \frac{-\$4000}{\$2000} = -2.0$$

Since the B/C ratio is less than 1, we reject the alternative with the extra cost, which in this case is the present condition. This is the correct and logical choice, as mentioned above. Thus, by considering the alternative with the larger total cost as the one to be justified, the interpretation of the B/C ratio will be consistent.

The steps for the B/C analysis are summarized below:

1. Calculate the total cost of each alternative.
2. Subtract the costs of the lower-investment alternative from those of the larger-cost alternative, which we consider the alternative to be justified. Use this value as C in the B/C ratio.
3. Calculate the total benefits of each alternative.
4. Subtract the benefits for the lower-cost alternative from the benefits for the larger-cost alternative, paying attention to algebraic signs. Use this value as B. Calculate the B/C ratio.
5. If B/C $\geq$ 1, the incremental investment is justified; select the larger-investment alternative. Otherwise, select the lower-cost alternative.

Example 9.3 illustrates this procedure.

Example 9.3

Two routes are under consideration for a new interstate highway segment. The northerly route N would be located about 5 km from the central business district and would require longer travel distances by local commuter traffic. The southerly route S would pass directly through the downtown area, and, although its construction cost would be higher, it would reduce the travel time and distance for local commuters. Assume that the costs for the two routes are as follows:

SEAT BELT

	Route N	Route S
Initial cost, $	10,000,000	15,000,000
Maintenance cost per year, $	35,000	55,000
Road-user cost per year, $	450,000	200,000

If the roads are assumed to last 30 years with no salvage value, which route should be selected on the basis of a B/C analysis using an interest rate of 5% per year?

Solution

Since most of the cash flows are already annualized, the B/C ratio will be expressed in terms of annual worth. The steps of the procedure above are used.

1. The costs in the B/C analysis are the initial construction and maintenance costs.

$$AW_N = 10,000,000(A/P,5\%,30) + 35,000 = \$685,500$$

$$AW_S = 15,000,000(A/P,5\%,30) + 55,000 = \$1,030,750$$

2. Route S has the larger AW of costs, so it is the alternative to be justified. The incremental cost value is

$$C = AW_S - AW_N = \$345,250 \text{ per year}$$

3. The benefits are derived from the road-user costs, since these are consequences to the public. The benefits for the B/C analysis are not the road-user costs themselves but the *difference* if alternative S is selected.

4. If route S is selected, the incremental benefit is the lower road-user cost each year. This is a positive benefit for route S, since it provides the larger benefits in terms of these reduced road-user costs.

$$B = \$450,000 - \$200,000 = \$250,000 \text{ per year for route } A$$

The B/C ratio is calculated by Equation [9.1].

$$B/C = \frac{\$250,000}{\$345,250} = 0.724$$

5. The B/C ratio is less than 1.0, indicating that the extra benefits associated with route S are not justified. Therefore, route N is selected for construction.

Note that there is no do-nothing alternative in this case, since one of the roads must be constructed.

Comment

If there are disbenefits associated with each route, the difference between the disbenefits is added to or subtracted from the $250,000 incremental benefits for route *S*, depending on whether the disbenefits for route *S* are less than or greater than the disbenefits for route *N*. That is, if the disbenefits for route *S* are less, the difference between the two is added to the $250,000 benefit for route *S*, since the disbenefits involved favor route *S*. Additional Example 9.5 illustrates this point.

   You should rework this example using the modified B/C method per Equation [9.2]. The annual maintenance costs are removed from the AW expressions of costs in the conventional B/C ratio and subtracted from the benefits.

Additional Example 9.5
Problems 9.10 to 9.16

9.4 SELECTION FROM ALTERNATIVES USING INCREMENTAL B/C ANALYSIS

In Section 8.1 we introduced the concepts of *mutually exclusive* alternatives and *independent* alternatives or proposals. Recall that for mutually exclusive alternatives, *only one* can be selected from among several and that it is necessary to compare the alternatives against each other as well as against the do-nothing alternative, when appropriate. For independent proposals, *more than* one can be chosen, and it is necessary only to compare the alternatives against the do-nothing alternative. *Simply compute the B/C value for each proposal and select all that have B/C ≥ 0.*

   As an example of independent proposals, consider that several flood-control dams could be constructed on a particular river and that adequate funding is available for all dams. The B/C ratios to be considered are those for each particular dam versus no dam. That is, the result could show that several dams along the river may be economically justified on the basis of reduced flood damage, recreation, etc. Independent proposals are treated further in Chapter 17, where a limitation to the total initial investment is imposed.

   However, if the dams are mutually exclusive alternatives, only one is selected for construction, and the B/C analysis must compare dams against one another. In order to use the conventional B/C ratio as an evaluation technique for mutually exclusive alternatives, an incremental B/C ratio must be computed in a fashion similar to that used for the ROR on the incremental investment (Chapter 8). The selected alternative must have an *incremental B/C ≥ 1.0 and require the largest justified initial investment.* However, in a B/C analysis it is generally convenient, although not necessary, to first compute an overall B/C ratio for each alternative, using the total PW or AW values determined in preparation for the incremental analysis. *Any alternatives that have an overall B/C < 1.0 can be eliminated immediately and need not be considered in the incremental analysis.* Example 9.4 presents a complete application of the incremental B/C ratio for mutually exclusive alternatives.

Example 9.4

Consider the four mutually exclusive alternatives described in Example 8.7 (Table 8–5). Apply incremental B/C analysis to select the best alternative for a MARR = 10% per year. Use a PW analysis.

Solution

The alternatives are first ordered from smallest to largest initial investment cost (*C, A, B, D*) and the PW values of annual cash flows are determined (Table 9–2). (Positive signs are used for all cost estimates.) Next calculate the overall B/C ratio and eliminate any alternative that has B/C < 1.0. Location *C* can be eliminated, since its overall B/C ratio is only 0.97. All other locations are initially acceptable and are further compared on an incremental basis. The incremental benefits and costs are determined on the basis of present worth.

Incremental benefit.   Increment in PW of cash flows between alternatives.

Incremental cost.   Increment in building cost between alternatives.

Table 9–2   Incremental B/C analysis for mutually exclusive alternatives, Example 9.4

Alternative	C	A	B	D
Building cost, $	190,000	200,000	275,000	350,000
Cash flow, $	19,500	22,000	35,000	42,000
PW of cash flow, $	183,826	207,394	329,945	395,934
Overall B/C ratio	0.97	1.03	1.20	1.13
Projects compared	—	—	B to A	D to B
Incremental benefit, $	—	—	122,551	65,989
Incremental cost, $	—	—	75,000	75,000
Incremental B/C ratio	—	—	1.63	0.88
Project selected	—	—	B	B

A summary of the incremental B/C analysis is included as Table 9–2, lower half. Using the initially acceptable alternative with the lowest investment cost as the defender (A) and the next-lowest acceptable alternative as the challenger (B), the incremental B/C ratio is 1.63 ($122,551/$75,000). This indicates that location B is the new defender, and location A is eliminated. Using D as the new challenger, the incremental analysis of D-to-B yields an incremental B/C = 0.88. Location D is removed since 0.88 < 1.0.

Since location B has an overall B/C > 1.0, an incremental B/C > 1.0, and is the largest justified investment, it is selected. This, of course, is the same conclusion of the ROR method in Table 8–6.

Comment

Note that alternative selection is not made on the overall B/C ratio, even though location B would still be selected, but only coincidentally in this case. The incremental investment must also be justified in order to select the best alternative.

Although a PW-based B/C ratio was utilized in this example, an AW- or FW-based ratio is equally correct; in fact, an AW-based ratio is generally easier for different alternative lives since the LCM need not be used, as it is for a PW-based ratio.

Additional Example 9.6
Problems 9.17 to 9.25

ADDITIONAL EXAMPLES

Example 9.5

SEAT BELT

B/C FOR TWO ALTERNATIVES, SECTION 9.3   Assume the same situation as in Example 9.3 for the routing of a new interstate highway segment, where the B/C analysis favored the northerly route N. However, this route will go through an agricultural region and the local farmers have complained about the great loss in revenue they and the economy will suffer. Likewise, the downtown merchants have complained about the southerly route because of the loss in revenue due to reduced merchandising ability, parking problems, etc. To include these concerns in the analysis, the state highway department has completed a study which predicts that the loss to state agriculture for route N will be about $500,000 per year and that route S will cause an estimated reduction in retail sales and rents of $400,000 per year. What effect does this new information have on the B/C analysis?

Solution

These loss estimates are considered disbenefits. Since the disbenefits of route S are $100,000 less than those of route N, this positive difference is added to the $250,000 benefits of route S to give it a total net benefit of $350,000. Now,

$$B/C = \frac{\$350,000}{\$345,250} = 1.01$$

Route S is slightly favored. In this case the inclusion of disbenefits has reversed the previous decision.

Example 9.6

B/C ANALYSIS OF MULTIPLE ALTERNATIVES, SECTION 9.4   The Corps of Engineers still wants to construct a dam on the Sacochsi River as proposed in Example 8.10. The construction and average annual dollar benefits (income) are repeated below. If a MARR of 6% per year is required and dam life is infinite for analysis purposes, select the one best location using the B/C method.

Site	Construction Cost, $ (millions)	Annual Income, $
A	6	350,000
B	8	420,000
C	3	125,000
D	10	400,000
E	5	350,000
F	11	700,000

Solution

We make use of the capitalized-cost equation, Equation [5.2], $A = Pi$, to obtain AW values for annual capital recovery of the construction cost, as shown in the first row of Table 9–3. The analysis between the mutually exclusive alternatives in the lower

Table 9–3   Use of incremental B/C ratio analysis for Example 9.6

	C	E	A	B	D	F
Capital recovery, $ (thousands)	180	300	360	480	600	660
Annual benefits, $ (thousands)	125	350	350	420	400	700
Comparison	C to do-no	E to do-no	A to E	B to E	D to E	F to E
Δ capital recovery, $ (thousands)	180	300	60	180	300	360
Δ annual benefits, $ (thousands)	125	350	0	70	50	350
Δ B/C ratio	0.70	1.17	0	0.39	0.17	0.97
Site selected	Do nothing	E	E	E	E	E

portion of Table 9–3 is based on the relation:

$$\text{Incremental B/C} = \Delta \text{ B/C} = \frac{\Delta \text{ annual benefits}}{\Delta \text{ capital recovery cost}}$$

Since only site E is justified and has the largest (incremental) investment, it is selected. None of the increments above E are justified.

Comment

Suppose that site G is added with a construction cost of $10 million and an annual benefit of $700,000. What site should G be compared with? What is the Δ B/C ratio? If you determine that a comparison of G to E is in order and that Δ B/C = 1.17 in favor of G, you are correct. Now site F must be incrementally evaluated with G, but since the annual benefits are the same ($700,000), the Δ B/C ratio is zero and the added investment is not justified. Therefore, site G is chosen.

CHAPTER SUMMARY

The benefit/cost ratio method is often used to evaluate projects for governmental entities, especially at the federal level. The evaluation criterion is the size of the number obtained when the present, future, or annual worths of the estimated benefits are divided by the similarly expressed costs. A number equal to or greater than 1 indicates that the alternative is acceptable.

For multiple alternatives, an incremental analysis is necessary when the alternatives are mutually exclusive, just as is necessary for the rate-of-return method. For independent proposals, the only B/C comparisons required are between the individual alternatives and the do-nothing alternative.

CASE STUDY #3, CHAPTER 9: FREEWAY LIGHTING

Introduction

A number of studies have shown that a disproportionate number of freeway traffic accidents occur at night. There are a number of possible explanations for this, one of which might be poor visibility. In an effort to determine whether freeway lighting was economically beneficial for reducing nighttime accidents, data were collected regarding accident frequency rates on lighted and unlighted sections of certain freeways. This case study is an analysis of part of that data.

Background

The Federal Highway Administration (FHWA) places value on accidents depending on the severity of the crash. There are a number of crash categories, the most severe of which is fatal. The cost of a fatal accident is placed at $2.8 million. The most common type of accidents is not fatal or injurious and involve only property damage. The cost of this type of accident is placed at $4500. The ideal way to determine whether or not lights reduce traffic accidents is through before and after studies on a given section of freeway. However, this type of information is not readily available, so other methods must be used. One such method compares night to day accident rates for lighted and unlighted freeways. If lights are beneficial, the ratio of night to day accidents will be lower on the lighted section than on the unlighted one. If there is a difference, the reduced accident rate can be translated into benefits which can be compared to the cost of lighting to determine its economic feasibility. This technique is used in the following analysis.

Economic Analysis

The results of one particular study conducted over a 5-year period are shown below. For illustrative purposes, only the property damage category will be considered in this example.

	Freeway accident rates, lighted and unlighted				
	Unlighted		**Lighted**	
Accident Type	**Day**	**Night**	**Day**	**Night**
Fatal	3	5	4	7
Incapaciting	10	6	28	22
Evident	58	20	207	118
Possible	90	35	384	161
Property damage	379	199	2069	839
Totals	540	265	2697	1147

SOURCE: Michael Griffin, "Comparison of the Safety of Lighting Options on Urban Freeways, " *Public Roads,* 58 (Autumn 1994), pp. 8–15.

The ratios of night to day accidents involving property damage for the unlighted and lighted freeway sections are $199/379 = 0.525$ and $839/2069 = 0.406$, respectively. These results indicate that the lighting was beneficial. To quantify the benefit, the accident-rate ratio from the unlighted section will be applied to the lighted section. This will yield the number of accidents that was prevented. Thus, there would have been $(2069)(0.525) = 1086$ accidents instead of the 839 if there had not been lights on the freeway. This is a difference of 247 accidents. At a cost of $4500 per accident, this results in a net benefit of

$$B = (247)(\$4500)$$
$$= \$1,111,500$$

To determine the cost of the lighting, it will be assumed that the light poles are center poles 67 meters apart, the bulb size is 400 watts, and the installation cost is $3500 per pole. Since these data were collected over 87.8 kilometers (54.5 miles) of lighted freeway, the installed cost of the lighting is

$$\text{Installation cost} = \$3500 \; \frac{87.8}{0.067}$$

$$= 3500(1310.4)$$

$$= \$4,586,400$$

The annual power cost based on 1310 poles is

$$\text{Annual power cost} = 1310 \text{ poles } (2 \text{ bulbs/pole})(0.4 \text{ kilowatts/bulb}) \times$$

$$(12 \text{ hours/day})(365 \text{ days/year}) \times$$

$$(\$0.08/\text{kilowatt-hour})$$

$$= \$367,219 \text{ per year}$$

These data were collected over a 5-year period. Therefore, the annualized cost (C) at $i = 6\%$ per year is

$$\text{Total annual cost} = \$4,586,400(A/P,6\%,5) + 367,219$$

$$= \$1,456,030$$

The B/C ratio is

$$B/C = \frac{\$1,111,500}{\$1,456,030} = 0.76$$

Since B/C < 1, the lighting is not justified on the basis of property damage alone. In order to make a final determination about the economic viability of the lighting, the benefits associated with the other accident categories would obviously also have to be considered.

Questions to Consider

1. What would the B/C ratio be if the light poles were twice as far apart as assumed above?
2. What is the ratio of night to day accidents for fatalities?
3. What would the B/C ratio be if the installation cost were only $2500 per pole?
4. How many accidents would be prevented on the unlighted portion of freeway if it were lighted? Consider the property damage category only.

5. Using only the category of property damage, what would the lighted night to day accident ratio have to be for the lighting to be economically justified?

PROBLEMS

9.1 What is the difference between disbenefits and costs?

9.2 Classify the following as either benefits, disbenefits, or costs:

(a) Less tire wear on automobiles and trucks because of smoother road surface.

(b) Loss of income by local businesses because of rerouting of traffic to the interstate highway.

(c) Adverse environmental impact due to a poorly controlled logging operation.

(d) Income to local lodges from an extended national park season.

(e) Cost of fish from government-operated hatchery to stock a trout stream.

(f) More recreational use of the lake because of better access roads.

(g) Decrease in property values because of closure of a national research lab.

(h) Expenditures associated with construction of a flood-control dam.

9.3 What is the difference between a conventional and a modified B/C ratio?

9.4 Where is the salvage value placed (numerator or denominator) in the conventional B/C ratio method? Why?

9.5 The first cost of grading and spreading gravel on a short rural road is expected to be $700,000. The road will have to be maintained at a cost of $25,000 per year. Even though the new road is not very smooth, it allows access to an area that previously could only be reached with off-road vehicles. This improved accessibility has led to a 200% increase in the property values along the road. If the previous market value of the property was $400,000, calculate the (a) conventional B/C ratio and (b) modified B/C ratio for the road, using an interest rate of 8% per year and a 20-year study period.

9.6 A small flood-control dam is expected to have an initial cost of $2.8 million and an annual upkeep cost of $20,000. In addition, minor reconstruction will be required every 5 years at a cost of $190,000. As a result of the dam, flood damage will be reduced by an average of $120,000 per year. Using an interest rate of 7% per year, determine the (a) conventional B/C ratio, (b) modified B/C ratio, and (c) B − C value. Assume the dam will be permanent.

9.7 The U.S. Bureau of Reclamation plans to install a pipe and to cover an agricultural drain in a colonia to reduce disease transmission and injury to local children. The project is expected to cost $1 million to build, $10,000 per year to maintain, and will have a useful life of 40 years. If the interest rate is 8% per year, what would the annual benefits have to be in order to justify the project?

9.8 A small water utility is trying to decide between installing a laboratory for conducting the required water analyses or sending the samples to a private lab. In order to equip the lab, an initial expenditure of $300,000 will be required. In addition, a full-time technician will have to be hired at a cost of $4000 per month. A total of 400 analytical tests are required each month. If the analyses are done in-house, the cost per sample will average $3, but if the samples are sent to an outside lab, the average cost will be $25. The equipment purchased for the lab is expected to have a useful life of 5 years. If the utility uses an interest rate of 0.5% per month, determine the B/C ratio for the project.

9.9 From the following data, determine the B/C ratio at $i = 7\%$ per year for a project which has a 20-year life.

Consequences to the People	Consequences to the Government
Benefits in year 0: $50,000	First cost: $400,000
Benefits in years 1–20: $10,000 in year 1 increasing by $1000 per year	Annual cost: $8000 per year
	Annual savings: $3000 per year
Disbenefits: $8000 per year	

9.10 Two methods are being evaluated for constructing a second-story floor onto an existing building. Method *A* will use lightweight expanded shale concrete on a metal deck with open web joists and steel beams. For this method, the costs will be $5300 for concrete, $2600 for metal decking, $2000 for joists, and $1200 for beams. Method *B* will be a reinforced concrete slab which will cost $2100 for concrete, $700 for rebars, $1000 for equipment rental, and $500 for extra labor costs. Special additives will be included in the lightweight concrete that will improve the heat transfer properties of the floor. If the energy costs for method *A* will be $100 per year lower than for method *B*, which one is more attractive at an interest rate of 9% per year over a 25-year study period? Use the B/C method.

9.11 The U.S. Bureau of Reclamation is evaluating two sites for injection of fresh water. At the east site, recharge basins could be used. They will cost $9 million to construct and $300,000 per year to operate and maintain. At this site, 380,000 cubic meters per year could be injected. The north site will cost $900,000 to develop injection wells. The annual M&O cost will be $6000, but at this site, only 50,000 cubic meters per year could be injected. If the value of the injected water is $0.40 per cubic meter, which alternative, if either, should be selected according to the B/C ratio method? Use a 20-year study period and an interest rate of 8% per year.

9.12 Select the better of the two alternatives shown below using an interest rate of 10% per year and the B/C ratio method. Assume one of the alternatives must be selected.

	Alternative X	Alternative Y
First cost, $	320,000	540,000
Annual M&O cost, $	45,000	35,000
Annual benefits, $	110,000	150,000
Annual disbenefits, $	20,000	45,000
Life, years	10	20

9.13 The water conservation department of a water utility is considering two alternatives for decreasing residential water consumption. Alternative 1 involves providing free water conservation kits to anyone who requests one. The kits will cost the utility $3 each and will likely reduce the water usage of a household that requests one by 2%. Administrative costs for this program are expected to be $10,000 per year. Alternative 2 involves on-site inspections of houses, again by request. This alternative will require the utility to hire two inspectors at a cost of $90,000 per year (for wages, fringe benefits, transportation, etc.). Through alternative 2, water usage will likely decrease by 5%. If the utilities cost of developing new water supplies is $0.20 per thousand liters and the average household consumes 700 liters per day, which alternative, if either, should be implemented at an interest rate of 6% per year? Assume 4000 households per year will request the kits and 800 will request inspections. Use the B/C ratio method.

9.14 Two routes are under consideration for a new interstate highway segment. The long route would be 25 kilometers in length and would have an initial cost of $21 million. The short transmountain route would be 10 kilometers long and would have an initial cost of $45 million. Maintenance costs are estimated at $40,000 per year for the long route and $15,000 per year for the short route. Regardless of which route is selected, the volume of traffic is expected to be 400,000 vehicles per year. If the vehicle operating expense is assumed to be $0.27 per kilometer, determine which route should be selected by (*a*) conventional B/C analysis and (*b*) modified B/C analysis. Assume an infinite life for each road, an interest rate of 6% per year, and that one of the roads will be built.

9.15 The Corps of Engineers is considering three sites for flood-control dams, designated as sites *A*, *B*, and *C*. The construction costs are $10 million, $12 million, and $20 million, respectively, and maintenance costs are expected to be $15,000, $20,000, and $23,000, respectively. In addition, expenditures of $20,000, $50,000, and $80,000, respectively, will be required every 10 years at each site. The present cost of flood damage is $2 million per year. If only the dam at site A is constructed, the flood damage will be reduced to $1.5 million per year. For dams B and C, the respective reductions in flood damage would be to $1.2 million and $0.77 million per year. Since the dams would be built on different branches of a larger river, either one or all of the dams could be constructed and the decrease in flood damages would be additive. If the interest rate is 5% per year, determine which ones, if any, should be built on the basis of their B/C ratios. Assume the dams will be permanent.

9.16 A state highway department is considering two types of surface coatings for a new road. An armor-coat surface will cost only $800,000 to install, but because of its relatively rough surface, the road users will have to spend more money for gasoline, tire wear, and automobile upkeep. The annual cost for these items is estimated to be $196,000. Additionally, disbenefits of $40,000 per year have been identified for this alternative. A smooth asphalt coating is an alternative also under consideration. This surface would have an initial cost of $2 million but the annual road-user cost will be only $75,000. The asphalt surface will have no disbenefits associated with it. If the life of either surface is expected to be 5 years, determine which should be selected on the basis of a B/C analysis, using an interest rate of 9% per year compounded annually.

9.17 Five methods could be used to recover grease from a rendering plant wastewater stream. The investment costs and incomes associated with each one are shown below. Assuming that all methods have a 10-year life with zero salvage value, determine which one should be selected using a minimum attractive rate of return of 15% per year and the modified B/C analysis method. Consider operating costs as an M&O cost in the modified B/C method.

	Method				
	1	2	3	4	5
First cost, $	15,000	19,000	45,000	33,000	48,000
Annual operating cost, $	10,000	12,000	9,000	11,000	13,000
Annual income, $	12,000	20,000	19,000	22,000	27,000

9.18 Which projects are selected in Problem 9.17 if they are not mutually exclusive?

9.19 Select the best mutually exclusive alternative using the B/C ratio method from the proposals shown below if the MARR is 10% per year and the projects will have a useful life of 15 years. Assume that the cost of the land will be recovered when the project is terminated. Treat maintenance costs as disbenefits.

	Proposal						
	1	2	3	4	5	6	7
Land cost, $	50,000	60,000	70,000	80,000	90,000	64,000	76,000
Construction cost, $	200,000	150,000	170,000	185,000	165,000	175,000	190,000
Annual maintenance, $	15,000	6,000	14,000	17,000	18,000	13,000	12,000
Annual income, $	52,000	49,000	68,000	50,000	81,000	70,000	5,000

9.20 Which projects are selected in Problem 9.19 if they are not mutually exclusive?

9.21 An oil and gas company is considering five sizes of pipe for a new pipeline. The costs for each size are shown below. Assuming that all pipes will last 25 years and the company's MARR is 8% per year, which size of pipe should be used according to the conventional B/C method?

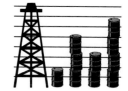

	Pipe Size, Millimeters				
	140	160	200	240	300
Initial investment, $	9,180	10,510	13,180	15,850	30,530
Installation cost, $	600	800	1,400	1,500	2,000
Annual operating cost, $	5,000	4,900	4,700	3,900	3,500

9.22 The two highway projects shown below are to be compared using the modified B/C method. Which one, if either, should be built? Use $i = 10\%$ per year.

	Alternative 2L	Alternative 3L
First cost, $	5,000,000	7,000,000
Annual maintenance cost, $	70,000	60,000
Annual benefits, $	175,000	450,000
Annual disbenefits, $	30,000	35,000
Life, years	∞	∞

9.23 A federal prison is expanding its existing facility to include two new buildings. Three alternatives are proposed to provide for wastewater treatment. Alternative 1 consists of setting up a package plant and discharging the treated wastewater to groundwater through recharge basins. The cost of this option is $160,000 for the plant, $25,000 for concrete work at the site, $20,000 for installation, and $50,000 for a building, piping, electrical work, etc. The annual M&O cost is expected to be $30,000 for a part-time operator, electricity, and parts replacement. Alternative 2 consists of constructing facultative stabilization ponds 200 meters away from the facility. The effluent will be discharged through six percolation beds. The cost of the ponds will be $128,000 for excavation, $304,000 for a plastic liner, and $16,000 for piping, manholes, etc. The annual M&O cost for this alternative is expected to be $10,000. Alternative 3 consists of constructing a force main to transport the wastewater to an existing gravity flow sewer line which is part of the city sewer system. This alternative will cost $190,000 for the pipe, lift station, backfilling, etc. The M&O cost for alternative 3 will be $12,000 per year. If the prison officials have identified annual benefits of $45,000, $22,000, and $14,000 for alternatives 1, 2, and 3, respectively, which one should be selected using the modified B/C method at an interest rate of 8% per year and a 30-year study period? Assume one alternative *must* be selected?

9.24 If the do-nothing alternative is a possibility in Problem 9.23, which alternative should be selected?

9.25 The government is considering the lining of the main canals of its irrigation ditches. The initial cost of lining is expected to be $3 million, with $25,000 per year required for maintenance. If the canals are not lined, a weed-control and dredging operation will have to be instituted, which will have an initial cost of $700,000 and a cost of $50,000 the first year, $52,500 the second, and amounts increasing by $2500 per year for 25 years. If the canals are lined, less water will be lost through infiltration so that additional land can be cultivated for agricultural use. The agricultural revenue associated with the extra land is expected to be $42,000 per year. Assume that the project life is 25 years and the interest rate is 6% per year. Use the (*a*) conventional B/C ratio and (*b*) modified B/C ratio to determine if the canals should be lined.

THREE

*I*n this level you will learn how to perform an analysis for the replacement or retention of currently owned assets and the procedure to determine the number of years to retain an asset so that it has a minimum equivalent annual cost.

The understanding of bonds and the calculation of their expected rate of return is presented. Also, inflationary effects on present-worth and future-worth computations are treated in association with basic cost indexing and cost-estimation techniques for components and entire systems. The fundamental principles and techniques for allocating indirect costs to business units, product lines, and departments complete this level.

10

Replacement Analysis

Keep

or Replace?

The result of an alternative evaluation process is the selection and implementation of a project, asset, or service which has some estimated and planned functional or economic life. In time, companies often find it necessary to determine how the in-place asset may be replaced, upgraded, or augmented. This analysis may be necessary before, at, or after the estimated life. The results of the analysis, which is commonly referred to as *replacement analysis* or a *replacement study,* provide answers to questions such as

- Since the asset has become technologically obsolete, what is the most economic choice—upgrade or complete replacement?
- Which of the identified alternatives should be accepted as a replacement?
- What is the most economic life estimate for an asset?
- Should I keep the asset for one more year before replacement? 2 more years? 3 more years?

The logic and computations of replacement analysis are discussed in this chapter. There may be significant income-tax consequences to replacement, especially a premature one. After-tax replacement analysis is presented in Chapter 15.

LEARNING OBJECTIVES

Purpose: Perform a replacement study of an in-use asset or project (the defender) and one or more alternatives (challengers).

This chapter will help you:

Reasons	1. Understand the basic reasons why a replacement study is performed.
Basic concept	2. Explain the basic concepts of and data used for a replacement study.
Study period	3. Select the better of defender or challenger alternatives using a specified study period.
Two approaches	4. Illustrate the difference between the conventional and cash-flow approaches to replacement analysis.
Economic service life	5. Determine the economic service life of an asset which minimizes the AW measure of worth.
One-additional year	6. Perform a replacement study using the one-additional-year analysis procedure.

10.1   WHY REPLACEMENT STUDIES ARE PERFORMED

The basic replacement study is designed to determine if a currently used asset should be replaced. The term replacement study is also used to identify a variety of economic analyses which compare a currently owned asset with augmentation by new, more advanced features; with custom upgrading of in-place equipment; or with retrofitting of existing, undersized, or oversized equipment.

Whether unplanned or anticipated, replacement is considered for one or more of several reasons. Some are

Reduced performance.   Because of the physical deterioration of parts, the ability to perform at an expected level of *reliability* (being available and performing correctly when needed) or *productivity* (performing at a given level of quality and quantity) is not present. This usually results in increased costs of operation, higher scrap and rework costs, lost sales, and larger maintenance expenses.

Altered requirements.   New requirements of accuracy, speed, or other specifications have been established. These requirements cannot be met by the existing equipment or system. Often the choice is between complete replacement or enhancement through retrofitting or augmentation.

Obsolescence.   International competition and the rapidly changing technology of automation, computers, and communications make currently used systems and assets perform acceptably but less productively than equipment coming onto the market. Replacement due to obsolescence is always a challenge, but management may want to undertake a formal analysis to determine if newly offered equipment may force the company out of current markets or may open new market areas. The ever-decreasing development cycle time to bring new products to market is often the reason for premature replacement studies, that is, studies performed before the estimated functional or economic life is reached.

10.2   BASIC CONCEPTS OF REPLACEMENT ANALYSIS

In most engineering economy studies two or more alternatives are compared. In a replacement study, one of the assets, referred to as the *defender,* is currently owned (or in place), and the alternatives are one or more *challengers.* For the analysis we take the *perspective (viewpoint) of the consultant or outsider;* that is, we assume that we currently own or use neither asset and we must select between the challenger alternative(s) and the in-place defender alternative. In order to acquire the defender, therefore, we must "invest" the going market value in this used asset. This estimated market or trade-in value becomes the first cost of the defender alternative. There will be new estimates for the remaining economic life, annual operating cost (AOC), and salvage value for the defender. All these values will likely differ from the original estimates. Because of the consultant's perspective, however, all estimates made and used previously should be neglected in the replacement analysis. Example 10.1 identifies the correct use of information for a replacement analysis.

Example 10.1

Paradise Isle, a tropical island hotel, purchased a state-of-the-art ice-making machine 3 years ago for $12,000 with an estimated life of 10 years, a salvage value of 20% of the purchase price, and an AOC of $3000 per year. Depreciation has reduced the first cost to a current book value of $8000.

A new model, selling for $11,000, has just been announced. The hotel manager estimates its life at 10 years, salvage value at $2000, and an AOC of $1800 per year. A trade-in amount of $7500 is offered by the salesperson for the 3-years old defender. Based on experiences with the current machine, revised estimates are: remaining life, 3 years; salvage value, $2000; and the same AOC of $3000.

If a replacement study is performed, what values are correct for P, n, SV, and AOC for each ice machine?

Solution

Taking the consultant's perspective, use only the most recent estimates.

Defender	Challenger
$P = \$7,500$	$P = \$11,000$
AOC $= \$3,000$	AOC $= \$1,800$
SV $= \$2,000$	SV $= \$2,000$
$n = 3$ years	$n = 10$ years

The original defender cost of $12,000, estimated salvage value of $2400, remaining 7 years of life, and book value of $8000 are not relevant to the replacement analysis of the defender versus the challenger.

Since the past is common to alternatives, past costs are considered irrelevant in a replacement analysis. This includes a *sunk cost*, which is an amount of money invested earlier that cannot be recovered now or in the future. This may occur due to changed economic, technological, or other conditions or ill-advised business decisions. You or I may personally experience a sunk cost when we purchase an item, say, some software, and discover very soon thereafter that it does not perform as expected and we can't get a refund. The purchase price is the amount of the sunk cost. In industry, a sunk cost also occurs when an asset is considered for replacement and the actual market or trade-in value is less than that predicted by the depreciation model used to write off the original capital investment or is less than the estimated salvage value. (A complete discussion of depreciation models is included in Chapter 13.) The sunk cost of an asset is computed as

$$\text{Sunk cost} = \text{present book value} - \text{present market value} \qquad \text{[10.1]}$$

If the result of Equation [10.1] is a negative number, there is no sunk cost involved. The present book value is the investment remaining after the total amount of depreciation has been charged; that is, the book value is the current book worth of the asset. For example, an asset purchased for $100,000 five years ago now has a depreciated book value of $50,000. A replacement study is being conducted and only $20,000 is offered as the trade-in amount toward the challenger. By Equation [10.1], a sunk cost of $50,000 - 20,000 = $30,000 is present.

In a replacement analysis the *sunk cost should not be included in the economic analysis.* The sunk cost actually represents a capital loss (discussed in Section 14.3) and is correctly reflected if included in the company's income statement and income tax computations for the year in which the sunk cost is incurred. However, some analysts try to "recover" the sunk cost of the defender by adding it to the first cost of the challenger. This is incorrect; it penalizes the challenger, making its first cost appear higher and thereby biasing the decision.

Often, incorrect estimates have been made about the utility, worth, or market value of an asset. This is quite possible, since estimates are made at one point in time about an uncertain future. The result may be a sunk cost when replacement is considered. Incorrect estimates and economic decisions of the past should not be allowed to incorrectly influence current economic studies and decisions.

In Example 10.1, a sunk cost is incurred for the defending ice machine if it is replaced. With a book value of $8000 and a trade-in offer of $7500, Equation [10.1] yields

$$\text{Sunk cost} = \$8000 - 7500 = \$500$$

The $500 should never be added to the first cost of the challenger. This would (1) penalize the challenger since the capital investment amount to be recovered each year would be larger due to the artificially increased first cost and (2) attempt to remove past, but likely nonavoidable, errors in estimation.

Problems 10.1 to 10.5

10.3 REPLACEMENT ANALYSIS USING A SPECIFIED STUDY PERIOD

The *study period* or *planning horizon* is the number of years selected for the economic analysis to compare the defender and challenger alternatives. When selecting the study period, one of two situations is typical: The anticipated remaining life of the defender equals or is shorter than the life of the challenger.

If the defender and challenger have equal lives, use any of the evaluation methods with the most recent data. Example 10.2 compares a defender and challenger with equal lives.

Example 10.2

Moore Transfer currently owns several moving vans which are deteriorating faster than expected. The vans were all purchased 2 years ago for $60,000 each. The company currently plans to keep the vans for 10 more years. Fair market value for a 2-year-old van is $42,000 and for a 12-year-old van is $8000. Annual fuel, maintenance, tax, etc., costs, that is, AOC, are $12,000 per year.

The replacement option is to lease on a yearly basis. The annual lease cost is $9000 (year-end payment) with annual operating costs of $14,000. Should the company lease its vans if the MARR is 12%?

Solution

Consider a 10-year planning horizon for a currently owned van and a leased van and perform an AW analysis to make the selection.

Defender	Challenger
P = $42,000	Lease cost = $9,000 per year
AOC = $12,000	AOC = $14,000
SV = $8,000	No salvage
n = 10 years	n = 10 years

The defender D has a $42,000 fair market value, so this represents its initial investment. The AW_D calculation is

$$AW_D = -P(A/P,i,n) + SV(A/F,i,n) - AOC$$

$$= -42,000(A/P,12\%,10) + 8000(A/F,12\%,10) - 12,000$$

$$= -42,000(0.17698) + 8000(0.05698) - 12,000$$

$$= -\$18,977$$

The AW_C relation for the challenger C is

$$AW_C = -\$9000 - 14{,}000 = -\$23{,}000$$

Clearly, the firm should retain ownership of the vans since AW_D is numerically larger than AW_C.

Comment

Recalculate the AW_C for the challenger if the lease payments are paid at the beginning of each year and the AOC occurs at the end of each year. You should get $AW_C = -\$24{,}080$, which is a larger AW cost. Of course, the decision to own the vans remains the same.

When a defender may be replaced with a challenger having an estimated life different from that of the defender's remaining life, the length of the study period must be determined. It is common practice to use a study period equal to the life of the longer-lived asset. Then the AW value for the shorter-lived asset will apply for it throughout the entire study period. This implies that the service performed by the shorter-lived asset can be acquired with the same AW value after its expected life. For example, if we compare a 10-year-life challenger with a 4-year-life defender, we assume for the replacement analysis that the service provided by the defender will be available for the same AW value for an additional 6 years. If this does not seem a reasonable assumption, an updated estimate of acquiring the equivalent service beyond year 4 should be included separately in the defender's cash flow and distributed over the 10-year study period.

Example 10.3

A municipality has owned and used a recyclable material sorter for 3 years. Based on recent computations, the asset has an AW value of $5200 per year for an estimated remaining life of 5 years. An upgraded replacement has a first cost of $25,000, estimated salvage value of $3800, projected life of 12 years, and an annual operating cost of $720. The city uses a MARR of 10% per year. If it plans to retain the new sorter for its full estimated life, should the city replace the old sorter?

Solution

Select a study period of 12 years to correspond with the challenger's life. Assume that the defender AW of $5200 is a good estimate of the equivalent annual cost to obtain the same level of recyclable material sorting after the remaining defender life of 5 years.

$$AW_D = -\$5200$$

$$AW_C = -25{,}000(A/P,10\%,12) + 3800(A/F,10\%,12) - 720$$

$$= -\$4211$$

Purchase of the new sorter is less costly by about $1000 per year.

Comment

If PW analysis is used for the study period of different-life assets, the analysis assumes the purchase of a similar shorter-lived asset when necessary, and, further, it assumes that the AW value will continue to be the same as in the previous life cycle. In this example, a defender-similar sorter would be purchased twice—at the end of 5 and 10 years. Present-worth values are

$$PW_D = -\$5200(P/A,10\%,12) = -\$35{,}431$$

$$PW_C = -\$4211(P/A,10\%,12) = -\$28{,}693$$

Of course, the decision is still in favor of the challenging new sorter.

International competition and swift obsolescence of in-place technologies are constant concerns. Skepticism and uncertainty of the future is often reflected in management's desire to impose *abbreviated study periods* upon all economic evaluations, knowing it may well be necessary to consider yet another replacement in the near future. This approach, though reasonable from a manager's perspective, forces the recovery of the initial investment and the required MARR over a shortened period of time compared to what may be longer lives of alternatives. In such analyses, the n values in all computations reflect the shortened study period, not the values estimated as alternative lives, which may be larger than the abbreviated study period. Example 10.4 illustrates the consequences of abbreviated study periods.

Example 10.4

Reconsider the situation in Example 10.3, but now use a shortened 5-year study period which corresponds with the remaining life of the defender. The city manager specifies 5 years because of concern over the progress being made in recycling technology, progress that has already called into question retention of operationally sound equipment. Assume that the challenger's estimated salvage value will remain at $3800 after 5 years.

Solution

The AW analysis is the same, except that a period of only $n = 5$ years is allowed for the challenger to recover the capital investment of $25,000 and a 10% return.

$$AW_D = -\$5200$$

$$AW_C = -\$25{,}000(A/P,10\%,5) + 3800(A/F,10\%,5) - 720$$

$$= -\$6693$$

Select the defender. In Example 10.3, the challenger had an AW advantage of approximately $1000; now the defender is less costly by nearly $1500 annually, thus reversing the decision made using a 12-year study period. Figure 10–1 compares the results of this example with those of Example 10.3 above. The defender's AW value remains

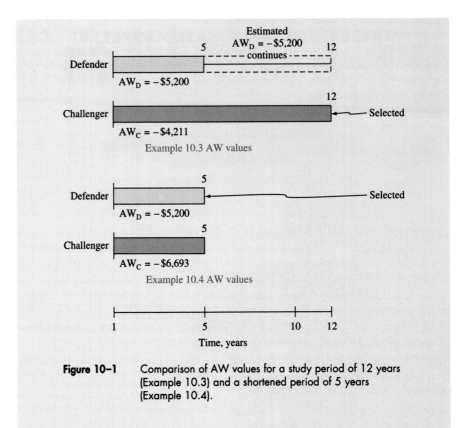

Figure 10-1   Comparison of AW values for a study period of 12 years (Example 10.3) and a shortened period of 5 years (Example 10.4).

at −$5200, but the shortened 5-year study period in this example significantly increases the challenger's AW value, enough to change the selected alternative.

Comment

The reason the decision is reversed in this example is quite simple. The challenger is given only 5 years to recover the same investment and a 10%-per-year return, whereas in the previous example 12 years is allowed. So, the AW value must increase.

A realistic alternative to the approach above is to recognize the unused value in the challenger by increasing the salvage value from $3800 to the estimated fair market value after 5 years of service, if such a value can be estimated. This would improve the AW value of the challenger.

By not allowing the full estimated life to be used in the previous example, the challenger is effectively preempted from selection. However, the decision to not consider use of the challenger after 5 years in the analysis is an example of management decision making. Selection of the study period is a difficult decision, one which must be based on information and sound judgment. The use of a shortened period may bias the economic decision in that the investment-recovery period for the challenger may be limited to significantly less than its estimated life. However, use of a longer horizon may also be detrimental due to the uncertainty of the future and accuracy of estimates. In this case, the direction of bias is less certain than in the case of a shortened horizon. A common practice is asset augmentation; that is, the defender is supplemented with a newly purchased asset to make it comparable in abilities with challenger attributes of speed, volume, and precision. The analysis for augmentation, is included as an additional example.

Additional Example 10.8
Problems 10.6 to 10.13

10.4   OPPORTUNITY-COST AND CASH-FLOW APPROACHES TO REPLACEMENT ANALYSIS

There are two equally correct and equivalent ways to consider the first cost of alternatives in replacement analysis. The first, called the *opportunity-cost approach,* or *conventional approach,* uses the defender's trade-in or current market value as the first cost of the defender and the initial cost of the replacement as the challenger's first cost. The term opportunity cost recognizes the fact that the owner forgoes an amount of capital equal to the trade-in value. This is the "cost" of the opportunity if the defender is selected. This is the approach of Example 10.1. For the defender the first cost reflects the highest value attainable through disposal via sale, trade-in, or scrap. This approach is cumbersome when there are multiple challengers each quoting a different trade-in value for the defender, because it requires a different defender first cost for comparison with each challenger.

The second approach—the *cash-flow approach*—recognizes the fact that when a challenger is selected, the defender's market value is a cash inflow to each challenger alternative; and that when the defender is selected, there is no actual capital outlay. However, to correctly use the cash-flow approach, *the defender and challenger must have the same life estimates.* Set the defender's first cost to zero and subtract the trade-in (market, sale, or scrap) value from the challenger's first cost. Again, it is important to remember that this approach can be used only when the defender and challenger lives are the same or when the comparison is over the same study period for all alternatives.

Example 10.5

This morning your boss asked you to analyze the following situation using the cash-flow approach and a MARR of 18%. Determine the most economical decision. A 7-year-old asset may be replaced with either of two new assets. Current data for each alternative are given below.

	Current Asset, Defender	Possible Replacements	
		Challenger 1	Challenger 2
First cost, $	—	10,000	18,000
Defender trade-in, $	—	3,500	2,500
Annual cost, $	3,000	1,500	1,200
Salvage value, $	500	1,000	500
Estimated life, years	5	5	5

Solution

The first cost is different for challenger 1 (C1) and challenger 2 (C2) for the cash-flow approach. Subtract the trade-in value from the respective challenger's first cost and compute the AW value over either the respective life of each alternative or a selected study period. The estimated life value of 5 years is the logical study period here. With this approach, the first cost of the defender is zero, since the asset is already owned and no actual initial investment is necessary if the defender is retained.

Defender: $AW_D = -3000 + 500(A/F,18\%,5) = -\$2930.11$

Challenger 1: $AW_{C1} = [-10,000 - (-3500)](A/P,18\%,5) - 1500$
$$+ 1000(A/F,18\%,5) = -\$3438.79$$

Challenger 2: $AW_{C2} = [-18,000 - (-2500)](A/P,18\%,5) - 1200$
$$+ 500(A/F,18\%,5) = -\$6086.70$$

Choose to retain the defender since it has the numerically largest AW value (lowest annual cost).

Comment

If the opportunity-cost approach is applied, two analyses are performed: D versus C1, with a defender first cost of $3500; and D versus C2, with a defender first cost of $2500. Resulting AW values are

D versus C1	D versus C2
$AW_D = -\$4049.34$	$AW_D = -\$3729.56$
$AW_{C1} = -\$4558.02$	$AW_{C2} = -\$6886.15$

As expected, the decision to retain the defender is made. It offers the smallest equivalent annual cost.

It is possible to use the opportunity-cost approach in a replacement study to determine what is commonly called the *replacement value (RV)* of the defender.

The RV is the market or trade-in value of the defender that must be exceeded in order to make the challenger more attractive economically. At the RV amount, the defender and challenger are economically equivalent. To determine RV, develop the relation $AW_D = AW_C$ with the defender first cost represented by RV, the unknown trade-in value. Solve the equation for RV. As an exercise, find RV for the defender in Example 10.2. Your answer should be $69,250. Since the fair market value of a 2-year-old moving van is estimated to be $42,000, the defender should be retained, which is the conclusion in Example 10.2.

Problems 10.14 to 10.22

10.5   ECONOMIC SERVICE LIFE

We may want to know the number of years that an asset should be retained in service to minimize its total cost, considering the time value of money, capital investment recovery, and annual operating and maintenance costs. This minimum-cost time is an n value and is referred to by several names including *economic service life,* minimum cost life, retirement life, and replacement life. Up to this point, the life of an asset has been assumed to be known or provided. This section explains how to determine an asset life (n value), that minimizes overall cost. This analysis is appropriate whether the asset is currently in use and replacement is considered or whether it is a new asset being considered for acquisition.

With each passing year of asset use, the following trends are usually observable. Figure 10–2 depicts these trends graphically.

*   The equivalent annual-worth value of annual operating cost (AOC) increases (identified as AW of AOC in Figure 10–2). The AOC term may also be referred to as operating and maintenance (O&M) costs.

*   The equivalent annual-worth value of the asset's initial investment or first cost decreases (the AW-of-investment curve in Figure 10–2).

*   The actual trade-in amount or salvage value decreases relative to the first cost. This effect is imbedded in the AW-of-investment curve.

These factors cause the asset's total AW curve to decrease for some years and to increase thereafter. The total AW curve is determined using the following relation over a number of years k.

$$\text{Total AW} = \text{AW of investment} + \text{AW of AOC} \qquad \text{[10.2]}$$

The minimum total AW value indicates the n value for the economic service life—the n value when replacement is the most economic. This should be the estimated asset life used in an engineering economy analysis, if only economics is considered.

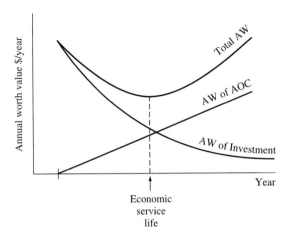

Figure 10–2   Components and shape of a total annual worth curve for an asset.

The approach to estimate n in the service life analysis uses the conventional AW computations of Chapter 6. Increase the life value index from 1 to k, where the longest possible life value is N, that is, $k = 1, 2, \ldots, N$. Start with Equation [10.2] and, for each k, determine AW_k.

$$AW_k = -P(A/P,i,k) + SV_k(A/F,i,k) - \left[\sum_{j=1}^{j=k} AOC_j(P/F,i,j)\right](A/P,i,k) \qquad \text{[10.3]}$$

where P = initial investment or asset first cost
         SV_k = salvage value after retaining the asset k years
     AOC_j = annual operating cost for year j ($j = 1, 2, \ldots, k$)

The economic service life is the k value at which AW_k indicates the smallest cost value. (*Remember:* Select the numerically largest AW_k, since costs have a minus sign.) The k value and AW amount are included in an economic analysis or a replacement analysis as the estimated life n and AW value, respectively.

Example 10.6

A 3-year-old asset is being considered for early replacement. Its current market value is $13,000. Estimated salvage values and annual operating costs for the next 5 years are given in Table 10–1, columns 2 and 3, respectively. What is the economic service life of this defender if a 10%-per-year return is required?

(1)	(2)	(3)	(4)	(5)	(6) = (4) + (5)
				Equivalent Annual	
Life, k, years	SV_k	AOC_j ($j = 1$ to 5)	Capital Recovery	Operating Costs	AW_k
1	$9000	−$2500	−$5300	−$2500	−$7800
2	8000	−2700	− 3681	−2595	−6276
3	6000	−3000	−3415	−2717	−6132
4	2000	−3500	−3670	−2886	−6556
5	0	−4500	−3429	−3150	−6579

Table 10–1 Computation of economic service life

Solution

Equation [10.3] yields AW_k for $k = 1, 2, \ldots, 5$. Table 10–1, column 4, provides the $13,000 investment recovery plus 10%-return value using the first two terms in Equation [10.3]. Column 5 gives the equivalent AOC for k years using the last term in the AW_k equation. The sum is AW_k shown in column 6. As an illustration, the computation for $k = 3$ is

$$AW_3 = -13,000(A/P,10\%,3) + 6000(A/F,10\%,3) - [2500(P/F,10\%,1)$$
$$+ 2700(P/F,10\%,2) + 3000(P/F,10\%,3)](A/P,10\%,3)$$
$$= -\$3415 - 2717$$
$$= -\$6132$$

The minimum AW cost is $6132 per year for $k = 3$, which indicates that 3 years should be the anticipated remaining life. If you plot column 6 amounts, the result will appear much like the convex-shaped total AW curve in Figure 10–2.

Comment

If several of the AW_K values are approximately equal, the total AW curve will be flat on the bottom, which indicates that the cost is relatively insensitive over several years. So, the economic service life is also less sensitive to a particular n value.

Realize that the method presented in this example is general. It can be utilized whether the economic service life is to be determined for an anticipated purchase or for an in-place asset being evaluated for retention or replacement.

Example 10.6 (Spreadsheet)

A 3-year-old asset is being considered for early replacement. Its current market value is $13,000. Estimated salvage values and annual operating costs for the next 5 years

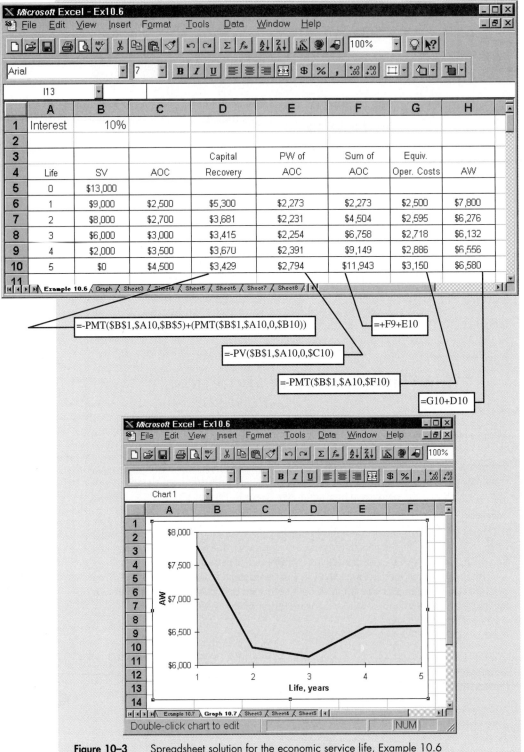

Figure 10–3 Spreadsheet solution for the economic service life, Example 10.6 (Spreadsheet).

are given in Table 10–1, columns 2 and 3, respectively. Use a spreadsheet to determine the economic service life of this defender if a 10%-per-year return is required?

Solution

The spreadsheet values in Figure 10–3, column H, tabulate the total AW. (Positive signs are used in the spreadsheet, since all values are cost estimates.) Column D is the capital recovery amount from Equation [10.3], first two terms, and the equivalent AW of AOC is in column G from the last term of Equation [10.3]. The function calculations shown are samples for $n = 5$ for each year 1 through 5.

The resulting total AW is plotted below the spreadsheet. The economic service life is 3 years (as in Table 10–1) with AW = $6132.

Using a spreadsheet system for the solution of the problems for this section is recommended.

Additional Example 10.9
Problems 10.23 to 10.27

10.6   REPLACEMENT ANALYSIS FOR ONE-ADDITIONAL-YEAR RETENTION

It is expected that an asset will normally be retained to the end of its economic service life (from the previous section) or its estimated useful life (if different). However, as the life of a currently owned asset passes, it deteriorates; a more attractive, modern, or upgraded model becomes available; or the original cost and revenue estimates are found to be significantly different from actual amounts. Then, a frequently asked question is: Should the asset be replaced or retained in service for 1, 2, 3, or more years? This is a good question if the asset has been in service the entire time expected; that is, n years of service have been given or the life has expired and there are seemingly more years of service possible from the asset. There are two options for each additional year: Select a challenger now, or keep the defender for one more year.

To make the replace-or-keep decision, it is not correct to simply compare only the equivalent defender cost and challenger cost over the remainder of the economic service life, the anticipated useful life, or some selected number of years beyond either of these two numbers. (Refer to any of these values as the *remaining life*.) Rather, the annual-worth procedure presented in Figure 10–4 is used to calculate AW_C and $C_D(1)$, where

  AW_C = challenger annual worth value
  $C_D(1)$ = defender cost estimate for the next year ($t = 1$)

If the defender 1-year cost $C_D(1)$ is less than the challenger annual worth AW_C, retain the defender one more year because its cost is less. (*Remember*: Use

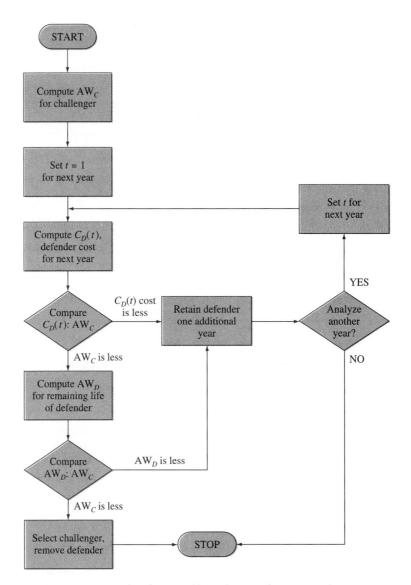

Figure 10–4 Procedure for one-additional-year replacement analysis.

minus signs to indicate costs.) If, for example, $C_D(1) = -\$100$ and $AW_C = -\$200$, the defender costs less. Figure 10–4 correctly indicates to retain the defender, because the cost of $100 is *less* than $200. If the $C_D(1)$ cost is more than AW_C, the challenger costs less for the next year. However, to be correctly selected, the challenger must also cost less than the defender's equivalent annual worth, AW_D, for its remaining life. Now, if the AW_D cost is less than AW_C, Figure 10–4 indicates that the defender should be retained for one additional year; otherwise select the challenger now.

It is possible to continue the analysis for years $t = 2, 3, \ldots$, one year at a time until either the challenger is selected or the defender remaining life is expended. Remember, only when the challenger cost is less than both the next-year defender cost *and* the defender annual worth is the challenger selected. This procedure assumes AW_C will remain the same.

Example 10.7

Engineering Models, Inc., usually keeps its company fleet cars for 5 years before replacement. Since discounted autos purchased exactly 2 years ago have deteriorated much more rapidly than expected, management wonders what is more economical: replace the cars this year with new autos; retain the cars for 1 more year and then replace them; retain for 2 more years then replace; or keep them for 3 more years until the end of their estimated lives. Perform a replacement analysis at $i = 20\%$ using these estimated costs.

	Currently-Owned Car (Defender)			Possible Replacement (Challenger)	
	Value at Beginning of Year	Annual Operating Cost			
Next year (3)	$3800	$4500	First cost	$8700	
Next year (4)	$2800	$5000	Annual cost	$3900 per year	
Last year (5)	$500	$5500	Life	5 years	
Remaining life	3		Salvage value	$1800	
Salvage after 3 more years	$500				

Solution

Following the procedure in Figure 10–4, compute the AW for the challenger over 5 years.

$$AW_C = -\$8700(A/P,20\%,5) + 1800(A/F,20\%,5) - 3900 = -\$6567$$

Calculate the defender's next-year ($t = 1$) cost using the next-year value of $2800 as the salvage value estimate.

$$C_D(1) = -\$3800(A/P,20\%,1) + 2800(A/F,20\%,1) - 4500 = -\$6260$$

Since the $C_D(1)$ cost is less than AW_C, retain the current cars for the next year.

After the next year is over, to determine if the autos should be kept yet another year (year 4) follow Figure 10–4 for $t = 2$. Now the salvage value of $2800 in $C_D(1)$ is the first cost for year $t = 2$. The 1-year cost $C_D(2)$ for the defender is now

$$C_D(2) = -\$2800(A/P,20\%,1) + 500(A/F,20\%,1) - 5000 = -\$7860$$

Now $C_D(2)$ costs more than $AW_C = -\$6567$, so we must compute the AW_D value for the remaining 2 years of the defender's life.

$$AW_D = -\$2800(A/P,20\%,2) + 500(A/F,20\%,2) - 5000 - 500(A/G,20\%,2)$$

$$= -\$6833$$

Since the challenger at $AW_C = -\$6567$ is also cheaper for the remaining 2 years, select it and replace the current cars (defenders) after the one additional year (year 3) of service. [If AW_D were less costly for this 2-year analysis, the defender is retained and a similar analysis is conducted for the last year (year 5) using $C_D(3)$ and AW_D for 1 year.]

Comment

If only the AW_D value for 3 years were used in the replacement analysis, the wrong decision would be made because the 3-year AW_D is slightly more costly than the 5-year AW_C.

$$AW_D = -\$3800(A/P,20\%,3) + 500(A/F,20\%,3) - 4500 - 500(A/G,20\%,3)$$

$$= -\$6606$$

$$AW_C = -\$6567$$

Here the decision is to select the challenger immediately, whereas the one-additional-year analysis has shown it to be more economical to retain the current fleet of cars one additional year.

Problems 10.28 to 10.32

ADDITIONAL EXAMPLES

Example 10.8

AUGMENTATION ANALYSIS USING A STUDY PERIOD, SECTION 10.3   Three years ago the City of Megapolis purchased a new fire truck. Because of expanded growth in a certain portion of the city, new fire-fighting capacity is needed now. An additional truck of the same capacity can be purchased now, or a double-capacity truck can replace the current fire truck. Estimates for each asset are presented below. Compare them at 12% per year using (*a*) a 12-year study period and (*b*) a 9-year period.

	Presently Owned	**New Purchase**	**Double Capacity**
First cost P, $	51,000 (3 years ago)	58,000	72,000
AOC, $	1,500	1,500	2,500
Trade-in, $	18,000	—	—
Salvage value, $	10% of P	12% of P	10% of P
Life, years	12	12	12

Solution

Identify plan A as the retention of the presently owned truck and augmentation with the new same-capacity vehicle. Define plan B as purchase of the double-capacity truck.

Plan A		Plan B
Presently Owned	**Augmentation**	**Double capacity**
$P = \$18,000$	$P = \$58,000$	$P = \$72,000$
AOC = \$1,500	AOC = \$1,500	AOC = \$2,500
SV = \$5,100	SV = \$6,960	SV = \$7,200
$n = 9$ years	$n = 12$ years	$n = 12$ years

(a) For a full-life 12-year study period of the defending plan,

$$AW_A = \text{(AW of presently owned)} + \text{(AW of augmentation)}$$
$$= [-18,000(A/P,12\%,9) + 5100(A/F,12\%,9) - 1500]$$
$$\quad + [-58,000(A/P,12\%,12) + 6960(A/F,12\%,12) - 1500]$$
$$= -4533 - 10,575$$
$$= -\$15,108$$

This computation assumes the equivalent annual cost for the current fire truck will continue at \$4533 for years 10 through 12.

$$AW_B = -72,000(A/P,12\%,12) + 7200(A/F,12\%,12) - 2500$$
$$= -\$13,825$$

Purchase the double-capacity truck (plan B) with an advantage of \$1283 per year.

(b) The analysis for an abbreviated 9-year study period is identical, except that $n = 9$ in each factor; that is, 3 fewer years are allowed for the augmentation and double-capacity trucks to recover the capital investment plus a 12%-per-year return. The salvage values remain the same since they are quoted as a percentage of P for all years.

$$AW_A = -\$16,447 \qquad AW_B = -\$15,526$$

Plan B is again selected; however, now the economic advantage is smaller at \$921 annually. If the study period were abbreviated more severely, at some point the decision would reverse. Set up this example (on your computer) and decrease the study period n until the decision reverses from plan B to A.

Example 10.9

ECONOMIC SERVICE LIFE, SECTION 10.5 Harold J. Beacon and Associates, Inc., an agribusiness consulting firm, can purchase an asset for \$5000 with a negligible salvage value. The annual operating costs are expected to follow a uniform arithmetic

gradient of $200 per year with a base amount of $300 in year 1. (*a*) Find the number of years that the asset should be retained from an economic viewpoint, if any return on the investment is neglected. (*b*) Derive a general closed-end expression for the economic service life for the situation in this example.

Solution

(*a*) The primary effects of $i = 0\%$ are to decrease the total annual cost values, to shorten the economic service life, and to make computations simpler since averages are determined with no time value of money calculations necessary. Table 10–2 presents the computation results for years $k = 1$ to 9. Columns 2 to 4 treat the AOC with the average total AOC presented in column 4 which is equal to the column 3 value divided by k. A sample computation of total annual cost is given below the table. The economic service life is 7 years, when the estimated total annual cost of $1614 is the smallest.

(*b*) Because of the regularity of this situation and the fact that $i = 0\%$, a formula can be derived to find the economic service life directly. Define total annual cost as TAC and use the approach of Equation [10.2].

$$\text{TAC} = \text{average first cost} + \text{average AOC}$$

Substituting n for k years, TAC for each n value may be expressed as

$$\text{TAC}_n = \frac{P}{n} + \frac{\sum_{j=1}^{j=n} \text{AOC}_j}{n}$$

where TAC$_n$ = total annual cost for n years of ownership

P/n = average of first cost over n years

AOC$_j$ = annual operating cost through year j ($j = 1, 2, \ldots, n$)

(1)	(2)	(3)	(4)	(5)	(6)
		Operating Cost, $		Average First Cost, $	Total Annual Cost, $
Year, k	Annual	Cumulative	Average		
1	−300	−300	−300	−5000	−5300
2	−500	−800	−400	−2500	−2900
3	−700	−1500	−500	−1667	−2167
4	−900	−2400	−600	−1250	−1850
5	−1100	−3500	−700	−1000	−1700
6	−1300	−4800	−800	−833	−1633
7	−1500	−6300	−900	−714	−1614*
8	−1700	−8000	−1000	−625	−1625
9	−1900	−9900	−1100	−555	−1655

Table 10–2 Computation of economic service life for $i = 0\%$, Example 10.9

* In column notation, total annual cost is computed as

$$(6) = (4) + (5) = (3)/k - 5000/k$$

For $k = 7$, this is $-\$1614 = -6300/7 - 5000/7$.

For the AOC series including an arithmetic gradient, use the expression for the sum of an arithmetic progression, with n = number of terms.

$$\text{Sum} = \frac{n}{2}[2(\text{first term}) + (n - 1)(\text{gradient})]$$

Now, we substitute.

$$\frac{\sum\limits_{j=1}^{j=n} \text{AOC}_j}{n} = B + \frac{n - 1}{2}G$$

where B = base amount of the series
G = amount of the uniform gradient

The total annual cost relation is now

$$\text{TAC}_n = \frac{P}{n} + B + \frac{n - 1}{2}G \qquad\qquad [10.4]$$

The general shape of the terms and TAC_n is shown in Figure 10.5. We can apply differential calculus to determine the optimal n value at which TAC_n will be a

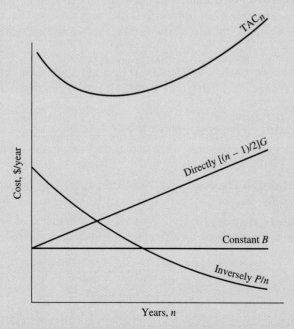

Figure 10–5   Total annual cost (TAC) relations for $i = 0\%$ used to determine economic service life.

minimum. Take the first derivative of Equation [10.4] and solve for an optimum service life value n^*.

$$\frac{d\text{TAC}_n}{dn} = -\frac{P}{n^2} + \frac{G}{2} = 0$$

$$n^* = \left(\frac{2P}{G}\right)^{1/2} \qquad\qquad \text{[10.5]}$$

Substituting $P = \$5000$ and $G = \$200$ for this example,

$$n^* = \left(\frac{10,000}{200}\right)^{1/2} = 7.07 \text{ years}$$

which is essentially the same as the value of 7 obtained in Table 10–2.

CHAPTER SUMMARY

Replacement analysis serves a vital role in engineering economy when a defender (in-place) asset and one or more challengers are compared. To perform the analysis, the evaluator takes the perspective of a consultant to the company—that neither asset is owned currently and that the two choices are to acquire the used asset or acquire a new asset.

There are two equivalent approaches that may be taken in a replacement study when determining the first cost P for the alternatives and when performing the analysis.

Cash-flow approach.   Recognize there is an actual cash-flow advantage to the challenger if the defender is traded. For the analysis, use the following:

Defender: First cost amount is zero.

Challenger: First cost is the actual cost minus the quoted trade-in value of the defender.

Important note.   The estimated lives of the defender, that is, its remaining life, and of the challenger must be equal to use this method.

Opportunity-cost approach.   Assume the defender trade-in amount is forgone and the defender service can be acquired as a used asset at its trade-in value.

Defender: First cost is the trade-in value.

Challenger: First cost is its actual cost.

When the defender remaining life and challenger life are unequal, it is necessary to preselect a study period for the analysis. The annual worth is commonly assumed to continue at the same calculated amount for an alternative with an *n* value less than the study period. If this assumption is not appropriate, perform the analysis using new estimates for the defender, the challenger, or both.

If the study period is abbreviated to be less than one or both of the alternative's life estimates, it is necessary to recover the first cost and the required return at the MARR in less time than normally expected. This will artificially increase the AW value(s). It is usually a management decision to use an abbreviated study period.

A sunk cost represents a previous capital investment which cannot be recovered completely or at all. When performing a replacement analysis, any sunk cost (for the defender) is never added to the challenger's first cost. This will unfairly bias the analysis against the challenger due to the resulting artificially higher AW value.

This chapter detailed the procedure to select the number of years to retain an asset using the economic service life criterion. The economically best *n* value occurs when the AW value resulting from Equation [10.3] is minimum at a specified rate of return. Though not usually correct, if interest is not considered ($i = 0$), computations are based upon simple averages.

When conducting a one-year-at-a-time or one-additional-year replacement analysis before or after the estimated life has been reached, the defender cost is calculated for one more year and compared to the challenger AW value. Use the logic of Figure 10–4.

CASE STUDY #4, CHAPTER 10   REPLACEMENT ANALYSIS FOR QUARRY EQUIPMENT

Equipment used to support the movement of raw material from the quarry to the rock crushers was purchased by Tres Cementos, SA, 3 years ago. When purchased, the equipment had $P = \$85,000$, $n = 10$ years, SV = \$5000, with an annual capacity of 180,000 metric tons. Additional equipment with a capacity of 240,000 metric tons per year is now needed. Such equipment can be purchased for $P = \$70,000$, $n = 10$ years, SV = \$8000.

However, a consultant has pointed out that the company can construct conveyor equipment to move the material from the quarry. This will cost an estimated \$115,000 with a life of 15 years and no significant salvage value. It will carry 400,000 metric tons per year. The company needs some way to move material to the conveyor in the quarry. The presently owned equipment can be used, but it will have excess capacity. If new smaller-capacity equipment is purchased, there will be a \$15,000 trade-in on the currently used equipment. The smaller-capacity equipment will require a capital outlay of \$40,000 with an estimated life of $n = 12$ years and an SV = \$3500. The capacity is 400,000 metric tons per year over this short distance. Monthly operating, maintenance, and insurance costs will average \$0.01 per ton-kilometer for the movers. Corresponding costs for the conveyor are expected to be \$0.0075 per metric ton.

The company wants to make 12% per year on this investment. Records show that the equipment must move raw material an average of 2.4 kilometers from the quarry to the crusher pad. The conveyor will be placed to reduce this distance to 0.75 kilometer.

Question 1. You have been asked to determine if the old equipment should be augmented with new equipment or if the conveyor equipment should be considered as a replacement. If replacement is more economical, which method of moving the material in the quarry should be used?

Question 2. Because of new safety regulations, the control of dust in the quarry and at the crusher site has become a real problem and implies that new capital must be invested to improve the environment for employees or large fincs may be imposed by the government. The Tres Cementos president has obtained an initial quote from a subcontractor which would take over the entire raw material movement operation being evaluated here for a base annual amount of $21,000 and a variable cost of 1 cent per metric ton moved. The 10 employees in the quarry operation would be employed elsewhere in the company with no financial impact upon the estimates for this evaluation. Should this offer be seriously considered if the best estimate is that 380,000 metric tons per year would be moved by the subcontractor? Identify any additional assumptions or estimates you had to make in order to adequately address this new question posed by the president.

PROBLEMS

10.1 Explain why the actual original cost of an asset is neglected when performing a replacement analysis.

10.2 An elaborate CD display purchased by The Music Company 4 years ago for $38,000 with an estimated salvage value of $1000 is being considered for replacement due to lagging CD sales. Book value is currently $20,000 with 5 years of economic life remaining. The owners wish to trade for a new, smaller display, which costs $14,000. Current estimates for the old display are that the trade-in value now is $25,000 and that it can last another 7 years. Annual upkeep costs average $150 for the old display. (*a*) Determine the values of *P*, *n*, SV, and AOC for the existing display if a replacement analysis is performed. (*b*) Is there a sunk cost involved? If so, what is its size?

10.3 The owner of a downtown men's clothing store is considering moving to leased space in an urban center. He purchased the downtown shop 5 years ago for $80,000 cash. He estimates the annual investment in property improvement to have averaged $1500 and believes the property should have a book value of the purchase price plus these improvements considered with an 8%-per-year interest factor. He will ask $115,000 for the shop if he sells at this time. The annual operating costs have averaged $4500.

If he stays downtown, he plans to retire in 10 years and will give the shop to a son-in-law. If the owner moves to the shopping center, he will sell the store now, and he must sign a 10-year lease for $16,600 per year with no additional yearly charges. He must pay a $7500 deposit now, but this amount is returned when the lease expires. (*a*) Determine the values of *P*, *n*, SV, and AOC for the two alternatives, and (*b*) determine the amount of the sunk cost, if one exists.

10.4 How is a replacement analysis different from and the same as an analysis performed using the approaches you have learned in the previous chapters on PW and AW analysis?

10.5 A small moving van can be swapped for this year's model at a trade-in value of 40% of the first cost at which it was purchased 3 years ago. The asset, purchased for $28,000, has been depreciated over 5 years with a current book value of $10,000 for tax purposes and depreciated over 12 years as its estimated economic life with a current book value of $15,500. Compute the sunk cost (*a*) for tax purposes and (*b*) for the company's decision makers.

10.6 The Canadian Touring Company (CTC) purchased 20 tour buses 3 years ago for $98,000 each. The president plans to contract for a major overhaul next year at a cost of $18,000 each. The president also estimates the following for each vehicle: an additional life of 7 years once the overhauls are complete, revised annual operating costs of $6000, and an $8000 salvage value.

However, the vice president for operations proposes a replacement with 25 new smaller touring buses, which offers a trade-in of $14,000 for each current vehicle and an out-of-pocket cost of $75,000 each. The vice president also estimates the AOC to be $2000 per year less, that the new buses will last for 8 years, and have a final market value of $5000 each if sold to church groups. If the management wants to make 12% per year on its investments, determine whether the president's overhaul plan or the vice president's replacement plan is more economical.

10.7 (*a*) Rework Problem 10.6 if the vice president for operations estimates that the new vehicles will last 14 years rather than 8 years, due to the use of much better materials. (*b*) By how many dollars did the increase in *n* from 8 to 14 years change the annual cost of the new vehicles?

10.8 A new member of the CTC Board of Directors has reviewed the proposals described in Problem 10.6. She proposes that an abbreviated 5-year study period be used to evaluate all proposals, since she hopes that the company will be sold at a large profit in the next 5 years. Will this shortening of the study period make a difference in the decision about whether to purchase new buses?

10.9 A newly formed freight railroad corporation has emerged from several marginally successful companies. There is a fleet of seven 3-to 5-year-old engines currently in use. They can be traded in for $800,000 (five engines) and $1 million (two engines) on six newer models with fuel savings and enhanced safety options which cost $1.5 million each. The AOC for the current engines is expected to continue at $600,000 and for the new engines at 5% of initial cost.

The new president wants all alternatives evaluated at 8%-per-year interest and over a 10-year period with no salvage value. The currently owned engines can be expected to last 6 more years and the new engines have an estimated life of 12 years. Use an AW analysis for (*a*) the abbreviated study period and (*b*) the respective estimated lives to make the decision between current and new engines. Did the abbreviated study period make the decision change?

10.10 Determine which alternative is better in Problem 10.3*a* if a MARR of 6% per year is expected by the owner.

10.11 Machine *A*, purchased 2 years ago, is wearing out more rapidly than expected. It has a remaining life of 2 years, annual operating costs of $3000, and no salvage value. To continue the function of this asset, machine *B* can be purchased now and a trade-in value of $9000 will be allowed for machine *A*. Machine *B* has $P = \$25,000$, $n = 12$ years, AOC = $4000, and SV = $1000. As an alternative, machine *C* can be bought to replace *A*. No trade-in will be allowed for *A*, but it can be sold for $7000. This new asset *C* will have $P = \$38,000$, $n = 20$ years, AOC = $2500, and SV = $1000. If plan I is the retention of *A*, plan II is the purchase of *B*, and plan III is the selling of *A* and the purchase of *C*, use AW analysis and a MARR of 8% to determine which plan is best.

10.12 Angstrom Technologies intends for the company to use the newest and finest equipment in its labs. Accordingly, the senior engineer has recommended that a 2-year-old piece of precision measurement equipment be replaced immediately. This engineer believes it can be shown that the proposed equipment is economically advantageous at a 15%-per-year return and a planning horizon of 5 years. (*a*) Perform the replacement analysis for a 5-year study period using the estimates below.

	Current	Proposed
Original purchase price, $	30,000	40,000
Current market value, $	15,000	—
Remaining estimated life, years	5	15
Estimated value in 5 years, $	7,000	10,000
Salvage after 15 years, $	—	5,000
Annual operating cost, $	8,000	3,000

(*b*) Is the decision the same as in part (*a*) if a 15-year period is used for the replacement analysis? What is the inherent assumption of this analysis for the defender when a 15-year study period is used?

10.13 Reread Problem 10.12(*a*) about Angstrom Technologies. The president believes he can negotiate a much better purchase price than the quoted amount of $40,000 for the proposed equipment. How much does the president have to negotiate down to economically justify the proposed measurement equipment? Use a 5-year study period.

10.14 (*a*) Describe the rationale for and explain the differences between the opportunity-cost (or conventional) and cash-flow approaches to replacement analysis. (*b*) Which of the two approaches more closely reflects the actual amount of money which flows in and out of the corporation for each alternative? Explain your answer.

10.15 Use the cash-flow approach for replacement analysis to select the current or proposed measurement equipment in Problem 10.12(*a*).

10.16 Describe a situation from your own experiences which fits the replacement analysis of this chapter and has at least two challenger alternatives to the defender. Demonstrate how the cash-flow approach is correctly applied for the first-cost amounts in your example if a replacement analysis were performed.

10.17 Rework Problem 10.6 using the cash-flow approach to replacement analysis.

10.18 Resolve Problem 10.11 using the cash-flow approach to replacement analysis. Assume that machine *A* can be refurbished to last for a total of another 12 years with a $20,000 rework in 2 years. Also, the estimate for SV for machine *C* in 12 years is $1000. Use a 12-year study period.

10.19 WWW Computers owns an asset used in hard disk drive construction. This asset has had unexpectedly high annual maintenance costs and could be replaced with one of two more modern versions. Model A-1 can be installed for a total cost of $155,000 with estimates of $n = 5$ years, AOC = $10,000, and SV = $17,500. Model B-2 has a first cost of $100,000 with $n = 5$ years, AOC = $13,000, and SV = $7000. If the presently owned asset is traded, it is valued at $31,000 by the A-1 manufacturer and $28,000 by the B-2 salesperson. The remaining useful life of the asset has been estimated to be 5 more years at an AOC of $34,000 and a negative salvage value of $2000 after the 5 years for hazardous waste disposal. (*a*) Use the cash-flow approach to determine which is the more economical decision at a required return of 16% per year. (*b*) If the WWW president and CEO decides to donate the current asset (and neglect any tax effect) and purchase one of the challenger models, use PW analysis to determine which has the lower present worth. Use the estimates stated above as appropriate.

10.20 A veterinarian currently has a diagnosis system in her lab that should last for 6 more years with costs of $24,000 this year and increasing by 10% per year for upkeep and supplies. A new model with special software options would cost $70,000, have an estimated nonobsolescence life of 6 years, an AOC of $22,000, and no salvage value. What is a fair trade-in value of the current system that will make replacement economical if a 5%-per-year return is expected?

10.21 What is the replacement value of the old CD display described in Problem 10.2 if the new one has $n = 10$, AOC = $30, and SV = $1500? Use $i = 15\%$. Compare this value with the $25,000 trade-in offer stated in Problem 10.2.

10.22 A construction company bought a 180,000 metric-ton capacity earth sifter 3 years ago at a cost of $55,000. The expected life at the time of purchase was 10 years with a $5000 salvage value and an AOC of $2700. A 480,000 metric-ton replacement sifter is being considered. This new sifter costs $40,000 with $n = 12$ years, SV = $3500, and AOC = $7200. (*a*) Compute the required trade-in value of the current sifter if the replacement is bought and $i = 12\%$. (*b*) Compute the required trade-in value if an abbreviated study period of 4 years is used. By what percent does this abbreviation in n increase the required trade-in value?

10.23 Paul Adams operates a landscape maintenance business during the summer months and a firewood-splitting service in the winter months. He purchased a commercial-grade gasoline-fueled wood splitter for $5800. Paul used $400 of business capital and financed the balance through his dad's bank at 5% per year for 3 years. The estimated values of the splitter for the next 6 years are $2200 after the first year of ownership, decreasing by $400 per year to year 5, after which the resale value remains at $600. Annual operating costs are expected to be $1000 the first year, increasing by 10% each year thereafter. Paul plans to keep the splitter at least 6 years. If money is worth 7% per year, how many years should the splitter be retained?

10.24 Rework Problem 10.23 assuming that money has no time value for the economic service life but that the loan still costs 5% per year. Compare the answers between the two problems and comment on the difference between them.

10.25 A heat-treating machine was purchased 5 years ago for $40,000 with an expected life of 10 years. The observed past and estimated future operating costs, maintenance costs, and salvage values are given below. If $i = 10\%$, determine how many more years the machine should be kept in service before it reaches its economic service life.

Year	Operating Cost, $	Maintenance Cost, $	Salvage Value, $
1	1,500	2,000	25,000
2	1,600	2,000	25,000
3	1,700	2,000	22,000
4	1,800	2,000	22,000
5	1,900	2,000	15,000
6	2,000	2,100	5,000
7	2,100	2,700	5,000
8	2,200	3,300	0
9	2,300	3,900	0
10	2,400	4,500	0

10.26 One year ago the Miller Paint Company purchased a machine, called EZ Clean, to thoroughly clean the wood surface of a wall and reduce the time necessary to prepare for painting. The machine cost $8000 and was expected to last 14 more years. The owner has already seen improved versions, so he would like to know the most economical life of his EZ Clean. If operating costs are $500 for the first year and are expected to increase by $100 per year, compute the economic service life for (*a*) 0% return, and (*b*) $i = 5\%$ per year and compare the answers.

10.27 A general manager wants to know the economic service life for currently owned machines. (*a*) Find this value at 20% per year if the first cost is $5000 per machine. (*b*) Plot the annual-worth curve and determine if it is insensitive over some range of n values.

Year	Trade-in Value, $	Estimated AOC, $
0	5000	—
1	3000	2000
2	1500	2500
3	1000	3000
4	500	3500
5	0	4000
6	0	5000

10.28 In Example 10.3, the current recyclable material sorter has $AW_D = \$-5200$ for another 5 years of service. The challenger, with $AW_C = \$-4211$ for a 12-year life, is selected because AW_C is the more favorable measure of worth. However, Helen, the city manager, wants to keep the current sorter another year before replacement. Make a recommendation to Helen, if additional study indicates that the value of the defender is

$3000 now with an anticipated value of $1800 one year from now. The projected operating cost for next year is $3500 and the minimum acceptable return is still 10% per year.

10.29 The service person who uses the pool sweeper on a daily basis told Clear Blue Pools' owner it should be replaced next year. A new sweeper costs $1800, will last 7 years with annual operating costs estimated at $100 the first year and $50 higher each year, and has no salvage value. The owner estimates he can sell the current sweeper to his brother for $400 this year (now), $300 next year, or $50 the third year. Clear Blue could keep the sweeper for a maximum of another 2 years, with operating costs increasing to $175 next year and $350 the following year. Should the owner trade now, next year, or 2 years from now if the new sweeper will have the same costs in the future as estimated now? Use $i = 12\%$ per year.

10.30 You and your spouse have to make the decision to keep your present work-car or purchase a low-mileage new one with no warranty. You don't wish to keep both cars, and you have just received an inheritance which allows you to pay cash for the automobile. The new car costs $20,000, can last you 7 years, has estimated annual maintenance costs of $200 the first year, increasing by $100 per year thereafter, and will trade for an estimated $1000 in 7 years. If you keep your car, the expected trade-in value and annual maintenance costs are as follows:

Additional Years Retained	Annual Maintenance Cost, $	Trade-in Value, $
1	1800	2500
2	1500	2000
3	1500	1500

(*a*) You will not consider keeping the current car for more than an additional 3 years, at which time you anticipate a $1000 sales price. If all other costs are considered equal for the two cars, use $i = 15\%$ to determine when to purchase a different car.

(*b*) If you both are absolutely sure you want to trade after one more year, what is the largest purchase price that will make the new car decision economically correct? Is this a reasonable amount given the $20,000 quote you already have?

10.31 Last year Dr. Morse made the decision to keep some existing medical diagnosis equipment used in her medical practice for another year (this year) in lieu of purchasing new, equivalent-capability equipment for $15,000.

(*a*) Using the data below that is now known, determine if the doctor made the correct economic decision at $i = 18\%$ per year.

Existing Equipment for Last Year

Trade-in value last year	$3000
Market value this year	$2000
Operating cost last year	$500

New Equipment if Purchased Last Year

P = $15,000	SV after 10 years = $1000
n = 10 years	AOC = $3000 (constant)

(*b*) A major price reduction from $15,000 to $8000 has just been announced for the same equipment. Since the AOC of the old equipment is rising substantially this year, Dr. Morse feels this is the year to trade up. Should the doctor trade this year or next year?

New Equipment

P = $8000	SV = $1000	AOC = $3000	n = 10 years

Existing Equipment, Current Estimates

This year	Value = $2000	AOC = $2000
Next year	Value = $ 500	AOC = $2500
Year after next	Value = 0	AOC = $2500

10.32 John must perform a replacement study on pressing equipment in an industrial laundry. The challenging asset has a computed AW_C = $-42,000 for its anticipated 10-year life. John has completed data collection on the defender with the following projected AOC and trade-in values for the next 5 years, after which the currently owned equipment must be replaced.

Additional Years Retained	AOC	Trade-in Value
1	$34,000	$28,000
2	30,000	22,000
3	30,000	15,000
4	30,000	5,000
5	30,000	0

If the current equipment is used for another 5 years, it will cost a net estimated $2000 to remove it from the plant. Perform one-additional-year replacement analysis at a 16%-per-year return to determine how many years to keep the pressing unit before replacing it.

Bonds

In Section 1.8 we introduced the term *capital* as the primary means for a company to generate new business and revenue through investment. There are essentially only two types of capital—debt and equity. Equity capital is developed from stock sales, primarily, and debt capital is developed by borrowing money directly. (Chapters 15 and 18 treat debt and equity financing in some detail.)

A time-tested method of raising debt capital is through the issuance of an IOU. One form of IOU is a bond. Bonds usually come from one of the following three sources: (1) federal (U.S.) government, (2) states and municipalities, and (3) corporations. Although these entities could probably raise capital in a number of other ways, bonds are usually issued when it would be difficult to borrow a large amount of money from a single source or when repayment is to be made over a long period of time. A major feature that differentiates bonds from other forms of financing is that bonds can be bought and sold in the open market by people other than the original issuer and lender. In this chapter, some of the types of bonds and their characteristics and economic analysis are discussed.

Purpose: Understand the types of bonds and their characteristics; perform an economic analysis of a bond.

	This chapter will help you:
Types of bonds	1. Describe the various types of bonds.
Bond interest	2. Calculate the interest payable or receivable for a specified bond.
Bond PW	3. Calculate the present worth of a bond.
Bond ROR	4. Calculate the nominal and effective rate of return (ROR) for a bond.

11.1   BOND CLASSIFICATIONS

A *bond* is a long-term note issued by a corporation or governmental entity for the purpose of financing major projects. In essence, the borrower receives money now in return for a promise to pay later, with interest paid between the time the money was borrowed and the time it was repaid. The bond interest rate is often called the *coupon rate*. Bonds can be classified and subclassified in literally hundreds of ways. For our purposes, we will consider four general classifications: *treasury securities, mortgage bonds, debenture bonds,* and *municipal bonds*. These types of bonds can be further subdivided, some of which will be discussed below. The types described in this section are summarized in Table 11–1.

Table 11–1	Classification and characteristics of bonds	
Classification	**Characteristics**	**Type**
Treasury securities	Backed by U.S. government	Bills (≤ 1 year)
		Notes (2–10 years)
		Bonds (10–30 years)
Mortgage	Bonds backed by mortgage or specified assets	First mortgage
		Second mortgage
		Equipment trust
Debenture	No lien to creditors	Convertible
		Nonconvertible
		Subordinated
		Junk
Municipal	Usually federal income tax free	General obligation
		Revenue
		Zero coupon
		Variable rate
		Put

Treasury securities are issued and backed by the U.S. government and therefore are considered to be the lowest-risk securities on the market. Their interest is commonly exempt from state and local income taxes. There are three types of treasury securities: Treasury bills, Treasury notes, and Treasury bonds. Treasury bills are issued with maturities of 3 months, 6 months, and 1 year, with interest payable at maturity. Treasury notes have maturities of 2 to 10 years, while Treasury bonds have 10- to 30-year maturities. Treasury notes and bonds pay interest semiannually. It is the interest rate paid by a Treasury bill that is referred to as a *safe investment* in the Chapter 1 discussion on MARR and a project's lowest hurdle rate (Section 1.8 and Figure 1–6).

A *mortgage bond* is one which is backed by a mortgage on specified assets of the company issuing the bonds. If the company is unable to repay the bondholders at the time the bonds mature, the bondholders have an option of foreclosing on the mortgaged property. Mortgage bonds can be subdivided into first-mortgage and second-mortgage bonds. As their names imply, in the event of foreclosure by the bondholders, the first-mortgage bonds take precedence during liquidation. The first-mortgage bonds, therefore, generally provide the lowest rate of return (less risk). Second-mortgage bonds, when backed by collateral of a subsidiary corporation, are referred to as *collateral bonds.* An *equipment trust bond* is one in which the equipment purchased through the bond serves as collateral. These types of bonds are generally issued by railroads for purchasing new locomotives and train cars.

Debenture bonds are not backed by any form of collateral. The reputation of the company is important for attracting investors to this type of bond. As further incentive for investors, debenture bonds sometimes carry a floating interest rate or are often *convertible* to common stock at a fixed rate as long as the bonds are outstanding. For example, a $1000 convertible debenture bond issued by the GoT Company may have a conversion option to 50 shares of GoT common stock. If the value of 50 shares of GoT common stock exceeds the value of the bond at any time prior to bond maturity, the bondholder has the option of converting the bond to common stock, thus reaping the financial return. Debenture bonds generally provide the highest rate of interest because of the increased risk associated with them. *Subordinated debentures* represent debt that ranks behind other debt (senior debt) in the event of liquidation or reorganization of the company. Because they represent even riskier investments, these bonds provide a higher rate of return to investors than regular debentures.

The fourth general type of bonds is *municipal bonds.* Their attractiveness to investors lies in their income-tax-free status. As such, the interest rate paid by the governmental entity is usually quite low. Municipal bonds can be either *general obligation bonds* or *revenue bonds.* General obligation bonds are issued against the taxes received by the governmental entity (i.e., city, county, or state) that issued the bonds and are backed by the full taxing power of the issuer. School bonds are an example of general obligation bonds. Revenue bonds are issued against the revenue generated by the project financed, as a water treatment plant or a bridge. Taxes cannot be levied for repayment of revenue bonds.

Three other common types of municipal bonds are zero-coupon, variable-rate, and put bonds. Zero-coupon bonds are securities for which no periodic interest payments are made. They are sold at a discount from their face value, which is paid at maturity. Variable-rate bonds have coupon rates that are adjusted at specified points in time (weekly, monthly, annually, etc.). Put bonds give the holder the option to cash in the bond on specified dates (one or more) prior to maturity.

In order to assist prospective investors, all *bonds* are rated by various companies according to the amount of risk associated with their purchase. One such rating is Standard and Poor's, which rates bonds from AAA (highest quality) to DDD (bond in default). In general, first-mortgage bonds carry the highest rating, but it is not uncommon for debenture bonds of large corporations to carry an AAA rating or ratings higher than first-mortgage bonds of smaller, less reputable companies. The term *junk bonds* refers to debenture bonds rated lower than BBB. Junk bonds are frequently issued when a corporation wants to raise enough money to purchase another company.

Problems 11.1 to 11.3

11.2   BOND TERMINOLOGY AND INTEREST

As stated in the preceding section, a bond is a long-term note (IOU) issued by a corporation or governmental entity for the purpose of obtaining needed capital for financing major projects. The conditions for repayment of the money obtained by the borrower are specified at the time the bonds are issued. These conditions include the bond face value, the bond interest rate or coupon rate, the bond interest-payment period, and the bond maturity date.

The bond *face value,* which refers to the denomination of the bond, is usually an even denomination starting at $100, with the most common being the $1000 bond. The face value is important for two reasons:

1. It represents the lump-sum amount that will be paid to the bondholder on the bond maturity date.

2. The amount of interest I paid per period prior to the bond maturity date is determined by multiplying the face value of the bond by the bond interest rate per period as follows:

$$I = \frac{\text{(face value)(bond interest rate)}}{\text{number of payment periods per year}} \qquad \text{[11.1]}$$

$$= \frac{Vb}{c}$$

Often a bond is purchased at a discount (less than face value) or a premium (greater than face value), but only the face value, not the purchase price, is used to compute the bond interest amount I. Examples 11.1 and 11.2 illustrate the computation of bond interest.

Example 11.1

A bicycle manufacturing company planning an expansion issued 4% $1000 bonds for financing the project. The bonds will mature in 20 years with interest paid semiannually. Mr. John Doe purchased one of the bonds through his stockbroker for $800. What payments is Mr. Doe entitled to receive?

Solution

The face value of the bond is $1000. Therefore, Mr. Doe will receive $1000 on the date the bond matures, 20 years from now. In addition, Mr. Doe will receive the semiannual interest the company promised to pay when the bonds were issued. The interest every 6 months will be computed using $V = \$1000$, $b = 0.04$, and $c = 2$ in Equation [11.1]:

$$I = \frac{1000(0.04)}{2} = \$20 \text{ every 6 months}$$

Example 11.2

Determine the amount of interest you will receive per period if you purchase a 6% $5000 bond which matures in 10 years with interest payable quarterly.

Solution

Since interest is payable quarterly, you would receive the interest payment every 3 months. The amount by Equation [11.1] is

$$I = \frac{5000(0.06)}{4} = \$75$$

Therefore, you will receive $75 interest every 3 months in addition to the $5000 lump sum after 10 years.

A *zero-coupon bond* does not pay periodic interest, so the coupon rate is zero. Because of this, they often sell at discounts of more than 75% of their face value so that their yield to maturity will be sufficient to attract investors. *Stripped bonds* are simply conventional bonds whose interest payments are sold separately from the face value. The stripped bond then behaves as if it were a zero-coupon bond.

Problems 11.4 to 11.8

11.3 BOND PRESENT-WORTH CALCULATIONS

When a company or government agency offers bonds for financing major projects, potential investors should determine how much they will have to pay in PW terms for a bond of a given denomination. The amount they pay now determines the rate of return on the investment. The standard relations are used to calculate the PW of the bond and its interest payments for a specified rate of return and a specific bond interest rate. These calculations are shown in Example 11.3.

Example 11.3

Jennifer Jones wants to make a nominal 8% per year compounded semiannually on a bond investment. How much should she be willing to pay now for a 6% $10,000 bond that will mature in 15 years and pays interest semiannually?

Solution

Since the interest is payable semiannually, Ms. Jones will receive the following interest payments:

$$I = \frac{10,000(0.06)}{2} = \$300 \text{ every 6 months}$$

The cash-flow diagram (Figure 11–1) for this investment helps us to write a present-worth relation to compute the value of the bond now, using a return rate of 4% per 6-month period, which is the same as the payment period for bond interest. Note in the following equation that I is simply an A value.

$$PW = 300(P/A,4\%,30) + 10,000(P/F,4\%,30) = \$8270.60$$

Thus, if Jennifer is able to buy the bond for $8270.60, she will receive a nominal 8%-per-year compounded semiannually on her investment. If she were to pay more than $8270.60 for the bond, the rate of return would be less than 8%, and vice versa.

Comment

It is important to note that the interest rate used in the present-worth calculation is the interest rate per period that Ms. Jones *wants to receive,* not the bond interest rate. Since she wants to receive a nominal 8% per year compounded semiannually, the interest rate per 6-month period is 8%/2 = 4%. The bond interest rate is used *only* to determine the amount of the bond interest payment.

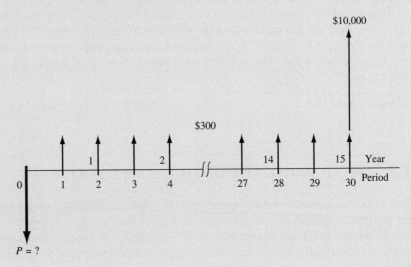

Figure 11–1   Cash flow for a bond investment, Example 11.3.

When the investor's compounding frequency is either more often or less often than the interest payment frequency of the bond, it is necessary to use the effective-interest-rate techniques of Chapter 3. Example 11.4 illustrates the calculations when the investor's compounding period is less than the bond interest period. If you wish to review nominal and effective interest rates, refer to Sections 3.1 through 3.3.

Example 11.4

Calculate the present worth of a 4.5% $5000 bond with interest paid semiannually. The bond matures in 10 years, and the investor desires to make 8% per year compounded quarterly on the investment.

Solution

First, determine the interest the investor receives.

$$I = \frac{5000(0.045)}{2} = \$112.50 \text{ every 6 months}$$

The present worth of the payments shown in Figure 11–2 can be determined in either of two ways:

1. Take each $112.50 interest payment back to year 0 separately with the P/F factor and add to the sum the PW of the $5000 in year 10. In this case, the interest rate is $8\%/4 = 2\%$ per quarter and the number of periods is 40, double those shown in Figure 11–2, since the interest payments are made semiannually while the desired rate of return is compounded quarterly. Thus,

$$PW = 112.50(P/F,2\%,2) + 112.50(P/F,2\%,4) + 112.50(P/F,2\%,6) + \cdots$$

$$+ 112.50(P/F,2\%,40) + 5000 \, (P/F,2\%,40)$$

$$= \$3788$$

2. Determine the effective interest rate compounded semiannually (the bond interest payment period) that is equivalent to the nominal 8% per year compounded quarterly (as stated in the problem); then use the P/A factor to compute the present worth of interest and add this amount to the present worth of the $5000 in year 10. The semiannual rate is $8\%/2 = 4\%$. Since there are two quarters per 6-month period, Table 3–3 (Chapter 3) indicates that the effective semiannual rate is $i = 4.04\%$. Alternatively, the effective semiannual rate can be computed from Equation [3.3]:

$$i = \left(\frac{1 + 0.04}{2}\right)^2 - 1 = 0.0404$$

The present worth of the bond can now be determined with calculations similar to those in Example 11.3.

$$PW = 112.50(P/A,4.04\%,20) + 5000(P/F,4.04\%,20) = \$3790$$

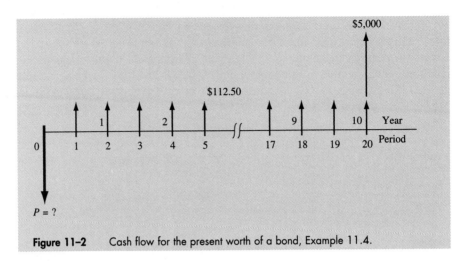

Figure 11-2   Cash flow for the present worth of a bond, Example 11.4.

In summary, the steps to calculate the present worth of a bond investment are the following:

1. Calculate the interest payment (I) per period, using the face value (V), the bond interest rate (b), and the number of interest periods (c) per year, by $I = Vb/c$.
2. Draw the cash-flow diagram of the bond receipts including bond interest and face value.
3. Determine the investor's desired rate of return per period. When the bond interest period and the investor's compounding period are not the same, it is necessary to use the effective-interest-rate formula to find the proper interest rate per bond payment period.
4. Add the present worths of all cash flows.

Additional Example 11.6
Problems 11.9 to 11.18

11.4   RATE OF RETURN OF A BOND INVESTMENT

To calculate the rate of return received on a bond investment, the procedures learned in this chapter and Chapter 7 are followed. That is, the procedures of Sections 11.1 and 11.2 of this chapter establish the timing and the magnitude of the income associated with a bond investment; the rate of return on the investment is then determined by setting up and solving a PW-based rate-of-return equation in the form of Equation [7.1], which is, $0 = -PW_D + PW_R$. Example 11.5 illustrates the general procedure for calculating the rate of return on a bond investment.

Example 11.5

In Example 11.1, Mr. John Doe paid $800 for a 4% $1000 bond that matures in 20 years with interest payable semiannually. What nominal and effective interest rates per year will Mr. Doe receive on his investment for semiannual compounding?

Solution

The income Mr. Doe will receive from the bond purchase is the bond interest every 6 months plus the face value in 20 years. The equation for calculating the rate of return using the cash flow of Figure 11–3 is

$$0 = -800 + 20(P/A,i^*,40) + 1000(P/F,i^*,40)$$

Solve via spreadsheet or manually to obtain $i^* = 2.87\%$ per period. The nominal interest rate per year is computed by multiplying the interest rate per period times the number of periods.

$$\text{Nominal } i = 2.87\%(2) = 5.74\% \text{ per year}$$

From Table 3–3 or Equation [3.3], the effective rate is 5.82% per year.

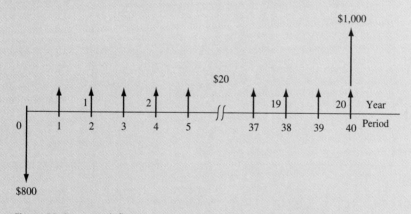

Figure 11–3   Cash flow to compute the rate of return for a bond investment, Example 11.5.

Additional Examples 11.7 and 11.8
Problems 11.19 to 11.28

ADDITIONAL EXAMPLES

Example 11.6

PW OF A BOND, SECTION 11.3   JP wants to invest in some 20-year 4% $10,000 mortgage bonds which pay interest semiannually. JP wants an ROR of 10% per year compounded semiannually. (*a*) If the bonds can be purchased through a broker at a discount price of $8375, should JP make the purchase? (*b*) If JP purchases a bond for $8375, what is the total gain in dollars?

Solution

(*a*) The interest every 6 months is

$$I = \frac{10,000(0.04)}{2} = \$200$$

For the nominal rate of 10%/2 = 5% per 6 months for 40 periods,

$$PW = 200(P/A,5\%,40) + 10,000(P/F,5\%,40) = \$4852$$

If the price is $8375 per bond, JP cannot even come close to 10% per year compounded semiannually. Do not buy these bonds!

(*b*) At $8375 per bond, we can find the dollar amount gained by computing the future worth after 40 periods. Assuming that all interest is reinvested at 10% per year compounded semiannually:

$$FW = 200(F/A,5\%,40) + 10,000 = \$34,159$$

Thus, JP will gain a total of $34,159 − $8375 = $25,784. However, as stated in part (*a*), the ROR will be much less than 10% per year.

Example 11.7

ROR OF A BOND, SECTION 11.4   In the preceding example, JP is naturally saddened by her inability to make 10% compounded semiannually if she pays $8375 for a 20-year 4%-per-year $10,000 bond. Compute (*a*) the actual nominal and effective returns per year and (*b*) the actual dollar gain if this rate is used for the reinvestment of the bond interest received.

Solution

(*a*) The rate-of-return equation is

$$0 = -8375 + 200(P/A,i^*,40) + 10,000(P/F,i^*,40)$$

Solution by spreadsheet or table interpolation shows that $i^* = 5.40\%$ per year nominally (2.70% semiannually) and that 5.47% is the effective annual rate.

(*b*) Calculate a nominal rate of 2.70% per semiannual period (from the effective rate of 5.40% per year compounded semiannually) and determine the future-worth value.

$$FW = 200(F/A,2.7\%,40) + 10,000 = \$24,180$$

which represents a gain of \$15,805 (\$24,180 − 8375).

Example 11.8

ROR OF A BOND, SECTION 11.4   An investor paid \$4240 for an 8% \$10,000 bond with interest payable quarterly. The bond was in default and, therefore, it paid no interest for the first 3 years after the investor bought it. If interest was paid for the next 7 years and then the investor was able to resell the bond for \$11,000, what rate of return did he make on the investment? Assume the bond is scheduled to mature 18 years after the investor bought it.

Solution

The bond interest received in years 4 through 10 was

$$I = \frac{(10,000)(0.08)}{4} = \$200 \text{ per quarter}$$

The rate of return *per quarter* can be determined by solving the PW equation developed on a per-quarter basis.

$$0 = -4240 + 200(P/A,i^* \text{ per quarter},28\ )(P/F,i^* \text{ per quarter},12)$$
$$+\ 11,000(P/F,i\% \text{ per quarter},40)$$

The equation is correct for $i^* = 4.1\%$ per quarter, which is a nominal 16.4% per year compounded quarterly.

Comment

The rate-of-return equation could also be written in terms of annual worth or future worth with the same result for i^*.

Example 11.8
(Spreadsheet)

Work Example 11.8 using a spreadsheet.

Solution

The solution is shown in Figure 11–4. The spreadsheet is set up to directly calculate an annual interest rate of 16.41% in cell E1. The quarterly bond interest receipts of \$200 are converted into equivalent annual receipts of \$724.24 using the PV function in cell E6. A quarterly rate could be determined initially on the spreadsheet, but this approach requires four times as many entries of \$200 each, compared to the six of \$724.24 entered here.

332

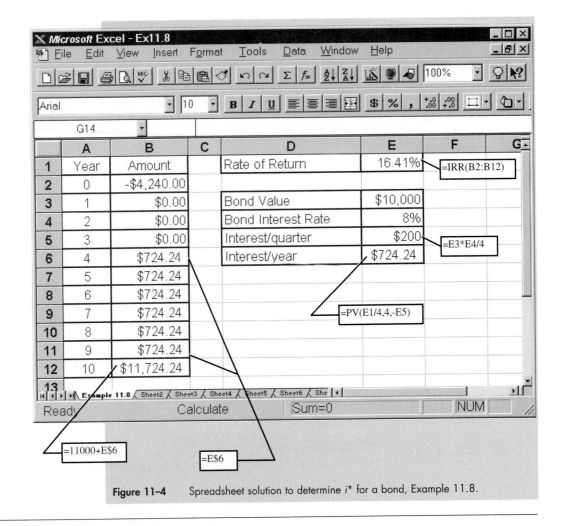

Figure 11-4   Spreadsheet solution to determine *i** for a bond, Example 11.8.

CHAPTER SUMMARY

A bond is a financial instrument issued by private or public entities for the sole purpose of raising debt capital. Bonds come in a variety of types, with many denominations and repayment schedules. As is true for all investments, the greater the risk to the purchaser, the greater is the reward in terms of rate of return. The two most common calculations associated with bond transactions are (1) the present worth of a bond to yield a specified rate of return, and (2) the rate of return associated with a specified cash-flow scenario.

PROBLEMS

11.1 What is meant by a convertible bond?

11.2 Rank the following bonds in terms of expected highest before-tax rate of return to the investor: general obligation, first mortgage, second mortgage, debenture, junk.

11.3   What is the difference between a U.S. Treasury note, a U.S. Treasury bill, and a U.S. Treasury bond?

11.4   What is the frequency and amount of the interest payments on a 6% $5000 bond for which interest is payable quarterly?

11.5   What are the interest payments and their frequency on an 8% $10,000 bond that pays monthly interest?

11.6   A debenture bond with an interest rate of 10% yielded $200 in interest every 6 months. What was the face value of the bond?

11.7   A collateral bond with interest payable quarterly was yielding $125 every 3 months in interest. If the face value of the bond was $8000, what was the bond interest rate?

11.8   A mortgage bond with an interest rate of 8% per year and a face value of $5000 is due 12 years from now. If the bond interest payments are $100, what is the bond interest payment period?

11.9   How much should you be willing to pay for a 7% $10,000 bond that is due 8 years from now if you want to make a nominal 8% per year compounded quarterly? Assume that the bond interest is payable quarterly.

11.10 What is the present worth of a 6% $50,000 bond that has interest payable monthly? Assume that the bond is due in 25 years and the desired rate of return is 12% per year compounded monthly.

11.11 If a manufacturing company needs to raise $2 million in capital to finance a small expansion, what total face value must the bonds have? The bonds will pay interest quarterly at a rate of 12% per year and will mature in 20 years. Assume that potential investors require a rate of return of a nominal 12% per year compounded quarterly and that the brokerage fees for handling the sale total $100,000.

11.12 A 9% $20,000 mortgage bond with interest payable semiannually was issued 4 years ago. The bond's maturity date is 20 years after issue. If the interest rate in the marketplace is 12% per year compounded semiannually, how much should an investor be willing to pay for (*a*) the stripped bond? (*b*) the bond coupons?

11.13 You desire to make a nominal 12% per year compounded quarterly. How much should you be willing to pay for a 9% $15,000 bond which has interest payable semiannually and is due in 20 years?

11.14 How much should you be willing to pay for a $4000 bond which has interest payable quarterly at 6%, if you desire to make an effective 10% per year compounded quarterly on the investment? The bond will mature in 10 years.

11.15 An investor purchased a convertible bond for $6000 four years ago. The bond has a face value of $10,000 and a bond interest rate of 8% payable semiannually. The investor can sell the bond now for $7500. If the investor keeps the bond, she thinks she can sell it for $8000 one year from now, $9100 two years from now, or get $10,000 upon maturity 3 years from now. What should she do if her MARR is 12% per year compounded semiannually?

11.16 An investor who was looking for protection against inflation was considering the purchase of inflation-adjusted bonds known as U.S. Treasury Inflation-Adjusted Securities (TIPS). With these bonds, the face values but not the interest payment is regularly adjusted to account for inflation. The investor could purchase either a $10,000 TIPS bond with a bond interest rate of 4% payable semiannually, or a $10,000 corporate bond with a bond interest rate of 8% per year payable semiannually. Both bonds mature 10 years after their purchase. The investor believes that there will be no inflation adjustment for the first 5 years, but after that, he expects the face value of the TIPS bond to increase by $1000 per year (i.e., years 6 through 10). Which bond should the investor purchase based on their present worth and an interest rate of 10% per year compounded semiannually?

11.17 Wet 'N' Wild Ship Building, Inc., must raise money for expanding its hull construction facility. If conventional bonds are issued, the bond interest rate will have to be 16% per year compounded semiannually. The face value of the bonds will be $7 million. If convertible bonds are issued, however, the bond interest rate will have to be only 7% per year compounded semiannually. What will the face value of the convertible bonds have to be to make the present worth of both issues the same at an interest rate of 20% per year compounded semiannually on the conventional bonds and 10% per year compounded semiannually on the convertible bonds? The conventional bonds mature in 10 years and the convertible bonds mature in 20 years.

11.18 Bonds purchased for $9000 have a face value of $10,000 and a bond interest rate of 10% per year payable semiannually. The bonds are due in 3 years. The company that issued the bonds expects a liquidity problem in 3 years and has advised all bondholders that if they will keep their bonds for another 2 years past the original due date, the bond interest for the extended 2-year period will be 16% per year payable semiannually. What is the present worth of the bonds if an investor would require a rate of return of 12% per year compounded semiannually?

11.19 An investor purchased a $5000 bond for $4000 two years ago. The bond interest rate is 6%, payable quarterly. If the investor plans to sell the bond now for $3500, what would be his rate of return on the investment?

11.20 You have been offered a 6% $20,000 bond at a 4% discount. If interest is paid quarterly and the bond is due in 15 years, what rate of return per quarter would you make if you bought the bond?

11.21 A $10,000 zero-coupon bond was issued when the prevailing market interest rate was 9%. If the bond sold for $3000 when it was issued, how many years into the future was the bond's maturity date?

11.22 Three years ago, Ms. Johnson purchased an 8% $10,000 bond which had interest payable semiannually for $7000. If the rate of return to maturity was a nominal 14% per year compounded semiannually when she bought it, how many years into the future was the bond's maturity date at that time?

11.23 In order to repave local streets, a city needed to acquire $3 million through a bond issue. At the time the bond issue was approved by the voters, the bond interest rate was set at 8% which was the same as the market interest rate. Between the time the bonds were approved and the time they were sold, however, the interest rate that the market required to attract investors went down to 6% per year. If the city puts the difference between 8% and 6% interest into an account which earns 5% per year interest, how much will it have when the bonds mature in 20 years?

11.24 A company is considering issuing bonds for financing a major construction project. The company is presently trying to determine whether it should issue conventional bonds or 'put' bonds (i.e., bonds which provide the holder the right to sell the bonds at a certain percentage of face value). If the company issues conventional bonds, it expects the bond purchasers to require an overall rate of return (to maturity) of 12% per year compounded quarterly. If put bonds are issued, however, investors would be satisfied with an 8% overall rate of return. The bond interest rate is 12% payable quarterly on a face value of $1,000,000 due 20 years from now. (*a*) How much more money could the company get now from the bond sale if it issues the put rather than the conventional bonds? (*b*) What rate of return per year will the company need to make on the extra money in order to have accumulated $1,000,000 to pay for the bonds in year 20?

11.25 At what bond interest rate will a $10,000 bond yield a nominal 12% per year compounded semiannually, if the purchaser pays $8000 and the bond becomes due in 12 years? Assume that the bond interest is payable semiannually.

11.26 At what bond interest rate will a $20,000 bond that has interest payable semiannually yield an effective 4% rate of return per 6 months, if the price of the bond is $18,000 and the bond becomes due in 20 years?

11.27 An investor purchased a $1000 convertible bond for $1200 from the Grow Company. The bond had an interest rate of 8% per year payable quarterly and was convertible to 20 shares of Grow common stock. If the investor kept the bond for 5 years and then converted it into common stock when the stock was selling for $51 per share, what was the nominal rate of return per year on his investment?

11.28 The Hi-Cee Steel Company issued $5 million worth of 20-year 14%-per-year, payable semiannually, callable bonds (i.e., bonds which could be called in and paid off at any time). The company agreed to pay a 10% premium on the face value if the bonds were called. Five years after the bonds were issued, the prevailing interest rate in the marketplace dropped to 10% per year. (*a*) What rate of return would the company make by calling the bonds and paying the $5.5 million? (*b*) Should the company call the bonds? (*c*) What rate of return will an investor make, if the bonds are called?

Inflation, Cost Estimation, and Indirect Cost Allocation

This chapter concentrates upon understanding how to consider inflation in time-value-of-money computations and in an engineering economy study. *Inflation* is a reality that we all deal with nearly every day.

The estimation of values to be utilized in an economic analysis, particularly cost-related amounts, is discussed here. Several techniques to estimate equipment expenses and process costs are presented based upon past data, indexes, and capacity relations. Other chapters which include sections on estimation techniques are Chapter 1 (cash flows, in general) and Chapter 18 (MARR and interest rates).

Finally, this chapter discusses the important area of *indirect cost allocation* for all types of indirect costs, especially those incurred in a manufacturing setting. Both the traditional method of cost allocation and the newer Activity-Based Costing (ABC) method are explained.

LEARNING OBJECTIVES

Purpose: Consider the elements of inflation, cost estimation and indirect cost allocation in an engineering economy analysis.

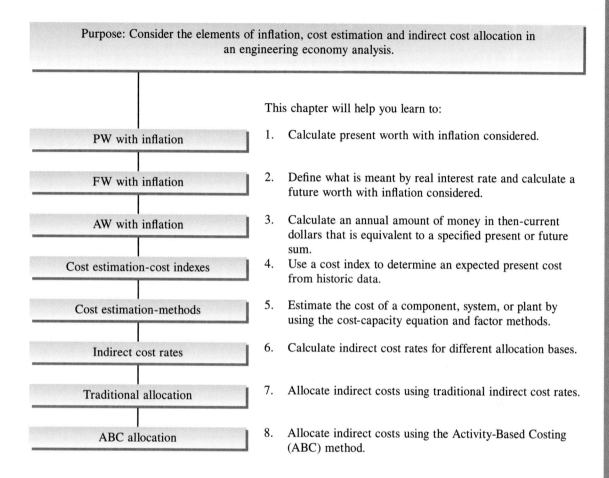

This chapter will help you learn to:

PW with inflation

1. Calculate present worth with inflation considered.

FW with inflation

2. Define what is meant by real interest rate and calculate a future worth with inflation considered.

AW with inflation

3. Calculate an annual amount of money in then-current dollars that is equivalent to a specified present or future sum.

Cost estimation-cost indexes

4. Use a cost index to determine an expected present cost from historic data.

Cost estimation-methods

5. Estimate the cost of a component, system, or plant by using the cost-capacity equation and factor methods.

Indirect cost rates

6. Calculate indirect cost rates for different allocation bases.

Traditional allocation

7. Allocate indirect costs using traditional indirect cost rates.

ABC allocation

8. Allocate indirect costs using the Activity-Based Costing (ABC) method.

12.1 INFLATION TERMINOLOGY AND PRESENT-WORTH COMPUTATION

Most people are very well aware of the fact that $20 now does not purchase the same amount as $20 did in 1995 or 1990 and purchases significantly less than in 1970. Why? This is inflation in action. Simply put, inflation is an increase in the amount of money necessary to obtain the same amount of product or service before the inflated price was present. Inflation occurs because the value of the currency has changed—it has gone down in value. The value of money has decreased and, as a result, it takes more dollars for fewer goods. This is a sign of *inflation.* In order to make comparisons between monetary amounts which occur in different time periods, the different-valued dollars must first be converted into constant-value dollars in order to represent the same buying power over time. This is especially important when future sums of money are considered, as is the case with all alternative evaluations.

Deflation is the opposite of inflation. The computations for inflation are equally applicable to a deflationary economy.

Money in one period of time, t_1, can be brought to the same value as money in another period of time, t_2, by using the generalized equation:

$$\text{Dollars in period } t_1 = \frac{\text{dollars in period } t_2}{\text{inflation rate between } t_1 \text{ and } t_2} \qquad \textbf{[12.1]}$$

Let dollars in period t_1 be called today's dollars and dollars in period t_2 be called future dollars or then-current dollars. If f represents the inflation rate per period and n is the number of time periods between t_1 and t_2, Equation [12.1] becomes

$$\text{Today's dollars} = \frac{\text{then-current dollars}}{(1 + f)^n} \qquad \textbf{[12.2]}$$

Another term for today's dollars is constant-value dollars. It is always possible to state future inflated amounts in terms of current dollars by applying Equation [12.2]. For example, if an item cost $5 in 1998 and inflation averaged 4% during the previous year, in constant 1997 dollars, the cost is equal to $5/(1.04) = $4.81. If inflation averaged 4% per year over the previous 10 years, the constant 1988 dollar equivalent is considerably less at $5/$(1.04)^{10} = $3.38.

There are actually three different rates used in this chapter; only the first two are interest rates: the real interest rate (i), the market interest rate (i_f), and the inflation rate (f).

Real or inflation-free interest rate *i*. This is the rate at which interest is earned when the effects of changes in the value of currency have been removed. Thus, the real interest rate presents an actual gain in buying power. This is the rate used in all the previous chapters in this book.

Market interest rate i_f.   As its name implies, this is the interest rate in the marketplace—the rate we hear about and commonly quoted every day. This rate is a combination of the real interest rate i and the inflation rate f, and, therefore, it changes as the inflation rate changes. It is also known as the inflated interest rate.

Inflation rate f.   As described above, this is a measure of the rate of change in the value of the currency.

A company's MARR, when adjusted for inflation, is correctly referred to as an inflation-adjusted MARR.

When the dollar amounts in different time periods are expressed as constant-value dollars per Equation [12.2], the equivalent present, future, or annual amounts are determined by using the real interest rate i in any of the Chapter 2 formulas. The calculations involved in this procedure are illustrated in Table 12–1 where an inflation rate of 4% per year is assumed. Column 2 indicates the increase in cost for each of the next 4 years for an item that has a cost of $5000 today. Column 3 shows the cost in future dollars, and column 4 presents the cost in constant-value (today's) dollars via Equation [12.2].

Observe from column 4 that when the future dollars of column 3 are converted into today's dollars, the cost is always $5000—as you should expect—the same as the cost at the start. This is predictably true when the costs are increasing by an amount *exactly equal* to the inflation rate. The actual cost of the item 4 years from now will be $5849, but in today's dollars the cost in 4 years will still amount to $5000.

Column 5 shows the present worth of the $5000 at the real interest rate $i = 10\%$ per year. In 4 years, PW = $3415. So, at $f = 4\%$ and $i = 10\%$, in 4 years $5000 today inflates to $5849, while $5000 four years from now has a PW of only $3415 today at a 10% return requirement.

Also, at a 4% interest rate, the amount $5849 has a current PW = $5849 $(P/F,4\%,4)$ = $5849 (0.8548) = $5000 since inflation and interest are exactly equal at 4%.

Table 12–1   Inflation calculations using today's dollars (rounded)

(1) Year, n	(2) Cost Increase Due to Inflation	(3) Cost in Future Dollars	(4) = (3)/(1.04)^n Future Cost in (Current) Today's Dollars	(5) = (4)(P/F,10%,n) Present Worth at $i = 10\%$
0		$5000	$5000	$5000
1	$5000(0.04) = $200	5200	$5200/(1.04)^1 = 5000$	4545
2	5200 (0.04) = 208	5408	$5408/(1.04)^2 = 5000$	4132
3	5408 (0.04) = 216	5624	$5624/(1.04)^3 = 5000$	3757
4	5624 (0.04) = 225	5849	$5849/(1.04)^4 = 5000$	3415

Figure 12–1 graphically presents the differences over a 4-year period of the constant-value amount of $5000, the future-dollar costs at 4% inflation, and the loss in present worth at 10% real interest caused by 4% inflation. The effect of compounded inflation and interest rates is large, as you can see by the shaded area.

An alternative method of accounting for inflation in a present-worth analysis involves adjusting the interest formulas themselves to account for inflation. Consider the P/F formula, where i is the real interest rate.

$$P = F\frac{1}{(1 + i)^n}$$

F (in future dollars) can be converted into today's dollars by using Equation [12.2].

$$P = \frac{F}{(1 + f)^n}\frac{1}{(1 + i)^n}$$

$$= F\frac{1}{(1 + i + f + if)^n} \qquad [12.3]$$

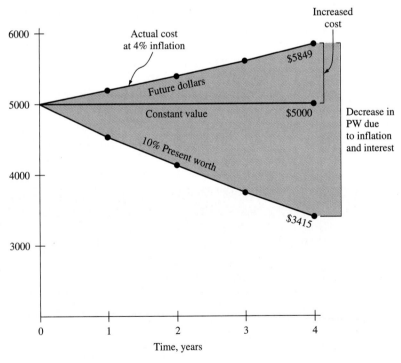

Figure 12–1   Comparison of constant-value dollars, future dollars, and present-worth dollars.

If the term $i + f + if$ in Equation [12.3] is defined as i_f, the equation becomes

$$P = F \frac{1}{(1 + i_f)^n} = F(P/F,i_f,n) \qquad [12.4]$$

The expression i_f is called the *inflated interest rate* and is defined as

$$i_f = i + f + if \qquad [12.5]$$

where i = real interest rate
     f = inflation rate
     i_f = inflated interest rate

For a real interest rate of 10% per year and an inflation rate of 4% per year, Equation [12.5] yields an inflated interest rate of 14.4%.

$$i_f = 0.10 + 0.04 + 0.10(0.04)$$
$$= 0.144$$

Table 12–2 details the use of i_f = 14.4% in the PW calculations for a current $5000 amount, which inflates to $5849 in future dollars 4 years hence as previously demonstrated in Table 12–1. As shown in column 4, the present worth of the item for each year is the same as that calculated in column 5 of Table 12–1.

The present worth of a series of cash flows—equal annual, uniform arithmetic gradient, or geometric (uniform percentage) gradient—can be found similarly. That is, either i or i_f is introduced into the P/A, P/G, or P_E factors, depending upon whether the cash flow is expressed in today's dollars or future dollars, respectively. If the series is expressed in today's dollars, then its PW is simply the discounted value using the real interest rate i. If the cash flow is expressed in future dollars, the amount today that would be equivalent to the

Table 12–2	Present-worth calculation using an inflated interest rate		
(1) Year, n	**(2)** Cost in Future Dollars	**(3)** $(P/F,14.4\%,n)$	**(4)** PW
0	$5000	1	$5000
1	5200	0.8741	4545
2	5408	0.7641	4132
3	5624	0.6679	3757
4	5849	0.5838	3415

inflated future dollars is obtained using i_f in the formulas. Alternatively, you can convert all future dollars into today's dollars and then use i.

Example 12.1

A former student of State College, who wishes to donate to her alma mater's Student Development Fund, has offered any one of the following three plans:

Plan A. $60,000 now.

Plan B. $15,000 per year for 8 years beginning 1 year from now.

Plan C. $50,000 three years from now and another $80,000 five years from now.

The only condition placed on the donation is that the college agree to spend the money on applied research related to the advancement of environmentally conscious manufacturing processes. From the college's perspective, it wants to select the plan which maximizes the buying power of the dollars received, so it has asked the engineering professor evaluating the plans to account for inflation in the calculations. If the college wants to earn a real 10% per year on its investments and the inflation rate is expected to average 3% per year, which plan should the college accept?

Solution

The quickest evaluation method is to calculate the present worth of each plan in today's dollars. For plans B and C, the easiest way to obtain the present worth is through the use of the inflated interest rate i_f. By Equation [12.5],

$$i_f = 0.10 + 0.03 + 0.10(0.03) = 0.133 \quad (13.3\%)$$

Compute the PW values with the appropriate use of Equation [12.4]:

$$PW_A = \$60,000$$

$$PW_B = \$15,000(P/A,13.3\%,8) = \$15,000(4.7508) = \$71,262$$

$$PW_C = \$50,000(P/F,13.3\%,3) + 80,000(P/F,13.3\%,5)$$

$$= \$50,000(0.68756) + 80,000(0.53561) = \$77,227$$

Since PW_C is the largest in today's dollars, select plan C.

Comment

The present worths of plans B and C could also have been found by first converting the cash flows into today's dollars using $f = 3\%$ and then using the real i of 10%. This procedure is more time-consuming, but the answers are the same.

Example 12.2

Calculate the present worth of a uniform series of payments of $1000 per year for 5 years if the real interest rate is 10% per year and the inflation rate is 4.5% per year, assuming that the payments are in terms of (a) today's dollars and (b) future dollars.

Solution

(a) Since the dollars are already expressed in today's dollars, the present worth is determined using the real $i = 10\%$.

$$PW = -1000(P/A,10\%,5) = -\$3790.80$$

(b) Since the dollars are expressed in future dollars, as shown in Figure 12–2a, use an inflated interest rate in the P/A factor.

$$i_f = i + f + if = 0.10 + 0.045 + (0.1)(0.045) = 14.95\%$$

$$PW = -1000(P/A,14.95\%,5) = -1000(3.3561) = -\$3356$$

The present worth can also be obtained by converting the future cash flows into today's dollars and then finding the present worth using the real interest rate of 10%. Refer to Figure 12–2b.

$$PW = -956.94(P/F,10\%,1) - 915.73(P/F,10\%,2) - 876.30(P/F,10\%,3)$$

$$-838.56(P/F,10\%,4) - 802.45(P/F,10\%,5)$$

$$= -\$3356$$

This is the same result obtained using i_f.

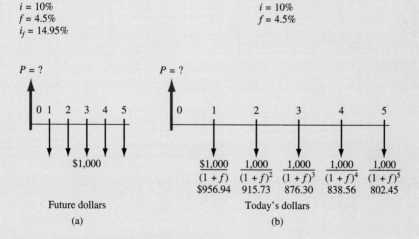

Figure 12–2   Cash flows for Example 12.2.

Comment

It is clear from this example that using the inflated interest rate i_f in the P/A factor is much simpler than converting the future dollar values into today's dollars and then applying the P/F factor.

Example 12.2
(Spreadsheet)

Use a spreadsheet to calculate the present worth of a uniform series of payments of $1000 per year for 5 years if the real interest rate is 10% per year and the inflation rate is 4.5% per year, assuming that the payments are in terms of (*a*) today's dollars and (*b*) future dollars.

Solution

Figure 12–3 includes the present-worth values in cells B7 and C7, respectively, for today's dollars using $i = 10\%$ and for future dollars using $i_f = 14.95\%$. The inflated interest rate i_f is calculated as indicated in cell C5 using Equation [12.5].

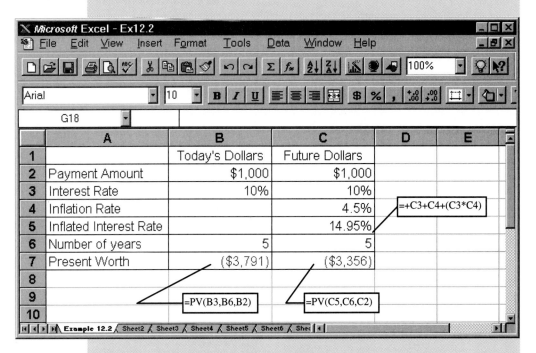

Figure 12–3   Spreadsheet solution of PW in today's and future dollars, Example 12.2.

To summarize, if future dollars are expressed in today's dollars (or have already been converted into today's dollars), the present worth should be calculated using the real interest rate i in the present-worth formulas. If the future dollars are expressed in then-current or future dollars, the inflated interest rate i_f should be used in the formulas.

Additional Examples 12.12 and 12.13
Problems 12.1 to 12.12

12.2   FUTURE-WORTH CALCULATIONS WITH INFLATION CONSIDERED

In future-worth calculations, the future sum of money in dollars can represent any one of four different amounts:

Case 1.   The *actual amount* of money that will be accumulated at time n.

Case 2.   The *buying power,* in terms of today's dollars, of the actual amount of dollars accumulated at time n.

Case 3.   The number of *future dollars required* at time n to maintain the same purchasing power as a dollar today; that is, no interest is considered.

Case 4.   The number of dollars required at time n to *maintain purchasing power and earn a stated real interest rate.* (Actually this makes case 4 and case 1 computations identical.)

It should be clear that for *case 1,* the actual amount of money accumulated is obtained by using a stated market interest rate, which we identify by i_f in this chapter because it includes inflation.

For *case 2,* the buying power of the future dollars is determined by using the market interest rate i_f to calculate F and then dividing by $(1 + f)^n$. Division by $(1 + f)^n$ deflates the inflated dollars. This, in effect, recognizes that prices increase during inflation so that \$1 in the future will purchase less goods than \$1 now. In equation form, case 2 is

$$F = \frac{P(1 + i_f)^n}{(1 + f)^n} = \frac{P(F/P,i_f,n)}{(1 + f)^n}$$ [12.6]

As an illustration, suppose \$1000 earns the market rate of 10%-per-year interest for 7 years. If the inflation rate for each year is 4%, the amount of money accumulated in 7 years, but with today's buying power, is

$$F = \frac{1000(F/P,10\%,7)}{(1.04)^7}$$

$$= \$1481$$

To understand the power of inflation, consider this. If inflation were nil (f approximates 0), in 7 years the \$1000 at 10% interest rate would grow to

$$F = 1000(F/P,10\%,7) = \$1948$$

This means that the buying power today and 7 years in the future are equal. Inflation at 4% annually removed \$467 in buying power.

Also for case 2, the future amount of money accumulated with today's buying power could equivalently be determined by calculating the real interest rate and using it in the F/P factor to compensate for the decreased purchasing power of the dollar. This real interest rate can be obtained by solving for i in Equation [12.5].

$$i_f = i + f + if$$
$$= i(1 + f) + f$$
$$i = \frac{i_f - f}{1 + f} \qquad [12.7]$$

This equation allows us to calculate the real interest rate from the market (inflated) interest rate. The real interest rate i represents the rate at which present dollars will expand with their *same buying power* into equivalent future dollars. The use of this interest rate is appropriate when calculating the future worth of an investment, especially a savings account or money market fund, when the effects of inflation must be considered. For the $1000 amount mentioned previously, from Equation [12.7]

$$i = \frac{0.10 - 0.04}{1 + 0.04} = 0.0577 \qquad (5.77\%)$$
$$F = 1000(F/P,5.77\%,7) = \$1481$$

The stated (market) interest rate of 10% per year has been reduced to less than 6% per year because of the erosive effects of inflation. An inflation rate larger than the interest rate, that is, $f > i_f$, leads to a negative real interest rate i in Equation [12.7].

Case 3 also recognizes that prices increase during inflationary periods, and, therefore, purchasing something at a future date will require more dollars than are required now for the same thing. Simply put, future (then-current) dollars are worth less, so more are needed. No interest rate is considered at all in this case. This is the situation present if someone asks, How much will a car cost in 5 years if its current cost is $15,000 and its price will increase by 6% per year? (The answer is $20,073.38.) No interest rate, only inflation, is involved. To find the cost F, substitute f directly for the interest rate in the F/P factor.

$$F = P(1 + f)^n = P(F/P,f,n)$$

Thus, if $1000 represents the cost of an item which is escalating in price exactly in accordance with the inflation rate of 4% per year, the cost 7 years from now will be

$$F = 1000(F/P,4\%,7) = \$1316$$

Calculations for *case 4* (maintaining purchasing power and earning interest) take into account both increasing prices (case 3) and the time value of money; that is, if real growth of capital is to be obtained, funds must grow at a rate equal to the real interest rate i plus a rate equal to the inflation rate f. Thus, to make

Table 12–3	Calculation methods for various future values	
Future Value Desired	**Method of Calculation**	**Example for $P = \$1000$, $n = 7$, $i_f = 10\%$, $f = 4\%$**
Case 1: Actual dollars accumulated	Use stated market rate i_f in equivalence formulas	$F = 1000(F/P,10\%,7)$
Case 2: Buying power of accumulated dollars in terms of today's dollars	Use market rate i_f in equivalence and divide by $(1+f)^n$ or Use real i	$F = \dfrac{1000(F/P,10\%,7)}{(1.04)^7}$ or $F = 1000(F/P,5.77\%,7)$
Case 3: Dollars required for same purchasing power	Use f in place of i in equivalence formulas	$F = 1000(F/P,4\%,7)$
Case 4: Future dollars to maintain purchasing power and to earn interest	Calculate i_f and use in equivalence formulas	$F = 1000(F/P,10\%,7)$

a *real rate of return of 5.77%* when the inflation rate is 4%, i_f is used in the formulas. Using the same $1000 amount,

$$i_f = 0.0577 + 0.04 + 0.0577(0.04) = 0.10 \qquad (10\%)$$

$$F = 1000(F/P,10\%,7) = \$1948$$

This calculation shows that $1948 7 years in the future will be equivalent to $1000 now with a real return of $i = 5.77\%$ per year and inflation of $f = 4\%$ per year.

   In summary, the calculations made in this section reveal that $1000 now at a market rate of 10% per year would accumulate to $1948 in 7 years; the $1948 would have the purchasing power of $1481 of today's dollars if $f = 4\%$ per year; an item with a cost of $1000 now would cost $1316 in 7 years at an inflation rate of 4% per year; and it would take $1948 of future dollars to be equivalent to the $1000 now at a real interest rate of 5.77% with inflation considered at 4%. Table 12–3 summarizes which rate is used in the equivalence formulas as a function of which interpretation of F is taken.

Example 12.3

Abbott Chemical wants to determine whether it should pay now or pay later for upgrading its production facilities. If the company selects action plan A, the necessary equipment will be purchased now for $20,000. However, if the company selects inaction plan I, the equipment purchase will be deferred for 3 years when the cost is expected to rise rapidly to $34,000. The non-inflation-adjusted (real) MARR is 12% per year and the inflation rate is estimated at 3% per year. Determine whether the company should purchase now or later (*a*) when inflation is not considered and (*b*) when inflation is considered.

Solution

(*a*) **Inflation not considered.** The non-inflation-adjusted MARR is $i = 12\%$ per year, and the cost of plan *I* is $34,000 three years hence. We can compute either P now or F 3 years from now and select the plan with the lower cost. For F,

$$F_A = -20{,}000\ (F/P{,}12\%{,}3) = -\$28{,}098$$

$$F_I = -\$34{,}000$$

Select plan *A* because it costs less; purchase now.

(*b*) **Inflation considered.** First, compute the inflation-adjusted MARR by Equation [12.5].

$$i_f = 0.12 + 0.03 + 0.12(0.03) = 0.1536$$

Use i_f to compute the F value for plan *A* to determine the future dollars necessary. Plan *I* still costs $34,000 in 3 years.

$$F_A = -20{,}000(F/P{,}15.36\%{,}3) = -\$30{,}704$$

$$F_I = -\$34{,}000$$

Plan *A* is still selected, because it requires less equivalent future dollars.

Comment

Suppose this example involved a company in a country with higher inflation, say, 12% annually, and a higher non-inflation-adjusted MARR was accordingly required, say, 18%. Now when inflation is taken into account, the more economic alternative is different from the one selected if inflation is ignored. See the results below.

No Inflation Considered	Inflation Considered
$i = 18\%$	$i_f = 0.18 + 0.12 + 0.18(0.12) = 32.2\%$
$F_A = -\$20{,}000(F/P{,}18\%{,}3) = -\$32{,}860$	$F_A = -\$20{,}000(F/P{,}32.2\%{,}3) = -\$46{,}209$
$F_I = -34{,}000$	$F_I = -34{,}000$
Select plan *A*	Select plan *I*

Most countries have inflation rates in the range of 2% to 8% per year, but *hyperinflation* is a problem in countries where political instability, overspending by the government, weak international trade balances, etc., are present. Hyperinflation rates may be very high, 10% to 50% per month, for example. In these cases, often the government redefines the currency in terms of the currency of other countries, controls banks and corporations, and controls the flow of capital into and out of the country in order to decrease inflation.

In a hyperinflated environment, everyone usually spends all their money immediately since the cost will be so much higher the next month, week, or day. To appreciate the disastrous effect of hyperinflation on a company's ability to keep up, rework Example 12.3*b* using an inflation rate of 10% per month, that is, 120% per year (not considering compounding of inflation). Your F_A amount should skyrocket, and plan *I* is a clear choice. Of course, in such an environment the $34,000 purchase price 3 years hence would obviously not be guaranteed, so the entire economic analysis is not reliable. Good economic decisions in a hyperinflated economy are very difficult to make using traditional engineering economy methods, since the estimated future values are very unreliable and the future availability of capital is uncertain.

Problems 12.13 to 12.19

12.3   CAPITAL-RECOVERY AND SINKING-FUND CALCULATIONS WITH INFLATION CONSIDERED

It is particularly important for capital-recovery calculations to include inflation because current capital dollars must be recovered with future inflated dollars. Since future dollars have less buying power than do today's dollars, it is obvious that more dollars will be required to recover the present investment. This suggests the use of the inflated or market interest rate in the A/P formula. For example, if $1000 is invested today at a real interest rate of 10% per year when the inflation rate is 8% per year, the annual amount of capital that must be recovered each year for 5 years in then-current dollars will be

$$A = 1000(A/P,18.8\%,5) = \$325.59$$

On the other hand, the decreased value of dollars through time means that investors may be willing to spend fewer present (higher-value) dollars to accumulate a specified amount of future (inflated) dollars by using a sinking fund; that is, an A value is computed. This suggests the use of a higher interest rate, that is, the i_f rate, to produce a lower A value in the A/F formula. The annual equivalent (when inflation is considered) of the same $F = \$1000$ five years from now in then-current dollars is

$$A = 1000(A/F,18.8\%,5) = \$137.59$$

For comparison, the equivalent annual amount to accumulaate $F = \$1000$ at a real $i = 10\%$, (before inflation is considered), is $1000(A/F,10\%,5) = \$163.80$ as illustrated in Example 12.4. Thus, when F is fixed, uniformly distributed future costs should be spread over as long a time period as possible so that inflation will have the effect of reducing the payment involved ($137.59 versus $163.80 here).

Example 12.4

What annual amount is required for 5 years to accumulate an amount of money with the same buying power as $680.58 today, if the market interest rate is 10% per year and inflation is 8% per year?

Solution

The actual number of future (inflated) dollars required in 5 years is

$$F = \text{(present buying power)}(1+f)^5 = 680.58(1.08)^5 = \$1000$$

Therefore, the actual amount of the annual deposit is calculated using the market (inflated) interest rate of 10%.

$$A = 1000(A/F,10\%,5) = \$163.80$$

Comment

Note that the real interest rate implied here is $i = 1.85\%$ as determined using Equation [12.7]. To put the above calculations in perspective, if the inflation rate is zero when the real interest rate is 1.85%, the future amount of money with the same buying power as $680.58 today is obviously $680.58. Then the annual amount required to accumulate this future amount in 5 years is $A = 680.58(A/P,1.85\%,5)$, which is $131.17. This is $32.63 lower than the $163.80 calculated above where $f = 8\%$. This difference is due to the fact that during inflationary periods, dollars deposited into sinking fund accounts (or savings accounts) have more buying power than the dollars returned at the end of the deposit period. In order to make up for the buying power than the dollars returned at the end of the deposit period. In order to make up for the buying power difference between higher-value dollars deposited and the lower-value dollars returned, more lower-value dollars are required. To maintain equivalent buying power, there is in this case, a difference of $32.63 required between $f = 0$ and $f = 8\%$ per year.

Quite simply, the logic discussed here explains why, in times of increasing inflation, lenders of money, such as credit card companies, mortgage companies, and banks, tend to further increase their market interest rates. People tend to pay off less of their incurred debt at each payment because they use any excess money to purchase additional items before the price is further inflated. Also, the lending institutions must have more dollars in the future to cover the expected higher costs of lending money. All of this is because of the spiraling effect of increasing inflation. Breaking this cycle is difficult to do at the individual level and much more difficult to alter at a national level. When unchecked, a hyperinflated economy can result.

Problems 12.20 to 12.24

12.4   COST INDEXES FOR ESTIMATION

Even a cursory study of recent world history reveals that the currency values of virtually every country are in a constant state of change. For engineers involved in project planning and design, this makes the difficult job of cost estimation even more difficult. One method of obtaining preliminary cost estimates is by looking at the costs of similar projects that were completed at some time in the past and updating these past cost figures. Cost indexes represent a convenient tool for accomplishing this.

A *cost index* is a ratio of the cost of something today to its cost at some time in the past. One such index that most people are familiar with is the Consumer Price Index (CPI), which shows the relationship between present and past costs for many of the things that "typical" consumers must buy. This index, for example, includes such items as rent, food, transportation, and certain services. However, other indexes are more relevant to the engineering profession because they track the costs of goods and services which are more pertinent to engineers. Table 12–4 is a listing of some of the more common indexes, and, as shown,

Table 12–4	Types and sources of various cost indexes
Type of Index	**Source**
Overall prices	
Consumer (CPI)	Bureau of Labor Statistics
Producer (wholesale)	U.S. Department of Labor
Construction	
Chemical plant overall	*Chemical Engineering*
Equipment, machinery, and supports	
Construction labor	
Buildings	
Engineering and supervision	
Engineering News Record overall	*Engineering News Record (ENR)*
Construction	
Building	
Common Labor	
Skilled Labor	
Materials	
EPA treatment plant indexes	Environmental Protection Agency, *ENR*
Large-city advanced treatment (LCAT)	
Small-city conventional treatment (SCCT)	
Federal highway	
Contractor cost	
Equipment	
Marshall and Swift (M&S) overall	Marshall & Swift
M&S specific industries	

some are applicable to a wide variety of goods and services (CPI) while others are more directed toward the needs of a specific user (federal highway index).

Generally, the indexes are made up of a mix of components which are assigned certain weights, with the components sometimes further subdivided into more basic items. For example, the equipment, machinery, and support component of the chemical plant cost index is subdivided further into process machinery, pipes, valves and fittings, pumps and compressors, and so forth. These subcomponents, in turn, are built up from even more basic items like pressure pipe, black pipe, and galvanized pipe. Table 12–5 shows the *Chemical Engineering* plant cost index, the *Engineering News Record* (*ENR*) construction cost index, and the Marshall and Swift (M&S) equipment cost index for the years 1970 through 1996, with the base period of 1957–1959 assigned a value of 100 for the *Chemical Engineering* (*CE*) plant cost index, 1913 = 100 for the *ENR* index, and 1926 = 100 for the M&S equipment cost index.

Table 12–5	Values for selected indexes		
Year	*CE* Plant Cost Index	*ENR* Construction Cost Index	M&S Equipment Cost Index
1975	182.4	2304.60	444.3
1976	192.1	2494.30	472.1
1977	204.1	2672.40	505.4
1978	218.8	2872.40	545.3
1979	238.7	3139.10	599.4
1980	261.2	3378.17	659.6
1981	297.0	3725.55	721.3
1982	314.0	3939.25	745.6
1983	316.9	4108.74	760.8
1984	322.7	4172.27	780.4
1985	325.3	4207.84	789.6
1986	318.4	4348.19	797.6
1987	323.8	4456.53	813.6
1988	342.5	4573.29	852.0
1989	355.4	4672.66	895.1
1990	357.6	4770.03	915.1
1991	361.3	4886.52	930.6
1992	358.2	5070.66	943.1
1993	359.2	5335.81	964.2
1994	368.1	5443.14	993.4
1995	381.1	5523.13	1027.5
1996	381.8	5618.08	1039.2

The general equation for updating costs through the use of any cost index over a period from time $t = 0$ (base) to another time t is

$$C_t = \frac{C_0 I_t}{I_0} \qquad\qquad \text{[12.8]}$$

where C_t = estimated cost at present time t
$\quad\quad\ C_0$ = cost at previous time t_0
$\quad\quad\ I_t$ = index value at time t
$\quad\quad\ I_0$ = index value at time t_0

Example 12.5 illustrates the use of the *ENR* index to estimate present costs from past values.

Example 12.5

In evaluating the feasibility of a major construction project, an engineer is interested in estimating the cost of skilled labor for the job. The engineer finds that a project of similar complexity and magnitude was completed 5 years ago when the *ENR* skilled labor index was 3496.27. The skilled labor cost for that project was $360,000. If the *ENR* skilled labor index now stands at 4038.44, what is the expected skilled labor cost for the new project?

Solution

The base time t_0 is 5 years ago. Using Equation [12.8], the present cost estimate is

$$C_t = \frac{C_0 I_t}{I_0} = \frac{(360,000)(4038.44)}{3496.27}$$

$$= \$415,825.55$$

Problems 12.25 to 12.32

12.5 COST ESTIMATING

While the cost indexes discussed above provide a valuable tool for estimating present costs from historical data, they become even more valuable when combined with some of the other cost-estimating techniques. One of the most widely used methods of obtaining preliminary cost information is through the use of *cost-capacity equations*. As the name implies, a cost-capacity equation relates the cost of a component, system, or plant to its capacity. Since many cost-capacity relationships plot as a straight line on log-log paper, one of the most common cost-prediction equations is

$$C_2 = C_1 \left(\frac{Q_2}{Q_1} \right)^x \qquad \text{[12.9]}$$

where C_1 = cost at capacity Q_1
   C_2 = cost at capacity Q_2
   x = exponent

The value of the exponent for various components, systems, or entire plants can be obtained or derived from a number of sources, including *Plant Design and Economics for Chemical Engineers, Preliminary Plant Design in Chemical Engineering, Chemical Engineers' Handbook,* technical journals (especially *Chemical Engineering*), the U.S. Environmental Protection Agency, professional or trade organizations, consulting firms, and equipment companies. Table 12–6 is a partial listing of typical values of the exponent for various units. When an exponent value for a particular unit is not known, it is common practice

Table 12–6	Sample exponent values for cost-capacity equations	
Component/System/Plant	**Size Range**	**Exponent**
Activated sludge plant	1–100 MGD	0.84
Aerobic digester	0.2–40 MGD	0.14
Blower	1,000–7,000 ft/min	0.46
Centrifuge	40–60 in	0.71
Chlorine plant	3,000–350,000 tons/year	0.44
Clarifier	0.1–100 MGD	0.98
Compressor	200–2,100 hp	0.32
Cyclone separator	20–8,000 ft^3/min	0.64
Dryer	15–400 ft^2	0.71
Filter, sand	0.5–200 MGD	0.82
Heat exchanger	500–3,000 ft^2	0.55
Hydrogen plant	500–20,000 scfd	0.56
Laboratory	0.05–50 MGD	1.02
Lagoon, aerated	0.05–20 MGD	1.13
Pump, centrifugal	10–200 hp	0.69
Reactor	50–4,000 gal	0.74
Sludge drying beds	0.04–5 MGD	1.35
Stabilization pond	0.01–0.2 MGD	0.14
Tank, stainless	100–2,000 gal	0.67

| NOTE: MGD = million gallons per day; hp = horsepower; scfd = standard cubic feet per day.

to use the average value of 0.6. The next example illustrates the use of Equation [12.9].

Example 12.6

The total construction cost for a stabilization pond to handle a flow of 0.05 million gallons per day (MGD) was $73,000 in 1987. Estimate the cost today of a pond 10 times larger. Assume the index (for updating the cost) was 131 in 1987 and is 225 today. The exponent from Table 12–6 for the MGD range of 0.01 to 0.2 is 0.14.

Solution

Using Equation [12.9], the cost of the pond in 1987 funds is

$$C_2 = 73{,}000\left(\frac{0.50}{0.05}\right)^{0.14}$$

$$= \$100{,}768$$

Today's cost can be obtained through the use of Equation [12.8] as follows:

$$C_t = \frac{(100{,}768)(225)}{131}$$

$$= \$173{,}075 \qquad \text{(today's dollars)}$$

A different, but widely used, simplified approach for obtaining preliminary cost estimates of process plants is called the *factor method*. While Equation [12.9] can be used for estimating both the costs of major items of equipment and the total plant costs, the factor method was developed only for obtaining total plant costs. The method is based on the premise that fairly reliable total plant costs can be obtained by multiplying the cost of the major equipment by certain factors. Since major-equipment costs are readily available, rapid plant estimates are possible if the appropriate factors are known. These factors are commonly referred to as Lang factors after Hans J. Lang, who first proposed the method in 1947.

In its simplest form, the factor method of cost estimation can be expressed as

$$C_T = hC_E \qquad\qquad\qquad \text{[12.10]}$$

where C_T = total plant cost
     h = overall cost factor or summation of individual cost factors
     C_E = summation of cost of major items of equipment

Note that h may be one overall cost factor (Example 12.7) or the sum of individual cost factors, as described later and illustrated in Examples 12.8 and 12.14.

In his original work, Lang showed that construction cost factors and overhead cost factors can be combined into one overall factor for various types of plants as follows: solid process plants, 3.10; solid-fluid process plants, 3.63; and fluid process plants, 4.74. These factors reveal that the total installed-plant cost is many times the purchase cost of the major items of equipment. Example 12.7 illustrates the use of overall cost factors.

Example 12.7

A solid-fluid process plant is expected to have a delivered-equipment cost of $565,000. If the overall cost factor for this type of plant is 3.63, estimate the plant's total cost.

Solution

The total plant cost is estimated by Equation [12.10].

$$C_t = 3.63(565,000)$$
$$= \$2,051,000$$

Subsequent refinements of the factor method have led to the development of separate factors for various elements of the direct and indirect costs. Direct costs are those which are specifically identifiable with a product, function, or activity. These costs usually include expenditures such as raw materials, direct labor, and specific equipment. Indirect costs are those not directly attributable to a single function but shared by several because they are necessary to perform the overall objective. Examples of indirect costs are general administration, taxes, support functions (such as purchasing), and security. The factors for both direct and indirect costs are sometimes developed from delivered-equipment costs and other times from installed-equipment costs. In this text, we will assume that all factors apply to delivered-equipment costs unless otherwise specified.

Furthermore for indirect costs, some of the factors apply to equipment costs while others apply to the total direct cost. In the former case, the indirect cost factors apply to equipment cost, just as do the direct cost factors. Therefore, the simplest procedure is to add the direct and indirect cost factors before multiplying by the delivered-equipment cost. In the latter case, the direct cost must be calculated first because the indirect cost factor must be applied to the direct cost rather than to the equipment cost. As with direct costs, we will assume that the indirect cost factors apply to the delivered-equipment cost. Additional Example 12.14 illustrates these calculations.

It should be pointed out that since the literature-reported values of the factors are decimal fractions of the total equipment costs, we must add 1 to their sum in order to obtain the total plant cost estimate from Equation [12.10]. It is *not necessary to add 1 to the overall plant cost factors* described earlier (the 1 is already included). Example 12.8 illustrates the use of direct and indirect cost factors for estimating the total plant cost.

Example 12.8

The delivered-equipment cost for a small chemical process plant is expected to be $2 million. If the direct cost factor is 1.61 and the indirect cost factor is 0.25, determine the total plant cost.

Solution

Since all factors apply to the delivered-equipment cost, they can be added to obtain the total cost factor. Remember that 1 must be added to the total because the factors are decimal values. Thus,

$$h = 1 + 1.61 + 0.25 = 2.86$$

From Equation [12.10], the total plant cost is

$$C_T = 2.86(2,000,000) = \$5,720,000$$

Comment

A more complicated use of cost factors is given in Additional Example 12.14.

Additional Example 12.14
Problems 12.33 to 12.38

12.6   COMPUTATION OF INDIRECT COST (OVERHEAD) RATES

Costs incurred in the production of an item or delivery of a service are tracked and assigned by a *cost accounting system.* For the manufacturing environment, it can be stated generally that the *statement of cost of goods sold* (see Appendix B) is one end product of this system. The cost accounting system accumulates material costs, labor costs, and indirect costs (also called overhead costs or factory expenses) by using *cost centers.* All costs incurred in one department or process line are collected under one cost-center title, for example, Department 3X. Since direct materials and direct labor are usually directly assignable to a cost center, the system need only identify and track these costs. Of course, this in itself is no easy chore, and the cost of the tracking system may prohibit collection of all direct cost data to the detail desired.

One of the primary and more difficult tasks of cost accounting is the allocation of *indirect costs* when it is necessary to allocate them separately to departments, processes, and product lines. The costs associated with property

taxes, service and maintenance departments, personnel, legal, quality, supervision, purchasing, utilities, software development, etc., must be allocated to the using cost center. Detailed collection of these data is cost-prohibitive and often impossible; thus, allocation schemes are utilized to distribute the expenses on a reasonable basis. A listing of possible bases is included as Table 12–7. Historically common bases are direct labor cost, direct labor hours, space, and direct materials.

Table 12–7	Indirect cost allocation bases
Cost Category	**Possible Allocation Basis**
Taxes	Space occupied
Heat, light	Space, usage, number of outlets
Power	Space, direct labor hours, direct labor cost, machine hours
Receiving, purchasing	Cost of materials, number of orders, number of items
Personnel, machine shop	Direct labor hours, direct labor cost
Building maintenance	Space occupied, direct labor cost
Software development	Cycle time, throughput, number of accesses
Quality control	Number of inspections

Most allocation is accomplished utilizing a predetermined *indirect cost rate,* or factory expense rate, computed using the general relation:

$$\text{Indirect cost rate} = \frac{\text{estimated indirect costs}}{\text{estimated basis level}} \qquad \text{[12.11]}$$

The estimated indirect cost is the amount allocated to a cost center. For example, if a company division has two producing departments, the total indirect cost allocated to a department is used as the numerator in Equation [12.11] to determine the department rate. Example 12.9 illustrates allocation when the cost center is a machine.

Example 12.9

EnviroTech, Inc., is computing indirect cost rates for the production of glass products. The following information is obtained from last year's budget for the three machines used in production.

Cost Source	Allocation Basis	Estimated Activity Level
Machine 1	Direct labor cost	$10,000
Machine 2	Direct labor hours	2,000 hours
Machine 3	Direct material cost	$12,000

Determine rates for each machine if the estimated indirect cost budget is $5000 per machine.

Solution

Applying Equation [12.11] for each machine, annual rates are

$$\text{Machine 1 rate} = \frac{\text{indirect budget}}{\text{direct labor cost}} = \frac{5000}{10{,}000}$$

$$= \$0.50 \text{ per direct labor dollar}$$

$$\text{Machine 2 rate} = \frac{\text{indirect budget}}{\text{direct labor hours}} = \frac{5000}{2000}$$

$$= \$2.50 \text{ per direct labor hour}$$

$$\text{Machine 3 rate} = \frac{\text{indirect budget}}{\text{material cost}} = \frac{5000}{12{,}000}$$

$$= \$0.42 \text{ per direct material dollar}$$

Comment

Once the product has been manufactured and actual direct labor costs and hours, and material costs are computed, each dollar of direct labor cost spent on machine 1 implies that $0.50 in indirect cost will be added to the cost of the product. Indirect expenses are added for machines 2 and 3 using the rates determined.

Additional Example 12.15
Problems 12.39 to 12.41

12.7 TRADITIONAL INDIRECT COST ALLOCATION AND VARIANCE

Once a period of time (a month, quarter, or year) has passed, the indirect cost rates and actual data for the relevant basis are used to determine the indirect cost *charge,* which is then added to other (direct) costs. This results in the total cost of production. These costs are all accumulated by *cost center,* as described in the previous section.

If the total indirect cost allocation budget is correct, the total indirect costs charged to all cost centers should equal this allocation. However, since some error in budgeting always exists, there will be some overallocation or underallocation relative to actual charges, which is termed *allocation variance*. Experience in indirect cost estimation assists in reducing the variance at the end of the accounting period. Example 12.10 illustrates indirect cost allocation and variance computation.

Example 12.10

Since we determined indirect cost rates for EnviroTech (Example 12.9), we can now compute the actual cost of production. Perform the computations using the actual data in Table 12–8. Also, calculate the variance for indirect cost allocation.

Table 12–8	Actual data used for indirect cost allocation		
Cost Source	**Machine Number**	**Actual Cost**	**Actual Hours**
Material	1	$ 3,800	
	3	19,550	
Labor	1	2,500	650
	2	3,200	750
	3	2,800	720

Solution

To determine actual production costs, start with the cost of goods sold (factory cost) relation given by Equation [C.1] in Appendix C, which is

Factory cost = direct material + direct labor + factory expense

To determine factory expense, which is another term for indirect cost, the rates from Example 12.9 are utilized:

Machine 1 indirect = (labor cost)(rate) = 2500(0.50)

= $1250

Machine 2 indirect = (labor hours)(rate) = 750(2.50)

= $1875

Machine 3 indirect = (material cost)(rate) = 19,550(0.42)

= $8211

Total charged indirect cost = $11,336

Factory cost is the sum of material and labor costs from Table 12–8 and the indirect cost charge for a total of $43,186.

Based on the indirect cost allocation of $5000 per machine, the variance for total indirect cost is

$$\text{Variance} = 3(\$5000) - \$11,336 = \$3664$$

This is an overallocation, since less was charged than allocated. The $15,000 budgeted for the three machines represents a 32.3% overestimate of indirect costs. This analysis may prompt a different indirect cost budget for EnviroTech in future years.

Once estimates of indirect costs are determined, it is possible to perform an economic analysis of the present operation versus a proposed or anticipated operation. Such a study is described in Additional Example 12.16.

Additional Example 12.16
Problems 12.42 to 12.45

12.8   ACTIVITY-BASED COSTING (ABC) FOR INDIRECT COSTS

As automation and manufacturing technologies have advanced, the number of direct labor hours necessary to manufacture a product has decreased substantially. Where once as much as 35% to 45% of the final product cost was represented in labor, now the labor component is commonly 5% to 10% of total manufacturing cost. However, the indirect or overhead cost may now be as much as 35% of the total manufacturing cost. The use of historically relied upon bases, such as direct labor hours, to allocate indirect cost is not accurate enough for automated environments. This has led to the development of new methods which supplement traditional cost accounting allocations that rely upon one form or another of Equation [12.11]. Also, allocation bases different from traditional ones are commonly utilized.

It is important from an engineering economy viewpoint to realize when traditional cost accounting systems should be augmented with better indirect cost accounting methods. A product which, by traditional methods, may have seemingly contributed a large portion to profit may actually be a loser when indirect costs are allocated more precisely. Companies which have a wide variety of products and produce some in small lots may find that traditional cost accounting methods have a tendency to underallocate the indirect cost to small-lot products. This may indicate that they are profitable, when in actuality they are losing money.

A commonly employed augmentation technique for indirect cost allocation is *Activity-Based Costing, ABC* for short. By design, its goal is to develop a cost center, called a *cost pool,* for each event, or *activity,* which acts as a *cost driver.* In other words, cost drivers actually *drive* the consumption of a shared resource and are charged accordingly. Cost pools are usually departments or functions—purchasing, inspection, maintenance, and software development. Activities are events such as purchase orders, reworks, repairs, computer package activations, machine setups, wait time, and engineering changes.

Some proponents of the ABC method recommend discarding the traditional cost accounting methods of a company and utilizing ABC exclusively. This is not a good approach, since ABC is not a complete cost system. The two systems work well together with the traditional methods allocating costs which have identifiable direct bases, e.g., direct labor, as in Example 12.10. The ABC method can then be utilized to further allocate support-service costs using activity bases such as those mentioned above.

The ABC methodology involves a two-step process:

1. **Define cost pools.**   Usually these are support functions.
2. **Identify cost drivers.**   These help trace costs to the cost pools.

As an illustration, a company which produces an industrial laser has three primary support departments identified as cost pools in step 1: *A*, *B*, and *C*. The annual support cost for the purchasing cost driver (step 2) is allocated to these departments based on the number of purchase orders each department issues to support its laser production functions. Example 12.11 illustrates the two-step process of ABC.

Example 12.11

A multinational aerospace firm uses traditional cost accounting to allocate manufacturing and management support costs for its European division. However, accounts such as business travel have historically been allocated on the basis of number of employees at the plants in France, Italy, Germany, and Greece.

The president recently stated that some products are likely generating much more management travel than others. The ABC system is chosen to augment the traditional method to more precisely allocate travel costs to major product lines at each plant.

(a)   First, let's assume that allocation of total observed travel expenses of $500,000 to the plants using a traditional basis of workforce size is sufficient. If total employment of 29,100 is distributed as follows, allocate the $500,000.

Paris, France plant	12,500 employees
Florence, Italy plant	8,600 employees
Hamburg, Germany plant	4,200 employees
Athens, Greece plant	3,800 employees

(b)   Now, let's assume that corporate management wants to know more about travel expenses based on product line, not merely plant location and workforce size. We will use the ABC method to allocate travel costs to major product lines assuming that annual plant support budgets indicate that the following percentages of total budget are expended for travel:

Paris	5% of $2 million
Florence	15% of $500,000
Hamburg	17.5% of $1 million
Athens	30% of $500,000

Further, the study indicates that in 1 year a total of 500 travel vouchers were processed by the management of the major five product lines produced at the four plants. The distribution may be summarized as follows:

Paris.   Product lines—1 and 2; number of vouchers—50 for line 1, 25 for 2.

Florence.   Product lines—1, 3, and 5; vouchers—80 for line 1, 30 for 3, 30 for 5.

Hamburg.   Product lines—1, 2 and 4; vouchers—100 for line 1, 25 for 2, 20 for 4.

Athens.   Product line—5; vouchers—140 for line 5.

Use the ABC two-step method to determine how the product lines drive travel costs at the plants.

Solution

(*a*)   Equation [12.11] takes the form

$$\text{Indirect cost rate} = \frac{\text{travel budget}}{\text{total workforce}}$$

$$= \frac{\$500,000}{29,100} = \$17.1821 \text{ per employee}$$

Using this traditional basis of rate times workforce size results in a plant-by-plant allocation.

Paris:	$17.1821(12,500) = $214,777
Florence:	$147,766
Hamburg:	$72,165
Athens:	$65,292

(*b*)   The ABC method is more involved since it requires the definition of the cost pool and its size (step 1) and the allocation to products using the cost driver (step 2). Also, the by-plant amounts will be different than in part (*a*) since completely different bases are being applied.

Step 1.   The cost pool is travel activity and the size of the cost pool is determined from the percentages of each plant's support budget devoted to travel. Using the travel expense information in the problem statement, a total cost pool of $500,000 is to be allocated to the five products. This number is determined from the percent-of-budget data as follows:

$$0.05(2,000,000) + \cdots + 0.30(500,000) = \$500,000$$

Step 2. The cost driver for the ABC method is the number of travel vouchers developed by the management unit responsible for each product line at each plant. The allocation will be to the products directly, not to the plants. However, the travel allocation to the plants can be determined afterward since we know which product lines are produced at each plant. For the cost driver of travel vouchers, the format of Equation [12.11] can be used to determine an ABC allocation rate.

$$\text{ABC cost allocation per travel voucher} = \frac{\text{total travel cost pool}}{\text{total number of vouchers}}$$

$$= \frac{\$500,000}{500}$$

$$= \$1000 \text{ per voucher}$$

Table 12–9 summarizes the vouchers and allocation by product and by city. Product 1 ($230,000) and product 5 ($170,000) drive the travel costs based on the ABC analysis. Comparison of the by-plant totals in Table 12–9 with the respective totals in part (*a*) indicates a substantial difference in the amounts allocated, especially to Paris, Hamburg, and Athens. This comparison verifies the president's suspicion that products, more than plants, drive travel requirements.

Table 12–9   ABC allocation of travel cost ($ in thousands), Example 12.11

| | Product | | | | | |
	1	2	3	4	5	Total
Paris	50	25				75
Florence	80		30		30	140
Hamburg	100	25		20		145
Athens					140	140
Total	$230	$50	$30	$20	$170	$500

Comment

Let's assume that product 1 has been produced in small lots at the Hamburg plant for a number of years. This analysis, when compared to the traditional cost allocation method in part (*a*), reveals a very interesting fact. In the ABC analysis, Hamburg has a total of $145,000 management travel dollars allocated, $100,000 of it from product 1. In the traditional analysis based on workforce size, Hamburg was allocated only $72,165—about 50% of the more-precise ABC analysis amount. This should point out to management the need to examine the manufacturing lot size practices at Hamburg and possibly other plants, especially when a product is currently manufactured at more than one plant.

ABC analysis is usually more expensive and time-consuming, but in many cases it can assist in understanding the economic impact of management decisions and determining the actual cost drivers for certain types of indirect costs. Often the combination of traditional and ABC analyses reveals areas where further economic analysis is warranted.

Problems 12.46 to 12.50

ADDITIONAL EXAMPLES

Example 12.12

PW WITH INFLATION, SECTION 12.1   A $50,000 bond which has a bond dividend rate of 10% per year payable semiannually is currently for sale. The bond is due in 15 years. If the rate of return requested by the investor is a nominal 8% per year compounded semiannually and if the inflation rate is expected to be 2.5% each 6-month period, what is the bond worth now (*a*) when inflation is not considered and (*b*) when inflation is considered?

Solution

(*a*) Without inflation, the dividend by Equation [11.1] is $I = [(50,000)(0.10)]/2 = \$2500$ per semiannual period. At a nominal 4% per 6 months for 30 periods, the PW is

$$PW = 2500(P/A,4\%,30) + 50,000(P/F,4\%,30) = \$58,645$$

(*b*) With inflation, use the inflated rate i_f in the P/F and P/A factors to determine the PW of the bond face value and 30 years of dividends.

$$i_f = 0.04 + 0.025 + (0.04)(0.025) = 0.0666 \text{ per semiannual period}$$

$$PW = 2500(P/A,6.66\%,30) + 50,000(P/F,6.66\%,30)$$

$$= 2500(12.8445) + 50,000(0.1445)$$

$$= \$39,338$$

Comments

The $19,307 difference in PW values illustrates the tremendous negative effect of inflation on fixed-income investments. On the other hand, the organizations which issue these instruments (bonds, in this case) are benefactors to the same extent, tempered by the fact that the bond dividend rate itself is an inflation-adjusted rate which must be higher when inflation is increasing.

Example 12.13

PW WITH INFLATION, SECTION 12.1   A business owner in a relatively high inflation country wishes to calculate an alternative's PW with costs of $35,000 now and $7000 per year for 5 years beginning 1 year from now with increases of 12% per year thereafter for the next 8 years. Use an interest rate of 15% per year and make the calculations (a) without inflation and (b) considering inflation at a rate of 11% per year.

Solution

(a) Figure 12–4 presents the cost cash flows. The total PW is found using $i = 15\%$ and Equation [2.18] for the geometric series which has its present worth PW_E in year 4.

$$PW = -35,000 - 7000(P/A,15\%,4)$$

$$- \left\{ \frac{7000[(1.12/1.15)^9 - 1]}{0.12 - 0.15} \right\} (P/F,15\%,4)$$

$$= -35,000 - 19,985 - 28,247$$

$$= -\$83,232$$

In the P/A factor, $n = 4$ because the $7000 cost in year 5 is the D term in Equation [2.18]. The expression in braces is PW_E in year 4 (Figure 12–4), which is discounted to time 0 with the $(P/F,15\%,4)$ factor.

(b) To consider inflation, calculate the inflated interest rate from Equation [12.5].

$$i_f = 0.15 + 0.11 + (0.15)(0.11) = 0.2765$$

$$PW = -35,000 - 7000(P/A,27.65\%,4)$$

$$- \frac{7000[(1.12/1.2765)^9 - 1]}{0.12 - 0.2765}(P/F,27.65\%,4)$$

$$= -35,000 - 7000(2.2545) - 30,945(0.3766)$$

$$= -\$62,436$$

This result demonstrates, once again, that in a high-inflation economy, when negotiating the amount of the payments to repay a loan, it is economically advantageous for the borrower to use future (inflated) dollars whenever possible to make the payments. The present value of future inflated dollars is significantly less when inflation is considered. And, the higher the inflation rate, the greater the discounting factors P/F and P/A.

Comment

You can check the result of the geometric series factor in part (a) by multiplying each of the amounts in years 5 through 13 by the appropriate P/F factor for $i = 15\%$. The same is correct in part (b) using the inflated interest rate.

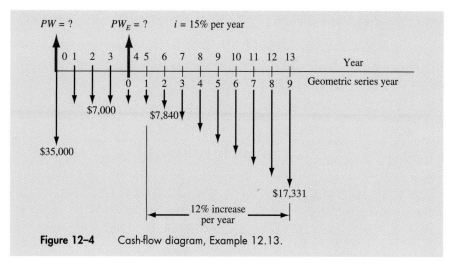

$PW = ?$ $PW_E = ?$ $i = 15\%$ per year

$35,000

$7,000

$7,840

$17,331

12% increase
per year

Year

Geometric series year

Figure 12–4   Cash-flow diagram, Example 12.13.

Example 12.14

PLANT COST ESTIMATION, SECTION 12.5   An activated sludge wastewater treatment plant is expected to have the following equipment purchase costs:

Equipment	Cost
Preliminary treatment	$20,000
Primary treatment	30,000
Activated sludge	14,000
Clarification	47,000
Chlorination	21,000
Digestion	60,000
Vacuum filtration	17,000
Total cost	$209,000

The multiplication factor for the cost of installation for piping, concrete, steel, insulation, supports, etc., is 0.49. The construction factor is 0.53 and the indirect cost factor is 0.21. Determine the total plant cost if (*a*) all cost factors are applied to the purchase cost of the equipment and (*b*) the indirect cost factor is applied to the total direct cost.

Solution

(*a*) Total equipment cost is $209,000. Since both the direct and indirect cost factors are applied to only the equipment cost, the overall cost factor is

$$h = 1 + 0.49 + 0.53 + 0.21 = 2.23$$

Thus, the total plant cost is

$$C_T = 2.23(209,000) = \$466,070$$

(b) Now the total direct cost is calculated first. The overall direct cost factor is

$$h = 1 + 0.49 + 0.53 = 2.02$$

The total direct cost is $2.02(209,000) = \$422,180$. Now apply the indirect cost factor to the total direct cost, after adding 1 to the 0.21 factor:

$$C_T = 1.21(422,180) = \$510,838$$

Comment

Note the difference in estimated plant cost when the indirect cost is applied to the equipment cost only in part (a) as compared to being applied to total direct cost in part (b). This illustrates the importance of determining exactly what the factors apply to before they are used.

Example 12.15

INDIRECT COST RATES, SECTION 12.6   J+L, Inc., produces several products, two of which are stereo speaker cabinets and dining tables. A total of $300,000 is allocated to indirect costs for next year. Management wants to determine indirect rates on the basis of direct labor hours for the two processing lines and one finishing line (a) individually and (b) using an overall or *blanket* rate. Table 12–10 presents departmental indirect allocation and estimated direct labor hours. Develop the rates.

Solution

(a) The rate for each department is computed using Equation [12.11]:

$$\text{Speaker processing} = \frac{145,000}{20,000} = \$7.25 \text{ per hour}$$

$$\text{Dining table processing} = \frac{145,000}{10,000} = \$14.50 \text{ per hour}$$

$$\text{Finishing} = \frac{10,000}{6000} = \$1.67 \text{ per hour}$$

(b) A blanket rate is found by computing

$$\text{Indirect rate} = \frac{\text{total indirect allocation}}{\text{total direct labor hours}}$$

$$= \frac{300,000}{36,000}$$

$$= \$8.33 \text{ per hour}$$

Table 12–10	Allocation information for Example 12.15		
Allocation	Speaker Processing	Dining Table Processing	Finishing
Indirect costs, $	145,000	145,000	10,000
Direct labor hours	20,000	10,000	6,000

Comment

An overall rate is easier to compute and use; however, it will not account for differences in the type of work accomplished in the individual departments.

Example 12.16

INDIRECT COST ALLOCATION, SECTION 12.7   For several years a company has purchased the motor and frame assembly of its major product line at an annual cost of $1.5 million. The suggestion to make the components in-house using existing departmental facilities has been made. For the three departments involved the indirect cost rates, estimated material, labor, and hours are quoted in Table 12–11. The allocated hours column is the time necessary to produce the motor and frame only.

Equipment must be purchased in order to make the products. The machinery has a first cost of $2 million, a salvage value of $50,000, and a life of 10 years. Perform an economic analysis for the 'make' alternative assuming a market rate of 15%-per-year is required.

Solution

For making the components in-house, the AOC is composed of labor, material, and indirect costs. Using the data of Table 12–11 we calculate the indirect cost allocation.

Department A: 25,000(10) = $250,000
Department B: 25,000(5) = 125,000
Department C: 10,000(15) = 150,000
                           $525,000

AOC = 500,000 + 300,000 + 525,000 = $1,325,000

Table 12–11	Production cost estimates for Example 12.16				
	Indirect Costs				
Department	Basis, Hours	Rate per Hour	Allocated Hours	Material Cost	Direct Labor Cost
A	Labor	$10	25,000	$200,000	$200,000
B	Machine	5	25,000	50,000	200,000
C	Labor	15	10,000	50,000	100,000
				$300,000	$500,000

The 'make' alternative annual worth is

$$\text{AW}_{\text{make}} = -P(A/P,i,n) + \text{SV}(A/F,i,n) - \text{AOC}$$

$$= -\$2,000,000(A/P,15\%,10) + 50,000(A/F,15\%,10) - 1,325,000$$

$$= -\$1,721,037$$

Currently,

$$\text{AW}_{\text{buy}} = -\$1,500,000$$

It is cheaper to continue to buy the motor and frame assembly.

CHAPTER SUMMARY

We learned about several new areas in this chapter—primarily cost-related areas. First, we learned about inflation, which is treated computationally like an interest rate i, but the inflation rate f makes the cost of the same product or service increase over time due to the decreased value of money. There are a variety of ways to consider inflation in engineering economy computations in terms of today's value and in terms of future or then-current value. A summary of important relations follows.

Inflated interest rate: $i_f = i + f + if$

Real interest rate: $i = (i_f - f)/(1 + f)$

PW of a future amount with inflation considered: $P = F(P/F,i_f,n)$

Future value of a present amount with the same buying power:
$F = P(F/P,i,n)$

Future amount to cover a current amount with no interest:
$F = P(F/P,f,n)$

Future amount to cover a current amount with interest: $F = P(F/P,i_f,n)$

Annual equivalent of a future dollar amount: $A = F(A/F,i_f,n)$

Annual equivalent of a present amount in future dollars: $A = P(A/P,i_f,n)$

Hyperinflation implies very high f values. Available funds are expended immediately because costs increase so rapidly that larger cash inflows cannot offset the fact that the currency is losing value. This can, and usually does, cause national financial disaster over extended periods of hyperinflated costs.

A cost index is a ratio of costs for the same item at two separate times. The cost is updated via Equation [12.8] using the index at the two time points. The Consumer Price Index (CPI) is an oft-quoted example of cost indexing.

Cost estimating may be accomplished with a variety of models. Two of them, the relations, and their best uses are

Cost-capacity method. Good for estimating present costs from historical data.

Factor method. Good for estimating total plant cost.

Traditional cost allocation uses an indirect cost rate determined for a machine, department, product line, etc. Bases such as direct labor cost, direct material cost, and direct labor hours are used. With increased automation in manufacturing, new and more-accurate techniques of indirect cost allocation have been developed. The activity-based costing (ABC) method is an excellent technique to augment the traditional allocation method.

The ABC method uses the rationale that cost drivers are activities—purchase orders, machine setups, reworks—which fundamentally *drive* the costs accumulated in cost pools, which are commonly departments or functions, such as, quality, purchasing, accounting, and maintenance. Improved understanding of how the company or plant actually accumulates indirect costs is a major by-product of implementing the ABC method.

CASE STUDY #5, CHAPTER 12   TOTAL COST ESTIMATES FOR OPTIMIZING COAGULANT DOSAGE

Background

There are several processes involved in the treatment of drinking water, but three of the most important ones are associated with removal of suspended matter, which is known as turbidity. Turbidity removal is brought about by adding chemicals that cause the small suspended solids to clump together (coagulation) forming larger particles which can be removed through settling (sedimentation). The few particles that remain after sedimentation are filtered out in sand, carbon, or coal filters (filtration).

In general, as the dosage of chemicals is increased, more 'clumping' occurs (up to a point), so there is increased removal of particles through the settling process. This means that fewer particles have to be removed through filtration, which obviously means that the filter will not have to be cleaned as often through backwashing. Thus, more chemicals means less backwash water and vice versa. Since backwash water and chemicals both have costs, a primary question is: What amount of chemicals will result in the lowest overall cost when the chemical coagulation and filtration processes are considered together?

Formulation

To minimize the total cost associated with coagulation and filtration, it is necessary to obtain the relationship between the chemical dosage and water turbidity after coagulation and sedimentation, but before filtration. This allows the chemical costs for different operating strategies to be determined. This cost relationship, derived using polynomial regression analysis, is shown in Figure 12–5 and is described by the equation:

$$T = 37.0893 - 7.7390F + 0.7263F^2 - 0.0233F^3 \qquad \text{[CS5.1]}$$

where T = settled water turbidity and F = coagulant dosage, milligrams/liter (mg/L).

Similarly, the backwash water data is described by the equation:

$$B = -0.549 + 1.697T \qquad \text{[CS5.2]}$$

where B = backwash water rate, m³/1000 m³ product water

By substituting Equation [CS5.2] into Equation [CS5.1] and multiplying by the unit water cost of \$0.06/m³, C_B, the cost of washwater versus turbidity, is found.

$$C_B = -0.002399F^3 + 0.0749F^2 - 0.798F + 3.791 \qquad \text{[CS5.3]}$$

where C_B = cost of washwater, \$/1000 m³ product water

The chemical cost C_C is \$0.183 per kilogram or

$$C_C = 0.183F \qquad \text{[CS5.4]}$$

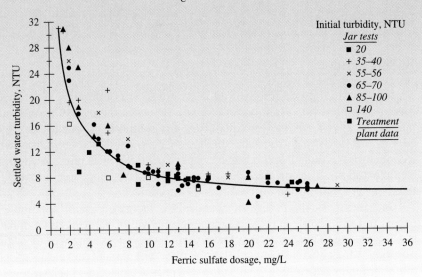

SOURCE: A. J. Tarquin, Diana Tsimis, and Doug Rittmann, "Water Plant Optimizes Coagulant Dosages," *Water Engineering and Management* 136, no. 5 (1989), pp. 43–47.

Figure 12–5   Nonlinear relation between coagulant dosage and sedimentation-basin effluent turbidity.

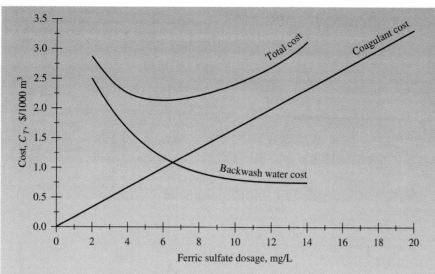

SOURCE: A. J. Tarquin, Diana Tsimis, and Doug Rittmann, "Water Plant Optimizes Coagulant Dosages," *Water Engineering and Management* 136, no. 5 (1989), pp. 43–47.

Figure 12–6   Total cost curve for coagulant dosage and backwash water.

The total cost of backwash water and chemicals, C_T, is obtained by adding the last two equations.

$$C_T = C_B + C_C \qquad\qquad \text{[CS5.5]}$$
$$= 0.002399F^3 + 0.0749F^2 - 0.615F + 3.791$$

Equations [CS5.3] through [CS5.5] are plotted in Figure 12–6.

Results

As Figure 12–6 shows, when a dosage of 6 mg/L is used, the total cost for coagulation and filtration is minimum at approximately $C_T = \$2.16$ per 1000 m^3 of product water. Prior to this analysis, the plant was using 12 mg/L. The costs at 6 and 12 mg/L are shown in Table 12.12 below. As an illustration, at the average flow rate of 189,250 m^3/day, a 23% savings results, which represents an annual dollar savings of over $44,000.

Table 12.12	Operational costs at 6 and 12 mg/L coagulant dosages				
Coagulant Dosage, mg/L	Coagulant Cost, $/1000 m^3	No. of Backwashes per Day*	Washwater Cost, $/1000 m^3	Total Cost, $/1000 m^3	Cost Savings
6	1.00	5.93	1.16	2.16	23%
12	2.00	4.12	0.81	2.81	

*Average amount of washwater per backwash is 305 m^3 and average flow rate per day is 94,625 m^3/day (25 MGD).

Questions to Consider

1. What effect does an increase in chemical cost have on the optimum dosage?
2. What effect does an increase in backwash water cost have on the optimum dosage?
3. What is the chemical cost at a dosage of 10 mg/L?
4. What is the backwash water cost at a dosage of 14 mg/L?
5. If the chemical cost changes to $0.21/kg, what will be the total coagulation and filtration cost?
6. At what chemical cost will the minimum total cost occur at 8 mg/L?

PROBLEMS

12.1 If the cost of a certain piece of equipment today is $20,000, what was its cost 5 years ago, if its price increased only by the inflation rate of 6% per year?

12.2 If an investor would be satisfied making a real rate of return of 4% per year, what rate of return would he have to make on his investments when the inflation rate is 16% per year?

12.3 If the market interest rate is 12% per year when the real rate of return is 4% per year, what inflation rate is built into the marketplace?

12.4 Determine the market interest rate that would be equivalent to a real interest rate of 1% per quarter and an inflation rate of 2% per quarter.

12.5 Calculate the present worth of $50,000 seven years from now when the real interest rate is 3% per year and the inflation rate is 2% per year (*a*) without consideration for inflation and (*b*) with inflation considered.

12.6 Find the present worth of $35,000 twenty years hence if the company's required real MARR is 20% per year and the inflation rate is 6% per year (*a*) without inflation and (*b*) with inflation considered.

12.7 Rework Problem 12.6, except assume the company's 20% MARR is inflation-adjusted.

12.8 A very appreciative girl was just told that her great-grandfather died and left her his entire investment account which contains $3 million. If he started the account 50 years ago with a single deposit and never added even a single dollar to the account after the initial deposit, how much did he deposit? Assume that the account earned interest at a rate of 5% per year and that the inflation rate during that time period averaged 2% per year.

12.9 How many dollars are required now to have the same buying power as $3 million 50 years ago, if the interest rate is 5% per year and the inflation rate averaged 2% per year?

12.10 Compare the machines shown below on the basis of their present worths using a real interest rate of 6% per year and an inflation rate of 3% per year.

	Machine X	Machine Y
First cost, $	31,000	43,000
Annual operating cost, $	18,000	19,000
Salvage value, $	5,000	7,000
Life, years	3	6

12.11 Compare the alternatives shown below on the basis of their capitalized costs with inflation considered. Use $i = 14\%$ per year compounded quarterly and $f = 2\%$ per quarter.

	Alternative V	Alternative W
First cost, $	$8,500,000	$50,000,000
Annual operating cost, $	8,000	7,000
Salvage value, $	5,000	2,000
Life, years	5	∞

12.12 An engineer is trying to decide which of two machines he should purchase to manufacture a certain part. He obtains estimates from two salespeople. Salesman *A* gives him the estimated costs in constant-value dollars (i.e., today's dollars), while saleswoman *B* gives him the estimated costs in then-current dollars. If the company's inflation-adjusted minimum attractive rate of return is 20% per year and the company expects inflation to be 10% per year, which product should the engineer purchase, assuming that the estimated costs are correct? Use a present-worth analysis.

	Salesman A (Today's Dollars)	Saleswoman B (Future Dollars)
First cost, $	60,000	95,000
Annual operating cost, $	25,000	35,000
Life, years	5	10

12.13 Assume $23,000 is invested now at an interest rate of 13% per year. (*a*) How much money will be accumulated in 7 years, if the inflation rate is 10% per year? (*b*) What would be the buying power of the accumulated amount with respect to today's dollars?

12.14 (*a*) What future amount of money in then-current dollars 6 years from now is equivalent to a present sum of $80,000 at a market interest rate of 18% per year and an inflation rate of 12% per year? (*b*) How many dollars would you have to have at that time just to keep up with inflation?

12.15 Calculate the number of (*a*) today's dollars and (*b*) then-current dollars in year 10 that will be equivalent to a present investment of $33,000 at a market interest rate of 15% per year and an inflation rate of 10% per year.

12.16 R-Gone Signs invests $3000 per year for 8 years beginning 1 year from now in a new production process. (*a*) How much money must be received in a lump sum in year 8 in then-current dollars in order for the company to recover its investment at a real rate of return of 6% per year and an inflation rate of 10% per year? (*b*) How much will the company need to receive just to keep up with inflation?

12.17 An investor purchased a 6%, $20,000 bond with interest payable annually for $16,000. The inflation rate during the time he owned the bond was 4% per year. If he decides to sell 3 years after he purchased it, what must be the selling price for him to make a real rate of return of 7% per year?

12.18 (*a*) In 1996, the Nobel prize awards were raised from $489,000 to $653,000. If the first award in 1901 was for $150,000 and inflation averaged 1.75% over that 95-year period, was the increase to $653,000 enough to offset the effects of inflation? (*b*) If the awarding foundation expects inflation to average 4% per year between 1996 and 2006, how much will the award have to be in 2006 to make it worth the same as in 1901?

12.19 A company involved in environmental restoration maintained a contingency fund of $10 million. The company kept the money in a stock market fund which earned 16% per year. The inflation rate during the 5-year period the company had the money invested was 5% per year. (*a*) How much money did the company have at the end of the 5-year period? (*b*) What was the buying power of the money in terms of dollars when the investment was originally made? (*c*) What was the company's real rate of return on its investment?

12.20 How much money must a recycling company make each year for 5 years to recover a $100,000 investment in a tire shredder if it wants to make a real rate of return of 4% per year when the inflation rate averages 3% per year?

12.21 A cellular telephone company is in the process of upgrading its transmission and receiver facilities. It expects to spend $2 million on the upgrade. The company expects to make a real rate of return of 12% per year on all investments. If the inflation rate is 3% per year, how much money must the company make each year to recover its investment in 6 years?

12.22 A biotech genetics research company is planning for a major expenditure on a new research facility. It expects to need $5 million of today's dollars so that it can start construction in 4 years. The inflation rate over this time period is expected to be 6% per year. (*a*) How many future dollars will it need at that time? (*b*) If the company plans to make annual deposits into a bond fund which earns 10% per year, what annual deposit is required to have the amount determined in part (*a*)?

12.23 An industrial firm is considering the purchase of equipment to automate one of its processes. It can purchase the necessary controllers for $40,000 now. Alternatively, it can make a $10,000 down payment and finance the balance at an interest rate of 10% per year. (*a*) If the company's inflation-adjusted MARR is 15% per year, should it pay now or pay later? (*b*) If the inflation rate is 4% per year, what real rate of return is built into the company's inflation-adjusted MARR?

12.24 Calculate the perpetual equivalent uniform annual worth in years 6 through ∞ of $50,000 now, $10,000 five years from now, and $5000 per year thereafter if the interest rate is 8% per year and the inflation rate is 4% per year for the first 5 years and 3% per year thereafter.

12.25 If a piece of equipment had a cost of $20,000 in 1985 when the M&S equipment index was 789.6, what would it be expected to cost when the index is 1150?

12.26 An item which had a cost of $7000 in 1988 was estimated to cost $13,000 in 2000. If the cost index which was used in the calculation had a value of 326 in 1988, what was its value in 2000?

12.27 A certain labor cost index had a value of 426 in 1960 and 1231 in 1997. If the labor cost for constructing a building was $160,000 in 1997, what would the labor cost have been in 1960?

12.28 Use the *ENR* construction cost index (Table 12–5) to update a cost of $325,000 in 1981 to a 1996 figure.

12.29 If one wanted 1980 to equal 100 for the *CE* plant cost index (Table 2–5), what would be the value of the index in 1995?

12.30 Use the *F/P* or *P/F* factor to calculate the average percent increase per year between 1975 and 1995 for the *CE* plant cost index.

12.31 Determine the value of the M&S equipment cost index in 2000 if it was 797.6 in 1986 and it increases by 4% per year.

12.32 A mass spectrometer can be purchased for $60,000 today, but the owner of a water analysis laboratory expects the cost to increase exactly by the equipment inflation rate over the next 10 years. (*a*) If the inflation rate is 0% per year for the next 3 years and 5% per year thereafter, how much will the spectrometer cost 10 years from now if the company's MARR is 15% per year? (*b*) If the applicable equipment cost index is at 1203 now, what will it be 10 years from now?

12.33 If a 15-horsepower pump has a cost of $2300, what would a 50-horsepower pump be expected to cost if the exponent for the cost-capacity equation is 0.69?

12.34 Mechanical dewatering equipment for a 1-million-liter-per-day (MLD) plant would cost $400,000. How much would the equipment cost for a 15-MLD plant if the exponent in the cost-capacity equation is 0.63?

12.35 The cost of site work for construction of a manufacturing facility which has a capacity of 6000 units per day was $55,000. If the cost for a plant with a capacity of 100,000 units per day was $3 million, what is the value of the exponent in the cost-capacity equation?

12.36 The equipment cost for a laboratory dedicated to immunological response research will cost $900,000. If the direct cost factor is 2.10 and the indirect cost factor is 0.26, what is the expected total cost of the laboratory?

12.37 Estimate the cost in 1996 of a 60-square-meter falling-film evaporator if the cost of a 30-square-meter unit was $19,000 in 1970. The exponent in the cost-capacity equation is 0.24. Use the M&S equipment index to update the cost.

12.38 Estimate the cost in 1996 of a 1000-horsepower steam turbine air compressor. A 200-horsepower unit cost $90,000 in 1975. The exponent in the cost-capacity equation is 0.29. Use the M&S equipment index to update the cost.

12.39 A company has a processing department with 25 machines. Because of the nature and use of three of these machines, each is considered a separate account for indirect cost accumulation. The remaining 22 machines are grouped under one account, #104. Machine operating hours are used as the allocation basis for all machines. A total of $50,000 is allocated to the department for next year. Use the data below to determine the indirect cost rate for each account.

Account Number	Indirect Cost Allocated	Estimated Machine Hours
#101	$10,000	600
#102	10,000	200
#103	10,000	800
#104	20,000	1,600

12.40 All the indirect costs are allocated by accounting for John's department. John has obtained records of allocation rates and actual charges for the prior months and estimates for this month (May) and next month. The basis of allocation is not indicated and the company accountant has no record of the basis used. However, the accountant says not to be concerned because the allocation rates have decreased each month. John has asked you to evaluate the allocation amounts.

You now have the following information from accounting.

		Indirect Cost	
Month	Rate	Allocated	Charged
February	$1.40	$2800	$2600
March	1.33	3400	3800
April	1.37	3500	3500
May	1.03	3600	
June	0.92	6000	

During the evaluation, you collect the following from departmental and accounting records.

	Direct Labor		Material	Departmental Space Square
Month	Hours	Costs	Costs	Feet
February	640	$2560	$5400	2000
March	640	2560	4600	2000
April	640	2560	5700	3500
May	640	2720	6300	3500
June	800	3320	6500	3500

(*a*) Determine the actual allocation basis used each month, and (*b*) comment on the accountant's comment about decreasing allocation rates.

12.41 A computer manufacturer serving the airline industry has five departments. Indirect cost allocations for 1 month are detailed below, along with space assigned, direct labor hours, and direct labor costs for each department which directly manufactures the machines.

		Actual Data for 1 Month		
Department	Indirect Cost Allocation	Space, Square Feet	Direct Labor Hours	Direct Labor Costs
Preparation	$20,000	10,000	480	$1,680
Subassemblies	15,000	18,000	1,000	3,250
Final assembly	10,000	10,000	600	2,460
Quality assurance	15,000	1,200		
Engineering	19,000	2,000		

Determine the manufacturing department allocation rates for redistributing the indirect cost allocation to quality assurance and engineering ($34,000) to the other departments. Use the following bases to determine the rates: (*a*) space, (*b*) direct labor hours, and (*c*) direct labor costs.

12.42 Compute the actual indirect cost charges and allocation variances for (*a*) each account and (*b*) total for all accounts in Problem 12.39 using the individual account allocation rates. The actual hours credited to each account are as follows: #101 has 700 hours; #102 has 350 hours; $103 has 650 hours; $104 has 1300 hours.

12.43 Indirect cost rates and bases for the six producing departments at Mutant, Inc., are listed below. (*a*) Use the recorded data to distribute indirect costs to the departments. (*b*) Determine the allocation variance relative to a total indirect cost allocation budget of $760,000.

Department	Allocation Basis*	Rate	Direct Labor Hours	Direct Labor Cost	Machine Hours
1	DLH	$2.50	5,000	$20,000	3,500
2	MH	0.95	5,000	35,000	25,000
3	DLH	1.25	10,500	44,100	5,000
4	DLC	5.75	12,000	84,000	40,000
5	DLC	3.45	10,200	54,700	10,200
6	DLH	1.00	29,000	89,000	60,500

| *DLH = direct labor hours; MH = machine hours; DLC = direct labor costs.

12.44 For Problem 12.41, determine the actual indirect cost charges using the rates you determined. For actual charges, use bases of direct labor hours for the preparation and subassembly departments and space for the final assembly department.

12.45 A computer manufacturer serving the airline industry has five departments. Indirect cost allocations for 1 month are detailed below, along with space assigned, direct labor hours, and direct labor costs for each department which directly manufactures the machines. (For reference, this is the same information presented in Problem 12.41 above, but you do not need to have worked it to complete this problem.)

| Department | Indirect Cost Allocation | Actual Data for 1 Month | | |
| | | Space, Square Feet | Direct Labor | |
			Hours	Costs
Preparation	$20,000	10,000	480	$1,680
Subassemblies	15,000	18,000	1,000	3,250
Final assembly	10,000	10,000	600	2,460
Quality assurance	15,000	1,200		
Engineering	19,000	2,000		

The company presently makes all the components required by the preparation department. The company is considering buying rather than making these components. An outside contractor has offered to make the items for $67,500 per month.

(*a*) If the costs for preparation for the particular month shown are considered good estimates for an engineering economy study, and if $41,000 worth of materials is charged to preparation, do a comparison of the make versus buy alternatives. Assume that the preparation department's share of the quality assurance and engineering department's costs is a total of $3230 per month.

(*b*) Another alternative for the company is to purchase new equipment for the preparation department and continue to make the components. The machinery will cost $375,000, have a 5-year life, no salvage value, and a monthly operating cost of $475. This purchase is expected to reduce costs in quality assurance and engineering by $2000 and $3000, respectively, and also reduce monthly direct labor hours to 200 and monthly direct labor costs to $850 for the preparation department. The redistribution of the indirect costs from quality assurance and engineering to the three production departments is on the basis of direct labor hours. If other costs remain the same, compare the present cost of making the components and the estimated cost if the new equipment is purchased. Select the more economic alternative. A market rate of (nominal) 12% per year compounded monthly is expected on capital investments.

12.46 Use Equation [12.11] and the bases listed in Table 12–7 to explain why a decrease in direct labor hours, coupled with an increase in indirect labor hours due to automation on a production line, may require the use of new bases to allocate indirect costs.

12.47 RelaxFull Resorts currently distributes its advertising costs for the four sites in Canada on the basis of size of resort budget. For this year, in round numbers, the budgets and allocation of $1 million advertising indirect costs are

	Site			
	A	**B**	**C**	**D**
Budget, $	2 million	3 million	4 million	1 million
Allocation, $	200,000	300,000	400,000	100,000

(a) Determine the allocation if the ABC approach is used. Define the cost pool as the $1 million in advertising costs, and make the activity the number of guests during the year, which are

Site	A	B	C	D
Guests	3500	4000	8000	1000

(b) Again use the ABC approach, but now make the activity be the number of guest-nights. The average number of lodging-nights for guests at each site are

Site	A	B	C	D
Length of Stay, Nights	3.0	2.5	1.25	4.75

(c) Comment on the distribution of advertising costs using the different methods. Can you identify any other activities (cost drivers) that might be considered for the Activity-Based Costing approach which may reflect a realistic allocation of the costs? What are they?

12.48 Indirect costs are allocated each calendar quarter to three processing lines using direct labor hours as the basis. New automated equipment has decreased direct labor and the time to produce a unit significantly, so the division manager plans to use cycle time per unit produced as the basis. However, the manager wants, initially, to determine what the allocation would have been had the cycle time been the basis prior to the automation. Use the data below to determine the allocation rate and actual indirect charges for the three different situations if the amount to be allocated in an average quarter is $400,000. Comment on changes in the amount of allocated indirect cost per processing line.

Processing Line	10	11	12
Direct labor hours per quarter	20,000	12,700	18,600
Cycle time per unit now, seconds	3.9	17.0	24.8
Cycle time per unit previously, seconds	13.0	55.8	28.5

12.49 This problem consists of three parts which build upon each other. The object is to compare and comment upon the amount of indirect cost allocated to the electricity-generating facilities located in two states for the different situations described in each part.

(*a*) Historically, Mesa Power Authority has allocated the indirect costs associated with its employee safety program to its plants in California and Arizona based on the number of employees. Information to allocate a $200,200 budget for this year follows:

State	Workforce Size
California	900
Arizona	500

(*b*) The head of the department of accounting recommends that the traditional method be abandoned and that Activity-Based Costing be used to allocate the $200,200 using expenditures on the safety program as the cost pool, and the number of accidents as the activities which drive the costs. Accident statistics indicate the following for the year.

State	Number of Accidents
California	425
Arizona	135

(*c*) Further study indicates that 80% of the safety program indirect costs are expended for employees in generation areas and the remaining 20% goes to office-area employees. Because of this apparent imbalance in expenditures, a split allocation of the $200,200 amount is proposed: 80% of total dollars allocated via ABC with the activities being the number of accidents occurring in generation areas and the cost pool being 80% of the total safety program expenditures; and,

20% of the total dollars using the traditional indirect-cost allocation method with a basis of number of office-area employees. The following data have been collected.

	Number of Employees		Number of Accidents	
State	**Generation Area**	**Office Area**	**Generation Area**	**Office Area**
California	300	600	405	20
Arizona	200	300	125	10

12.50 Develop your own application of the ABC approach to allocating indirect costs. Identify the cost pool and cost driver for your example. Develop numerical estimates and show how ABC would be used for your example.

FOUR

*T*he basics of corporate taxes and
individual income taxes are treated in this section. The essentials of depreciation
methods (capital-recovery models) are used to develop asset depreciation schedules and, later, to perform an after-tax engineering economy study.

The second chapter of this level covers the simple formulas for computing
taxable income, capital gains and losses, income tax amounts for individuals
and corporations, and other tax-related amounts. The final chapter examines the
fundamentals of an after-tax analysis using the alternative evaluation techniques of PW, AW, and ROR.

CHAPTER

chapter

13

Depreciation and Depletion Models

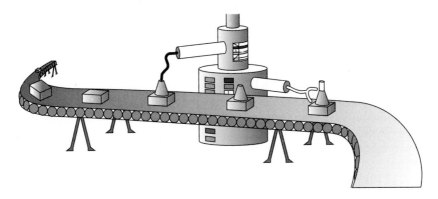

The capital investments of a corporation in tangible assets—equipment, computers, vehicles, buildings, and machinery—are commonly recovered by a company through a process called *depreciation*. The process of depreciating an asset, also referred to as *capital recovery,* accounts for the loss of an asset's value because of age, wear, and obsolescence over its useful life. Even though an asset may be in excellent working condition, the fact that it is worth less through time is taken into account in economic evaluation studies. An introduction to basic depreciation methods is followed by the *Modified Accelerated Cost Recovery System* (MACRS), which is the standard currently used in the United States.

Why is depreciation so important to engineering economy? Depreciation is a tax-allowed deduction included in income tax calculations via the relation

$$\text{Taxes} = (\text{income} - \text{deductions})(\text{tax rate})$$

Since depreciation is an allowable deduction for businesses (along with salaries and wages, materials, rent, etc.), it lowers income taxes. Income taxes are discussed further in the two chapters following this one.

This chapter concludes with an introduction to two methods of *depletion,* which are used to recover economic interest in deposits of natural resources to include minerals, ores, and timber.

Purpose: Use an approved model to recover the capital investment in an asset or natural resource.

This chapter will help you:

Depreciation terms	1. Understand the basic terminology of capital recovery using depreciation.
Straight line	2. Utilize the straight-line model of depreciation.
Declining balance	3. Utilize the declining balance-model of depreciation.
MACRS	4. Utilize the Modified Accelerated Cost Recovery System (MACRS) of depreciation.
Recovery period	5. Select the recovery period of an asset for MACRS depreciation.
Switching	6. Understand the principle of switching from one depreciation model to another.
MACRS and switching	7. Compute MACRS depreciation rates using depreciation model switching (optional section).
Capital expense	8. Calculate the amount of capital expense allowed when depreciating an asset.
Depletion	9. Utilize the cost and percentage models of depletion for natural resources.

13.1 DEPRECIATION TERMINOLOGY

Some terms commonly used in depreciation are defined here. The terminology is applicable to corporations, as well as individuals who own depreciable assets.

Depreciation is the reduction in value of an asset. Depreciation models use government-approved rules, rates, and formulas to represent the current value on the company books. The depreciation amount, D_t, usually computed annually, does not necessarily reflect the actual usage pattern of the asset during ownership. Annual depreciation charges are tax deductible for U.S. corporations.

First cost or *unadjusted basis* is the installed cost of the asset including purchase price, delivery and installation fees, and other depreciable direct costs incurred to prepare the asset for use. The term unadjusted basis, or simply basis, and symbol B are used when the asset is new, with the term adjusted basis used after some depreciation has been charged.

Book value represents the remaining, undepreciated investment on the books after the total amount of depreciation charges to date have been subtracted from the basis. The book value, BV_t, is usually determined at the end of each year, which is consistent with our end-of-year convention.

Recovery period is the depreciable life, n, of the asset in years for depreciation (and income tax) purposes. This value may be different than the estimated productive life because of government laws which dictate allowable recovery periods and depreciation.

Market value is the estimated amount realizable if an asset were sold on the open market. Because of the structure of depreciation laws, the book value and market value may be substantially different. For example, a commercial building tends to increase in market value, but the book value will decrease as depreciation charges are taken. However, a computer workstation may have a market value much lower than its book value due to rapidly changing technology.

Depreciation rate or *recovery rate* is the fraction of the first cost removed by depreciation each year. This rate, d_t, may be the same each year, which is called the straight-line rate, or different for each year of the recovery period. A depreciation rate without reference to the year is identified by the letter d.

Salvage value is the estimated trade-in or market value at the end of the asset's useful life. The salvage value, SV, expressed as an estimated dollar amount or as a percentage of the first cost, may be positive, zero, or negative due to dismantling and carry-away costs. (Currently, U.S. government-approved depreciation models assume a salvage value of zero, even though the actual salvage value may be positive.)

Personal property, one of the two types of property for which depreciation is allowed, is the income-producing, tangible possessions of a corporation used to conduct business. Included are most manufacturing and service industry property—vehicles, manufacturing equipment, materials-handling devices, computers, telephone switching equipment, office furniture, refining process equipment, and much more.

Real property includes real estate and any improvements and similar types of property—office buildings, manufacturing structures, warehouses, apartments, and other structures. *Land itself is considered real property, but it is not depreciable.*

Half-year convention assumes that assets are placed in service or disposed of in midyear, regardless of when these events actually occur during the year. This convention is utilized in this text. There are midmonth and midquarter conventions as well.

There are several models approved for depreciating assets, with the straight-line (SL) model being the most commonly used, historically. However, accelerated models such as the declining-balance (DB) model are attractive because the book value decreases to zero (or to the salvage value) more rapidly than by the straight-line method, as shown by the general book-value curves in Figure 13–1.

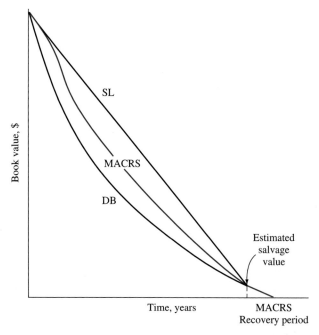

Figure 13–1   General shape of book-value curves for different depreciation models.

In 1981 and 1986, respectively, the U.S. government initiated efforts to standardize accelerated depreciation. The *Accelerated Cost Recovery System (ACRS)* and the *Modified Accelerated Cost Recovery System (MACRS)* were announced in 1981 and 1986, respectively, as the prime capital-recovery models. MACRS continues as the only approved depreciation method in the United States. Prior to these fundamental changes, it was acceptable to utilize the classical straight-line, declining-balance, and sum-of-year digits methods to reduce the book value to the anticipated salvage value.

Tax law revisions are occurring rapidly. The depreciation rules and rates used here may differ slightly from those in effect when you study this material or in the country in which you are studying engineering economy. However, the general principles and relations for the depreciation models presented here are universal.

Problems 13.1 to 13.4

13.2   STRAIGHT-LINE (SL) DEPRECIATION

The straight-line model is a method of depreciation used as the standard of comparison for most other methods. It derives its name from the fact that the book value decreases linearly with time because the depreciation rate is the same each year at 1 over the recovery period. Therefore, $d = 1/n$. The annual depreciation is determined by multiplying the first cost minus the estimated salvage value by the depreciation rate d, which is the same as dividing by the recovery period n. In equation form,

$$D_t = (B - SV)d$$

$$= \frac{B - SV}{n} \qquad \text{[13.1]}$$

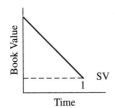

where t = year ($t = 1, 2, \ldots , n$)
D_t = annual depreciation charge
B = first cost or unadjusted basis
SV = estimated salvage value
d = depreciation rate (same for all years)
n = recovery period or expected depreciable life

Since the asset is depreciated by the same amount each year, the book value after t years of service, BV_t, will be equal to the unadjusted basis B minus the annual depreciation times t.

$$BV_t = B - tD_t \qquad \text{[13.2]}$$

Earlier we defined d_t as a depreciation rate for a specific year t. However, the SL model has the same rate for all years, that is,

$$d_t = \frac{1}{n} \qquad \qquad \textbf{[13.3]}$$

The relations above are illustrated by Example 13.1.

Example 13.1

If an asset has a first cost of $50,000 with a $10,000 estimated salvage value after 5 years, (a) calculate the annual depreciation and (b) compute and plot the book value of the asset after each year using the straight-line depreciation model.

Solution

(a) The depreciation each year can be found by Equation [13.1].

$$D_t = \frac{B - SV}{n} = \frac{50{,}000 - 10{,}000}{5}$$

$$= \$8000 \text{ per year for 5 years}$$

(b) The book value after each year t is computed using Equation [13.2].

$$BV_t = B - tD_t$$

$$BV_1 = 50{,}000 - 1(8000) = \$42{,}000$$

$$BV_2 = 50{,}000 - 2(8000) = \$34{,}000$$

....

$$BV_5 = 50{,}000 - 5(8000) = \$10{,}000 = SV$$

A plot of BV_t versus t is given in Figure 13–2.

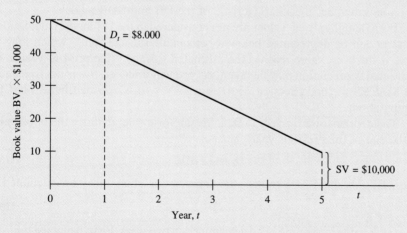

Figure 13–2   Book-value plot of an asset depreciated using the straight-line model, Example 13.1.

Problems 13.5 to 13.7

13.3   DECLINING-BALANCE (DB) DEPRECIATION

The declining-balance method, also known as the uniform- or fixed-percentage method, is an accelerated write-off model. Simply put, the annual depreciation charge is determined by multiplying the book value at the beginning of each year by a uniform percentage, which we shall call d, in decimal equivalent form. For example, if the uniform-percentage rate is 10% (that is, $d = 0.10$), the depreciation write-off for any given year would be 10% of the book value at the beginning of that year. The depreciation charge is largest in the first year and decreases each succeeding year.

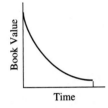

The maximum percentage depreciation permitted (for tax purposes) is double the straight-line rate. When this rate is used, the method is known as *double-declining balance (DDB)*. Thus, if an asset had a useful life of 10 years, the straight-line recovery rate is $1/n = 1/10$, and the uniform rate for DDB is $d = 2/10$ or 20% of book value. Since d represents the uniform depreciation rate, the formula for calculating the maximum rate, d_{max}, is two times the straight-line rate, or

$$d_{max} = \frac{2}{n} \qquad \text{[13.4]}$$

This is the rate used for the DDB method. Another commonly used percentage for the DB method is 150% of the straight-line rate, where $d = 1.50/n$.

The actual depreciation rate for each year t, relative to the first cost, is

$$d_t = d(1 - d)^{t-1} \qquad \text{[13.5]}$$

For DB or DDB depreciation, the *estimated salvage value is not subtracted from the first cost* when calculating the annual depreciation charge. It is important that you remember this characteristic of the DB and DDB models.

Even though salvage values are not considered in the DB model calculations, no asset can be depreciated below a reasonable salvage value, which may be zero. If the book value reaches the estimated salvage value prior to year n, no additional depreciation may be taken. As you will discover, this is not true under the MACRS method (Section 13.4), since SV = 0 is assumed in all MACRS computations.

The depreciation for year t, D_t, is the uniform rate, d, times the book value at the end of the previous year.

$$D_t = (d)BV_{t-1} \qquad \text{[13.6]}$$

If the BV_{t-1} value is not known, the depreciation charge may be calculated as

$$D_t = (d)B(1 - d)^{t-1} \qquad \text{[13.7]}$$

The book value in year t may be determined in two ways. First, using the uniform rate d and the first cost B.

$$BV_t = B(1 - d)^t \qquad \text{[13.8]}$$

Also, BV_t for any depreciation model can always be determined by subtracting the current depreciation charge from the previous book value, that is,

$$BV_t = BV_{t-1} - D_t \qquad \text{[13.9]}$$

The book value in declining-balance methods never goes to zero. There is an implied SV after n years, which is equal to the BV in year n, as calculated from Equation [13.8], that is,

$$\text{Implied SV} = BV_n = B(1 - d)^n \qquad \text{[13.10]}$$

If the implied SV is less than the estimated SV, the asset will be totally depreciated before the end of its expected life.

It is also possible to determine an implied uniform depreciation rate by using the estimated SV amount. For SV > 0,

$$\text{Implied } d = 1 - \left(\frac{SV}{B}\right)^{1/n} \qquad \text{[13.11]}$$

The allowed range on d is $0 < d < 2/n$. In all DB models, d is stated or it may be calculated by Equation [13.11], if an SV > 0 is estimated. And for the DDB model, $d = d_{max} = 2/n$. Examples 13.2 and 13.3 illustrate the DDB and DB models.

Example 13.2

Assume that an asset has a first cost of $25,000 and an estimated $4000 salvage value after 12 years. Calculate its depreciation and book value for (a) year 1 and (b) year 4. (c) Calculate the implied salvage value after 12 years for the DDB model.

Solution

First compute the DDB depreciation rate d.

$$d = \frac{2}{n} = \frac{2}{12} = 0.1667 \text{ per year}$$

(a) For the first year, the depreciation and book value are calculated using Equations [13.7] and [13.8],

$$D_1 = (0.1667)25,000(1 - 0.1667)^{1-1} = \$4167.50$$
$$BV_1 = 25,000(1 - 0.1667)^1 = \$20,832.50$$

(b) For year 4, Equations [13.7] and [13.8] with $d = 0.1667$ result in

$$D_4 = 0.1667(25,000)(1 - 0.1667)^{4-1} = \$2411.46$$
$$BV_4 = 25,000(1 - 0.1667)^4 = \$12,054.40$$

(c) From Equation [13.10], the implied salvage value after 12 years is

$$\text{Implied SV} = 25,000(1 - 0.1667)^{12} = \$2802.57$$

Since the estimated salvage value of $4000 is larger, than $2802.57, the asset will be completely depreciated before its expected 12-year life is reached. Therefore, once

BV_t reaches $4000, no further depreciation charges are allowed; in this case $BV_{10} = 4036.02. By Equation [13.6], D_{11} would be $672.80, making $BV_{11} = 3362.22, which is less than the estimated SV of $4000. So, in years 11 and 12 the depreciation amounts are $D_{11} = 36.02 and $D_{12} = 0$, respectively.

Comment

Two important facts to remember about the DB and DDB methods are that the salvage value is not subtracted from the first cost when the depreciation is calculated, and, when the book value reaches the estimated salvage value, no additional depreciation may be taken.

Example 13.3

The Rush Mining Company has purchased a computer-controlled ore-grading unit for $80,000. The unit has an anticipated life of 10 years and a salvage value of $10,000. Use the declining-balance method to develop a schedule of depreciation and book values for each year.

Solution

An implied depreciation rate is determined by Equation [13.11] using the SV of $10,000.

$$d = 1 - \left(\frac{10{,}000}{80{,}000}\right)^{1/10} = 0.1877$$

Note that $0.1877 < 2/n = 0.2$, so this DB model does not exceed twice the straight-line rate. Table 13–1 presents the D_t values using Equation [13.6] and the BV_t values

Table 13–1	D_t and BV_t values using declining-balance depreciation, Example 13.3	
Year, t	D_t	BV_t
0	—	$80,000
1	$15,016	64,984
2	12,197	52,787
3	9,908	42,879
4	8,048	34,831
5	6,538	28,293
6	5,311	22,982
7	4,314	18,668
8	3,504	15,164
9	2,846	12,318
10	2,318	10,000

from Equation [13.9] rounded to the nearest dollar. For example, in year $t = 2$,

$$D_2 = (d)BV_1 = 0.1877(64,984) = \$12,197$$

$$BV_2 = 64,984 - 12,197 = \$52,787$$

Because we round off to even dollars, \$2312 is calculated for depreciation in year 10, but \$2318 is deducted to make $BV_{10} = SV = \$10,000$ exactly.

Problems 13.8 to 13.14

13.4   MODIFIED ACCELERATED COST RECOVERY SYSTEM (MACRS)

During the 1980s in the United States, major changes in capital recovery systems were enacted at the federal government level. The Economic Recovery Act of 1981 introduced the Accelerated Cost Recovery System (ACRS), and the 1986 Tax Reform Act modified it to *MACRS,* which is the asset depreciation law-of-the-land at the time of this writing. Both systems dictate statutory depreciation rates for all personal and real property while taking advantage of accelerated methods of capital recovery.

Many aspects of MACRS are more specific to depreciation accounting and income tax computations than they are to the evaluation of investment alternatives, so only the major elements which materially affect engineering economy analysis are discussed here. Additional information and derivation of the MACRS depreciation rates are discussed in Sections 13.6 and 13.7.

In general, MACRS computes the annual depreciation using the relation

$$D_t = d_t B \qquad\qquad [13.12]$$

where the depreciation rate d_t is provided by the government in tabulated form and updated periodically. The book value in year t is determined in the standard ways, by subtracting the year's depreciation amount from the previous year's book value,

$$BV_t = BV_{t-1} - D_t \qquad\qquad [13.13]$$

or by subtracting the total depreciation for years 1 through $(t - 1)$ from the first cost, that is,

$$BV_t = \text{first cost} - \text{sum of accumulated depreciation}$$

$$= B - \sum_{j=1}^{j=t} D_j \qquad\qquad [13.14]$$

Therefore, the first cost B is always completely depreciated, since MACRS assumes that the estimated $SV = 0$, even though there may be a positive SV that is realizable.

The MACRS recovery periods are standardized to the values of 3, 5, 7, 10, 15, and 20 years for personal property. The real property recovery period for structures is commonly 39 years, but it is possible to justify a 27.5-year recovery. Section 13.5 explains how to determine an allowable MACRS recovery period.

For annual depreciation computations on personal property the first cost (unadjusted basis) is multiplied by the MACRS rate using Equation [13.12]. The d_t values for $n = 3, 5, 7, 10, 15,$ and 20 are included in Table 13-2. These values are used often throughout the remainder of this text, so you may wish to tab this page. (Explanation of how these rates are determined is included in Section 13.7.)

MACRS depreciation rates incorporate the DDB method ($d = 2/n$) and switch to SL depreciation during the recovery period as an inherent component for *personal property* depreciation. The MACRS rates usually start with the DDB rate ($d_t = 2/n$) and switch to the SL rate ($d_t = 1/n$) when the SL method offers a faster write-off.

For *real property*, MACRS utilizes the SL method for $n = 39$ throughout the recovery period. The annual percentage depreciation rate is $d = 1/39 = 0.02564$. However, MACRS forces partial recovery in years 1 and 40. The rates in percentage amounts are

Year 1	$100d_1 = 1.391\%$
Years 2–39	$100d_t = 2.564\%$
Year 40	$100d_{40} = 1.177\%$

Table 13-2   Depreciation rates, d_t, applied to the first cost B for the MACRS method

Year	Depreciation Rate (%) for Each MACRS Recovery Period in Years					
	$n = 3$	$n = 5$	$n = 7$	$n = 10$	$n = 15$	$n = 20$
1	33.33	20.00	14.29	10.00	5.00	3.75
2	44.45	32.00	24.49	18.00	9.50	7.22
3	14.81	19.20	17.49	14.40	8.55	6.68
4	7.41	11.52	12.49	11.52	7.70	6.18
5		11.52	8.93	9.22	6.93	5.71
6		5.76	8.92	7.37	6.23	5.29
7			8.93	6.55	5.90	4.89
8			4.46	6.55	5.90	4.52
9				6.55	5.91	4.46
10				6.55	5.90	4.46
11				3.28	5.91	4.46
12					5.90	4.46
13					5.91	4.46
14					5.90	4.46
15					5.91	4.46
16					2.95	4.46
17–20						4.46
21						2.23

You have probably noticed that all MACRS depreciation rates are presented for 1 year longer than the stated recovery period n. Also, note that the extra-year rate is usually about one-half of the previous year's rate. This is because a built-in *half-year convention* is imposed by the MACRS system. This convention assumes that all property is placed in service at the midpoint of the tax year of installation. Therefore, only 50% of the first-year DB depreciation applies for tax purposes. This removes some of the accelerated depreciation advantage and requires that one-half year of depreciation be taken in year $n + 1$. We utilize the half-year convention in all MACRS examples and problems. When this material was prepared there was also an allowed midquarter convention and a midmonth convention (which is built into the previously stated real property rates for $n = 39$). We won't concern ourselves with this level of detail, nor need it be considered in routine engineering economy studies.

Example 13.4

Consumer1st, a nationwide franchise for environmental engineering services, has acquired new workstations and 3D modeling software for its 100 affiliate sites at a cost of $4000 per site. The estimated salvage for each system after 3 years is expected to be 5% of the first cost. Assume that the franchise owner wants to compare the depreciation for a 3-year MACRS model with that for a 3-year DDB model. The franchise owner is curious about the total depreciation over the next 2 years.

(*a*) Determine which model offers the larger total depreciation after 2 years.

(*b*) Determine the book value for each model after 2 years and at the end of the recovery period.

Solution

The total basis is $B = \$400,000$ and the estimated SV $= 0.05(400,000) = \$20,000$. The MACRS rates for $n = 3$ are taken from Table 13–2, and the depreciation rate for DDB, Equation [13.4], is $d_{max} = \frac{2}{3} = 0.6667$. Table 13–3 presents the depreciation amounts and book values. Year 3 depreciation for DDB would be $44,444(0.6667) = \$29,629$, except that this would make the BV$_3$ < \$20,000, the estimated SV. So, only the remaining available amount of $24,444 may be utilized.

	MACRS			DDB	
Year	Rate	Depreciation	Book Value	Depreciation	Book Value
0		$ 0	$400,000		$400,000
1	0.3333	133,320	266,680	$266,667	133,333
2	0.4445	177,800	88,880	88,889	44,444
3	0.1481	59,240	29,640	24,444	20,000
4	0.0741	29,640	0		

Table 13–3   MACRS and DDB depreciation for $B = \$400,000$ and $n = 3$ years

(*a*) The 2-year total accumulated depreciation values from Table 13–3 are

MACRS: $D_1 + D_2 = \$133,320 + 177,800 = \$311,120$

DDB: $D_1 + D_2 = \$266,667 + 88,889 = \$355,556$

The DDB depreciation is larger. Remember that Consumer1[st] does not have the choice in the United States of the DDB model as applied here. MACRS has to be used.

(*b*) After 2 years the book value for DDB at $44,444 is 50% of the MACRS book value of $88,880. At the end of recovery, which is 4 years for MACRS (due to the built-in half-year convention) and 3 years for DDB, the MACRS value is $BV_4 = 0$ and the DDB value is $BV_3 = \$20,000$. This occurs because MACRS depreciation always removes the entire B value, regardless of the estimated salvage value.

Example 13.4
(Spreadsheet)

Develop the solution for Example 13.4 using a spreadsheet system.

Solution

Figure 13–3 presents the solution using the DDB function on a spreadsheet. Some comments around the spreadsheet explain the functions used to develop the MACRS and DDB schedules.

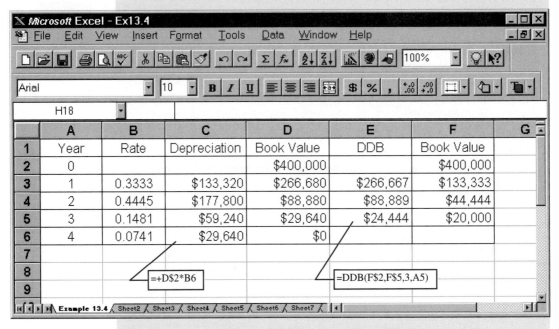

Figure 13–3   Spreadsheet using computed amounts for MACRS and a built-in model for DDB, Example 13.4 (Spreadsheet).

Note that the DDB value for D_3 is correct at \$24,444 in cell E5. The spreadsheet function automatically checks to ensure that the book value does not go below the estimated SV of \$20,000. Spreadsheets are excellent for depreciation schedules, especially when the function is built in. MACRS is not in all the spreadsheet systems.

The solutions to the problem using Figure 13–3 are calculated identically to those in Example 13.4.

(a) The 2-year total depreciation values are

MACRS, add cells C3 + C4:   \$133,320 + 177,800 = \$311,120

DDB, add cells E3 + E4:   \$266,667 + 88,889 = \$355,556

The DDB depreciation is larger.

(b) Book values after 2 years are

MACRS, cell D4:   \$88,880

DDB, cell F4:   \$44,444

The book values at the end of the recovery periods are in cells D6 and F5.

Comment

If you have not been developing the spreadsheets yourself, now is an excellent time to do so. You may wish to set up a *shell spreadsheet* for use with depreciation problems in this and future chapters.

MACRS, and its predecessor ACRS, have simplified depreciation computations considerably, but they have also removed much of the flexibility in model selection. Since alterations to the current methods of capital recovery are inevitable in the United States, current methods of tax-related depreciation computations may be different at the time you study this material. Since all economic analyses utilize future estimates, they may be performed in many instances more rapidly and often about as accurately by utilizing the classical straight-line model described in Section 13.2.

Another classical accelerated depreciation model, sum-of-year-digits, is now used very seldom. It is summarized in the appendix to this chapter.

Problems 13.15 to 13.19

13.5   DETERMINING THE MACRS RECOVERY PERIOD

The expected useful life of property is estimated in years and used as the n value in depreciation computations. Since depreciation is a tax-deductible amount, most large corporations and individuals want to minimize the n value. The advantage of a recovery period shorter than the anticipated useful life is compounded by the use of accelerated depreciation models which write off more of the first cost (or basis B) in the initial years. There are tables provided by government agencies which assist in determining the life and recovery period.

The U.S. government requires that all depreciable property be classified into a *property class* which identifies its MACRS-allowed recovery period. Table 13–4 gives example of assets and the MACRS n values. Virtually all property considered in an economic analysis has an n value presented in Table 13–2, where the MACRS rates are detailed.

Table 13–4 provides two MACRS n values for each property. The first is the *general depreciation system (GDS)* value, which we will commonly use in our examples and problems. The depreciation rates used in Section 13.4 correspond to the n values for the GDS column and provide the fastest write-off allowed, since the rates utilize the 200% DB method (that is DDB) or the 150% DB model

Table 13–4	Example MACRS recovery periods for various property descriptions		
		MACRS **n Value, Years**	
Examples (Personal and Real Property)		**GDS**	**ADS Range**
Special manufacturing and handling devices, tractors, racehorses		3	3–5
Computers and peripherals, duplicating equipment, autos, trucks, buses, cargo containers, some manufacturing equipment		5	6–9.5
Office furniture; some manufacturing equipment; railroad cars, engines, tracks; agricultural machinery; petroleum and natural gas equipment; *all property not in another class*		7	10–15
Equipment for water transportation, petroleum refining, agriculture product processing, durable-goods manufacturing, ship building		10	15–19
Land improvements, docks, roads, drainage, bridges, landscaping, pipelines, nuclear power production equipment, telephone distribution		15	20–24
Municipal sewers, farm buildings, telephone switching buildings, power production equipment (steam and hydraulic), water utilities		20	25–50
Residential rental property (house, mobile home)		27.5	40
Nonresidential real property attached to the land, but not the land itself		39	40

with a switch to SL depreciation. Notice that any asset not identifiable with a stated class is automatically assigned a 7-year recovery period under GDS.

The far-right column of Table 13–4 lists the *alternative depreciation system (ADS)* recovery period range of *n* values. This alternative method allows the use of *SL depreciation over a longer recovery period* than the GDS. The half-year convention applies, and any salvage value is neglected in the SL calculations, as it is in regular MACRS. The use of ADS is generally a choice left to a company, but it is required for some special asset situations. Since it takes longer to depreciate the asset, and since the SL model is required (thus removing the advantage of accelerated depreciation), ADS is usually not included in an economic analysis. This electable SL option is, however, sometimes chosen by businesses which are young and do not need the tax benefit of accelerated depreciation during the first few years of operation and capital-asset ownership. If ADS is selected, tables of d_t rates are printed and available.

Problems 13.20 to 13.23

13.6   SWITCHING BETWEEN DEPRECIATION MODELS

Switching between depreciation models may assist in accelerated reduction of the book value. It also maximizes the present value of accumulated and total depreciation over the recovery period. Therefore, switching usually increases the tax advantage in years where the depreciation is larger. (The approach you will learn now is an inherent part of the U.S. government MACRS system.)

Switching from a DB model to the SL method is the most common switch because it usually offers a real advantage, especially if the DB model is the DDB. General rules of switching are summarized here.

1. Switching is recommended when the depreciation for year *t* by the currently used model is less than that for a new model. The selected depreciation D_t is the larger amount.

2. Regardless of the depreciation models used, the book value can never go below the estimated salvage value. We assume the estimated SV = 0 in all cases; switching within MACRS always uses an estimated SV of zero.

3. The undepreciated amount, that is, BV_t, is used as the new adjusted basis to select the larger D_t for the next switching decision.

4. When switching from a DB model, the estimated salvage value, not the DB-implied salvage value, is used to compute the depreciation for the new method.

5. Only one switch can take place during the recovery period.

In all situations, the criterion is to *maximize the present worth of the total depreciation,* PW_D. The combination of depreciation models which results in the largest present worth is the best strategy.

$$PW_D = \sum_{t=1}^{t=n} D_t(P/F,i,t) \qquad \textbf{[13.15]}$$

This logic minimizes tax liability in the early part of an asset's recovery period. (Sections 14.5 and 14.6 include more discussion on the tax effects of depreciation.)

Virtually all switching occurs from a rapid write-off model to the SL model. The most advantageous is the DDB-to-SL switch. This switch is predictably advantageous if the implied salvage value computed by Equation [13.10] exceeds the salvage value estimated at the purchase time; that is, switch if

$$BV_n = B(1 - d)^n > \text{estimated SV} \qquad \textbf{[13.16]}$$

Since we assume that the actual SV will be zero, and since BV_n *will be greater than zero,* a switch to SL is advantageous. Depending upon the values of d and n, the switch may be best in the later years or last year of the recovery period, which removes the implied SV amount inherent to the DDB model.

The procedure to switch from DDB to SL depreciation is

1. For each year t, compute the two depreciation charges.

    For DDB: $\qquad\qquad\qquad\qquad D_{\text{DDB}} = (d)BV_{t-1} \qquad \textbf{[13.17]}$

    For SL: $\qquad\qquad\qquad\qquad D_{\text{SL}} = \dfrac{BV_{t-1}}{n-t+1} \qquad \textbf{[13.18]}$

2. Select the larger depceciation value so that the depreciation for each year $t = 1, 2, \ldots, n$ is

    $$D_t = \max[D_{\text{DDB}}, D_{\text{SL}}] \qquad \textbf{[13.19]}$$

3. Compute the present worth of total depreciation, PW_D, using Equation [13.15].

It is acceptable, though not usually financially advantageous, to state that a switch will take place in a particular year, for example, a mandated switch from DDB to SL in year 7 of a 10-year recovery period. This approach is usually not taken, but the switching technique will work correctly for any depreciation model involved with switching in any year $t < n$.

Example 13.5

M-E CyberSpace, Inc., has purchased a $100,000 computer-controlled on-line document imaging system with an estimated useful life of 8 years. Compute the annual capital recovery and compare the present worth of total depreciation for (*a*) the straight-line method, (*b*) the DDB method, and (*c*) DDB-to-SL switching. Use a market rate of $i = 15\%$ per year.

Solution

We use the GDS column in Table 13–4, which indicates that the recovery period is 5 years. (*Note:* If MACRS recovery periods are not imposed, this example should use an 8-year recovery period.)

(a) For $B = \$100,000$, $SV = 0$, and $n = 5$, Equation [13.1] is used to determine the annual SL depreciation.

$$D_t = \frac{100,000 - 0}{5} = \$20,000$$

Since D_t is the same for all years $t = 1, 2, \ldots, 5$, the P/A factor replaces P/F in Equation [13.15] to compute PW_D.

$$PW_D = 20,000(P/A,15\%,5) = 20,000(3.3522) = \$67,044$$

(b) For the DDB, $d = 2/5 = 0.40$. Equation [13.6] or [13.7] generates the results in Table 13–5. The value $PW_D = \$69,915$ exceeds the $\$67,044$ value for SL depreciation above. As is predictable, the DDB model maximizes PW_D for the $\$100,000$ investment.

Table 13–5   DDB model depreciation and present-worth computations for Example 13.5*b*; $B = \$100,000$, $n = 5$, and $d = 0.40$

Year, t	D_t	BV_t	$(P/F,15\%,t)$	Present Worth of D_t
0		$100,000		
1	$40,000	60,000	0.8696	$34,784
2	24,000	36,000	0.7561	18,146
3	14,400	21,600	0.6575	9,468
4	8,640	12,960	0.5718	4,940
5	5,184	7,776	0.4972	2,577
	$92,224			$69,915 = PW_D

(c) Use the DDB-to-SL switching procedure.

1. The DDB values for D_t in Table 13–5 are repeated in Table 13–6 for comparison with the D_{SL} values from Equation [13.18]. The D_{SL} values change each year since the adjusted basis BV_{t-1} is different. Only in year $t = 1$ is $D_{SL} = \$20,000$, the same as computed in part (*a*). For illustration, let's compute a D_{SL} value for years 2 and 4. For $t = 2$, $BV_1 = \$60,000$ by the DDB method and

$$D_{SL} = \frac{60,000 - 0}{5-2+1} = \$15,000$$

For $t = 4$, $BV_3 = \$21,600$ by the DDB method and

$$D_{SL} = \frac{21,600 - 0}{5-4+1} = \$10,800$$

Table 13–6	Depreciation and present worth allowing DDB-to-SL switching, Example 13.5c				
Year, t	**DDB Model**		**SL Depr., D_{SL}**	**Selected D_t**	**P/F Factor for t**	**Present Worth of D_t**
	D_{DDS}	BV_t				
0	—	$100,000				
1	$40,000	60,000	$20,000	$40,000	0.8696	$34,784
2	24,000	36,000	15,000	24,000	0.7561	18,146
3	14,400	21,600	12,000	14,400	0.6575	9,468
4	8,640	12,960	10,800	10,800*	0.5718	6,175
5	5,184	7,776	12,960	10,800	0.4972	5,370
	$92,224			$100,000		PW$_D$ = $ 73,943

| * Indicates switch from DDB to SL depreciation.

2. The column 'Selected D_t' shown in Table 13–6 includes a switch to SL depreciation in year 4 with $D_4 = D_5 = \$10,800$. The $D_{SL} = \$12,960$ in year 5 would apply only if the switch occurred in year 5.

   Total depreciation with switching is $100,000 compared to the DDB amount of $92,224. Book-value plots with and without switching are presented in Figure 13–4.

3. With switching, PW$_D$ = $73,943, which is an increase in depreciation present worth over both the SL and DDB models.

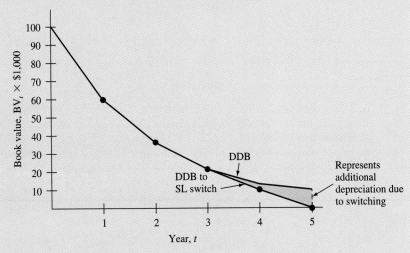

Figure 13–4   Comparison of book values for DDB and DDB-to-SL switching in year 4, Example 13.5.

In MACRS, recovery periods of 3, 5, 7, and 10 years include DDB with half-year-convention switching to SL. When the switch to SL takes place, which is usually in the last 1 to 3 years of the recovery period, any remaining basis is charged off in year $n + 1$ so that the book value reaches zero. Usually 50% of the applicable SL amount remains after the switch has occurred. For recovery periods of 15 and 20 years, 150% DB with the half-year convention and the switch to SL applies.

The present worth of depreciation PW_D is calculated using Equation [13.15]. It will always indicate which model is the most advantageous. Remember, however, currently in the United States only the MACRS rates (Table 13–2) for the general depreciation system recovery periods utilize the DDB-to-SL switch. The alternative-MACRS rates for the alternative depreciation system have longer recovery periods and impose the SL model for the entire recovery period, that is, no switching occurs.

Example 13.6

In Example 13.5c the DDB-to-SL switching model was applied to a $100,000 asset resulting in $PW_D = \$73,943$ at $i = 15\%$. Use MACRS to depreciate the same asset using a 5-year recovery period and compare PW_D values.

Solution

Table 13–7 summarizes the computations for depreciation (using Table 13–2 rates), book value, and present worth of depreciation. The PW_D values for the four different models are

DDB-to-SL switching	$73,943
Double declining balance	$69,916

Table 13–7   Depreciation and book value using MACRS, Example 13.6

t	d_t	D_t	BV_t
0	—	—	$100,000
1	0.20	$20,000	80,000
2	0.32	32,000	48,000
3	0.192	19,200	28,800
4	0.1152	11,520	17,280
5	1.1152	11,520	5,760
6	0.0576	5,760	0
	1.000	$100,000	

$$PW_D = \sum_{t=1}^{t=6} D_t(P/F,15\%,t) = \$69,016$$

| MACRS | $69,016 |
| Straight line | $67,044 |

Based on the smaller PW_D value for MACRS compared to DDB and DDB-to-SL switching, MACRS provides a slightly less accelerated write-off. This is, in part, because the half-year convention disallows 50% of the first-year DDB depreciation (which amounts to 20% of the first cost) compared to other DB-based models. Also, the recovery period extends into year 6 for MACRS, thus reducing PW_D further.

Problems 13.24 to 13.33

13.7 DETERMINATION OF MACRS RATES (optional section)

The depreciation rates for MACRS incorporate, in some form, the DB-to-SL switching model for all GDS recovery periods from 3 to 20 years. In the first year, some adjustments have been made to compute the MACRS rate. The adjustments vary and are not usually considered in detail in economic analyses. The half-year convention is always imposed, and any remaining book value in year n is removed in year $n + 1$. The value SV = 0 is assumed for all MACRS schedules.

Since different DB percentages are used for different n values, the following summary may be used to determine D_t and BV_t values. The symbols D_{DB} and D_{SL} are used to identify DB and SL depreciation, respectively.

For n = 3, 5, 7, and 10 Use DDB depreciation with the half-year convention switching to SL depreciation in year t when $D_{SL} \geq D_{DB}$. Use the switching rules of Section 13.6, and add one-half year to the life when computing D_{SL} to account for the half-year convention. The yearly depreciation rates are

$$d_t = \begin{cases} \dfrac{1}{n} & t = 1 \\[2ex] \dfrac{2}{n} & t = 2, 3, \ldots \end{cases}$$ [13.20]

Annual depreciation values for each year t applied to the adjusted basis are computed as

$$D_{DB} = (d_t)BV_{t-1}$$ [13.21]

$$D_{SL} = \frac{BV_{t-1}}{n-t+1.5}$$ [13.22]

When the switch to SL depreciation takes place—usually in the last 1 to 3 years of the recovery period—any remaining book value is removed in year $n + 1$.

This is commonly 50% of the SL amount applicable after the switch has occurred.

For n = 15 and 20   Use 150% DB with the half-year convention and the switch to SL when $D_{SL} \geq D_{DB}$. Until SL depreciation is more advantageous, the annual DB depreciation is computed as usual, that is, $D_{DB} = (d_t)BV_{t-1}$ where

$$d_t = \begin{cases} \dfrac{0.75}{n} & t = 1 \\ \dfrac{1.50}{n} & t = 2, 3, \ldots \end{cases} \qquad \textbf{[13.23]}$$

Example 13.7

A weaving machine for simulated-suede material with a MACRS 5-year recovery period has been purchased for $10,000. (*a*) Use the computed MACRS amounts (that is, don't use Table 13–2) to obtain the annual depreciation and book value. (*b*) Determine the resulting annual depreciation rates and compare them with the MACRS rates in Table 13–2 for $n = 5$. (*c*) Compute the present worth of depreciation at $i = 15\%$ per year.

Solution

(*a*)   With $n = 5$ and the half-year convention, use the DDB-to-SL switching procedure. Equations [13.20] through [13.22] are applied in conjunction with the switching procedure of Section 13.6 to obtain the results in Table 13–8.
     The switch to SL depreciation, which occurs in year 4 when both depreciation values are equal, is indicated using Equations [13.21] and [13.22] for $t = 4$.

$$D_{DB} = 0.4(2880) = \$1152$$

$$D_{SL} = \frac{2880}{5-4+1.5} = \$1152$$

The depreciation of $576 in year 6 is the result of the half-year convention. The amount $BV_5 = \$576$ is removed as one-half the SL amount, that is, $D_6 = 0.5(1152) = \$576$. The DB amount for $t = 5$, $D_{DB} = \$691.20$, is shown for illustration purposes only; it need not be computed since the SL values will always exceed DB values once the switch is made.

(*b*)   The actual rates are computed by dividing the Selected D_t column values by the first cost of $10,000. The rates are

t	1	2	3	4	5	6
d_t	0.20	0.32	0.192	0.1152	0.1152	0.0576

Table 13-8	Depreciation using computed MACRS rates for $n = 5$ and $B = \$10,000$				

Years, t	DDB		SL Depr., D_{SL}	Selected D_t	BV_t
	d_t	D_{DB}			
0	—	—	—	—	$10,000
1	0.2	$2,000.00	$1,818.18	$ 2,000	8,000
2	0.4	3,200.00	1,777.78	3,200	4,800
3	0.4	1,920.00	1,371.43	1,920	2,880
4	0.4	1,152.00	1,152.00	1,152	1,728
5	0.4	691.20	1,152.00	1,152	576
6	—	—	576.00	576	0
				$10,000	

These are exactly the same values presented in Table 13–2 for the MACRS recovery period of 5 years.

(c)   Equation [13.15] and the Selected D_t values in Table 13–8 are used to compute

$$PW_D = \sum_{t=1}^{t=6} D_t(P/F,15\%,t) = \$6901.61$$

It is easier to use the rates in Table 13–2 to calculate annual MACRS depreciation with Equation [13.12], $D_t = d_t B$, than to determine each MACRS rate using the switching logic above. But the logic behind the MACRS rates is described here for those interested. The annual MACRS rates may be derived using the applicable uniform rate, d, for the DB method. The subscripts DB and SL have been inserted along with the year t. For the first year $t = 1$,

$$d_{DB,1} = \frac{1}{n} \qquad\qquad [13.24]$$

For summation purposes only, we introduce the subscript i ($i = 1, 2, \ldots, t$) on d. Then the depreciation rates for years $t = 2, 3, \ldots, n$ are

$$d_{DB,t} = d\left(1 - \sum_{i=1}^{i=t-1} d_i\right) \qquad\qquad [13.25]$$

$$d_{SL,t} = \frac{\left(1 - \sum_{i=1}^{i=t-1} d_i\right)}{n-t+1.5} \qquad\qquad [13.26]$$

Also, for the year $n + 1$, the MACRS rate is one-half the SL rate of the previous year n.

$$d_{SL,n+1} = \frac{1}{2(d_{SL,n})} \qquad\qquad [13.27]$$

The results of Equations [13.25] and [13.26] are compared each year to determine which is larger and when the switch to SL depreciation should occur. The MACRS rates in Table 13–2 are the resulting d_t values.

Example 13.8

Verify the MACRS rates in Table 13–2 for a 3-year recovery period. The depreciation rates in percent are 33.33, 44.45, 14.81, and 7.41 for years 1, 2, 3, and 4, respectively.

Solution

The uniform rate for DDB with $n = 3$ is $d = 2/3 = 0.6667$. Using the half-year convention in year 1 and Equations [13.24] through [13.27], the results are

d_1. $\qquad\qquad d_{DB,1} = 0.5d = 0.5(0.6667) = 0.3333$

d_2.   The cumulative depreciation rate is 0.3333.

$$d_{DB,2} = 0.6667(1 - 0.3333) = 0.4445 \qquad \text{(larger value)}$$

$$d_{SL,2} = \frac{1 - 0.3333}{3 - 2 + 1.5} = 0.2267$$

d_3.   Cumulative depreciation rate is $0.3333 + 0.4445 = 0.7778$.

$$d_{DB,3} = 0.6667(1 - 0.7778) = 0.1481$$

$$d_{SL,3} = \frac{1 - 0.7778}{3 - 3 + 1.5} = 0.1481$$

Both d_3 values are the same; switch to straight-line depreciation.

d_4.   This rate is 50% of the last SL rate.

$$d_4 = 0.5(d_{SL,3}) = 0.5(0.1481) = 0.0741$$

The derived rates are 0.3333, 0.4445, 0.1481, and 0.0741, which are the same as those in Table 13–2.

Comment

The DDB and SL rates for year 3 are equal because 0.5 year is added to the denominator for $d_{SL,3}$, making the multipliers equal for both methods.

Problems 13.34 to 13.37

13.8   CAPITAL EXPENSE (SECTION 179) DEDUCTION

There are many exceptions, allowances, and details in depreciation accounting and capital-recovery law. By way of illustration, we cover one commonly used by small U.S. businesses. This allowance is designed to be an incentive for investment in capital directly used in business or trade. The allowance is historically called the *Section 179 deduction,* after the government [Internal Revenue Service (IRS)] code section which defines it.

Rather than depreciate the entire basis of a capital investment, it is acceptable to expense up to $18,500 of the first cost during the first year of service, provided the first cost (unadjusted basis) is reduced by the amount expensed prior to calculating the depreciation schedule. The remaining amount is depreciated over the regular recovery period. (The $18,500 amount does change with time. For example, it was $10,000 until the mid-1990s and $17,500 through 1996. The approved limits are $18,000 in 1997; $18,500 in 1998; $19,000 in 1999; $20,000 in 2000; $24,000 in 2001; and $25,000 in 2003.)

The deduction is additionally limited. Any amount invested in excess of $200,000 reduces the current expensing limit dollar for dollar. Thus, any investment over $218,500 eliminated the Section 179 deduction allowance completely in 1998. As an example, if a 3-year recovery property asset costing $50,000 has an $18,500 179-deduction allowance in year 1, the first-year MACRS depreciation will be $0.3333(50,000 - 18,500) = \$10,499$. Therefore, $10,499 is depreciated in year 1, and $18,500 is expensed by Section 179 in year 1. The remaining $21,001 is recovered in years 2, 3, and 4.

Of course, a corporation can claim only a single Section 179 expense amount each year, so the allowance is very limited in its overall impact, especially upon large corporations. It is usually not considered in engineering economy studies which involve large capital investments.

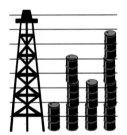

Example 13.9

Oil-field equipment purchased under the MACRS 7-year depreciation rules totals $205,000. Compute the total of first-year depreciation and Section 179 allowance, and the allowed MACRS depreciation for year 2. Did the Section 179 deduction help the owner of the equipment?

Solution

The maximum Section 179 deduction is reduced by $5000 since the investment exceeds the $200,000 limit by $5000. Using a Section 179 limit of $18,500, the 7-year MACRS rates in Table 13–2 apply to the remaining basis of $191,500 as calculated here.

Section 179 expense allowance: $18,500 - 5000 = \$13,500$

Total amount to be depreciated: $\$205,000 - 13,500 = \$191,500$

MACRS, year 1: $0.1429(191,500) = \$27,365$

The total is $13,500 (Section 179) plus $27,365 (MACRS), or $40,865.

MACRS, year 2: $0.2449(191,500) = \$46,898$

Had the Section 179 deduction not been used, MACRS depreciation would have been $29,294 and $50,204, for a total of $79,498 over 2 years. With the Section 179 deduction, a total of $87,763, or 10.4% more, has been removed from the books in 2 years. Yes, the Section 179 deduction did help.

Problems 13.38 and 13.39

13.9   DEPLETION METHODS

Up to this point, we have computed depreciation for an asset which has a value that can be recaptured by purchasing a replacement. Depletion, though similar to depreciation, is applicable only to natural resources. When the resources are removed, they cannot be replaced or repurchased in the same manner as can a machine, computer, or structure. Therefore, depletion is applicable to natural deposits removed from mines, wells, quarries, geothermal deposits, forests, and the like. There are two methods of depletion—*cost depletion* and *percentage depletion.*

Cost depletion, sometimes referred to as factor depletion, is based on the level of activity or usage, not time as in depreciation. It may be applied to most types of natural resources. The cost-depletion factor for year t, p_t, is the ratio of the first cost of the property to the estimated number of units recoverable.

$$p_t = \frac{\text{initial investment}}{\text{resource capacity}} \qquad \text{[13.28]}$$

The annual depletion charge is p_t times the year's usage or activity volume. *Accumulated cost-based depletion cannot exceed the total first cost of the resource.* If the capacity of the property is reestimated some year in the future, a new cost-depletion factor is computed based upon the undepleted amount and the new capacity estimate.

Example 13.10

Miller Lumber Co. has negotiated the rights to cut timber on privately held forest acreage for $350,000. An estimated 175 million board feet of lumber are harvestable.

(*a*) Determine the depletion amount for the first 2 years if 15 million and 22 million board feet are removed.

(*b*) If after 2 years the total recoverable board feet is reestimated to be 225 million from time $t = 0$, compute the new cost-depletion factor for years 3 and later.

Solution

(a) Use Equation [13.28] to compute p_t in dollars per million board feet for $t = 1$, 2, . . .

$$p_t = \frac{\$350,000}{175} = \$2000 \text{ per million board feet}$$

Multiply the p_t rate by the annual harvest to obtain depletion of $30,000 in year 1 and $44,000 in year 2. Continue using p_t until a total of $350,000 is depleted.

(b) After 2 years, a total of $74,000 has been depleted. A new p_t value must be based on the remaining $350,000 - 74,000 = \$276,000$ of undepleted investment. Additionally, with the new estimate of 225 million board feet, a total of $225 - 15 - 22 = 188$ million board feet remain. For years $t = 3, 4, \ldots$, the cost-depletion factor is now

$$p_t = \frac{\$276,000}{188} = \$1468 \text{ per million board feet}$$

Percentage depletion, the second depletion method, is a special consideration given for natural resources. A constant, stated percentage of the resource's gross income may be depleted each year provided it does not exceed 50% of the company's taxable income. So, annually the depletion amount is calculated as

$$\text{Percentage depletion amount} = \text{percentage} \qquad \text{[13.29]}$$
$$\times \text{ gross income from property}$$

Using percentage depletion, total depletion charges may exceed first cost with no limitation. The U.S. government does not generally allow percentage depletion to be applied to oil and gas wells (except small independent producers) or timber.

The depletion amount each year may be determined using either the cost method or the percentage method, as allowed by law. Usually, the percentage-depletion amount is chosen because of the possibility of writing off more than the original cost of the venture. However, the law also requires that the cost-depletion amount be chosen if the percentage depletion is smaller in any year. So, calculate both depletion amounts; cost depletion ($Depl) and percentage depletion (%Depl) and apply the following logic each year.

$$\text{Annual depletion} = \begin{cases} \%\text{Depl} & \text{if } \%\text{Depl} \geq \$\text{Depl} \\ \$\text{Depl} & \text{if } \%\text{Depl} < \$\text{Depl} \end{cases} \qquad \text{[13.30]}$$

The annual percentage depletions for certain natural deposits are listed below. These percentages are changed from time to time when new depletion legislation is enacted.

Deposit	Percentage
Sulfur, uranium, lead, nickel, zinc, and some other ores and minerals	22
Gold, silver, copper, iron ore, geothermal deposits	15
Oil and gas wells (special cases)	15
Coal, lignite, sodium chloride	10
Gravel, sand, peat, some stones	5
Most other minerals, metallic ores	14

Example 13.11

An old gold mine purchased for $750,000 has an anticipated gross income of $1.1 million per year for years 1 to 5 and $0.85 million per year after year 5. Assume that depletion charges do not exceed 50% of taxable income. Compute annual depletion amounts for the mine. How long will it take to recover the initial investment?

Solution

A 15% depletion applies to gold. Depletion amounts are

Years 1 to 5:   0.15(1.1 million) = $165,000

Years thereafter: 0.15(0.85 million) = $127,500

At this rate, the cost of $750,000 will be recovered in approximately 4.5 years of operation.

$$\frac{\$1.1 \text{ million}}{\$165,000} = 4.55 \text{ years}$$

Problems 13.40 and 13.41

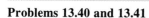

CHAPTER SUMMARY

The majority of this chapter has been devoted to learning the concepts and calculations of depreciation models. *Depreciation* does not result in actual cash flow directly. It is a book method by which the capital investment in tangible property is recovered. The annual depreciation amount is tax-deductible, which can result in actual cash flow changes.

Some important points about the straight-line, declining-balance, and MACRS models are presented here. Common relations for each model are summarized in the table of formulas (Table 13–9).

Table 13–9	Summary of common depreciation model relations		
Model	**MACRS**	**SL**	**DDB**
Uniform depreciation rate, d	Not defined	$\dfrac{1}{n}$	$\dfrac{2}{n}$
Annual rate, d_t	Table 13–2	$\dfrac{1}{n}$	$d(1-d)^{t-1}$
Annual depreciation, D_t	$d_t B$	$\dfrac{B-SV}{n}$	$(d)BV_{t-1}$
Book value, BV_t	$BV_{t-1}-D_t$	$B-tD_t$	$B(1-d)^t$

Straight Line (SL)

- It writes off capital investment linearly over n years.
- The estimated salvage value is always considered.
- This classical, nonaccelerated depreciation model is used to compare the rate of capital recovery with rates for accelerated models.

Declining Balance (DB)

- The model accelerates depreciation compared to the straight-line model.
- The book value is reduced each year by a uniform percentage.
- The most common rate is twice the SL rate, which is called double-declining-balance depreciation.
- It has an implied SV, which may be lower than the estimated SV.

Modified Accelerated Cost Recovery System (MACRS)

- It is the approved depreciation system in the United States.
- It automatically switches from DDB or DB to SL depreciation.
- It always depreciates to zero; that is, it assumes SV = 0.
- Recovery periods are specified by property classes.
- Depreciation rates are tabulated for general use.
- The actual recovery period is 1 year longer due to the imposed half-year convention.
- MACRS straight-line depreciation is an option, but recovery periods are longer than for regular MACRS.

The *rules for switching* between depreciation models *maximize the PW of total depreciation*, PW_D. Compute the D_t each year and switch (once) to the model with the larger depreciation. Usually the DDB-to-SL switch in the last few years of recovery maximizes PW_D. The MACRS rates include switching from DDB or DB to SL to maximize PW_D, but the half-year convention reduces the accelerated write-off somewhat.

A business, especially a small one, may include the *capital expense (Section 179) deduction* in its economic analysis. Though dollar-limited, when used in combination with depreciation, the first cost may be recovered more rapidly.

Cost and *percentage depletion* models recover investment in natural resources. The annual cost-depletion factor is applied to the amount of resource removed. No more than the initial investment can be recovered with cost depletion. Percentage depletion can recover more than the initial investment.

PROBLEMS

13.1 Contact the local office of the U.S. Internal Revenue Service (or the relevant office for your country of residence) and ask them to describe or provide you with a document or Internet address which describes the depreciation system authorized currently for (*a*) individual taxpayers or (*b*) businesses and corporations. Prepare a summary of your understanding of this system. Be prepared to explain the system to your classmates, as guided by your instructor for this material.

13.2 Earth Harvest Nursery paid $152,000 for an asset in 1990 and installed it at a cost of $3000. The asset's expected life was 10 years with a salvage value of 10% of the original purchase price. The asset was sold at the end of 1998 for $43,000.

 (*a*) Define the values needed to develop an annual depreciation schedule at purchase time.

 (*b*) What are the values for the actual life, the market value in 1998, and the book value at sale time if 75% of the first cost has been removed via an accelerated depreciation model?

13.3 A $100,000 piece of testing equipment was installed and depreciated for 5 years. Each year the end-of-year book value decreased at a rate of 10% of the book value at the beginning of the year. The system was sold for $24,000 at the end of the 5 years.

 (*a*) Compute the amount of the annual depreciation.

 (*b*) What is the actual depreciation rate for each year?

 (*c*) Plot the book value for each of the 5 years.

 (*d*) At the time of sale, what is the difference between the book value and the market value?

13.4 Explain why the recovery period of an asset used for depreciation purposes may be different from the expected-life value used when performing an engineering economy study.

13.5 Smoother and Sons has just purchased a machine for $325,000 with an additional $25,000 charge for installation onto a truck for mobility. The expected life is 30 years with a salvage value of 10% of the purchase price. For classical straight-line depreciation, determine the first cost, salvage value, annual depreciation amounts, and book value after 20 years.

13.6 A commercial van costing $12,000 has a life of 8 years with a $2000 salvage value. (*a*) Calculate the straight-line depreciation amount and book value for each year. (*b*) What is the rate of depreciation? Explain the meaning of this rate.

13.7 A new special-purpose computer workstation has $B = \$50,000$ with a 4-year recovery period. Use a computerized spreadsheet to tabulate (and plot, if requested by the instructor) the values for SL depreciation, accumulated depreciation, and book value for each year if (*a*) there is no salvage value and (*b*) SV = $10,000.

13.8 For the declining-balance method of depreciation, explain the differences between the three rates: uniform percentage rate d, d_{max}, and the annual recovery rate d_t.

13.9 (*a*) Use a spreadsheet system to work Problem 13.6(*a*) using the DDB model. Plot the book value for SL and DDB depreciation on a single graph. (Manual solution works also.) (*b*) Calculate the DDB annual depreciation rate for all years 1 through 8.

13.10 A building costs $320,000 to construct. It has a 30-year life with an estimated sales value of 25% of the construction cost. Calculate and compare the annual depreciation charge for years 4, 18, and 25 using (*a*) straight-line depreciation and (*b*) DDB depreciation.

13.11 Jeremy and Sheila have a word processing business for which they have purchased new computer equipment for $25,000. They do not expect the computers to have a positive salvage or trade-in value after the anticipated 5-year life. For possible reference in the future, they want book value schedules for several different models: SL, DB, 150% SL, 175% SL, and DDB. They believe that a fair uniform depreciation rate is 25% annually for the DB model. Use your spreadsheet system or manual computations to develop the schedules.

13.12 Newly purchased tree-planting equipment has an installed value of $82,000 with an estimated trade-in value of $10,000 after 18 years. For the years 2, 10, and 18, compute the annual depreciation charge using DDB depreciation and DB depreciation.

13.13 Declining-balance depreciation at a rate of 1.5 times the straight-line rate is to be used for automated process-control equipment with $B = \$175,000$, $n = 12$, and expected SV $= \$32,000$. (*a*) Compute the depreciation and book value for years 1 and 12. (*b*) Compare the expected salvage value and the book value remaining using the 150% DB model.

13.14 A newly installed electronic desktop printing system has $B = \$25,000$, a useful life of $n = 10$ years, and no salvage value. On a single graph, plot the book-value curves for the following situations: classical straight-line depreciation over the useful life and DDB depreciation for $n = 7$ years.

13.15 Give three specific examples each of personal property and real property that must be depreciated by the MACRS method.

13.16 Claude is the accountant for a business client with a new $30,000 personal-property asset to be depreciated using MACRS over 7 years. If the salvage value is expected to be $2000, compare the values and plots of book value for MACRS depreciation and classical SL depreciation over 7 years.

13.17 An asset, purchased for $20,000, has a 10-year useful life and an excellent resale value, so an SV of 20% of first cost is estimated. An abbreviated recovery period of 5 years is allowed by MACRS. Prepare the MACRS annual depreciation schedule, and compare these results with DDB depreciation over $n = 10$ years.

13.18 An automated assembly robot that cost $450,000 installed has a depreciable life of 5 years and no salvage value. An analyst in the financial management department used classical SL depreciation to determine end-of-year book values for the robot when the original economic evaluation was performed. You are now performing a replacement analysis after 3 years of service and realize that the robot should have been depreciated using MACRS with $n = 5$ years. What is the amount of the error in the book value for the classical SL method?

13.19 (*a*) Develop the MACRS depreciation schedule for a rental house purchased by Daryl Enterprises for $150,000. (*b*) What is the effect of the half-year convention that MACRS imposes; that is, how would the depreciation schedule change if the convention were not imposed?

13.20 Bowling Mania, Inc., has just installed $100,000 worth of depreciable equipment that represents the latest in automated scoring and graphics intended to make the bowler enjoy the sport more fully. A $15,000 salvage value is estimated. The alley's accountant can depreciate using MACRS for a 7-year recovery period or opt for the ADS alternate system over 10 years using the straight-line model. The SL rates require the half-year convention, that is, only 50% of the regular annual rate applies for years 1 and 11. (*a*) Construct the book value curves for both models on one graph. (*b*) After 3 years of use, what is the percentage of the $100,000 basis removed for each model?

13.21 If $B = \$20,000$, use the MACRS rate for $n = 5$ years and the SL alternative (with the half-year convention) for 9 years to develop two annual depreciation schedules.

13.22 The financial analyst for a large real-estate ownership/management corporation has just purchased the first residence for her own fledging business in rental property ownership. She is in the unique situation to make decisions at the same time on MACRS depreciation recovery periods of real property (accelerated GDS periods or the SL alternative using ADS periods) for a large corporation and her own small business. Prepare two lists of the elements which she should consider for these different business environments.

13.23 The president of a small business wants to understand the difference in the yearly recovery rates for classical SL, MACRS, and the SL alternative to MACRS. Prepare a single graph showing the annual recovery rates (in percent) for the three models versus the year of the recovery period. Use $n = 3$ years for your illustration.

13.24 Explain why the combination of shortened recovery rates and higher depreciation rates in the initial years of an asset's life may be financially advantageous to a corporation.

13.25 Use the estimates of Problem 13.16 for the $30,000 asset and compare the present worth of depreciation values for $i = 10\%$.

13.26 An asset has a first cost of $45,000, a recovery period of 5 years, and a $3000 salvage value. Use the switching procedure from DDB to SL depreciation and calculate the present worth of depreciation at $i = 18\%$ per year.

13.27 If $B = \$45,000$, $SV = \$3000$, and $n = 5$-year recovery period, use $i = 18\%$ per year to maximize the present worth of depreciation using the following methods: DDB-to-SL switching (as requested in Problem 13.26) and MACRS. Given that MACRS is the required depreciation system in the United States, comment on your results.

13.28 Above Ground Industries has a new milling machine with $B = \$110,000$, $n = 10$ years, and $SV = \$10,000$. Determine the depreciation schedule and present worth of depreciation at $i = 12\%$ per year using the 175% DB method for the first 5 years and switching to the classical SL method for the last 5 years. (Use a spreadsheet system to solve this problem.)

13.29 Let $B = \$12,000$, $n = 8$ years, SV = $\$800$, and $i = 20\%$ per year. (*a*) Develop the depreciation schedule and depreciation PW for the DDB method. (*b*) Allow switching to classical SL depreciation and develop the new depreciation schedule and PW value. (*c*) Determine the MACRS schedule and PW value for a 7-year recovery life. (*d*) Plot the three book-value curves on the same graph.

13.30 The Electric Company has erected a large portable building with a first cost of $\$155,000$ and an anticipated salvage of $\$50,000$ after 25 years. (*a*) Should the switch from DDB to SL depreciation be made? (*b*) For what values of the uniform depreciation rate in the DB method would it be advantageous to switch from DB to SL depreciation at some point in the life of the building?

13.31 For the situation in Problem 13.21, compare the PW values of depreciation for the regular MACRS and the alternative SL models if the rate of return is 10%.

13.32 In the United States, the MACRS system of depreciation has significantly simplified the development of depreciation schedules and the allowance for depreciation for income tax purposes. Explain this simplification relative to the other depreciation models you have learned in this chapter.

13.33 If you reside outside the United States, research the current tax laws for depreciation (or capital recovery) for your country and compare them with the classical depreciation methods of SL and DB. Also, compare them with the MACRS system. In your comparison, highlight any significant differences in depreciation philosophy, recovery periods, and annual recovery rates.

13.34 Verify the 5-year recovery period rates for MACRS given in Table 13–2. Start with the DDB model in year 1 and switch to SL depreciation when it offers a larger recovery rate.

13.35 A video recording system was purchased 3 years ago at a cost of $\$30,000$. A 5-year recovery period and MACRS depreciation have been used to write off the basis. The system is to be prematurely replaced with a trade-in value of $\$5000$. What is the difference between the book value and the trade-in value?

13.36 Use the computations in Equations [13.20] through [13.22] to determine the MACRS annual depreciation for the following asset data: $B = \$50,000$ and a recovery period of 7 years.

13.37 The 3-year MACRS recovery rates are 33.33%, 44.45%, 14.81%, and 7.41%, respectively. (*a*) What are the corresponding rates for the MACRS straight-line model alternative with the half-year convention imposed? (*b*) Compare the present worth of depreciation for these two sets of rates if $B = \$80,000$ and $i = 15\%$ per year.

13.38 A1 Janitorial Supply purchased a new supply truck for $36,500. The owner wants to compute the 3-year MACRS depreciation schedule and take advantage of the maximum amount of capital expensing allowed. Perform the analysis. (Use the current year's rules, or the year 2000 rules, as guided by your instructor.)

13.39 In 1998, the Beauty Supply Company acquired a $215,800 mixing machine which it capital expensed and depreciated using the 5-year recovery and MACRS method. (*a*) Determine the capital expensing and depreciation schedule. (*b*) Compare these amounts with those for MACRS only.

13.40 A company owns gold mining operations in the United States, Australia, and South Africa. The Colorado U.S. mine has the taxable income and sales summarized below. Determine the annual percentage depletion for the gold mine.

Year	Taxable Income	Sales, Ounces	Sales, $/Ounce
1	$500,000	2,000	375
2	500,000	4,500	390
3	400,000	3,000	385

13.41 An international concrete corporation has operated a quarry for the past 5 years. During this time the following tonnage has been extracted each year: 50,000; 42,000; 58,000; 60,000; and 56,000 tons. The mine is estimated to contain a total of 2.0 million tons of usable stones and gravel. The quarry land had an initial cost of $2.2 million. The company had a per-ton gross income of $15 for the first 2 years, $20 for the next 2 years, and $23 for the last year.

(*a*) Compute the depletion charges each year using the larger of the values for the two depletion methods.

(*b*) Compute the percent of the initial cost that has been written off in these 5 years using the depletion charges in part (*a*).

(*c*) If the quarry operation is reevaluated after the first 3 years of operation and estimated to contain another 1.5 million tons, rework parts (*a*) and (*b*).

CHAPTER 13 APPENDIX: SUM-OF-YEAR-DIGITS (SYD) DEPRECIATION

The SYD method is a classical accelerated-depreciation technique which removes much of the basis in the first one-third of the recovery period; however, write-off is not as rapid as DDB or MACRS. This technique, though not incorporated into the current MACRS method, may be used in an engineering economy analysis, especially in the depreciation of multiple-asset accounts (group and composite depreciation).

The mechanics of the method involve initially finding S, the sum of the year digits from 1 through the recovery period n. The depreciation charge for any given year is obtained by multiplying the basis of the asset less any salvage value $(B - SV)$ by the ratio of the number of years remaining in the recovery period to the sum of year digits, S.

$$D_t = \frac{\text{depreciable years remaining}}{\text{sum of year digits}} (\text{basis} - \text{salvage value}) \qquad \text{[13.A1]}$$

$$= \frac{n-t+1}{S}(B - SV)$$

where S is the sum of year digits 1 through n.

$$S = \sum_{j=1}^{j=n} j = \frac{n(n+1)}{2} \qquad \text{[13.A2]}$$

Note that the depreciable years remaining must include the year for which the depreciation charge is desired. That is why the 1 has been included in the numerator of Equation [13.A1]. For example, to determine the depreciation for the fourth year of an asset which has an 8-year life, the numerator of Equation [13.A1] is $8 - 4 + 1 = 5$ and $S = 36$.

The book value for any given year t is calculated as

$$BV_t = B - \frac{t(n-t/2 + 0.5)}{S}(B - SV) \qquad \text{[13.A3]}$$

The rate of depreciation d_t, which decreases each year for the SYD method, follows the multiplier in Equation [13.A1].

$$d_t = \frac{n-t+1}{S} \qquad \text{[13.A4]}$$

Example 13.A1

Calculate the SYD depreciation charges for years 1, 2, and 3 for electro-optics equipment with $B = \$25{,}000$, $SV = \$4000$, and an 8-year recovery period.

Solution

The sum of year digits is $S = 36$, and the depreciation amounts for the first 3 years by Equation [13.A1] are

$$D_1 = \frac{8 - 1 + 1}{36(25{,}000 - 4000)} = \$4667$$

$$D_2 = \frac{7}{36(21{,}000)} = \$4083$$

$$D_3 = \frac{6}{36(21{,}000)} = \$3500$$

Figure 13–A1 is a plot of the book values for an \$80,000 asset with $SV = \$10{,}000$ and $n = 10$ years using all the depreciation methods we have learned. The MACRS, DDB, and SYD curves track closely except for year 1 and years 9 through 11. As an exercise, you may wish to use your spreadsheet to compute the D_t and BV_t values to confirm the results of Figure 13–A1.

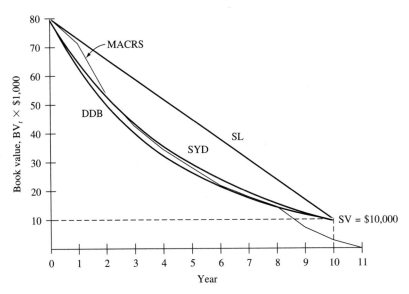

Figure 13–A1   Comparison of book values for an asset using SL, SYD, DDB, and MACRS depreciation.

APPENDIX PROBLEMS

13.A1 A music company's demonstration piano has a first cost of $12,000, an estimated salvage value of $2000, and a recovery period of 8 years. Use the SYD method to tabulate annual depreciation and book value.

13.A2 Earth-moving equipment having a first cost of $182,000 is expected to have a life of 18 years. The salvage value at that time is expected to be $15,000. Calculate the depreciation charge and book value for years 2, 7, 12, and 18 using the sum-of-year-digits method.

13.A3 If $B = \$12,000$, $n = 6$ years, and SV is estimated at 15% of B, use the SYD method to determine the (*a*) book value after 3 years, (*b*) rate of depreciation in year 4, and (*c*) depreciation amount in year 4 using the rate from part (*b*).

chapter

14

Basics of Income Taxes

Form **1040**	Department of the Treasury—Internal Revenue Service **U.S. Individual Income Tax Return**	19**96**	(5)	IRS

For the year Jan. 1–Dec. 31, 1996, or other tax year beginning,1995, ending

Label
(See instructions on page 11.)

Use the IRS label. Otherwise, please print or type.

L A B E L   H E R E

Your first name and initial	Last name
If a joint return, spouse's first name and initial	Last name

Home address (number and street). If you have a P.O. box, see page 11.

City, town or post office, state, and ZIP code. If you have a foreign address, see page 11.

Presidential Election Campaign ▶ Do you want $3 to go to this fund? . . . .

This chapter will help you understand *income taxes for corporations.* You will also learn some basics about *income taxes for individual taxpayers;* that's you and your family. The basic tax treatment of capital gains and losses is explained. And comparisons of how different depreciation models and shortened recovery periods affect income taxes are included.

Individuals and corporations pay taxes at the federal, state, and local levels of government in a variety of areas—income, property, school, value-added, sales, use, services, and others. By and large, corporations are taxed on income generated in the process of doing business, while individuals pay taxes on salaries, wages, royalties, and investment proceeds.

When an economic analysis is performed, it is reasonable to ask if the analysis should be on a *before-tax* or *after-tax* basis. For a tax-exempt organization (e.g., university, state, religious, not-for-profit foundation, or not-for-profit corporation), after-tax analysis is not necessary. For taxed organizations (corporations and partnerships) the after-tax analysis may or may not result in a different decision than that from a before-tax analysis. Though the alternative chosen may be the same, after-tax analysis gives much better estimates of cash flows and the anticipated rate of return for an alternative. For these reasons, many analysts prefer an after-tax analysis. Chapter 15 presents more detail on the inclusion of income tax considerations in engineering economic analyses.

We present here only a simplified version of income taxes. More detail is available from the U.S. Internal Revenue Service and its publications, some of which are listed at the end of the chapter.

LEARNING OBJECTIVES

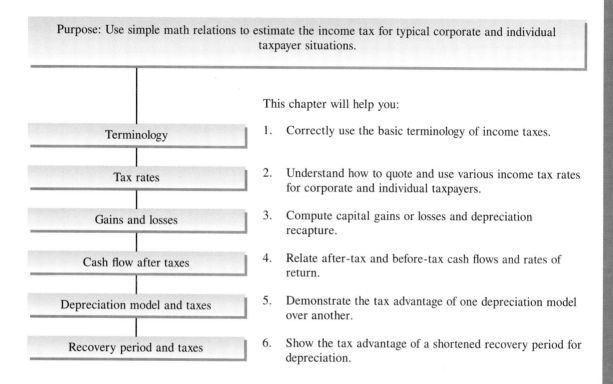

Purpose: Use simple math relations to estimate the income tax for typical corporate and individual taxpayer situations.

This chapter will help you:

Terminology	1.	Correctly use the basic terminology of income taxes.
Tax rates	2.	Understand how to quote and use various income tax rates for corporate and individual taxpayers.
Gains and losses	3.	Compute capital gains or losses and depreciation recapture.
Cash flow after taxes	4.	Relate after-tax and before-tax cash flows and rates of return.
Depreciation model and taxes	5.	Demonstrate the tax advantage of one depreciation model over another.
Recovery period and taxes	6.	Show the tax advantage of a shortened recovery period for depreciation.

Income taxes are the subject of frequent changes, because they may be "political hot spots" or they may offer economic leverage to speed up or slow down the economy. Usually, only the specific tax rates are altered, but from time to time, governments change some of the basic mechanisms and allowances for both corporations and individual taxpayers. We utilize tax rates and a tax structure characteristic of the mid-to-late 1990s in the United States in our examples and problems.

The general approach to tax treatment and computations that you will learn here is consistent over time and in many countries. Therefore, you will be familiar with the fundamental treatment of taxes when you perform an engineering economy analysis.

425

14.1 BASIC TERMINOLOGY FOR INCOME TAXES

Some basic income tax terms for corporations (and some terms for individual taxpayers) follow.

Gross income, GI, is the total of all income from revenue-producing sources, including all items listed in the revenue section of an income statement. (Refer to Appendix B for the basics of accounting reports.) The term *adjusted gross income* is used when certain allowable adjustments are made to gross income. We assume the two gross income amounts are equal. For individuals, gross income primarily consists of wages, salaries, interest, dividends, royalties, and capital gains.

Operating expenses, E, include all corporation costs incurred in the transaction of business. These are the operating costs for the engineering economy alternative. Most individual taxpayers don't have expenses for income tax calculations; they receive selected exemptions and deductions.

Taxable income, TI, is the dollar amount upon which taxes are calculated. For corporations, expenses and depreciation are subtracted from gross income to obtain taxable income.

$$TI = \text{gross income} - \text{expenses} - \text{depreciation}$$

Tax rate is a percentage, or decimal equivalent, of taxable income owed in taxes. Historically the tax rate, T, has been graduated for U.S. corporations and individuals—rates are higher as TI increases. In general, taxes are computed as

$$\text{Taxes} = \text{taxable income} \times \text{applicable tax rate} = (TI)T$$

Net profit or *net income* results, in general, when corporate income taxes are subtracted from taxable income.

Operating losses are experienced by a corporation in years when there is a net loss rather than a net profit. Special tax considerations are made in an attempt to balance the lean and fat years. Anticipation of operating losses, and thus the ability to take them into account in an economy study, is not practical; yet, the tax treatment of past losses may be relevant in some economy studies.

Capital gain is an amount of taxable income incurred when the selling price of a depreciable asset or property exceeds the original purchase price. At sale time,

$$\text{Capital gain} = \text{selling price} - \text{purchase price}$$

If the result is positive, a gain is recorded. If the sales date occurs within a given time of the purchase date, the capital gain is referred to as *short-term gain (STG);* if the ownership period is longer, the gain is a *long-term gain (LTG)*. Tax law sets and changes the required ownership period, commonly 1 year or 18 months. We use 1 year here.

Capital loss is the opposite of a capital gain. If the selling price is less than the book value, the capital loss is

$$\text{Capital loss} = \text{book value} - \text{selling price}$$

If the result is positive, a loss is reported. The terms *short-term loss (STL)* and *long-term loss (LTL)* are determined in a fashion similar to capital gains.

Depreciation recapture occurs when depreciable property is sold for an amount greater than the current book value. The excess is depreciation recapture, DR, and is taxed as ordinary taxable income. There is depreciation to recapture if $DR > 0$ by the following computation at sale time.

$$DR = \text{selling price} - \text{book value} \qquad \textbf{[14.1]}$$

If the selling price happens to exceed the purchase price (or first cost, B), a capital gain is also incurred, and all previous depreciation is considered recaptured and fully taxable. This is an unlikely event for most assets. Depreciation recapture is computed using a salvage (book) value of zero for assets disposed of after the recovery period.

The U.S. federal tax laws are based upon taxation of *income*. A variety of different bases are utilized for state taxation, and still other bases are used by the tax systems of other countries. The basis may be total sales, total value of property, winnings from gambling, value of imports, or another type of base. States and countries which have no income tax use means different from *income* taxes to develop revenue. Virtually no government entity survives for long without some form of taxation.

Problem 14.1

14.2   FUNDAMENTAL TAX RELATIONS FOR CORPORATIONS AND INDIVIDUALS

The following is a summary of frequently used relations for corporate and individual income taxes. These relations are developed to help you understand the fundamental elements of income tax computations. As such, they do not include terms for the sale of capital assets or capital expenditures (first costs). The cash flow generated by an asset sale can affect taxes. This consideration is introduced into the corporate taxable income relations in the next section and in Chapter 15.

General Relations to Estimate Annual *Corporate* Income Taxes		General Relations to Estimate Annual *Individual* Income Taxes	
Gross income		*Gross income*	
GI = business revenue + other income	[14.2]	GI = salaries and wages + interest and dividends + other income	[14.5]
Taxable income		*Taxable income*	
TI = gross income − operating expenses − depreciation	[14.3]	TI = gross income − personal exemption − itemized or standard deductions	[14.6]
Income tax		*Income tax*	
Taxes = (TI)(tax rate)	[14.4]	Taxes = (TI)(tax rate)	[14.7]

As you can conclude, there are some significant differences in how corporations and individuals are treated from a tax viewpoint. Corporations are able to reduce annual GI by all their legitimate business (operating) expenses, while individuals have stated amounts for personal exemptions and specific deductions. Personal exemptions are yourself, spouse, children, and others who depend on you for primary financial support. Each exemption reduces TI by some set amount, such as $2500. Most of your economic analysis will use corporate tax relations, but you should be aware of the computational differences in Equations [14.2] through [14.7].

The annual tax rate T is based upon the principle of *graduated tax rates,* which means that corporations and individuals pay higher rates for larger taxable incomes. Table 14–1 presents T values for corporations, and Table 14–2 details individual tax rates. The portion of each new dollar of TI is taxed at what is called the *marginal tax rate.* As an illustration, scan the tax rates in

Table 14-1	Corporate income tax rate schedule (1997) (mil = $ in millions)				
(1) TI Limits	(2) TI Range	(3) Tax Rate, T	(4) = (2)T Maximum Tax for TI Range	(5) = Sum of (4) Maximum Tax Incurred	
$1–$50,000	$ 50,000	0.15	$ 7,500	$ 7,500	
$50,001–75,000	25,000	0.25	6,250	13,750	
$75,001–100,000	25,000	0.34	8,500	22,250	
$100,001–335,000	235,000	0.39	91,650	113,900	
$335,001–10 mil	9.665 mil	0.34	3.2861 mil	3.4 mil	
Over $10–15 mil	5 mil	0.35	1.75 mil	5.15 mil	
Over $15–18.33 mil	3.33 mil	0.38	1.267 mil	6.417 mil	
Over $18.33 mil	Unlimited	0.35	Unlimited	Unlimited	

Table 14-2	Individual income tax rate schedule for single and married filing jointly (1997)	
(1) Tax Rate, T	(2) Taxable Income, $ Filing Single	(3) Filing Married and Jointly
0.15	0–23,350	0–39,000
0.28	23,351–56,550	39,001–94,250
0.31	56,551–117,950	94,251–143,600
0.36	117,951–256,500	143,601–256,500
0.396	Over 256,500	Over 256,500

Table 14–2 for individuals. A single person with an annual TI of $20,000 has a marginal rate of 15% (taxes = $3000). However, a single filer with TI = $30,000 pays 15% for the first $23,350 and 28% on the remainder of TI.

Taxes = 0.15(23,350) + 0.28(30,000 − 23,350) = $5365

The graduated tax-rate system gives small businesses and individuals with smaller taxable incomes a slight advantage. You can see from the tables that the marginal rates are in the mid to upper thirty percentages for TI values above (roughly) $100,000, while smaller TIs have rates in the 15% to 28% range. Often a *one-value federal tax rate* is used to avoid the detail of graduated tax rates. To reduce dependence on changing tax law, problems will include a stated one-value federal tax rate, or Table 14–1 rates will be explicitly called for. Example 14.1 illustrates further the application of graduated tax rates.

Each year the government reviews and/or alters the TI ranges in Tables 14–1 and 14–2 to account for inflation and other factors. This action is called *indexing*. The tax rates are also altered when new tax law is approved.

Because the marginal tax rates change with TI, it is not possible to quote directly the percent of TI paid in income taxes. It is helpful to compute a percent of total TI paid. This is called the *average tax rate* and is calculated as

$$\text{Average tax rate} = \frac{\text{total taxes paid}}{\text{total taxable income}} = \frac{\text{taxes}}{\text{TI}} \qquad [14.8]$$

For the single-person filer mentioned above with TI = \$30,000, the average tax rate is \$5365/30,000 = 0.179 or 17.9%.

As mentioned in the introduction, there are federal, state, and local taxes in many areas. For the sake of simplicity, the tax rate used in an economy study is often a single-figure *effective tax rate, T_e*, which accounts for federal, state, and local taxes. Commonly used effective tax rates are in the range of 35% to 50%. One reason to use the effective tax rate is that state taxes are deductible for federal tax computation. The effective tax rate as a decimal fraction is

$$\text{Effective tax rate} = T_e = \text{state rate} + (1 - \text{state rate})(\text{federal rate}) \qquad [14.9]$$

Since graduated tax rates at the federal and state levels make this computation difficult when marginal rates are used, it is common to use an average graduated tax rate to estimate T_e. This is illustrated in Example 14.1*b*.

Example 14.1

For a particular year, the software division of Intelligent Highway, Ltd., has a gross income of \$2,750,000 with expenses and depreciation totaling \$1,950,000. (*a*) Compute the company's exact federal income taxes. (*b*) Estimate total federal and state taxes if the state tax rate is 8% and a 34% federal one-value figure applies. (*c*) Estimate the average tax rate for the year for both federal taxes only and for total taxes.

Solution

(*a*) Compute the TI by Equation [14.3] and the exact tax using Table 14–1.

$$\text{TI} = 2,750,000 - 1,950,000 = \$800,000$$

$$\text{Taxes} = (\text{TI range})(\text{marginal tax rate}) \qquad [14.10]$$

$$= (50,000)0.15 + (25,000)0.25 + (25,000)0.34$$

$$+ (235,000)0.39 + (800,000 - 335,000)0.34$$

$$= \$7500 + 6250 + 8500 + 91,650 + 158,100$$

$$= \$272,000$$

A faster approach uses the amount in column 5 of Table 14–1 that is closest to the total TI and adds the tax for the next TI range.

$$\text{Taxes} = 113,900 + (800,000 - 335,000)0.34 = \$272,000$$

(*b*) Equation [14.9] determines the effective tax rate.

$$T_e = 0.08 + (1 - 0.08)(0.34) = 0.3928$$

Equation [14.4], with T_e substituted for T, estimates the taxes.

$$\text{Taxes} = (\text{TI})T_e = (800,000)(0.3928) = \$314,240$$

Do not compare these two tax amounts since the result in part (*a*) does not include state taxes.

(*c*) Use Equation [14.8] for both cases.

Federal only: $\text{Average tax rate} = \dfrac{272,000}{800,000} = 0.34 \qquad (34\%)$

Federal and state: $\text{Average tax rate} = \dfrac{314,240}{800,000}$

$$= 0.3928 \qquad (39.28\%)$$

The combined federal and state average rate of 39.28% is predictable, since we computed it in part (*b*). Relative to the federal average of 34%, the 39.28% reflects an effective increase of 5.28% of the quoted 8% state rate. This occurs because state taxes are deductible in the federal tax computation.

Example 14.2

Sheila Amos and Carl Baker submit a married-filing-jointly return to the IRS. During the year their two jobs provided them with a combined income of $82,000. They adopted their first child during the year, and they plan to use the standard deduction of $7000 applicable for the year. Dividends and interest amounted to $3550, and an investment vehicle—a stock mutual fund—reported capital gains of $550. Personal exemptions are $2500 each. (*a*) Compute their exact federal tax liability. (*b*) Compute their average tax rate. (*c*) What percent of their total income is consumed by federal taxes?

Solution

(*a*) Use Equations [14.5] and [14.6] to calculate GI and TI. Sheila and Carl have three personal exemptions and the standard deduction of $7000.

$$\text{Gross income} = \text{salaries} + \text{interest and dividends} + \text{capital gains}$$

$$= \$82,000 + 3550 + 550 = \$86,100$$

$$\text{Taxable income} = \text{gross income} - \text{exemptions} - \text{deductions}$$

$$= \$86,100 - 3(2500) - 7000$$

$$= \$71,600$$

Table 14–2 indicates the 28% marginal rate for a TI of $71,600. By Equation [14.7] and columns 1 and 3 of Table 14–2, federal taxes are

$$\text{Taxes} = (39,000)0.15 + (71,600 - 39,000)0.28$$

$$= 5850 + 9128$$

$$= \$14,978$$

(b) Using Equation [14.8],

$$\text{Average tax rate} = \frac{14,978}{71,600} = 0.209 \quad (20.9\%)$$

This indicates that about 1-in-5 dollars of taxable income is paid to the U.S. government.

(c) Of the total $82,000 in income that Sheila and Carl earned, $14,978/82,000 = 0.1826$ or 18.26% went to federal taxes.

Comment

There has been discussion in the U.S. Congress for some years about changing from a graduated tax structure to a flat-tax structure, especially for individual taxpayers. There are many, many ways to legislate taxes, and the amount to be chosen for the flat-rate is usually a real controversy.

For example, the flat-tax structure may allow no standard or itemized deductions and have only the personal exemption allowance. In this example, were there a flat-tax rate of, say, 20% on gross income reduced only by the three personal exemptions, the computations would be

$$\text{Gross income} = \$86,100$$

$$\text{Flat-rate taxable income} = 86,100 - 3(2500) = \$78,600$$

$$\text{Flat-rate taxes} = (78,600)0.20 = \$15,720$$

In this specific case, a 20% flat-tax rate would require that slightly more taxes be paid by this family—$15,720 versus $14,978.

Problems 14.2 to 14.10

14.3   CAPITAL GAINS AND LOSSES FOR CORPORATIONS

All the tax implications discussed here are the result of disposing of a depreciable asset before, at, or after its recovery period. The key is the size of the salvage amount (selling price or trade-in value) relative to the book value at disposal time and relative to the first cost of the asset. The computations and corporate tax treatment of capital gains, capital losses, and depreciation recapture presented here are summarized in Figure 14–1.

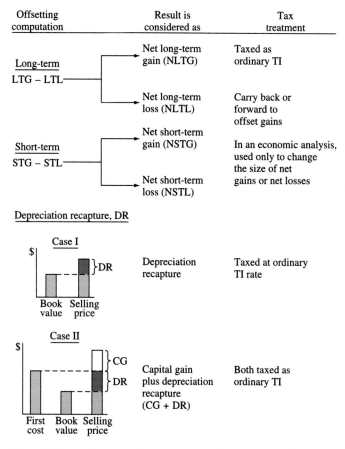

Figure 14–1   Summary of computation and tax treatment of capital gains and losses and depreciation recapture.

The capital gain and loss definitions presented in Section 14.1 are used in an economic analysis for depreciable assets when an after-tax study is conducted. Examples are when disposal plans for an asset or real property are predetermined and when an after-tax asset replacement study is undertaken (covered in Chapter 15). Prior to the major 1986 U.S. tax law alterations, long-term gains were taxed at a rate lower than short-term gains. This may again become tax law at some future time, but for now, all capital gains are taxed at the applicable tax rate. The distinction between short-term and long-term is explicitly maintained in case the tax laws are altered and preferential rates are reinstated.

The final result of capital sales is a *net capital gain or loss. Losses do not reduce taxable income directly,* because long-term losses are only allowed to offset long-term gains. Similarly, short-term losses offset short-term gains. If there is a resulting net gain (long or short), it is treated as ordinary taxable income. Any leftover net losses (above net gains) in a year may be carried back for 3 years or forward for 5 years. Net losses do create *tax savings* for the company, however, in the year of occurrence provided there are gains in other areas against which the loss can be utilized, as Example 14.3 illustrates.

Short-term gains and losses may be important for tax computation purposes as illustrated in Example 14.4. However, it may be corporate practice to not include them in the analysis of major economic investments, because 1 year is a short time compared to the expected life of most alternatives. But, if the size of the capital loss is estimated, it is definitely used to reduce the size of anticipated capital gains, thus generating a tax savings in the amount of the net capital loss times the (effective) tax rate.

Depreciation recapture, DR, may be included in after-tax studies. Two cases may occur, as Figure 14–1 shows graphically. Regardless of their nature, the total amount is taxed as ordinary taxable income.

Case I.   Selling price, that is, realized salvage value in year t, exceeds the book value in year t. This is the computational result of Equation [14.1]. The DR is taxed at the ordinary rate. An example would be selling a building for more than the current book value.

Case II.   Selling price exceeds the first cost. Now there are DR and capital gain (CG) components in year t.

$$\text{DR} = \text{first cost} - \text{book value} = B - BV_t \qquad \text{[14.11]}$$

$$\text{Capital gain} = \text{selling price} - \text{first cost} \qquad \text{[14.12]}$$

Equation [14.3] for TI can now be expanded to include the cash flow for an asset sale, which can result in a capital loss, depreciation recapture, and possibly a capital gain.

$$\text{TI} = \text{gross income} - \text{operating expenses} - \text{depreciation} \qquad \text{[14.3]}$$

$$+ \text{ depreciation recapture} + \text{net capital gains}$$

$$- \text{ net capital losses}$$

Remember, if there is a net capital loss, the assumption is made that a capital gain will be present in the same year elsewhere in the corporation to serve as the offset.

Example 14.3

A start-up company, CAI Control Systems, Inc., expects to realize a gross income of $500,000 and combined business expenses and depreciation of $300,000 in the next tax year. Using an effective tax rate of 35%, compute the expected income taxes for the following:

(a) Anticipated capital equipment replacements with long-term gains of $25,000 and long-term losses of $10,000.

(b) Specific anticipated long-term gains of $25,000 and long-term losses of $40,000. The company will likely have gains in other areas.

Solution

(a) The net long-term gain (NLTG) of $25,000 - 10,000 = $15,000 is taxed as ordinary income (see Figure 14–1). Calculate TI using the relevant terms in Equation [14.3].

$$TI = GI - expenses - depreciation + net capital gain$$

$$= \$500,000 - 300,000 + 15,000 = \$215,000$$

$$Taxes = (TI)T_e$$

$$= (\$215,000)0.35$$

$$= \$75,250$$

(b) The net long-term loss (NLTL) of $40,000 - 25,000 = $15,000 is not available to directly reduce TI for the year. However, it can be used to offset gains in other parts of the company, or the net loss may be carried back 3 years or forward 5 years to offset gains in other tax years. Since other gains are anticipated this year, there will be an effective *tax savings,* because taxes will not be paid on some other gains due to this offsetting. Therefore,

$$Taxes = (TI)T_e$$

$$= (500,000 - 300,000 - 15,000)0.35$$

$$= \$64,750$$

Because of other gains during the year, the amount of tax savings included in the $64,750 amount is

$$Tax\ savings = (15,000)0.35 = \$5250$$

Comment

In economy studies, net capital losses are usually treated as tax-savings generators and the full benefit of the incurred loss is taken in the year it occurs.

Example 14.4

How will taxable income of a corporation be affected if the following gains and losses have been incurred in one tax year?

LTG = $40,000   STG = $75,000

LTL = $5,000   STL = $90,000

Solution

Offsetting gains and losses results in NLTG = \$35,000 and NSTL = \$15,000. The net result is a long-term gain of \$20,000. The TI value will be increased by \$20,000 when taxes are computed.

Comment

If long-term gains were taxed at a preferentially lower rate, as mentioned earlier, the \$20,000 would be taxed at this lower rate.

We have mentioned that capital losses can be carried forward for several tax years. Another important tax advantage for corporations is the provision that allows an *operating loss* (defined in Section 14.1) to be carried backward and forward until the loss is completely exhausted. The number of years allowed for carry-back and carry-forward may vary (e.g., 3 and 7 years, respectively), but the amount of operating loss claimed in any 1 year cannot exceed taxable income. Since only the amount of the loss is recoverable, this and all carry-back and carry-forward laws present a complex question of strategy, that is, when to apply the tax advantage.

Additional Example 14.8 (Spreadsheet)
Problems 14.11 to 14.15

14.4   CASH FLOW AND RATE OF RETURN: BEFORE AND AFTER TAXES

The terms *CFBT* and *CFAT* are used to represent annual cash flows before taxes and after taxes, respectively, in an engineering economy study. Relationships between these two terms, and some pertinent relations from previous sections which directly affect corporate income taxes, are

$$\text{CFBT} = \text{gross income} - \text{operating expenses} = \text{GI} - E \qquad \textbf{[14.13]}$$

$$\text{TI} = \text{CFBT} - \text{depreciation} + \text{net capital gains} \qquad \textbf{[14.14]}$$
$$- \text{ net capital losses} + \text{depreciation recapture}$$

$$\text{Taxes} = (\text{TI})T \qquad \textbf{[14.15]}$$

$$\text{CFAT} = \text{CFBT} - \text{taxes} \qquad \textbf{[14.16]}$$

Equation [14.14] is the same as Equation [14.3] using the term CFBT and including any capital asset sales which affect taxes. The CFBT relation does not include the capital expense, that is, the first cost of the alternative being evaluated, since this does not affect taxes directly. However, the way in which the capital investment is financed can affect taxes. This dimension is introduced in the next chapter along with debt and equity financing for capital funds.

The amounts in the relations above are all *actual* tax-related cash flows, except for the depreciation term in Equation [14.14]. Depreciation is a *book amount* since it represents the reduction in value of depreciable property, but it is not actual cash outflow. However, since depreciation is tax-deductible, it does change actual cash flow by reducing income taxes, which are actual cash outflows. When an after-tax analysis is performed and the after-tax PW, AW, or ROR is computed, the CFAT values are used in all the computations.

If the after-tax ROR is important in an economic analysis, but the after-tax analysis details are not, it is common to increase (or inflate) the before-tax ROR to incorporate either the effective or the marginal tax rate. If a single-value tax rate is expressed in decimal form, an approximation of the tax effect on the project ROR is

$$\text{After-tax ROR} = (\text{before-tax ROR})(1 - \text{tax rate}) \qquad \textbf{[14.17]}$$
$$= (\text{before-tax ROR})(1 - T)$$

The value for T can be the applicable marginal tax rate (Table 14–1) or the effective tax rate, T_e, from Equation [14.9]. For example, assume a company's federal and state effective tax rate is 40% and an after-tax ROR of 12% is the market MARR. The before-tax ROR that is implied by a 12% after-tax return is estimated by solving Equation [14.17] for the before-tax rate.

$$0.12 = (\text{before-tax ROR})(1 - 0.40)$$

$$\text{Before-tax ROR} = \frac{0.12}{0.60} = 0.20 \qquad (20\%)$$

This analysis applies to corporations and individuals alike.

Example 14.5

In addition to corporate management positions, the Espinosas operate a direct-mail business out of their home. The results during one tax year are

Business revenue	= $60,000
Equipment depreciation	= $8000
Business expenses	= $18,500
Business investment income	= $5000
State tax rate	= 4% of TI

The rate of return represented by the business investment before taxes is 11%. Estimate the following: (*a*) after-tax cash flow generated by the business and (*b*) after-tax return on the Espinosas' business investment.

Solution

(*a*) CFAT is computed using the following: Equation [14.2] to determine gross income; Table 14–1 for the corporate tax rate; Equation [14.9] for the effective

tax rate; and Equations [14.13] through [14.16] to estimate CFAT. Figure 14–2 graphically profiles the development from CFBT to CFAT for the actual cash flows.

$$CFBT = (GI) - \text{operating expenses}$$

$$= (\text{revenue} + \text{investment income}) - \text{expenses}$$

$$= (\$60,000 + 5000) - 18,500$$

$$= \$46,500$$

$$TI = CFBT - \text{depreciation}$$

$$= \$46,500 - 8000$$

$$= \$38,500$$

$$T_e = 0.04 + (1 - 0.04)(0.15)$$

$$= 0.184$$

$$\text{Taxes} = (TI)T_e$$

$$= (\$38,500)0.184$$

$$= \$7084$$

$$CFAT = CFBT - \text{taxes}$$

$$= \$46,500 - 7084$$

$$= \$39,416$$

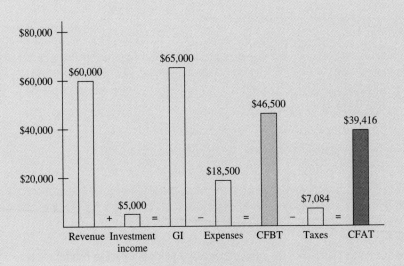

Figure 14–2   Development of cash flow after taxes for Example 14.5.

The business netted an actual cash flow of $39,416 after taxes. Note that in Figure 14–2 there is no depreciation column, since depreciation is not an actual cash flow. It is only used as a deduction in computing taxable income and taxes.

(b) Use 11% for the before-tax return and 0.184 for the effective tax rate in Equation [14.17].

$$\text{After-tax investment return} = 0.11(1 - 0.184)$$

$$= 0.09 \quad (9\%)$$

Taxes reduced the return realized on the business investment from 11% to 9%.

An effective way to increase after-tax cash flow and a project's rate of return is to take advantage of allowed tax credits. These credits, by definition, directly reduce taxes; they don't merely increase the deductions used in computing taxes.

The *investment tax credit (ITC)* has been used at times in the United States to encourage investment in equipment in capital-intensive industries by authorizing direct tax credits of 6% to 10% of an asset's first cost, without reducing the basis for depreciation purposes. That is, the ITC reduces income taxes directly while depreciation is only a deduction to taxable income.

This advantage for ITC was removed by the Tax Reform Act of 1986, which repealed the ITC because it was considered inflationary and unnecessary due to the introduction of MACRS. Since the ITC has been utilized historically to stimulate the economy, you should know how it functions. It may be reinstated in some form at a future time.

A 10% ITC allows 10% of the first cost as a tax credit, which may be applied in the year of asset purchase. The tax credit claimed in a year is usually limited by the company's income tax (total or a percentage of it) or some specified amount for each year. Additionally, this tax credit did not reduce the total depreciable amount of the asset. However, previous tax laws required that the ITC be taken on the reduced basis if a Section 179 capital expense deduction (Section 13.8) was also claimed.

Problems 14.16 to 14.20

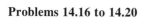

14.5   EFFECT OF DIFFERENT DEPRECIATION MODELS ON TAXES

The amount of taxes incurred is affected by the depreciation model. Accelerated methods, such as MACRS, result in less taxes in earlier years of the recovery period due to the larger reduction in taxable income. The criterion of *minimizing the present worth of total taxes* is used to evaluate the tax effect of depreciation models. That is, for the recovery period n, choose the depreciation model with the minimum present-worth value for taxes, PW_{tax}, where

$$PW_{tax} = \sum_{t=1}^{t=n} (\text{taxes in year } t)(P/F,i,t) \qquad [14.18]$$

Let's compare any two different depreciation models. Assume the following: (1) constant single-value tax rate, (2) gross income exceeds each annual depreciation amount, (3) capital recovery reduces book value down to the same salvage value (commonly zero), and (4) same recovery period in years. Then, for all depreciation models, the following are correct:

1. The total taxes paid are *equal* for all depreciation models.
2. The present worth of taxes, PW_{tax}, is *less* for accelerated depreciation models.

As we learned in Chapter 13, MACRS is the prescribed depreciation model, and the only alternative is MACRS straight-line depreciation (with an extended recovery period). The accelerated write-off of MACRS will always provide a smaller PW_{tax} value compared to less accelerated models. If the DDB model were available directly, rather than embedded in the MACRS computations, DDB would usually not be as good as MACRS. This is because the DB and DDB models do not reduce the book value to zero as do the MACRS models. This is illustrated in Example 14.6.

The selection of the depreciation model which minimizes PW_{tax} is the equivalent of selecting the model which maximizes the present worth of total depreciation, PW_D, discussed in Section 13.6.

Example 14.6

An after-tax analysis for a new $50,000 machine proposed for a fiber-optics manufacturing line is in process. The CFBT for the machine is estimated at $20,000. If a recovery period of 5 years applies, use the present-worth-of-taxes criterion, an effective tax rate of 35%, and a return of 8% per year to compare the following: classical straight-line, DDB, and MACRS depreciation. Use a consistent 6-year period for comparison purposes.

Solution

Table 14–3 presents a summary of annual depreciation, taxable income, and taxes for each model. For classical straight-line depreciation with $n = 5$, $D_t = \$10,000$ for 5 years and $D_6 = 0$ (column 3), the CFBT of $20,000 is fully taxed at 35% in year 6 for comparison with other models.

The DDB percentage of $d = 2/n = 0.40$ is applied for 5 years. The implied salvage value is $\$50,000 - 46,112 = \$3888$, so not all $50,000 is tax-deductible. The taxes incurred using DDB are $\$3888(0.35) = \$1361$ larger than for the classical SL model.

Table 14–3 Comparison of taxes and present worth of taxes for different depreciation models

(1) Year, t	(2) CFBT	Straight Line			Double-Declining Balance			MACRS		
		(3) D_t	(4) TI	(5) = 0.35(4) Taxes	(6) D_t	(7) TI	(8) = 0.35(7) Taxes	(9) D_t	(10) TI	(11) = 0.35(10) Taxes
1	+20,000	$10,000	$10,000	$ 3,500	$20,000	$ 0	$ 0	$10,000	$10,000	$ 3,500
2	+20,000	10,000	10,000	3,500	12,000	8,000	2,800	16,000	4,000	1,400
3	+20,000	10,000	10,000	3,500	7,200	12,800	4,480	9,600	10,400	3,640
4	+20,000	10,000	10,000	3,500	4,320	15,680	5,488	5,760	14,240	4,984
5	+20,000	10,000	10,000	3,500	2,592	17,408	6,093	5,760	14,240	4,984
6	+20,000	0	20,000	7,000	0	20,000	7,000	2,880	17,120	5,992
Totals		$50,000		$24,500	$46,112		$25,861*	$50,000		$24,500
PW_{tax}				$18,386			$18,549			$18,162

| * Larger than other values since there is an implied salvage value of $3888 not recovered by the DDB model.

MACRS writes off the $50,000 in 6 years using the rates of Table 13–2. Total taxes are $24,500 for MACRS, the same as for classical SL depreciation over the 6 years.

The annual taxes (columns 5, 8, and 11 in Table 14–3) are accumulated year by year for each model in Figure 14–3. Note the pattern of the curves, and notice the lower tax values relative to the SL model in year 2 for MACRS and in years 1 and 2 for DDB. These lower tax values cause the PW_{tax} value for SL depreciation to be larger.

The present-worth values for total taxes using Equation [14.18] are

SL: $PW_{tax} = 3500(P/A,8\%,6) + 7000(P/F,8\%,6) = \$18,386$

DDB: $PW_{tax} = 2800(P/F,8\%,2) + \cdots + 7000(P/F,8\%,6) = \$18,549$

MACRS: $PW_{tax} = 3500(P/F,8\%,1) + \cdots + 5992(P/F,8\%,6) = \$18,162$

The MACRS PW_{tax} value is the smallest at $18,162. The DDB amount would be smaller than classical SL depreciation if all of the $50,000 first-cost were depreciated.

These computations indicate that accelerated models, under the conditions stated, have a smaller income tax implication using the present-worth-of-tax criterion. This criterion will result in the same depreciation model as the before-tax criterion of maximizing the present-worth of depreciation, PW_D, as discussed in Section 13.6 (Switching between Depreciation Models).

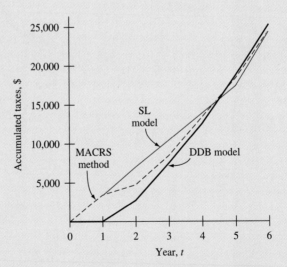

Figure 14–3   Taxes incurred by different depreciation models for a 6-year comparison period, Example 14.6.

Problems 14.21 to 14.26

14.6   EFFECT OF DIFFERENT RECOVERY PERIODS ON TAXES

Let's now make slightly different assumptions about depreciation models in order to compare recovery periods: (1) constant single-value tax rate, (2) gross income exceeds each annual depreciation amount, (3) capital recovery reduces book value down to the same salvage value (commonly zero), and (4) comparison of results for the same depreciation model. We can demonstrate that a shorter recovery period will offer a tax advantage over a longer period using the criterion of minimizing the PW_{tax} value computed by Equation [14.18]. Comparison of taxes for different n values will indicate that

1. The total taxes paid are *equal* for all n values.
2. The present worth of taxes, PW_{tax}, is *less* for smaller n values.

Example 14.7 demonstrates these conclusions using the classical straight-line model, but they may be demonstrated for any depreciation model.

Example 14.7

Grupo Grande Maquinaría, a diversified manufacturing corporation based in Mexico, maintains parallel records for depreciable assets in its U.S. operations. This is common for multinational corporations. One set is for corporate use, and it reflects the estimated useful life of assets. The second set is for U.S. government purposes—specifically for depreciation.

The company just purchased an asset with $B = \$90,000$ and an estimated life of 9 years; however, a shorter recovery period of 5 years is allowed by U.S. tax law. Demonstrate the tax advantage for the smaller n if CFBT = \$30,000 per year, an effective tax rate of 35% applies, invested money is returning 5% per year after taxes, and classical SL depreciation is allowed. Neglect the effect of any salvage value.

Solution

Complete the SL computations of Section 13.2, and then compute and compare the present worth of total taxes using Equation [14.18] for both n values.

$n = 9$ years.

$$D_t = \frac{90,000}{9} = \$10,000$$

$$TI = 30,000 - 10,000 = \$20,000 \text{ per year}$$

$$Taxes = 20,000(0.35) = \$7000 \text{ per year}$$

$$Total \ taxes = \$7000(9) = \$63,000$$

$$PW_{tax} = 7000(P/A,5\%,9) = \$49,755$$

$n = 5$ years. We use the same comparison period of 9 years, but depreciation occurs only during the first 5 years.

$$D_t = \begin{cases} \dfrac{90{,}000}{5} = \$18{,}000 & t = 1 \text{ to } 5 \\[2ex] 0 & t = 6 \text{ to } 9 \end{cases}$$

Apply Equation [14.14] for TI, followed by Equation [14.15] for taxes.

$$\text{Taxes} = \begin{cases} (30{,}000 - 18{,}000)0.35 = \$4200 & t = 1 \text{ to } 5 \\ (30{,}000)0.35 = \$10{,}500 & t = 6 \text{ to } 9 \end{cases}$$

$$\text{Total taxes} = \$4200(5) + 10{,}500(4) = \$63{,}000$$

$$\text{PW}_{\text{tax}} = 4200(P/A,5\%,5) + 10{,}500(P/A,5\%,4)(P/F,5\%,5)$$

$$= \$47{,}356$$

A total of $63,000 in taxes is paid for both the 9- and 5-year periods. However, the more rapid write-off for $n = 5$ results in a present-worth tax savings of nearly $2400 ($49,755 − 47,356).

Problems 14.27 to 14.30

ADDITIONAL EXAMPLE

**Example 14.8
(Spreadsheet)**

TAXES AND CAPITAL GAINS, SECTIONS 14.2 AND 14.3   Biotec-1, a medical testing company, is considering the purchase of cell-analysis machines. Details for the two contenders are

	Analyzer 1	Analyzer 2
First cost, $	150,000	225,000
Expenses per year, $	30,000	10,000
Recovery period, years	5	5

(a) Use an effective tax rate of 35% and MACRS depreciation to determine the tax advantage, if any, for the analyzers during the first 3 years. A gross income of $100,000 is expected for both analyzers.

(b) If the analyzers were to be sold during the third year of ownership, compute the income tax effect in year 3 due to the sale. Assume the partially depreciated cell analyzers become technological standards and have elevated third-year selling prices of $130,000 (analyzer 1) and $240,000 (analyzer 2). Also, calculate the net proceeds from the sale of each asset.

Solution

(*a*) Figure 14–4 presents the spreadsheet with the following column values.

> **Column B cells.** MACRS rates for $n = 5$ from Table 13–2.
>
> **Column C cells.** Annual MACRS depreciation using Equation [13.12], $d_t B$.

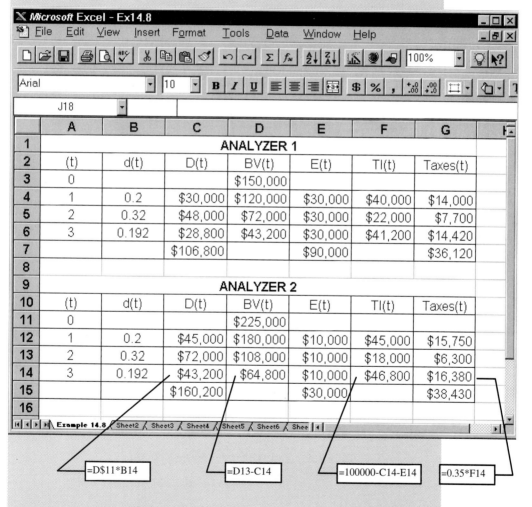

Figure 14–4 After-tax MACRS analysis of two cell-analysis machines, Example 14.8 (Spreadsheet).

Column D cells.   Book value, Equation [13.13], $BV_t = BV_{t-1} - D_t$.

Column E cells.   Anticipated expenses each year, E_t.

Column F cells.   Taxable income by Equation [14.3],
$TI_t = GI_t - D_t - E_t$.

Column G cells.   Income taxes using Equation [14.4],
$(TI_t)T_e = (TI_t)0.35$.

For the 3 years, total taxes for analyzer 1 ($36,120) are lower than those for analyzer 2 ($38,430). So, analyzer 1 has a $2310 tax advantage.

(b) *Analyzer 1* has depreciation recapture. According to Figure 14–1, case I, and Figure 14–4 which indicates that $BV_3 = \$43,200$, DR and the resulting estimated taxes are

$$DR = \$130,000 - 43,200 = \$86,800$$

$$\text{Taxes in year 3} = (TI_3 + DR)0.35 = (41,200 + 86,800)0.35$$

$$= \$44,800$$

Year 3 taxes will rise from $14,420 (Figure 14–4) to $44,800. Net proceeds consider the taxes on recaptured depreciation.

$$\text{Taxes on DR} = (\$86,800)0.35 = \$30,380$$

$$\text{Net proceeds from the sale} = \$130,000 - 30,380 = \$99,620$$

For *analyzer 2,* a similar analysis yields a capital gain plus depreciation recapture, since the selling price exceeds the first cost of $225,000. Referring to Figure 14–1, case II, and applying Equations [14.11] and [14.12], we have

$$DR = \text{first cost} - \text{book value}$$

$$= \$225,000 - 64,800$$

$$= \$160,200$$

$$\text{Net capital gain} = \text{selling price} - \text{first cost}$$

$$= \$240,000 - 225,000$$

$$= \$15,000$$

Using updated Equations [14.3] and [14.4],

$$\text{Taxes in year 3} = (TI_3 + DR + CG)0.35$$

$$= (46,800 + 160,200 + 15,000)0.35$$

$$= \$77,700$$

Year 3 taxes will rise from $16,380 (Figure 14–4) to $77,700. For net proceeds,

$$\text{Taxes on DR} + \text{CG} = (160,200 + 15,000)0.35 = \$61,320$$

$$\text{Net proceeds from the sale} = \$240,000 - 61,320 = \$178,680$$

Conclusion

For both analyzers, the net proceeds are around 75% of the selling price.

CHAPTER SUMMARY

Income tax computations for individual taxpayers and corporations take the same general form, and the graduated tax rate structure applies to both cases. But there are significant differences when computing gross income, taxable income, and income taxes.

Marginal tax rates are tabulated, and they are graduated with increasing TI. Two additional single-value tax rates used in engineering economy studies are the single-value effective tax rate and the average tax rate.

Corporate *capital losses* can offset *capital gains* and, therefore, reduce income taxes indirectly, provided there are gains available. If net gains result, they are considered as ordinary taxable income. Net losses are considered *tax savings* in economy studies. Figure 14–1 summarizes tax treatment for capital gains and losses and depreciation recapture at the time a capital asset is sold.

Accelerated depreciation models, such as MACRS and DDB, and abbreviated recovery periods can be shown to have considerable tax advantages. The criterion to *minimize the present worth of total taxes,* PW_{tax}, is used to conclude that

- Total taxes paid are the same for all models and all *n* values.
- Accelerated models have a smaller PW_{tax} value, because of larger depreciation in early years.
- Smaller *n* values have a smaller PW_{tax} value, because the first cost is written off in less time resulting in more depreciation in less years.

READING REFERENCES

Some references for individual taxpayers.

- *Your Federal Income Tax (for Individuals)*, U.S. Internal Revenue Service Publication 17, published annually.

- *The Ernst & Young Tax Saver's Guide,* John Wiley and Sons, Inc., New York, published annually.

- *J. K. Lasser's Your Income Tax,* Macmillan, New York, published annually.

- *Package 1040-5—Forms and Instructions,* U.S. Internal Revenue Service, published annually.

Some references for understanding corporate taxes.

- *Tax Information on Corporations,* U.S. Internal Revenue Service Publication 542, published annually.

- *U.S. Master Tax Guide,* Commerce Clearing House, Chicago, published annually.

- *Sales and Other Dispositions of Assets,* U.S. Internal Revenue Service Publication 544, published annually.

PROBLEMS

Note to instructors and students: When current tax rates and law are different from those presented here and the new ones are used to solve the following problems, minor adjustments may be required in the problem statement and solution. Obviously, the answers will also change from those in Appendix C.

14.1 The situations listed below were recorded for a corporation during the past year. For each situation, state which of the following is involved: gross income, taxable income, depreciation recapture, capital gain, capital loss, or operating loss.

(*a*) A machine was purchased and had a first-year depreciation of $9600.

(*b*) The company estimates that it will report a $75,000 net loss on its tax return.

(*c*) An asset with a book value of $8000 was retired and sold for $8450.

(*d*) The asset in part (*a*) will have a $4200-per-year interest cost to pay off the loan taken to purchase it.

(*e*) An asset that had a life of 8 years has been owned for 14 years and has a final book value of zero. It was sold this year for $275.

(*f*) The cost of goods sold in the past year was $468,290.

(*g*) Revenue from international sales was $1.8 million, of which $0.8 million is the cost for foreign licenses.

14.2 Summarize the main differences in income-tax-related formulas and tax rates for corporate and individual taxpayers.

14.3 Two small businesses have the following data on their tax returns:

	Company 1	Company 2
Sales revenue, $	1,500,000	820,000
Interest revenue, $	31,000	25,000
Expenses, $	754,000	591,000
Depreciation, $	48,000	54,000

(a) If both do business in the state of No-Taxes, compute the federal income tax for the year, using the federal tax rates.

(b) What percent of their sales revenues do the companies pay in income taxes?

(c) Compute the taxes using an effective rate of 34% for the entire TI. What percentage change in taxes is caused by graduated tax rates?

14.4 International Car Wholesalers will have a $250,000 taxable income this year. If an advertising campaign is initiated, TI is estimated to increase to $300,000. Neglect state and local taxes. After computing taxes using the federal graduated tax rates, determine the following:

(a) Average federal tax rate on TI = $250,000.

(b) Marginal federal tax rate on only the additional taxable income.

(c) Average federal tax rate on the entire $300,000 of TI.

(d) After-tax profit on the additional $50,000 taxable income.

14.5 Carl Hold and Associates has a gross income of $4.3 million for the year. Depreciation and expenses amount to $2.45 million. If the combined state and local tax rate amounts to 6.5% and an effective federal rate of 35% is estimated, compute the income taxes using the effective-tax-rate formula.

14.6 WB Contractors reported a TI of $80,000 last year. If the state tax rate is 8%, compute the (a) federal average tax rate, (b) overall effective tax rate, and (c) total taxes to be paid by the company based on the effective tax rate.

14.7 The taxable income for a small partnership business is $150,000 this year. A single-value tax rate of 39% is used by the owners. New capital investment is being considered. The equipment will cost $40,000, have a life of 5 years, be salvaged for an estimated $5000, and be written off

using classical straight-line depreciation. The purchase will increase taxable income by $10,000 and expenses by $1000 for the year. Compute the change in income taxes for the year if the purchase is made.

14.8 Julietta has worked as an engineer for several companies and now has an annual salary of $78,000. She has capital gains and earned interest of $5500 this year. Her total deductions are $13,800.

(*a*) Calculate her income taxes as a person filing single.

(*b*) Determine what percentage of her annual salary goes toward federal income taxes.

(*c*) What is the average tax rate paid for this year?

14.9 An executive with a university, who has a TI of $70,000 and files a tax return as filing married and jointly, tells a friend that he is in the 28% income tax bracket, so his taxes will be $19,600 this year. His tax advisor tells him he is incorrect and that his taxes will be lower.

(*a*) Compute his taxes correctly and explain the difference between the two tax amounts.

(*b*) If the $70,000 TI were computed correctly, what flat-tax rate would be required to impose the amount of taxes computed in part (*a*)?

14.10 Compare the income taxes due for the following two situations—a single person filing and a married couple filing jointly. No exemptions other than the individuals themselves are claimed.

	Filing Status	
	Single	Married and Jointly
Gross income, $	60,000	60,000
Personal exemptions, $	2,500	5,000
Deductions	7,000	7,000

14.11 The following capital gains and losses are recorded for 1 year for a small U.S. exporting company.

Long-term gain = $28,000

Long-term loss = $5000

Short-term gain = $2000

Taxable income is $380,000 before these capital asset sale results are considered. Incorporate the capital gains and losses into TI and compute the income taxes. The state tax rate is 5% and the effective federal tax rate is 34%.

14.12 Calculate the capital gains and losses and any depreciation recapture for each asset transaction described below and then use them to determine income taxes. Sales revenue for the year is $180,000 with expenses and accumulated depreciation totaling $39,400.

(*a*) A 3-year-old MACRS-depreciated asset was sold for 68% of first cost, which was $50,000. The asset had no salvage value and a MACRS recovery period of 7 years.

(*b*) A machine that was only 5 months old was replaced because of rapid technological advances. The asset had $B = \$10,000$, $SV = \$1000$, $n = 3$ years, and was depreciated by the MACRS method. The trade-in deal allowed the company $5000 on the replacement. Allow only 50% of the MACRS depreciation for the 5-month period.

(*c*) Land purchased 4 months ago for $18,000 was sold for a 10% profit.

(*d*) A 23-year-old asset was sold for $500. When purchased, the asset was entered on the books with $B = \$18,000$, $SV = \$200$, $n = 20$ years. Classical straight-line depreciation was used for the entire life of the machine.

14.13 Mountain Spring Water, Inc., purchased new refrigeration equipment in January with $B = \$40,000$, a MACRS recovery period of 5 years, and an estimated salvage value of $5000. The purchase has increased taxable income by $10,000 and expenses by $1000 for the year. In December of the same year, due to lagging sales the unit was sold at a sacrifice amount of $28,000. Tax law requires that an asset be retained at least 1 year to have a long-term capital gain or loss. Explain how the sale in December will affect income taxes for the year. What is the amount of the possible effect on taxes, if the effective tax rate is 34%?

14.14 The president of the Virtual-Health Club, which operates at 450 sites in North America and Europe, has placed your consulting company on retainer for the next 5 years to assist in making capital investment decisions on equipment purchases at all sites. You have been asked to prepare a 5-minute presentation accompanied with a written summary of how your company will treat capital gains, capital losses, depreciation expenses, and depreciation recapture when performing after-tax engineering economic analyses. Prepare the version of the written summary and presentation outline (if your instructor wants it) that you would offer to the president.

14.15 Three years ago Chic Fashions purchased land and assets which have recently been transferred to a subsidiary of the corporation. Use the information below to determine where capital gains or losses or depreciation recapture have occurred. Determine the amount of each effect.

Asset	Purchase Price	Recovery Period	Current Book Value	Sales Price
Land	$100,000	—	$100,000	$105,000
Machine 1	50,500	5	15,500	17,500
Machine 2	10,000	3	1,000	11,000

14.16 A vice president for finance wants to estimate the required annual cash flow before taxes necessary to have CFAT = $1,500,000. The effective federal tax rate is 40% and the state tax rate is 8%. The vice president knows that $1 million of depreciable asset value will be removed from the company books this year. Estimate CFBT.

14.17 Wholesome Grocers purchased new forklifts for $100,000 each at the end of last year. Tabulate the CFBT and CFAT for each of 6 years of ownership using an effective tax rate of 40% for the estimated annual cash flow and depreciation amounts below. Assume the gross income is the estimated increase generated by the use of these more-effective forklifts in moving pallets of grocery products. The forklifts are expected to be sold after 6 years for $4000.

Year	Gross Income	Operating Expenses	MACRS Depreciation	Sale of Forklifts
1	$30,000	$5,000	$20,000	
2	40,000	5,000	32,000	
3	35,000	6,000	19,200	
4	25,000	6,000	11,520	
5	20,000	7,000	11,520	
6	15,000	7,000	5,760	$4,000

14.18 Compute the required before-tax return for Problem 14.5 if an after-tax return of 8% per year is expected.

14.19 If a company president tells a friend that she expects to make a market rate of return of 15% per year on all corporate investments before taxes, and 8% per year after taxes, what percent of income is she assuming to be committed to income taxes?

14.20 A consultant has worked on new technology justification for the textile industry recently. In both a small business and a large corporation he performed economic evaluations which have an average 21%-per-year before-tax return. If the stated MARR for new projects in both companies is 12% per year after taxes, determine if management at both companies would accept the projects, provided the before-tax return is used to approximate the after-tax return. Effective incremental tax rates are 48% for the large corporation and 34% for the small company.

14.21 A transportation leasing company bought new trucks for $150,000 and expects to realize a CFBT of $100,000 for each of 3 years. The trucks have a recovery period of 3 years. Assume an effective tax rate of 40% and a market interest rate of 15% per year. Show the advantage of accelerated depreciation methods in terms of tax present worth for the MACRS method versus the classical SL method. Since MACRS takes an added year to fully depreciate the basis, assume no CFBT beyond year 3, but include any negative tax as a tax savings.

14.22 An international company based in the United States purchased computer printer A which generates an estimated CFBT of $65,000 for the next 6 years. The company's division in South America purchased five of printer B which generate the same CFBT. Money has a market rate of 12% per year in both countries. The United States requires MACRS depreciation for printer A, and classical SL depreciation is allowed for printer B. Neglect any capital gains, losses, or depreciation recapture at sale time, and assume that any negative income tax is a tax savings to the company. Use the full 6 years as the evaluation period for both cases. (*a*) What is the difference in the present worth of taxes paid? (*b*) Examine the sequence of income tax amounts for each year for the two cases. Calculate total taxes for the 6 years. Why are these two totals not equal? (*c*) Note differences in the profiles of annual tax amounts and explain how and why the tax burden is distributed differently in the two countries.

	Printer A	Five of Printer B
Total first cost, $	250,000	260,000
Total salvage value, $	25,000	25,000
Total annual CFBT, $	65,000	65,000
Depreciation method	MACRS	Classical SL
Tax rate, %	50	50
Recovery period, years	5	5

14.23 Resolve Problem 14.22 if asset A is depreciated by the DDB method and assets B are depreciated by the DB method using a rate of 150% of the SL rate. Consider the estimated salvage values and use a 5-year period for comparison.

14.24 An asset costing $45,000 has a life of 5 years, a salvage value of $3000, and an anticipated CFBT = $15,000 per year. Determine the depreciation schedule for classical SL and for switching from DDB to SL to maximize depreciation. (The switching method was used in Problem 13.26.) Use $i = 18\%$ and an effective tax rate of 50% to determine how the present worth of taxes decreases when switching is allowed. Assume that the asset is sold for $3000 in year 6 and that any negative TI or capital loss at sale time generates a tax savings.

14.25 Construct the cash-flow diagram for annual income taxes, and then calculate the present worth of taxes for a $9000, 5-year recovery-period asset. The CFBT is estimated at $10,000 the first 4 years and $5000 thereafter as long as the asset is retained. The effective tax rate is 40% and money is worth 10% per year. Use the MACRS method of depreciation.

14.26 It is possible to calculate the effective tax savings in year t due to depreciation alone. If TS_t identifies tax savings, the relation is

$$TS_t = \text{(effective tax rate)(depreciation)} = T_e(D_t)$$

(*a*) Determine the present worth of tax savings, PW_{TS}, for an asset using MACRS with $B = \$45,000$, $n = 3$ years, $i = 8\%$, and $T_e = 0.35$. (*b*) Explain how the value of PW_{TS} may be used to select one depreciation method over another? How does this criterion compare with the PW_{tax} criterion used in Section 14.5?

14.27 Daryl's company purchased a depreciable asset which cost $6000, is expected to last 4 years, and can produce a CFBT of $3000 for 4 years only. The asset can be MACRS depreciated using a recovery period of either 3 or 5 years. If the tax rate is 50% and $i = 5\%$ per year, use the minimum PW_{tax} criterion to select the recovery period. Assume that any annual depreciation in excess of cash flow before tax considerations is a tax advantage to the corporation in that year. (*Note:* You might wish to use your spreadsheet to solve this problem.)

14.28 You plan to purchase a $900,000 materials handling system for a processing line. Assume you have the choices of classical SL depreciation ($n = 3$ years) and the DDB method ($n = 5$ years). Which method and time period would you choose? Use either the criterion of PW_{tax} or PW_{TS} (as discussed in Problem 14.26) to make the decision. The effective tax rate is 45% and assume i is 12% per year.

14.29 Web Edge Microprocessor Co. has just bought an asset for $88,000 with an expected life of 10 years. For tax purposes, the company is allowed (*a*) to use a recovery period of 5 or 10 years and straight-line depreciation with the half-year convention, or (*b*) to use a period of 5 years and MACRS depreciation. If $i = 10\%$ per year; $T_e = 52\%$ for local, state, and federal taxes; and the anticipated CFBT is $25,000 per year for only 10 years, determine the recovery period and depreciation method to minimize the time value of taxes. Consider all operating losses to be a tax advantage in the year incurred.

14.30 MACRS depreciation ($n = 5$) and the SL alternative to MACRS ($n = 8$) are being considered by a new small-business president for an asset with $B = \$10,000$ and no expected salvage value. Use CFBT = \$4000 to select the better method based on the present worth of taxes. Let $T_e = 0.35$ and $i = 20\%$ per year. Use the half-year convention.

chapter

15

After-Tax Economic Analysis

SCHEDULE C (Form 1040)	Profit or Loss From Business
Department of the Treasury Internal Revenue Service (T)	(Sole Proprietorship) ▶ Partnerships, joint ventures, etc., must file Form 1065. ▶ Attach to Form 1040 or Form 1041. ▶ See Instructions for Schedule C

Name of proprietor *Susan J. Brown*

Principal business or profession, including product or service (see page C-1)
Retail, ladies' apparel

Business name. If no separate business name, leave blank.
Milady Fashions

In this chapter we will incorporate the effects of income taxes into an engineering economic analysis. First we will use the basic relations of income tax in the previous chapter, coupled with information about *debt and equity financing,* to estimate annual *net cash flow (NCF)* after taxes for the alternative. Then the NCF values are utilized to compute the present worth, annual worth, or rate of return for a single project and for two competing alternatives.

As you already realize, tax laws change over time and are different from one country to another. Only the significant economic impacts of tax laws should be considered in an engineering economy analysis.

LEARNING OBJECTIVES

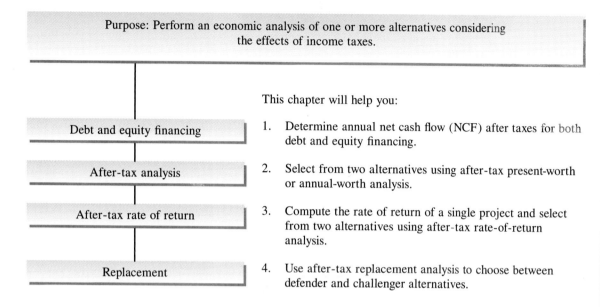

Purpose: Perform an economic analysis of one or more alternatives considering the effects of income taxes.

Debt and equity financing	This chapter will help you: 1. Determine annual net cash flow (NCF) after taxes for both debt and equity financing.
After-tax analysis	2. Select from two alternatives using after-tax present-worth or annual-worth analysis.
After-tax rate of return	3. Compute the rate of return of a single project and select from two alternatives using after-tax rate-of-return analysis.
Replacement	4. Use after-tax replacement analysis to choose between defender and challenger alternatives.

15.1 ESTIMATING NET CASH FLOW AFTER TAXES CONSIDERING DEBT AND EQUITY FINANCING

In Chapter 1 we introduced the terms cash flow and net cash flow. In fact, Equation [1.7] is

$$\text{Net cash flow} = \text{receipts} - \text{disbursements}$$

$$= \text{cash inflows} - \text{cash outflows}$$

Net cash flow (NCF), simply put, is the resulting, actual amount of cash that flows into the company (inflow, so the net is positive) or out of the company (outflow, so the net is negative) during a period of time, usually 1 year. After-tax analysis of cash flow implies that the net cash-flow amounts are used in all computations to determine PW, AW, ROR, or whatever measure of worth is of interest to the analyst. In fact, after-tax net cash flow is the same as the cash flow after taxes (CFAT) amount with some additional cash-flow terms. Let's explore exactly what makes up annual net cash-flow values, as we build upon what we learned in Chapter 14. Equations [14.13] through [14.16] show the computations for CFAT, where taxes reduce cash flow, but depreciation does not since it is not an actual cash outflow. Now, let's concentrate on an alternative's net cash flow by introducing the first cost of the asset, which is a capital expense in the amount P and commonly occurs in year 0 of the alternative's life. Secondly, we must introduce any salvage value, SV, which is a positive cash flow in year n. If we include the negative cash flow resulting from taxes, with any tax consequences of the SV accounted for in TI, the annual net cash flow for an after-tax engineering economy analysis, in general, is

$$\text{NCF} = -\text{capital expense} + \text{gross income} - \text{operating expenses}$$

$$+ \text{salvage value} - \text{taxes}$$

$$= -P + \text{GI} - E + \text{SV} - \text{TI}(T) \qquad \text{[15.1]}$$

There are several other inflows and outflows that directly affect cash flow and taxes. Primarily they result from the manner in which capital investments are financed through the use of debt or equity financing. We will now discuss how debt and equity financing are treated taxwise, and later we will learn how each type of financing affects measures of worth such as PW, AW, and IRR.

If a company uses funds borrowed from outside its own resources to acquire an asset, it is called debt financing. For our purposes, *debt financing (DF)* includes loans and bonds. We learned something about different loan repayment plans in Example 1.9, and Chapter 11 covered the details of bonds. *Loans* require the company to repay the principal over some stated time period, plus periodic *interest* on the principal. *Bonds* require the company to repay the face value after a stated number of years, plus periodic *dividends* on the bond's face value. Various types of loan and bond cash flows affect taxes and net cash flows differently as follows:

Type of Debt	Cash Flow Involved	Tax Treatment	Effect on Net Cash Flow
Loan	Receive principal	No effect	Increases it
Loan	Pay interest	Deductible	Reduces it
Loan	Repay principal	Not deductible	Reduces it
Bond	Receive face value	No effect	Increases it
Bond	Pay dividend	Deductible	Reduces it
Bond	Repay face value	Not deductible	Reduces it

Note that *only loan interest and bond dividends are tax-deductible.* We will use the symbol DE_I to identify the sum of these two. To develop a relation which explains the tax impact on cash flow of debt financing, start with the fundamental net cash-flow relation, that is, receipts minus disbursements. Identify receipts from debt financing as

$$DF_R = \text{loan principal receipt} + \text{bond sale receipt} \qquad \textbf{[15.2]}$$

Define debt financing disbursements as

$$DF_D = \text{loan interest payment} + \text{bond dividend payment} \qquad \textbf{[15.3]}$$
$$+ \text{ loan principal repayment}$$
$$+ \text{ bond face-value repayment}$$

It is common that a loan or bond sale, not both, is involved in a single asset purchase. The two terms in the first line of Equation [15.3] represent DF_I mentioned earlier. After developing this same logic for equity financing, we will utilize Equations [15.2] and [15.3] to calculate TI and taxes.

If a company uses its own resources for capital investment, it is called *equity financing (EF)*. This includes (1) the use of a corporation's own funds (such as retained earnings), which is referred to as investment of working capital; (2) the sale of corporation stock; and (3) the sale of corporate assets, including whole business units, to raise equity capital. There are no direct tax advantages considered in economic analyses for equity financing. Corporate working capital applied and stock dividends paid will reduce cash flow, but neither will reduce TI.

Therefore, to explain the impact on cash flow of equity financing, again start with the fundamental net cash-flow relation—receipts minus disbursements. Equity financing disbursements, defined as EF_D, are the portion of the first cost of an asset covered by a corporation's own resources.

$$EF_D = \text{corporate-owned funds} \qquad \textbf{[15.4]}$$

Any equity financing receipts are

$$EF_R = \text{sale of corporate assets} + \text{stock sale receipts} \qquad \textbf{[15.5]}$$

In Equation [15.4], stock dividends are a part of disbursements, but they are small in comparison to other disbursements and their timing depends on the financial success of the corporation overall, so they are neglected here. In fact, stock sales and stock dividends are not usually considered specifically in an analysis, because the money flows into and out of general corporate accounts. We include this level of detail here only to help in the understanding of how equity financing elements affect taxes and measures of worth.

Note that for debt financing, the loan principal and bond sale amounts are considered to be receipts in Equation [15.2], since they are *cash inflows*. However, for equity financing the use of corporate-owned funds in Equation [15.4] is *cash outflow*, since the company is spending its own funds to finance the capital investment.

Now, incorporate the DF and EF terms into Equation [15.1] in order to estimate annual NCF. The capital expense is equal to the amount of corporate-owned funds committed to the alternative's first cost, that is, $P = EF_D$. So, EF_D replaces P as the capital expense term in Equation [15.1]. (The NCF amounts will vary each year, but the year subscript t on every term has been omitted.)

$$\text{NCF} = -\text{equity-financed capital expense} + \text{gross income}$$
$$-\text{operating expenses} + \text{salvage value} - \text{taxes}$$
$$+ \text{debt financing receipts minus disbursements}$$
$$+ \text{equity financing receipts}$$
$$= -EF_D + GI - E + SV - TI(T) + (DF_R - DF_D) + EF_R \quad \textbf{[15.6]}$$

Recalling from Equation [15.3] that DF_D includes DF_I, which is the tax-deductible portion of debt financing, we can use the effective tax rate T_e times taxable income to compute taxes.

$$\text{Taxes} = (TI)(T_e) \qquad\qquad\qquad \textbf{[15.7]}$$
$$= (\text{gross income} - \text{operating expenses} - \text{depreciation}$$
$$-\text{loan interest and bond dividends})(T_e)$$

where DF_I is loan interest plus bond dividends.

These relations are quite easy to use when the investment involves no debt financing, that is, 100% equity financing or when 100% debt financing is used, since only the relevant terms in Equation [15.6] have nonzero values.

The approach taken above, and the resulting Equation [15.6], is called *generalized cash-flow analysis*. Though there are many more-detailed terms which could be included in a generalized cash-flow relation, this gives you an idea of how complex alternatives may be financially evaluated after taxes.

If Equation [15.7] includes a negative TI value, we assume the resulting negative tax is a *tax savings* and will offset taxes for the same year in other income-producing areas of the corporation. The effect is to increase the NCF value for the same year. This simplifying procedure is commonly used in lieu of confusing, and quite detailed, carry-forward and carry-back tax allowances mentioned at the end of Section 14.3.

Example 15.1

In order to reduce operating costs, a packaging machine incorporating fuzzy-logic software has been proposed for a small mail-order company. The first cost of $50,000 will be covered from company funds.

Since the start-up company is not seeking the advantage of accelerated depreciation, it will select a 5-year MACRS ADS-life from Table 13–4, which allows straight-line depreciation over 6 years with the half-year convention in years 1 and 6. Financial estimates for years $t = 1$ to 6 are

$$\text{Annual gross income} = 28{,}000 - (1000)(t)$$

$$\text{Operating expenses} = 9500 + (500)(t)$$

(a) Use an effective tax rate of 35% to determine NCF values for the 6 years. There is no salvage value expected for the packaging machine.

(b) Determine the NCF in year 6 if the machine is sold for $5000 after 6 years.

Solution

(a) The company plans to use equity financing i.e., its own funds. The NCF relation in Equation [15.6] simplifies greatly; the debt financing term is eliminated, and only EF_D—corporate-owned funds disbursement—is involved in the purchase year. Table 15–1 shows the gross income, operating expenses, capital expenses, depreciation, TI, taxes, and resulting NCF values. Depreciation rates for the 5-year MACRS straight-line model are $d_t = 0.10$ in years 1 and 6, and 0.20 in other years. Column 4 indicates the annual D_t values.

Since there is no debt financing, taxes shown in column 6 are computed as

$$\text{Taxes}_t = (TI)T_e = (GI_t - E_t - D_t)(0.35)$$

Equation [15.6] estimates annual NCF with the debt financing terms omitted

Table 15–1 Tabulation of after-tax net cash flow with 100% equity financing, Example 15.1

Year	(1) Gross Income	(2) Operating Expenses	(3) Capital Expense	(4) MACRS SL Depreciation	(5) Taxable Income	(6) Taxes	(7) NCF
0	—	—	$50,000	—	—	—	$-50,000
1	$27,000	$10,000	—	$5,000	$12,000	$4,200	+12,800
2	26,000	10,500	—	10,000	5,500	1,925	+13,575
3	25,000	11,000	—	10,000	4,000	1,400	+12,600
4	24,000	11,500	—	10,000	2,500	875	+11,625
5	23,000	12,000	—	10,000	1,000	350	+10,650
6	22,000	12,500	—	5,000	4,500	1,575	+7,925

and only the EF_D term included for the 100% equity-financed purchase price of $50,000 in year 0.

$$NCF_t = -EF_{D_t} + GI_t - E_t - \text{taxes}_t \qquad [15.8]$$

In column notation for Table 15–1, TI and NCF are

$$TI = \text{column } 1 - \text{column } 2 - \text{column } 4$$

$$NCF = -\text{column } 3 + \text{column } 1 - \text{column } 2 - \text{column } 6$$

(b) When the packaging machine is sold for $5000, the TI and taxes must account for $5000 in depreciation recapture. Also, the $5000 salvage cash flow must be added to the NCF. Use Equation [14.3] to compute TI (subscript for year 6 omitted).

$$TI = GI - E - D + DR$$

$$= \$22,000 - 12,500 - 5000 + 5000 = \$9500$$

$$\text{Taxes} = \$9500(0.35) = \$3325$$

From Equation [15.1], the NCF for year 6 is

$$NCF = GI - E + SV - \text{taxes}$$

$$= \$22,000 - 12,500 + 5000 - 3325$$

$$= \$11,175$$

This is an increase in NCF of $3250 over the $7925 in Table 15–1 where the salvage value is zero.

Example 15.2

The owner of Fifth Ave. Cleaners plans to invest in new dry cleaning equipment for $15,000 with the following estimates.

    Gross income = $7000 per year   Expenses = $1000 per year

    Effective tax rate = 35%   MACRS 5-year depreciation

Tabulate and compare NCF amounts for two alternate financing options: (a) All $15,000 is from company funds (100% equity-financed) and (b) one-half is via bank loan at 6%-per-year interest, and one-half is from company funds (50% debt–50% equity financing). Assume that the 6% is simple interest on the initial loan principal and repayment will be in five equal payments of accrued interest and principal. Which method offers a smaller total NCF to the company over the MACRS depreciation period, if the time value of money is neglected in this analysis?

Solution

(*a*) Table 15–2 includes the computations for the 100% equity financing option. Column 2 shows that GI expenses are $6000 per year; MACRS-5-year depreciation is in columns 4 and 5; and column 6 shows TI using Equation [15.6] with no debt financing interest or dividends. NCF is calculated as

$$\text{NCF} = -\text{capital expense} + (\text{GI} - \text{expenses}) - \text{taxes}$$

$$= -\text{column 3} + \text{column 2} - \text{column 7}$$

The equity-financed capital expense of $15,000 is a cash outflow only in year 0. The total NCF for the 100% equity financing option is $13,650 with no time value of money considered.

(*b*) The 50% debt–50% equity financing option is more complex as indicated in Table 15–3, where two new columns on debt financing have been added for debt financing. There is no bond financing, only loan interest and principal repayment. From Equations [15.3], [15.6], and [15.7], respectively, we write the following equations, with the table column relation.

$$\text{DF}_D = \text{loan } \textit{interest } \text{payment} + \text{loan } \textit{principal} \text{ repayment}$$

$$= \$7500(0.06) + \frac{7500}{5}$$

$$= \$450 + 1500$$

$$= \text{column 3} + \text{column 4}$$

where only the $450 is loan interest, DF_I, and it is tax-deductible. The loan amount of $7500 is not shown in Table 15–3, column 4 in year 0, because the loan principal receipt and disbursement for 50% of the asset first cost both occur in the same year, we assume. The NCF in column 10 is

$$\text{NCF} = -\text{EF}_D + (\text{gross income} - \text{expenses}) - \text{taxes} - (\text{DF}_D)$$

$$= -\text{column 5} + (\text{column 2}) - \text{column 9} - (\text{column 3} + \text{column 4})$$

The 50% equity-financed capital expense in year 0 is now only $\text{EF}_D = \$7500$ for the company. The taxes are

$$\text{Taxes} = [(\text{gross income} - \text{expenses}) - \text{depreciation} - \text{DF}_I]T_e$$

$$= [(\text{column 2}) - \text{column 7} - \text{column 3}]\,0.35$$

From column 10 of Table 15–3, for the 50% debt–50% equity financing option, the total NCF is $12,188.

Conclusion

The total NCF is larger for the 100% equity option, because there is no cash outflow for interest associated with a loan. However, if the Fifth Ave. Cleaners owner does not have the full $15,000 in equity funds up front, the 50% debt option has a total NCF that is only about $1500 less. Debt financing should be considered a viable option for the owner.

Table 15-2 Calculation of NCF for 100% equity-financed asset, Example 15.2a

(1) Year	(2) Gross Income − Expenses	(3) Capital Expense	(4) MACRS Depr. Rate	(5) Depreciation	(6) Taxable Income	(7) Taxes 0.35(TI)	(8) NCF
0	$ 0	$15,000	0	—	—	—	−$15,000
1	6,000		0.2	$ 3,000	$3,000	$1,050	4,950
2	6,000		0.32	4,800	1,200	420	5,580
3	6,000		0.192	2,880	3,120	1,092	4,908
4	6,000		0.1152	1,728	4,272	1,495	4,505
5	6,000		0.1152	1,728	4,272	1,495	4,505
6	6,000		0.0576	864	5,136	1,798	4,202
			1.0000	$15,000		$7,350	$13,650

Table 15-3 Calculation of NCF for 50% equity-financed and 50% debt-financed asset, Example 15.2b

(1) Year	(2) Gross Income − Expenses	(3) Debt Financing Interest	(4) Debt Financing Principal	(5) Capital Expense	(6) MACRS Depr. Rate	(7) Depreciation	(8) Taxable Income	(9) Taxes 0.35(TI)	(10) NCF
0	$ 0	$ 0	$ 0	$7,500	0	—	—	—	−$7,500
1	6,000	450	1,500		0.2	$ 3,000	$2,550	$ 893	3,158
2	6,000	450	1,500		0.32	4,800	750	263	3,788
3	6,000	450	1,500		0.192	2,880	2,670	935	3,116
4	6,000	450	1,500		0.1152	1,728	3,822	1,338	2,712
5	6,000	450	1,500		0.1152	1,728	3,822	1,338	2,712
6	6,000				0.0576	864	5,136	1,798	4,202
			$7,500		1.0000	$15,000		$6,563	$12,188

Example 15.2
(Spreadsheet)

Set up the debt-equity financing situation in Example 15.2 on a spreadsheet in order to perform an analysis which determines the impact of various debt and equity funding percentages on the total NCF. Examine the range 100% debt–no equity to 10% debt–90% equity, at the debt financing percentages of 100, 90, 70, 50, 30, and 10.

Solution

We can use the column definition of Table 15–3 to build a spreadsheet frame which allows the debt-equity percentages to take on any value from 0% to 100%. Figure 15–1 shows spreadsheets for 70% debt–30% equity and 90% debt–10% equity. Some of the functions for the spreadsheet cells are detailed in the margins.

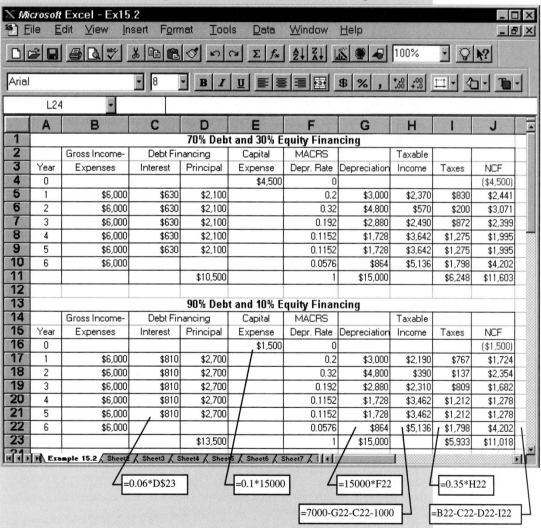

Figure 15–1   Calculation of total NCF amount for various debt-equity financing percentages of a $15,000 investment, Example 15.2 (Spreadsheet).

> Total NCF (column J in Figure 15–1) results for the various percentage splits obtained from the spreadsheets are
>
> | Debt-equity | 100%–0% | 90–10% | 70%–30% | 50%–50% | 30%–70% | 10%–90% |
> |---|---|---|---|---|---|---|
> | Total NCF, $ | 10,725 | 11,018 | 11,603 | 12,188 | 12,773 | 13,358 |
>
> For each 10% decrease in debt financing, the total NCF increases by $292.50. For example, from 100% to 90% debt, the total NCF difference is $11,018 - 10,725 = $293 (rounded). Larger debt financing will always decrease the resulting NCF, since more of the corporation's income is used to pay the interest.

If only equity financing is involved, we can rewrite the NCF relation, Equation [15.1], in a shorter form by utilizing the *tax savings* term (TI) $(1 - T)$, which is the portion of TI not absorbed by taxes. For 100% equity funds, $EF_D = P$, the capital expense in Equation [15.1]. If we write the depreciation times tax rate term, DT as $D - D(1-T)$, we derive a shortened form for NCF.

$$NCF = -P + GI - E - \text{taxes} = -P + GI - E - TI(T)$$

$$= -P + GI - E - (GI - E - D)(T)$$

$$= -P + GI(1-T) - E(1-T) - DT$$

$$= -P + (GI - E)(1-T) + D - D(1-T)$$

$$= -P + D + (GI - E - D)(1-T)$$

$$= -P + D + TI(1-T)$$

$$= -\text{capital expense} + \text{depreciation} + TI(1-T) \qquad \textbf{[15.9]}$$

Therefore, in Example 15.2*a*, Table 15–2, NCF for year 1 may be calculated as

$$NCF = 0 + 3000 + 3000(1-0.35) = \$4950$$

The same is true for each year 0 through 6.

But you must be careful. *If there is any debt financing whatsoever,* as in Example 15.2*b*, this shortened relation will not work. Let's try it in Table 15–3 for year 1.

$$NCF = 0 + 3000 + 2550(1-0.35) = \$4657.50 \neq \$3158$$

Why doesn't it work? This "tax savings" approach in Equation [15.9] neglects the fact that loan interest is tax-deductible and that NCF is determined using a different relation.

Problems 15.1 to 15.10

15.2   AFTER-TAX PRESENT WORTH AND ANNUAL WORTH

If the required after-tax MARR is established (at market rate as discussed in Chapter 12), the NCF values are used to compute the PW or AW for a project just as in previous chapters. When both positive and negative NCF amounts are involved, a resulting PW or AW < 0 indicates that the after-tax MARR is not met.

For mutually exclusive alternative comparison, use the guidelines below to select the better alternative. (This is the same logic as that developed in Chapter 5 for alternative selection.)

- If alternative PW or AW ≥ 0, the required after-tax MARR is met or exceeded; the alternative is financially viable.

- *Select the alternative with the PW or AW value that is numerically larger.*

If only cost estimates are included for an alternative, consider the tax savings that the AOC or operating expense generates to be positive NCF. Then use the same guideline to select an alternative.

Example 15.3

Paul has estimated the NCF values presented below. Use an AW analysis and a 7%-per-year after-tax MARR to select the better plan.

Plan X		Plan Y	
Year	NCF	Year	NCF
0	−$28,800	0	−$50,000
1–6	5,400	1	14,200
7–10	2,040	2	13,300
10	2,792	3	12,400
		4	11,500
		5	10,600

Solution

Develop the AW relations using the NCF values over each plan's life. Select the numerically larger value at $i = 7\%$ per year.

$$AW_X = [-28{,}800 + 5400(P/A,7\%,6) + 2040(P/A,7\%,4)(P/F,7\%,6)$$

$$+ \ 2792(P/F,7\%,10)](A/P,7\%,10)$$

$$= \$421.78 \hspace{3cm} \textbf{[15.10]}$$

$$AW_Y = [-50,000 + 14,200(P/F,7\%,1) + 13,300(P/F,7\%,2)$$
$$+ 12,400(P/F,7\%,3) + 11,500(P/F,7\%,4)$$
$$+ 10,600(P/F,7\%,5)](A/P,7\%,5)$$
$$= \$327.01 \hspace{3cm} \text{[15.11]}$$

Both plans are financially viable since each AW > 0. Select plan X because AW_X is larger.

Comment

You might want to use the net present-value or present-worth function on your spreadsheet system to calculate PW and/or AW directly.

$$PW_X = \$2962.99 \text{ over the 10 years}$$

$$AW_X = PW_X(A/P,7\%,10) = \$421.78$$

$$PW_Y = \$1340.81 \text{ over 5 years}$$

$$= \$2296.79 \text{ over LCM of 10 years}$$

$$AW_Y = PW_Y(A/P,7\%,5) = \$327.01$$

Plan X is selected again because of its larger PW value.

Problems 15.11 to 15.16

15.3   RATE-OF-RETURN COMPUTATIONS USING THE NCF SEQUENCE

The PW or AW relations are used to estimate the rate of return of the NCF values using the same procedures of Chapters 7 and 8. For a *single project,* set the PW or AW of the NCF sequence equal to zero and solve for the $i*$ value using the easier method.

Present worth:
$$0 = \sum_{t=1}^{t=n} NCF_t(P/F,i*,t) \hspace{3cm} \text{[15.12]}$$

Annual worth:
$$0 = \left[\sum_{t=1}^{t=n} NCF_t(P/F,i*,t) \right](A/P,i*,n) \hspace{2cm} \text{[15.13]}$$

Spreadsheet analysis using the internal rate-of-return function can make it easy to solve for $i*$ for relatively complex NCF sequences. However, multiple roots may exist as discussed in Section 7.4 when the NCF sequence has more than one sign change.

Example 15.4

A fiber-optics manufacturing company has spent $50,000 for a 5-year-life machine which has a projected $20,000 annual CFBT and an effective incremental tax rate of 40%. Compute the after-tax rate of return. Assume annual depreciation is $10,000 for computation purposes.

Solution

The NCF in year 0 is $EF_D = \$-50,000$ of equity-financed capital expense. For years $t = 1$ through 5, use Equation [15.9] to estimate the NCF as

$$NCF = \text{depreciation} + TI(1-T)$$
$$= D + (CFBT - D)(1-T)$$
$$= 10,000 + (20,000 - 10,000)(1-0.4)$$
$$= \$16,000$$

Since the NCF values for years 1 through 5 have the same value, the PW form of Equation [15.12] used to estimate $i*$ is

$$0 = -50,000 + 16,000(P/A,i*,5)$$

$$(P/A,i*,5) = 3.125$$

Solution gives $i* = 18.03\%$ as the after-tax rate of return.

Comment

As discussed in Section 14.4, it is always possible to use an exaggerated before-tax rate of return to estimate the effect of taxes without all the detailed computations of NCF. Solution of Equation [14.17] for the after-tax rate estimate yields

$$\text{Before-tax rate} = \frac{(\text{after-tax rate})}{1-T}$$

$$= \frac{0.1803}{1-0.40} = 0.3005 \quad (30.05\%)$$

The actual before-tax $i*$ using CFBT = $20,000 for the 5 years is 28.65% from the relation

$$0 = -50,000 + 20,000(P/A,i*,5)$$

Comparison shows that the tax effect is slightly overestimated if a MARR of 30.05% is used in the before-tax analysis.

For multiple alternatives, use a PW or AW relation to compute the *return on the incremental NCF series* for the larger-initial-investment alternative relative to the smaller one. The equations, which are of the same form used in Sections 8.4 and 8.5, are summarized in Table 15–4. These equations are developed for the incremental NCF values, represented by the Δ (delta)

Table 15-4	Guidelines for computation of the after-tax return using incremental analysis on ΔNCF values (B has the larger initial investment)		
Method Used to Estimate i^*_{B-A}	**Equal-life Alternatives**		**Different-life Alternatives**
Present worth (PW)	Set PW = 0 for ΔNCF for n years		Set PW = 0 for ΔNCF for least common multiple of years
Equation used:	$$\sum_{t=1}^{n} \Delta NCF_t(P/F,i^*,t) = 0$$		[15.14]
Annual worth (AW)	Set AW = 0 for ΔNCF for n years		Set difference of two AW relations = 0 over the unequal lives
Equations used: (PW of ΔNCF)$(A/P,i^*,n) = 0$ [15.15]		$AW_B - AW_A = 0$ [15.16]	

symbol. For each year t and for alternatives B and A, calculate the series

$$\Delta NCF = NCF_B - NCF_A$$

Solution of any Table 15–4 equation for the interest rate provides the breakeven incremental after-tax return i^*_{B-A} between the two alternatives, which is compared with the after-tax (market) MARR. Refer to Sections 8.3 through 8.5 for a review of incremental rate-of-return analysis. The complete procedure for after-tax return analysis of two alternatives is

1. Order the alternatives by decreasing initial investment. Consider the larger-investment as alternative B.
2. Decide whether to use a PW or AW relation to compute the incremental after-tax return. Select the appropriate equation from Table 15–4.
3. Calculate the ΔNCF values for PW or AW (equal lives) analysis, or determine the AW relation for different lives.
4. Calculate the after-tax return i^*_{B-A} of the ΔNCF series. The $B-A$ subscript may be omitted for simplicity.
5. Compare the resulting i^* value with the MARR. Select B if $i^* >$ MARR. Otherwise, select alternative A.

Figure 15–2 demonstrates this procedure for PW analysis. The top graph shows i^* as the breakeven rate between the actual NCF series of two alternatives. If i^* is larger than the MARR (marked as $MARR_1$), select B, which will have the larger PW. Correspondingly, the bottom graph indicates the incremental breakeven rate, i^*_{B-A}. Again, select B if $i^*_{B-A} >$ MARR, because the larger investment in B is justified. If the required return, shown as $MARR_2$, is above the incremental breakeven rate, the larger investment for B is not justified and A is selected.

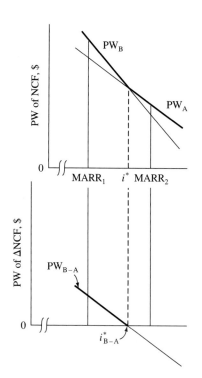

Figure 15–2   Selection of an alternative at different MARR values using after-tax PW analysis of actual NCF series (top graph) and incremental NCF series (bottom graph).

In all cases, remember that the acceptance of alternative B means that the incremental investment over A is justified only when the incremental breakeven rate exceeds the MARR.

In choosing the method to estimate i^*_{B-A} from Table 15–4 via manual trial and error, the PW method is usually easier for equal lives and the AW method is better for different lives, due to the number of computations involved.

For *spreadsheet analysis* the steps above are simplified because there are no trial-and-error computations. However, most spreadsheet systems use PW computations to find an internal rate of return, so the ΔNCF series must be determined for the least common multiple of years to obtain the correct i^*_{B-A} value. In brief, the procedure for spreadsheet analysis is

1. Order the alternatives by initial investment. Label the larger-investment alternative as B.

2. Enter the NCF amounts for each alternative for PW analysis for the least common multiple of years as in Equation [15.14].

3. Set up the work sheet to calculate the ΔNCF values for each year.

4. Enter the function to estimate the after-tax incremental return i^*_{B-A}. The $B-A$ subscript may be omitted for simplicity.

5. Compare the resulting i^* value with MARR. Select B if $i^* >$ MARR. Otherwise, select alternative A.

Example 15.5
(Spreadsheet)

Use the estimated NCF values for plans X and Y in Example 15.3 to select the better alternative if an after-tax MARR of 6% per year is required.

Solution

Since the lives are different, Table 15–4 indicates that either PW analysis over a 10-year period or AW analysis over respective lives may be performed. Choose the PW relation, Equation [15.14], and use the spreadsheet-analysis procedure. Plan Y has the larger initial investment.

The PW values for each plan are plotted in Figure 15–3. The accompanying spreadsheet shows the NCF and the computed ΔNCF series for 10 years. Reinvest-

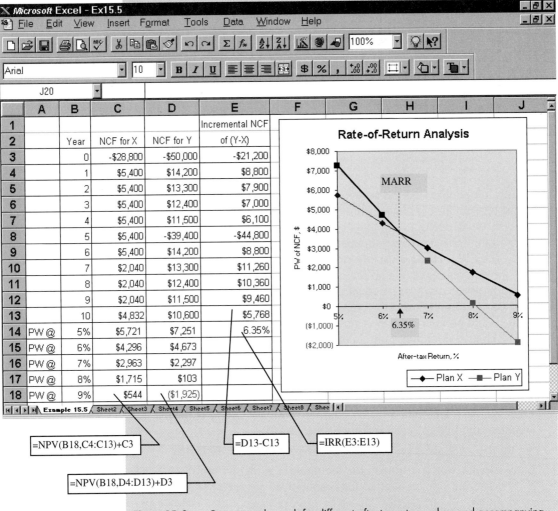

Figure 15–3   Present-worth graph for different after-tax return values and accompanying breakeven incremental rate-of-return analysis, Example 15.5.

ment in plan Y at the end of year 5 results in the cash outflow of $-\$50,000 + 10,600 = -\$39,400$. The form of Equation [15.14] to estimate breakeven $i^*_{Y-X} = 6.35\%$ (cell E14 in Figure 15–3) is

$$\sum_{t=0}^{t=10} \Delta NCF_t(P/F,i^*_{Y-X},10) = 0$$

Since $6.35\% >$ MARR $= 6\%$, select plan Y. As Figure 15–3 indicates, any MARR less than 6.35% favors plan Y because the PW_Y is larger. For any MARR exceeding 6.35%, select plan X.

Comment

If AW analysis is used, develop the two NCF series for the respective lives of Y and X and place them into the form of Equation [15.16].

$$(AW_Y \text{ over 5 years}) - (AW_X \text{ over 10 years}) = 0$$

Solve for i^*_{Y-X} by trial and error.

Example 15.6

Johnson Enterprises must decide between two alternatives in its Asian plant: system 1—a single robot assembly system for printed wiring boards will require a $100,000 investment now; and system 2—a combination of two robots for a total of $130,000. Management intends to implement one of the plans. This aggressive electronics manufacturer expects a 20% after-tax return on technology investments. Select one of the systems, if the following series of cost NCF values have been estimated for the next 4 years.

	Year				
	0	**1**	**2**	**3**	**4**
System 1 NCF, $	−100,000	−35,000	−30,000	−20,000	−15,000
System 2 NCF, $	−130,000	−20,000	−20,000	−10,000	−5,000

Solution

Only cost cash flows are involved as indicated by the minus signs. Considering the initial investment in year 0, system 2 is the alternative with the incremental investment which must be justified. Since lives are equal, select PW analysis to estimate i^* and determine the series shown below for system 2 minus system 1 NCF values. All cash flows have been divided by $1000. For example, in year 0, $[(-130,000) - (-100,000)]/1000 = -30$.

Year	0	1	2	3	4
ΔNCF for 2 − 1, $	−30	+15	+10	+10	+10

Equation [15.14] is set up to estimate an after-tax return i^*_{2-1}.

$$-30 + 15(P/F,i^*,1) + 10(P/A,i^*,3)(P/F,i^*,1) = 0$$

Solution gives an after-tax return of 20.10%, which just exceeds the 20% required return. The extra investment in system 2 is marginally justified.

Additional Example 15.8 illustrates the differences in after-tax returns for various debt-equity funding proposals on one project.

Additional Example 15.8
Problems 15.17 to 15.25

15.4   AFTER-TAX REPLACEMENT ANALYSIS

When a currently owned asset (the defender) is challenged by a new asset, the effects of income taxes may be considered. Accounting for all the tax details in *after-tax replacement analysis* is sometimes neither time- nor cost-effective. However, it is important from a tax perspective to account for any capital gain or loss or significant depreciation recapture which may occur if the defender is replaced. Also important is the future tax advantage stemming from deductible operating and depreciation expenses. Either the cash-flow approach (for equal-life alternatives only) or the opportunity cost approach (Section 10.4) may be used. The following example considers the impact of taxes on replacement analysis.

Example 15.7

The Midcontinent Power Authority purchased new coal extraction equipment 3 years ago for $600,000. Management has discovered it is technologically outdated now. A new-equipment alternative (the challenger) has been identified. If a generous trade-in of $400,000 is offered for the current equipment, perform AW analyses using (*a*) a before-tax MARR of 10% per year and (*b*) a 7%-per-year after-tax MARR. Assume an effective tax rate of 34%. As a simplifying assumption, use classical straight-line depreciation with SV = 0 for both alternatives. The following additional data are applicable.

	Defender (*D*)	Challenger (*C*)
Current basis, $	400,000	1,000,000
Annual costs, $	100,000	15,000
Recovery period, years	8 (originally)	5

Solution

(a) For the before-tax analysis, at this point in time $n = 5$ years for the defender, and the trade-in offer of $400,000 becomes the first cost amount of $-\$400,000$ for the AW_D equation. We use the *cash-flow approach* learned in Section 10.3, since the lives are equal; remove the $400,000 from the challenger's first cost. Now, complete the AW analysis.

$$AW_D = -\$100,000$$

$$AW_C = -600,000(A/P,10\%,5) - 15,000$$

$$= -\$173,280$$

We select the numerically larger AW value; the currently owned equipment (defender) is favored by $73,280.

(b) For the after-tax analysis, we must perform the tax-related computations before developing the AW relations. For the *retain-the-defender* alternative, using the cash-flow approach to replacement analysis, the defender after-tax annual costs and depreciation are

Annual costs: $100,000 as estimated.

SL depreciation: $600,000/8 = $75,000 continues for 5 more years.

Annual tax savings: (Annual costs + depreciation) $(T_e) =$ $(100,000 + 75,000)(0.34) = \$59,500$.

Actual annual after-tax expenses: $100,000 - 59,500 = \$40,500$.

The defender cash-flow diagram, presented in Figure 15–4 (top graph), using the cash-flow approach, shows no first cost and only the annual costs for 5 years. The after-tax AW expression of costs, which is the NCF for the *defender,* is

$$AW_D = \text{actual annual expenses} = -\$40,500$$

For the *accept-the-challenger* alternative, the after-tax analysis is more complex. When the old equipment is traded, there will be a 34% tax on the depreciation recapture. This raises the equivalent first cost of the challenger above $1,000,000, and the $400,000 trade-in amount reduces the first cost equivalently. (No capital gain or loss is incurred.)

Current defender SL book value: $600,000 - 3(75,000) = \$375,000$.

Depreciation recapture upon trade-in: $400,000 - 375,000 = \$25,000$.

Depreciation recapture tax: $(25,000)(0.34) = \$8500$.

Equivalent challenger after-tax investment now: $-\$1,000,000 + 400,000 - 8500 = -\$608,500$.

Annual classical SL depreciation: $1,000,000/5 = \$200,000$.

Annual tax savings: (Annual costs + depreciation) $(T) =$ $(15,000 + 200,000)(0.34) = \$73,100$.

Actual annual after-tax NCF: $\$-15,000 + 73,100 = \$58,100$.

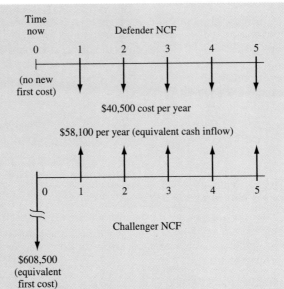

Time
now   Defender NCF

0   1   2   3   4   5

(no new
first cost)

$40,500 cost per year

$58,100 per year (equivalent cash inflow)

0   1   2   3   4   5

Challenger NCF

$608,500
(equivalent
first cost)

Figure 15–4   After-tax replacement analysis cash-
flow diagrams for Example 15.7 using
the cash-flow approach to replacement
analysis.

The annual expenses have now effectively become a net cash inflow (plus sign) for the AW analysis, rather than a cash outflow. The cash-flow diagram is presented in Figure 15–4 (bottom graph). The after-tax AW for the *challenger* is

$$AW_C = \text{(equivalent challenger investment)}(A/P,7\%,5) + \text{annual NCF}$$

$$= -\$608,500(A/P,7\%,5) + 58,100$$

$$= -\$90,307$$

Select the defender with a cost advantage, that is, equivalent NCF advantage, of $90,307 - 40,500 = \$49,807$ per year. Even though both analyses favor the defender, the advantage is less after taxes ($73,280 versus $49,807). Since the technological adequacy of the defender is at issue, management may decide to replace it anyway, thus basing the decision on a combination of economic and noneconomic factors.

Comments

If the trade-in offer had been less than the current defender book value of $375,000, the resulting capital loss would be treated as a tax savings for the accept-the-challenger alternative, using the cash-flow approach. The tax savings amount then decreases the equivalent first-cost amount below $600,000. For example, if the trade-in were for $300,000, a tax savings of $(\$375,000 - 300,000)(0.34) = \$25,500$ is removed to make the challenger basis $-\$1,000,000 + 400,000 + 25,500 = -\$574,500$.

From Section 10.3, it is equally correct to analyze these equal-life alternatives using the *opportunity cost approach* with the correct conclusion to select the defender by the same dollar amount as calculated above. The defender first cost is −$400,000 (before taxes) and −$391,500 after performing the tax-related adjustments. The AW amounts you should verify are

Before-tax analysis: $AW_D = -\$205,520$ $AW_C = -\$278,800$

After-tax analysis: $AW_D = -\$135,983$ $AW_C = -\$185,790$

Problems 15.26 to 15.29

ADDITIONAL EXAMPLE

Example 15.8

AFTER-TAX RATE OF RETURN, SECTION 15.3   Return to Example 15.2 (Fifth Ave. Cleaners) and estimate the after-tax return for the two funding options: (a) 100% equity financing and (b) 50% equity–50% debt financing. Compare the results and explain why debt financing increases the return.

Solution

(a) For *100% equity financing* include the NCF series in column 8 of Table 15–2, in the general PW form of Equation [15.12], to solve for $i*$.

$$0 = -15,000 + 4950(P/F,i*,1) + \cdots + 4202(P/F,i*,6)$$

Solution results in $i* = 23.39\%$.

(b) For the 50% debt–50% equity split, the NCF values in column 10 of Table 15–3 are inserted into the following relation to yield $i* = 37.01\%$.

$$0 = -7500 + 3158(P/F,i*,1) + \cdots + 4202(P/F,i*,6)$$

Comparison indicates that 50%-debt financing significantly increases the rate of return on company investment by over 13 percentage points, in part because less of the firm's capital funds are committed to the investment—$7500 for 50%–50% financing versus $15,000 for 100% equity. Additionally, the loan rate of 6% is significantly less than the $i*$ value of 37.01% using 50% debt capital. As the loan rate increases, the advantage of debt financing will diminish relative to 100% equity financing.

Comment

Let's examine this situation. Why not use 100% debt financing for all capital investments and maximize the return to the business? Please venture a guess now. Ask yourself: What if every time you wanted to make a purchase you had to borrow from someone—parent, bank, or credit union? How stable would your credit be over time? What kind of financial reputation would you have when you asked for more? What happens if something goes wrong? We will let you speculate until Section 18.7 where we discuss debt-equity financing more thoroughly.

If spreadsheet analysis is used as in Example 15.2 (Spreadsheet), the internal rate-of-return function may be applied to the far-right columns of Tables 15–2 and 15–3 to determine the i^* values of 23.39% and 37.01%, respectively.

CHAPTER SUMMARY

The bottom-line effects of debt and equity financing may be significantly different on after-tax net cash flow (NCF) depending upon the debt and equity financing percentages of the investment. This is due to the differences in tax-deductible allowances in debt and equity financing. A heavily debt-laden corporation tends to have a smaller tax burden; however, too much reliance on debt financing can reduce creditworthiness and the ability to raise new capital through additional debt financing.

A general NCF relation which considers all forms of debt and equity financing receipts and disbursements is

$$\begin{aligned} \text{NCF} = {} &-\text{capital expense} + \text{gross income} - \text{expenses} + \text{salvage value} \\ &- \text{taxes} + \text{debt financing receipts minus disbursements} \\ &+ \text{equity financing receipts} \end{aligned}$$

This relation is the foundation for a generalized cash-flow analysis with relations available for each of the debt-and equity-financed components.

The appropriate tax rules are applied to determine annual NCF for an alternative. A measure of worth—PW, AW, or ROR—is determined in the same fashion we learned in earlier chapters.

An after-tax area that can be complicated is replacement analysis of a defender and a challenger. For the retain-the-defender alternative, there may be depreciation recapture, and possibly a capital gain or capital loss involved. Either the cash-flow approach for equal-life assets or the opportunity cost approach may be used to adjust the first-cost amounts while considering tax implications.

Often the decisions of before-tax and after-tax analyses are the same. However, the considerations made for the after-tax computations give the analyst a much clearer estimate of the monetary impact of taxes.

PROBLEMS

15.1 Explain the basic differences in the contents and uses between the relation for CFAT presented in Equation [14.16] and for NCF in Equation [15.1].

15.2 In your own words describe the different impacts on the annual net cash-flow values over the life of an asset when debt rather than equity financing is used to fund the initial capital investment to purchase the asset.

15.3 Four years ago Flexinet Beverages purchased an asset for $20,000 with an estimated SV of zero. Depreciation was $5000 per year and the following annual gross incomes and operating expenses were recorded. Tabulate the annual net cash flows after an effective 50% tax rate was applied.

	Year of Ownership			
	1	2	3	4
Gross income, $	8,000	15,000	12,000	10,000
Operating expenses, $	2,000	4,000	3,000	5,000

15.4 Four years ago Flexinet Beverages purchased an asset for $20,000 with an estimated SV of zero. Depreciation was charged using MACRS for a 3-year recovery period. Tabulate the annual net cash flows after an effective 50% tax rate was applied, if the following annual gross incomes and operating expenses were recorded under the assumption (*a*) that the asset was discarded with no financial impact after 4 years and (*b*) that the asset was salvaged for $2000 after 4 years of ownership. (*Note:* The numbers below are the same as in Problem 15.3, if you wish to compare the NCF values.)

	Year of Ownership			
	1	2	3	4
Gross income, $	8,000	15,000	12,000	10,000
Operating expenses, $	2,000	4,000	3,000	5,000

15.5 An investment company just purchased an apartment complex for $3,500,000 using all equity capital. Annual gross income before taxes of $480,000 is expected for the next 8 years, after which the owner hopes to sell the property for the currently appraised value of $4,530,000. The applicable tax rate is 52%, the estimated annual operating expenses are $200,000, and any capital gain on the sale will be taxed at an estimated 40%. Tabulate the after-tax net cash flows for the 8 years of ownership, if the property will be straight-line depreciated over a 20-year life using a salvage value of zero. (Neglect the half-year convention in depreciation for simplification purposes.)

15.6 June's corporation purchased new water sampling equipment for use in contract work for $40,000 and depreciated it over a 5-year recovery period using MACRS. It had no estimated salvage value and produced annual CFBT of $20,000 which was taxed at an effective 40%.

(a) If all funds for the purchase were obtained from corporate retained earnings, and the machine was donated after 6 years, determine the net cash-flow values for years 0 through 6.

(b) Determine the NCF values if the machine was sold prematurely after 3 years for $20,000 and no replacement was acquired, but the CFBT for year 4 only was $10,000.

15.7 Determine the NCF values for the situation in Problem 15.6 (b) if 60% of the equipment cost was financed by a loan. You have been given data which shows that $2500 in annual interest was paid. The entire principal of the loan was repaid in year 5, and interest continued through year 5 even though the machine was sold in year 3.

15.8 An asset was acquired by AAA Floor Covering. Use the following to tabulate net cash-flow values and their sum if the asset is actually salvaged after 6 years for $3075.

$B = \$10,000$

$n = 5$ years

$SV = 0$

Classical straight-line depreciation

Gross income $-$ expenses $= \$5000$

$T = 48\%$

Assume the income and expenses continued at the same level for the year after full depreciation.

15.9 (a) Resolve Problem 15.8 if MACRS depreciation and a 5-year recovery period are used. (b) Is there any difference in the total NCF values for SL and MACRS depreciation? Why?

15.10 Use the asset data below to determine annual net cash-flow values if a $5000 loan assisted in acquiring the $10,000 asset. Assume the loan interest was computed at an annual rate of 3% on the initial principal and repayment of the principal and interest took place in five equal install-ments of $1150 each. The actual salvage value after 5 years was $2000.

$B = \$10,000$

$n = 5$ years

$SV = \$3075$

Classical straight-line depreciation

Gross income $-$ expenses $= \$5000$

$T = 48\%$

15.11 An aerospace company manager must choose between two machines which will bend metal on the production floor.

	Machine A	Machine B
First cost, $	24,000	15,000
Actual salvage value, $	6,000	3,000
Gross income − expenses, $	4,000	2,000
Life, years	12	12

The machines have an anticipated useful life of 12 years, but the MACRS alternative model requires straight-line depreciation over 10 years with a zero salvage value. An effective tax of 50% applies and an after-tax MARR of 10% is desired.

(a) Compare the two machines using present-worth analysis. In order to simplify the analysis, neglect the half-year convention for depreciation.

(b) Can the manager expect either machine to return the after-tax MARR? How did you make this conclusion?

15.12 Ms. Montoya wants to economically evaluate the following machines. Prepare the evaluation for her.

Machine A: $B = \$15,000$ $SV = \$3000$ $AOC = \$3000$ $n = 10$

Machine B: $B = \$22,000$ $SV = \$5000$ $AOC = \$1500$ $n = 10$

(a) Use classical straight-line depreciation, $T_e = 50\%$, and an after-tax MARR of 7% per year to select the more economical machine.

(b) Is the answer different if MACRS depreciation for a 5-year recovery period applies, but the machine is retained in use for a total of 10 years and then sold for the estimated SV?

15.13 Choose between the two alternatives detailed below if the after-tax MARR is 8% per year, MACRS depreciation is used, and $T_e = 40\%$.

	Alternative A	Alternative B
First cost, $	10,000	15,000
Actual salvage value, $	0	2,000
Gross income − expenses, $	1,500	600
Recovery period, years	3	3

The estimates for gross income − expenses are made for only 3 years. They are zero when each asset is sold in year 4.

15.14 Rework Problem 15.13 if one-half of the first cost of each asset is debt-financed and the owner agrees to make three equal loan repayments of $1870.55 for alternative *A* and $2805.83 for alternative *B*. For each loan, one-third of the principal amount is repaid each year with the remainder of the payment used to cover loan interest.

15.15 Dr. Smitherson plans to purchase new dental equipment for cleaning patients' teeth using capital funds from his own business. The purchase price is $10,000, the salvage value is estimated at $1000 after 6 years, and the MACRS alternative of SL depreciation with $n = 5$ years is chosen for depreciation. The annual increase in gross income is expected to be $5000, and annual operating expenses should average $700. The effective tax rate is 38%. Will Dr. Smitherson realize an after-tax market MARR of 15% over the 6-year period?

15.16 A manager in a canned-food processing plant in South America must decide between two labeling machines. Their respective costs are

	Machine *A*	Machine *B*
First cost, $	15,000	25,000
Annual operating cost, $	1,600	400
Salvage value, $	3,000	6,000
Life, years	7	10

Regardless of which machine is selected, a $10,000 loan (U.S. $ equivalent) will be necessary for the purchase. Repayment of this loan will be in five equal annual installments of $2700 each ($2000 principal, $700 interest). If the food-processing company is in the 52% tax bracket, can use classical straight-line depreciation, and requires an after-tax MARR of 6% per year, determine which labeling machine is more economical.

Note: You may want to use your spreadsheet system to solve the following rate-of-return problems.

15.17 Determine the after-tax rate of return for Problem 15.5.

15.18 Calculate the after-tax rate of return for (*a*) Problem 15.6*b* and (*b*) Problem 15.7. Explain the difference in the answers.

15.19 (*a*) What is the after-tax return for the situation in Problem 15.10 where some debt financing is used to fund the purchase? (*b*) What difference does the debt financing make in the after-tax return compared to that for 100% equity financing described in Problem 15.8?

15.20 A tractor and associated farm equipment purchased for $78,000 by Stimson Farms generated an average of $26,080 annually in before-tax cash flow during its 5-year life, which represents a return of 20%. However, the corporate tax expert said the effective NCF was only $15,000 per

year. If the corporation hopes for a return of 10% per year after taxes on all investments, for how many more years should the equipment remain in service to realize the 10% return?

15.21 (*a*) Tabulate the NCF values in Problem 15.12 and estimate the break-even rate of return using PW after-tax analysis. (*b*) Plot the PW values and select the better plan at each of the following after-tax MARR values: 6%, 9%, 10%, and 13% per year.

15.22 Determine the after-tax return for each machine in Problem 15.12 if gross income is $5000 per year.

15.23 In Example 15.4, $B = \$50,000$, $n = 5$, CFBT $= \$20,000$, and $T = 40\%$ for the fiber-optics cable manufacturer. Straight-line depreciation is used to compute $i* = 18.03\%$. Assume the owner wants an after-tax return of 20%. If the tax rate remains at 40%, estimate the value of the (*a*) first cost, (*b*) salvage value, and (*c*) annual SL depreciation at which this will occur. When determining any one of the values above, assume that the other two parameters retain the values specified in Example 15.4.

15.24 Compute the after-tax return for the owners of an offshore diving company. They can purchase equipment to handle a special job for $2500. The equipment will have no salvage value and will last 5 years. They will receive revenue of $1500 in year 1 of ownership and an estimated $300 in revenue each additional year. The effective tax rate is 45%. (*a*) Use classical SL depreciation. (*b*) Use MACRS depreciation.

15.25 If the offshore diving equipment in Problem 15.24*a* is not purchased, another firm will provide the service for the owners and they will make 5% per year after taxes on the $2500 already invested. If this is done, what percent of the $1500 in revenue must they realize in year 1 to make the same return offered by the purchase alternative with the additional $300 in revenue for years 2 through 5?

15.26 Gene-splicing equipment placed in service 6 years ago has been depreciated from $B = \$175,000$ to no book value using the MACRS rates. The system can be continued in use or replaced by a newer technology system with the following data:

$$B = \$40,000 \qquad n = 5 \text{ years} \qquad SV = 0$$

The new system will require $7000 in annual costs, but only $2000 of this is tax-deductible. Assume classical straight-line depreciation is again allowed and will be applied for the challenger.

    The currently owned system could be sold to a new upstart firm for $15,000. But if retained it would be used for 5 more years with an AOC of $6000 and a one-time $9000 upgrade now. The upgrade investment would be depreciated with $n = 3$ years and no salvage value.

(*a*) Use a 5-year planning horizon, an effective tax rate of 40%, and an after-tax MARR of 12% per year to perform the replacement analysis.

(b) Assume the company owner decides to purchase the new machine now regardless of the replacement analysis recommendation. The owner believes the company can sell the new equipment in 5 years for an amount which will make the current decision between the defender and challenger indifferent. Compute the required salvage value in 5 years for the defender and challenger to be equally desirable now. How does this amount compare with the challenger's first cost?

15.27 Perform an after-tax replacement analysis for Bob if his effective tax rate is 40% and his after-tax MARR is 8%. He bought a car 1 year ago for $33,000 to be used exclusively in his business. It is MACRS-depreciated with $n = 3$ years, and the annual expenses average $6000. A salesperson is attempting to sell Bob a new car as a replacement for the same amount of $33,000 while offering a $20,000 trade-in on his 1-year-old car. Bob estimates he would use MACRS with a 3-year recovery period and then sell this replacement for exactly the book value after exactly 3 years. If expenses for the new car average only $5000 per year, should Bob trade business cars?

15.28 Rework Example 15.7 under the assumptions that the trade-in value of the defender is only $200,000 and that the defender's remaining useful life is 10 years after which its salvage value estimate is $75,000 if sold on the international market. Salvage value is not considered in computing depreciation.

15.29 (a) Compare the two plans detailed below using an effective tax rate of 48% and an after-tax return of 8%. (b) Is the decision different from the before-tax result at $i = 15\%$?

	Defender	Challenger
First cost, $	28,000	15,000
AOC when purchased, $	—	1,500
Actual AOC, $	1,200	—
Expected salvage when purchased, $	2,000	3,000
Trade-in value, $	18,000	—
Depreciation	classical SL	classical SL
Life, years	10	8
Years owned	2	—

*T*he following chapters treat a full range of economic-analysis tools. First, the commonly used approach of breakeven analysis for a single project and for two or more alternatives, is detailed. This same chapter includes the description and correct usage of the payback period for an investment.

The decision to fund one or more projects from several independent proposals is discussed using the capital-budgeting approach. Then the setting of the MARR for an analysis is examined for debt capital financing, equity financing, and a mixture of debt-equity financing.

Sensitivity and risk are considered further, and simple probabilistic analysis makes it possible to determine expected values for cash flows. The decision tree approach is used to select the best sequence of alternatives from all the feasible paths. The last chapter further explores decision making under risk and introduces the use of simulation into engineering economy.

CHAPTER

16

Breakeven Analysis and Payback Period

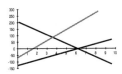

Some typical questions about alternatives and proposed projects are

- How many of these do I have to make in order to break even?

- How long do we have to keep this machine to recover our investment?

- Is there a rate of return at which I can select either alternative?

The answers are determined by estimating the number of units or years or the rate of return at which revenues and costs are equal.

The first question is best addressed using *breakeven analysis.* The number of units produced or percentage utilization of plant capacity is computed using relations for revenue and cost estimates for each alternative. We will learn both linear and nonlinear breakeven analyses.

In the case of the second question—the time that an investment may take to pay for itself—breakeven analysis is a correct approach. Another way to obtain an answer may be to determine the *payback period.* Usually a specified rate of return is used in the analysis. Payback is not *per se* an alternative evaluation technique, yet it is useful as a supplemental decision-making tool.

The ROR at which two PW or AW values are equal answers the third question. We use the same methods here as in rate of return analysis on incremental cash flows and rate-of-return breakeven analysis.

LEARNING OBJECTIVES

Purpose: For one or more alternatives, determine the level of activity necessary to break even, or the number of years needed to realize full investment payback.

One-alternative breakeven	This chapter will help you: 1. Determine the breakeven point for a single alternative.
Two-alternative breakeven	2. Calculate the breakeven point between two alternatives and use it to select one alternative.
Payback period	3. Compute the payback period for a single alternative and identify the shortcomings of payback analysis for alternative selection.

16.1   BREAKEVEN VALUE OF A VARIABLE

It is often necessary to determine the quantity of a variable at which revenues and costs are equal in order to estimate the amount of profit or loss. This quantity, called the *breakeven point, Q_{BE},* is determined using relations for revenue and cost estimation as a function of different quantities Q of a particular variable. The size of Q may be expressed in units per year, percentage of capacity, hours per month, and many other dimensions. We will commonly use units per year for illustration.

Figure 16–1a presents different shapes of the revenue relation, which we identify as R. A linear revenue relation is commonly assumed, but a nonlinear relation is sometimes more realistic, since it can model an increasing per-unit revenue with larger volumes—nonlinear curve 1 in Figure 16–1a. Also, a nonlinear R can recognize that, while additional revenues are possible, decreasing per-unit prices usually prevail at higher quantities—nonlinear curve 2.

Costs, which may be linear or nonlinear, are usually comprised of two components—fixed and variable—as indicated in Figure 16–1b.

> **Fixed costs (FC).**   Include costs such as buildings, insurance, fixed overhead or indirect costs, some minimum level of labor, and capital recovery.
>
> **Variable costs (VC).**   Include costs such as direct labor, materials, indirect and support labor, contractors, marketing, advertisement, and warranty.

The fixed-cost component is usually constant for all values of the variable, so it does not vary with differing production levels or workforce size. Even if no units are produced, fixed costs are incurred, because the plant must be maintained and some employees paid. Of course, this situation could not last for long before the plant would have to shut down to reduce fixed costs. Fixed costs are reduced through improved equipment, information system and workforce utilization, less costly fringe-benefit packages, subcontracting some functions, and so on.

Variable costs change with production level, workforce size, and other variables. It is usually possible to decrease variable costs through better product design, manufacturing efficiency, and sales volume.

There is often a predictable relation between the FC and VC components. For example, a product manufactured on outdated equipment and in outmoded facilities may have a large fixed cost due to high maintenance costs, ineffective processes, and lost production time. Also, the variable cost may be high due to high scrap rates; poor use of employee time; and the inability of the process to utilize improved input materials, software systems, and worker techniques.

When FC and VC are added, they form the total-cost relation, TC. Figure 16–1b illustrates the TC relation for linear fixed and variable costs. Figure 16–1c shows a general TC curve for a nonlinear VC in which unit variable costs decrease as the quantity level rises.

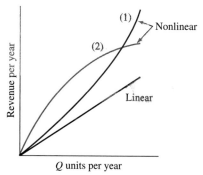

(a) Revenue relations–(1) increasing and
(2) decreasing revenue per unit

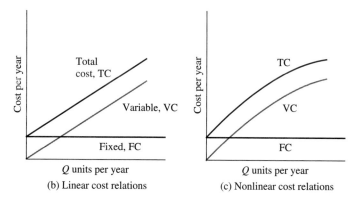

(b) Linear cost relations   (c) Nonlinear cost relations

Figure 16–1   Linear and nonlinear revenue and cost relations used in breakeven analysis.

At some quantity of the variable the revenue and total-cost relations will intersect to identify the breakeven point, Q_{BE} (Figure 16–2). If $Q > Q_{BE}$, there is a predictable profit, but if $Q < Q_{BE}$, there is loss, provided the relations continue to estimate correctly as Q changes in value. If the variable cost per unit is reduced, the TC line will also be lowered (see Figure 16–2) and the breakeven point will decrease in amount; that is, it will take less to break even. This is an advantage because the smaller the value of Q_{BE}, the greater the profit for a given amount of revenue. For linear models of R and VC, the greater the actual quantity sold, the larger the profit.

If nonlinear R or TC models are used, there may be more than one breakeven point. Figure 16–3 presents this situation for two breakeven points. The maximum profit is obtained now by operating at a quantity Q_P which occurs between the two breakeven points and when the distance between the R and TC relations is greatest.

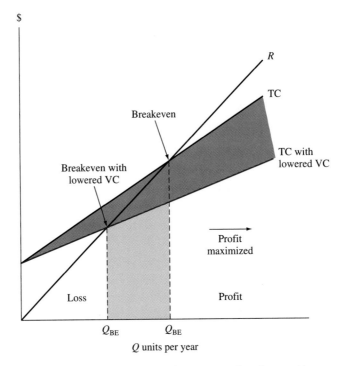

Figure 16-2   Effect on the breakeven point when the variable cost per unit is reduced.

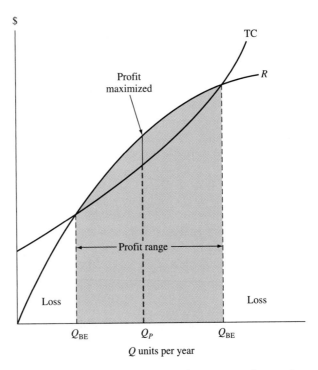

Figure 16-3   Breakeven points and maximum-profit point for a nonlinear analysis.

Of course, no static R and TC relations—linear or nonlinear—are able to estimate exactly the revenue and cost amounts for a product or service over an extended period of time. But it is possible to estimate breakeven points which may be excellent target points for planning purposes.

Example 16.1

A truck disc-brake-components manufacturing line has historically operated at about 80% of line capacity with an output of 14,000 units per month. Decreasing international demand has lowered production to 8000 units per month for the foreseeable future. Use the information below (a) to determine where the reduced level (8000 units per month) will place production relative to the linear breakeven point and the profit and (b) to estimate the variable cost per brake, v, necessary to break even at the 8000-unit level, if revenue per unit, r, and fixed costs remain constant.

Fixed cost: FC = \$75,000 per month

Variable cost: v = \$2.50 per unit

Revenue: r = \$8.00 per unit

Solution

First, develop linear relations for revenue R and total cost TC. Let Q be the quantity in units per month.

$$R = rQ = 8.00(Q) \quad (\$ \text{ per month})$$

$$VC = vQ = 2.50(Q) \quad (\$ \text{ per month})$$

$$TC = FC + VC = 75{,}000 + 2.50(Q) \quad (\$ \text{ per month})$$

To determine the breakeven point, $Q = Q_{BE}$, set revenue equal to total cost, that is, $R = TC$, and solve for Q_{BE}.

$$rQ = FC + VC = FC + vQ$$

$$Q_{BE} = \frac{FC}{r - v} \qquad \text{[16.1]}$$

(a) Solve Equation [16.1].

$$Q_{BE} = \frac{75{,}000}{8.00 - 2.50} = 13{,}636 \text{ units per month}$$

Figure 16–4 indicates that at 14,000 units (80% of capacity) this plant has been producing just above breakeven which occurs at 78% of capacity. (This is a very high production requirement to attain breakeven.)

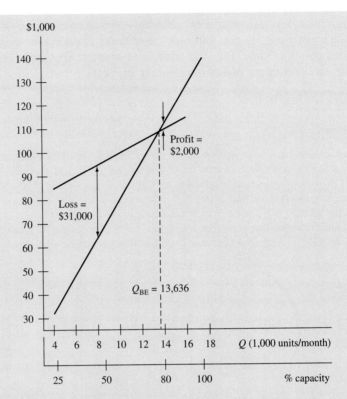

Figure 16–4   Breakeven graph for Example 16.1.

To estimate total profit, subtract total cost from revenue for $Q = 8000$ units.

$$\text{Profit} = R - \text{TC} = rQ - (\text{FC} + vQ)$$

$$= (r - v)\,Q - \text{FC} \qquad\qquad \textbf{[16.2]}$$

$$= (8.00 - 2.50)8000 - 75,000$$

$$= -\$31,000 \qquad .$$

There is a loss of $31,000 per month. Even at the historic output level of 14,000 units per month, Equation [16.2] indicates only a $2000 profit (Figure 16–4).

(b)   One way to determine v if Q_{BE} were 8000 is to state Equation [16.2] at a profit of zero and $Q = 8000$. Solve for v.

$$0 = (8 - v)\,8000 - 75,000$$

$$v = \frac{-11,000}{8000} = -\$1.375 \qquad (\$ \text{ per unit})$$

> Since v is less than zero, it is impossible to break even at 8000 units without reducing the fixed costs and/or increasing the revenue per unit. This plant has some serious problems.

Problems 16.1 to 16.6

16.2 BREAKEVEN POINT BETWEEN TWO ALTERNATIVES

In most economic analyses, one or more of the cost components vary as a function of the number of units. We commonly express the cost relations in terms of the quantity (or other variable) and calculate the value at which the alternatives break even. To find the breakeven point, *the variable must be common to both alternatives,* such as operating cost or production cost. Figure 16–5 illustrates this concept for two alternatives with linear cost relations. The fixed cost, FC, of alternative 2, which may be the annual worth of the initial investment, is greater than that of alternative 1. However, alternative 2 has a smaller variable cost as indicated by its lower slope. The intersection of the two TC lines locates the breakeven point. Thus, if the number of units of the common variable is expected to be greater than the breakeven amount, alternative 2 is selected, since the total cost of the operation will be lower. Conversely, an anticipated level of operation below the breakeven point favors alternative 1.

Instead of plotting the total costs of each alternative and estimating the breakeven point graphically, it may be easier to calculate the breakeven point numerically using engineering economy expressions for the PW or AW. Although the total cost may be expressed as either PW or AW, the latter is usually preferred when the variable units are expressed on a yearly basis. Additionally, AW calculations are simpler when the alternatives under consideration have

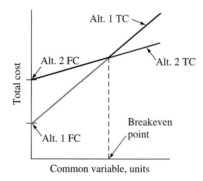

Figure 16–5   Breakeven between two alternatives involving linear cost relations.

different lives. The following steps may be used to determine the breakeven point and to select an alternative:

1. Clearly define the common variable and state its dimensional units.
2. Use AW or PW analysis to express the total cost of each alternative as a function of the common variable.
3. Equate the two cost relations and solve for the breakeven value of the variable.
4. If the anticipated variable level is below the breakeven value, select the alternative with the higher variable cost (larger slope). If the level is above the breakeven point, select the alternative with the lower variable cost. Refer to Figure 16–5.

Example 16.2

A small aerospace company is evaluating two alternatives: the purchase of an auto-matic-feed machine and a manual-feed machine for a product's finishing process. The auto-feed machine has an initial cost of $23,000, an estimated salvage value of $4000, and a predicted life of 10 years. One person will operate the machine at a cost of $12 an hour. The expected output is 8 tons per hour. Annual maintenance and operating cost is expected to be $3500.

The alternative manual-feed machine has a first cost of $8000, no expected salvage value, a 5-year life, and an output of 6 tons per hour. However, three workers will be required at $8 an hour each. The machine will have an annual maintenance and operation cost of $1500. All invested capital is expected to generate a market return of 10% per year before taxes.

(a) How many tons per year must be finished in order to justify the higher purchase cost of the auto-feed machine?

(b) If a requirement to finish 2000 tons per year is anticipated, which machine should be purchased?

Solution

(a) Use the steps above to calculate the breakeven point.

1. Let x represent the number of tons per year finished.
2. For the auto-feed machine the annual variable cost is

$$\text{Annual VC} = \frac{\$12}{\text{hour}} \frac{1 \text{ hour}}{8 \text{ tons}} \frac{x \text{ tons}}{\text{year}}$$

$$= 1.5x$$

The VC is developed in dollars per year, as required for AW analysis. The AW expression for the auto-feed machine is

$$\text{AW}_{\text{auto}} = -23{,}000(A/P,10\%,10) + 4000(A/F,10\%,10) - 3500 - 1.5x$$

$$= \$-6992 - 1.5x$$

Similarly, the annual variable cost and AW for the manual-feed machine are

$$\text{Annual VC} = \frac{\$8}{\text{hour}} (3 \text{ operators}) \frac{1 \text{ hour}}{6 \text{ tons}} \frac{x \text{ tons}}{\text{year}}$$

$$= 4x$$

$$\text{AW}_{\text{manual}} = -8000(A/P,10\%,5) - 1500 - 4x$$

$$= \$-3610 - 4x$$

3. Equate the two cost relations and solve for x, the breakeven amount in tons per year.

$$\text{AW}_{\text{auto}} = \text{AW}_{\text{manual}}$$

$$-6992 - 1.5x = -3610 - 4x$$

$$x = 1353 \text{ tons per year}$$

4. If the output is expected to exceed 1353 tons per year, purchase the auto-feed machine, since its VC slope of 1.5 is smaller than the manual-feed VC slope of 4.

(*b*) Using the logic of the steps detailed above, select the auto-feed machine since $2000 > 1353$ from part (*a*). If the breakeven point had not been determined, it is still possible to select an alternative by substituting the expected production level of 2000 tons per year into the AW relations above. The results are $\text{AW}_{\text{auto}} = \$-9992$ and $\text{AW}_{\text{manual}} = \$-11,610$. The purchase of the auto-feed machine is justified by its numerically larger AW value.

Comment

Rework this example using the PW relations to satisfy yourself that either method results in the same breakeven value. Remember to select the alternative with the smaller slope (i.e., lower variable cost) when the common variable units are above the breakeven point.

Example 16.3

The Jack n' Jill Toy Company currently purchases the metal parts which are required in the manufacture of certain toys, but there has been a proposal that the company make these parts. Two machines will be required for the operation: Machine *A* will cost $18,000 and have a life of 6 years and a $2000 salvage value; machine *B* will cost $12,000 and have a life of 4 years and a $-500 salvage value (carry-away cost). Machine *A* will require an overhaul after 3 years costing $3000. The annual operating cost for machine *A* is expected to be $6000 per year and for machine *B* $5000 per year. A total of four operators will be required for the two machines at a cost of $12.50 per hour per operator. In a normal 8-hour period, the operators and two machines can produce parts sufficient to manufacture 1000 toys. The purchase price of the metal parts is expected to be $0.60 per toy.

Use a MARR of 15% per year to determine the following.

(a) Number of toys to manufacture each year to justify the purchase.

(b) The maximum capital expense justifiable to purchase machine A, assuming all other estimates for machines A and B are as stated. The company expects to produce and sell 125,000 units per year.

Solution

(a) Use steps 1 to 3 above to determine the breakeven point in units per year.

1. Define x as the number of toys produced per year.

2. There are variable costs for the operators and fixed costs for the two machines. The annual VC is

$$\text{VC} = (\text{cost per unit})(\text{units per year})$$

$$= \frac{4 \text{ operators}}{1000 \text{ units}} \frac{\$12.50}{\text{hour}} (8 \text{ hours})x$$

$$= 0.4x$$

The fixed annual costs for machines A and B are the AW amounts.

$$\text{AW}_A = -18,000(A/P,15\%,6) + 2000(A/F,15\%,6)$$
$$- 6000 - 3000(P/F,15\%,3)(A/P,15\%,6)$$

$$\text{AW}_B = -12,000(A/P,15\%,4) - 500(A/F,15\%,4) - 5000$$

3. Equating the annual costs of the purchase option $(0.60x)$ and the manufacture option yields

$$-0.60x = \text{AW}_A + \text{AW}_B - \text{VC}$$

$$= -18,000(A/P,15\%,6) + 2000(A/F,15\%,6) - 6000$$

$$- 3000(P/F,15\%,3)(A/P,15\%,6) - 12,000(A/P,15\%,4)$$

$$- 500(A/F,15\%,4) - 5000 - 0.4x \qquad \text{[16.3]}$$

$$-0.2x = -20,352.43$$

$$x = 101,762 \text{ units per year}$$

A minimum of 101,762 toys must be produced each year to justify the manufacture proposal.

(b) Substitute 125,000 for the variable x and P_A for the to-be-determined first cost of machine A (currently $\$-18,000$) in Equation [16.3]. Solution yields $P_A = \$-35,588$, which is the amount justified as the first cost for machine A if 125,000 toys are produced and other costs are as estimated. This justifiable first cost is substantially larger than the estimated first cost of $18,000 because the expected production of 125,000 toys per year is larger than the breakeven amount of 101,762.

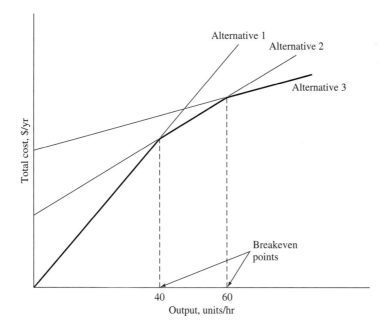

Figure 16–6   Breakeven points for three alternatives.

Even though the preceding examples deal with only two alternatives, the same type of analysis can be performed for three or more alternatives. It is necessary to compare the alternatives by pairs to find their respective breakeven points. The results reveal the ranges through which each alternative is more economical. For example, in Figure 16–6, if the output is expected to be less than 40 units per hour, alternative 1 should be selected. Between 40 and 60 units per hour, alternative 2 is more economical; and above 60 units per hour alternative 3 is favored.

If the variable cost relations are nonlinear, analysis is more complicated. If the costs increase or decrease uniformly, mathematical expressions that allow direct determination of the breakeven point can be developed.

Problems 16.7 to 16.18

16.3   DETERMINATION AND INTERPRETATION OF PAYBACK PERIOD

The *payback period, n_p,* for an asset or alternative is the estimated time, usually in years, it will take for the revenues and other economic benefits to *recover the initial investment and a stated return.* The payback period should never be used as the sole measure of worth to select an alternative. Rather, it is used to provide supplemental information about an alternative. Payback analysis is performed using before-tax or after-tax cash flows. It is best performed using a required return of $i\%$ on the initial investment, but in practice it is often determined with no return requirement, that is, at $i = 0\%$. To find the payback period at a stated return, determine the years n_p (not necessarily an integer) using the expression

$$0 = P + \sum_{t=1}^{t=n_p} \text{NCF}_t(P/F,i,t) \qquad \text{[16.4]}$$

where NCF_t is the estimated net cash flow at the end of each year. If the cash flows are expected to be the same each year, the P/A factor may be substituted and the relation is

$$0 = P + \text{NCF}(P/A,i,n_p) \qquad \text{[16.5]}$$

After n_p years, the cash flows will recover the investment and a return of $i\%$. If, in reality, the asset or alternative is active for more than n_p years, a larger return may result; but if the useful life is less than n_p years, there is not enough time to recover the investment and the return. It is very important to realize that in payback analysis *all net cash flow occurring after n_p years is neglected*. Since this is significantly different from the approach of PW, AW, or ROR analysis, where all cash flows for the estimated alternative life are included in the economic analysis, payback analysis can unfairly bias alternative selection. So use payback analysis only as a supplemental technique to the regular analysis methods.

When $i > 0$ is used to estimate the payback period via Equation [16.4] or [16.5], the n_p value does provide a sense of the risk taken if the alternative is undertaken. For example, if a company plans to produce a product under contract for only 3 years and the payback period for the equipment which must be purchased is computed as 10 years, the company should not undertake the contract. Even in this situation, the 3-year payback period is only supplemental information, not a good substitute for a complete economic analysis using PW or AW computations.

As mentioned above, you may experience the common (but not recommended) practice of determining n_p while neglecting any effects of a required return; that is, use $i = 0\%$ in Equation [16.4] and compute the *no-return payback period* using the relation

$$0 = P + \sum_{t=1}^{t=n_p} \text{NCF}_t \qquad \text{[16.6]}$$

For a uniform net cash-flow sequence, the no-return payback from Equation [16.5] is simply

$$n_p = \frac{P}{\text{NCF}} \qquad \text{[16.7]}$$

The resulting payback is too easily interpreted incorrectly. Computation of n_p from these relations is of little value to an economic decision. Use of the no-return payback period to make alternative decisions is incorrect because

1. The required return is neglected since the time value of money is omitted.
2. All cash flows which occur after the computed payback period which may contribute to the return of the investment are neglected.
3. The selected alternative may be different from that selected by an economic analysis based on PW or AW computations.

Example 16.4

The board of directors of AA International has just approved an $18 million world-wide financial services contract. The services are expected to generate new net annual revenues of $3 million. The contract has a potentially lucrative repayment clause to AA International of $3 million at any time that the contract is canceled by either party during the 10 years of the agreed-to contract period. (a) If the MARR is 15%, compute the payback period. (b) Determine the no-return payback period and compare it with the answer in part (a).

Solution

(a) The net cash flow each year is $3 million with an additional $3 million in any year that the contract is canceled. Assume that this single $3 million payment (call it CV for cancellation value) could be received at any time within the 10-year contract period. Equation [16.5] is slightly altered to include CV.

$$0 = -P + \text{NCF}(P/A,i,n) + \text{CV}(P/F,i,n)$$

In $1,000,000 terms

$$0 = -18 + 3(P/A,15\%,n) + 3(P/F,15\%,n)$$

The 15% payback period is n_p = 15.3 years. During the period of 10 years, the estimated revenues will not deliver the required return.

(b) If the assumption is made that AA International requires no return on its $18 million investment, Equation [16.6] would result in n_p = 5 years as follows (in million $):

$$0 = -18 + 5(3) + 3$$

There is a very significant difference in a MARR = 15% versus 0%. At 15% this contract would have to be in force for 15.3 years, while the no-return payback period is only 5 years. The fact of longer required ownership is always present for $i > 0$ for the obvious reason that the time value of money is considered.

Comment

The results in part (b) substantiate the general guideline that the use of a payback period with i = 0% may be appropriate when investment capital is in very short

supply and management wants recovery of only the invested capital in a short period of time. This is especially true when high project risk is present. The payback
calculation provides the number of years required to recover the invested dollars. But
from the points of view of engineering economic analysis and time value of money,
as mentioned earlier, no-return payback analysis is not a reliable method of alternative selection.

If two or more alternatives are present and payback analysis is used to
indicate that one may be better than the other(s), the second shortcoming of
payback analysis (neglect of cash flows after n_p) may lead to an economically
incorrect decision. When cash flows which would occur after n_p are neglected,
it is possible to favor short-lived assets even when longer-lived assets produce a
higher return. In these cases, PW or AW analysis should always be the primary
decision technique. Comparison of short- and long-lived assets in Example 16.5
illustrates this incorrect use of payback analysis.

Example 16.5

Two equivalent pieces of farm equipment are being considered for purchase. Machine
2 is expected to be versatile and technologically advanced enough to provide income
longer than Machine 1.

	Machine 1	Machine 2
First cost, $	12,000	8,000
Annual income, $	3,000	1,000 (years 1–5), 3,000 (years 6–15)
Maximum life, years	7	15

Mr. James used a return of 15% per year and a microprocessor-based economic
analysis package in the local agricultural extension office. The software utilized
Equations [16.4] and [16.5] to recommend machine 1 because it has a shorter payback period of 6.57 years at $i = 15\%$. The computations are summarized here.

Machine 1: $n_p = 6.57$ years, which is less than the 7-year life.

Equation used: $0 = -12,000 + 3000(P/A,15\%,n)$

Machine 2: $n_p = 9.52$ years, which is less than the 15-year life.

Equation used: $0 = -8000 + 1000(P/A,15\%,5)$
$$+ 3000(P/A,15\%,n_p-5)(P/F,15\%,5)$$

Recommendation: Select machine 1.

Now, use a 15% AW analysis to compare the machines and comment on any difference in the decision.

Solution

For each alternative, consider the cash flows for all years during the estimated (maximum) life.

$$AW_1 = -12,000(A/P,15\%,7) + 3000 = \$116$$

$$AW_2 = -8000(A/P,15\%,15) + 1000 + 2000(F/A,15\%,10)(A/F,15\%,15)$$
$$= \$485$$

Machine 2 is selected since its AW value is numerically larger than that of machine 1 at 15%.

The result is the opposite of the payback period decision. The AW analysis accounts for the increased cash flows for machine 2 in the later years. As illustrated in Figure 16–7, payback analysis neglects all cash-flow amounts which may occur after the payback time has been reached.

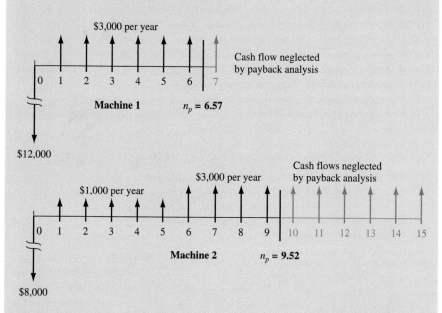

Figure 16–7   Illustration of payback periods and cash-flow estimates, Example 16.5.

Comment

This is a good example of why to use payback analysis only as supplemental risk assessment information. Often a shorter-lived alternative evaluated by payback analysis may appear to be more attractive, when the longer-lived alternative has estimated cash flows later in life which make it more economically attractive using the PW, AW, or ROR value. Therefore, always rely on PW, AW, or ROR analysis to correctly compare alternatives.

Additional Example 16.6 illustrates a very elementary way in which the computations for breakeven and payback analysis may be combined to serve as a supplemental tool in decision making.

Additional Example 16.6
Problems 16.19 to 16.25

ADDITIONAL EXAMPLE

Example 16.6

BREAKEVEN AND PAYBACK ANALYSES, SECTIONS 16.2 AND 16.3   As an energetic, top-selling, not-afraid-of-risks member of the marketing and sales staff at Positive Motion, Inc., Robin, who hopes to be considered in the near future (within 3 years) for director of marketing, needs a "big splash" in sales during this time. Robin took engineering economy in college and now wants to use breakeven and payback analyses to determine how many units of an improved product must be sold over some period of time. It may be a hard sell to convince management that the added investment and costs are worth it in new projected sales.

After consulting with some friends in the engineering and manufacturing departments, Robin has made the following estimates about costs and revenues for the improved model.

Fixed costs: $80,000 in new capital investment now, $1000 in annual upkeep expenses.

Variable costs: $8 per unit manufactured.

Revenues: Twice the unit variable cost for up to 5 sales years and only one-half the cost after 5 years.

Recognizing the pitfalls of payback analysis, Robin, who has decided this is a risk-assessment situation, wants to perform no-return payback analysis with the understanding that any significant cost and revenue cash flows after the payback period will be neglected. (*a*) If Robin estimates new sales at 5000 units per year, find the payback period. Using the results as a basis, should Robin go for it? (*b*) Determine the breakeven sales requirement for the five payback period values of 1 through 5 years. Again, using the results, should Robin go for it?

Solution

Define X as the units sold per year and n_p as the payback period. There are two unknowns and one relation, so it is necessary to establish values of one variable and solve for the other. The following approach will be used. Establish annual cost and revenue relations with no time value of money considered; determine the values of n_p for $X = 5000$ in part (*a*); and use n_p values of 1 through 5 to find the breakeven amounts for part (*b*). We'll pay special attention to the n_p values of 3 years (Robin's aspiration year for director) and 5 years (when the revenue estimates decrease substantially).

Fixed costs:
$$\frac{80,000}{n_p} + 1000$$

Variable costs:
$$8X$$

Revenues:
$$\begin{cases} 16X & \text{for years 1 through 5} \\ 4X & \text{for years 6 on} \end{cases}$$

(a) Set revenues equal to total costs, and determine if $n_p \leq 5$ with revenues at $16X$.

$$\text{Revenues} = \text{total costs}$$

$$16X = \frac{80,000}{n_p} + 1000 + 8X$$

$$n_p = \frac{80,000}{8X - 1000}$$

$$= \frac{10,000}{X - 125} \qquad \text{[16.8]}$$

At $X = 5000$ units per year, the payback period is

$$n_p = \frac{10,000}{4875} = 2.05 \text{ years}$$

Since $n_p < 5$ years, the result is very encouraging. With the horizon of 3 years, Robin should accept the risk involved and go for it.

(b) Solve Equation [16.8] for X, substitute n_p values of 1 through 5, and plot the different breakeven values (Figure 16–8).

$$X = \frac{10,000}{n_p} + 125$$

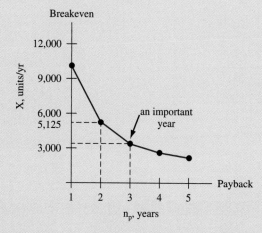

Figure 16–8   Breakeven sales volumes for different payback periods. Example 16.6.

As an illustration for $n_p = 2$ years, the breakeven sales volume is 5125. As in part (*a*), Robin's 3-year horizon is acceptable to go for it, but the decision is quite sensitive to sales estimates as Figure 16–8 clearly indicates. All other estimates occurring as expected, if sales are in the 3000 range, the payback is above 3 years. Robin will have to take the risk and the reward estimates into account in making the decision.

Example 16.6
(Spreadsheet)

BREAKEVEN ANALYSIS, SECTION 16.3   Solve part (*b*) of Example 16.6 using a spreadsheet. Plot the breakeven values for years 1 through 5, and use the graph to determine if Robin has a chance for promotion in 3 years if new sales are 5000 per year.

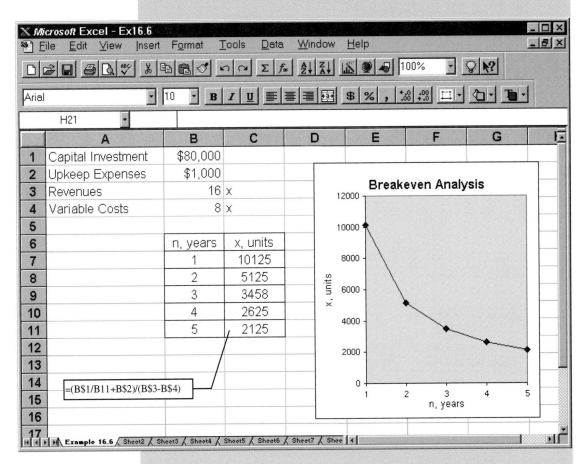

Figure 16–9   Spreadsheet solution for breakeven sales values, Example 16.6*b*.

Solution

Set revenues equal to costs with X identifying the number of units sold per year and n identifying the number of years. Through 5 years, the equation is developed with a revenue of $16 per unit.

$$16X = \frac{80,000}{n} + 1000 + 8X$$

Solve for x in terms of n. (See the relations in the solution of Example 16.6 above for details.)

$$X = \frac{10,000}{n} + 125$$

Figure 16–9 includes a spreadsheet with the values of X in cells C7 through C11 for different n values and a plot of X versus n. If sales are 5000, the n value is slightly larger than 2 years, so Robin has a good chance for promotion.

CHAPTER SUMMARY

The breakeven point for one alternative or between two alternatives is usually expressed in terms such as units per year or hours per month. At the breakeven amount, Q_{BE}, there is indifference to accept or reject the alternative. Use the following decision guideline:

Single alternative   (Refer to Figure 16–2)

Estimated quantity *larger* than Q_{BE} → accept alternative

Estimated quantity *smaller* than Q_{BE} → reject alternative

For two or more alternatives, determine the breakeven value of the common variable X. Use the following guideline to select an alternative:

Two alternatives   (Refer to Figure 16–5)

Estimated level of X is *below* breakeven → select alternative with the higher variable cost (larger slope)

Estimated level of X is *above* breakeven → select alternative with the lower variable cost (smaller slope)

Payback period is the number of years to recover an investment at a stated return. It is primarily used as supplemental to a complete PW, AW, or ROR analysis. Payback does provide a sense of the risk involved, especially when capital is scarce and short investment-recovery times are sought. Payback is *not* a good alternative selection method because it

- Neglects all cash flows which may occur after the computed payback period and which may enhance the rate of return on the initial investment.

- May indicate selection of an alternative different from that based on a PW, AW, or ROR computation over the entire estimated alternative life.

A quick way to estimate initial-investment payback is via no-return payback in which $i = 0\%$. Again, the result is merely supplemental to the engineering economy analysis and should never be used as the sole basis for alternative decisions. In addition to the fallacies above, no-return payback

- Completely neglects any return on investment, since the time value of money is disregarded.

A deceptive outcome of incorrectly used payback comparison of two alternatives can be that the shorter-lived one should seemingly be selected, when, in fact, cash flows estimated after the payback year for the longer-lived alternative make it much more attractive economically. This is why payback is considered supplemental information.

CASE STUDY #6, CHAPTER 16: **ECONOMIC EVALUATION OF ULTRALOW-FLUSH TOILET PROGRAM**

Introduction

In many cities in the southwestern part of the United States, water is being withdrawn from subsurface aquifers faster than it is being replaced. The attendant depletion of groundwater supplies has forced some of these cities to take actions ranging from restrictive pricing policies to mandatory conservation measures in residential, commercial, and industrial establishments. Beginning in 1991, a city undertook a project to encourage installation of ultralow-flush toilets in existing houses. In order to evaluate the cost effectiveness of the program, an economic analysis was conducted.

Background

The heart of the toilet replacement program involved a rebate of 75% of the cost of the fixture (up to $100 per unit), providing the toilet used 1.6 gallons of water per flush or less. There was no limit on the number of toilets any individual or business could have replaced.

Procedure

In order to evaluate the water savings achieved (if any) through the program, monthly water use records were searched for 325 of the household participants, representing a sample size of approximately 13%. Water consumption data were obtained for 12 months before and 12 months after installation of the ultralow-flush toilets. If the house changed ownership during the evaluation period, that account was not included in the evaluation. Since water consumption increases dramatically during the hot summer months for lawn watering, evaporative cooling, car washing, etc., only the winter months of December, January, and February were used to evaluate water consumption before and after installation of the toilet. Before any calculations were made, high-volume water users (usually businesses) were screened out by eliminating all records whose average monthly consumption exceeded 50 CCF (one CCF = 100 cubic feet = 748 gallons). Additionally, accounts which had monthly averages of 2 CCF or less (either before or after installation) were also eliminated because it was believed that such low consumption rates probably represented an abnormal condition, such as a house for sale which was vacant during part of the study period. The 268 records that remained after the screening procedures were then used to quantify the effectiveness of the program.

Results

Water Consumption   Monthly consumption before and after installation of the ultralow-flush toilets was found to be 11.2 CCF and 9.1 CCF, respectively, for an average reduction of 18.8%. When only the months of January and February were used in the before and after calculations, the respective values were 11.0 CCF and 8.7 CCF, resulting in a water savings rate of 20.9%.

Economic Analysis   The following table shows some of the program totals through the first $1^3/_4$ years of the program.

Program summary	
Number of households participating	2466
Number of toilets replaced	4096
Number of persons	7981
Average cost of toilet	$115.83
Average rebate	$76.12

The results in the previous section indicated monthly water savings of 2.1 CCF. For the average program participant, the payback period, n_p, in years with no interest considered is calculated using Equation [16.7].

$$n_p = \frac{\text{net cost of toilets} + \text{installation cost}}{\text{net annual savings for water and sewer charges}}$$

The lowest rate block for water charges is $0.76 per CCF. The sewer surcharge is $0.62 per CCF. Using these values and a $50 cost for installation, the payback period is

$$n_p = \frac{(115.83 - 76.12) + 50}{(2.1 \text{ CCF/month} \times 12 \text{ months})(0.76 + 0.62)/\text{CCF}}$$

$$= 2.6 \text{ years}$$

Less expensive toilets or lower installation costs would reduce the payback period accordingly, while consideration of the time value of money would lengthen it.

From the standpoint of the utility which supplies water, the cost of the program must be compared against the marginal cost of water delivery and wastewater treatment. The marginal cost c may be represented as

$$c = \frac{\text{cost of rebates}}{\text{volume of water not delivered} + \text{volume of wastewater not treated}}$$

Theoretically, the reduction in water consumption would go on for an infinite period of time, since replacement will never be with a less efficient model. But for a worst-case condition, if is assumed the toilet would have a "productive" life of only 5 years, after which it would leak and not be repaired. The cost to the city for the water not delivered or wastewater not treated would be

$$c = \frac{\$76.12}{(2.1 + 2.1 \text{ CCF/month})(12 \text{ months})(5 \text{ years})}$$

$$= \frac{\$0.302}{\text{CCF}} \text{ or } \frac{\$0.40}{1000 \text{ gallons}}$$

Thus, unless the city can deliver water and treat the resulting wastewater for less than $0.40 per 1000 gallons, the toilet replacement program would be considered economically attractive. For the city, the operating costs alone, that is, without the capital expense, for water and wastewater services that were not expended were about $1.10 per 1000 gallons, which far exceeds $0.40 per 1000 gallons. Therefore, the toilet-replacement program was clearly very cost effective.

Case Study Exercises

1. For an interest rate of 8% and a toilet life of 5 years, what would the participant's payback period be?

2. Is the participant's payback period more sensitive to the interest rate used or to the life of the toilet?

3. What would the cost ($ per CCF) to the city be if an interest rate of 6% per year were used with a toilet life of 5 years?

4. From the city's standpoint, is the success of the program sensitive to (*a*) the percentage of toilet cost rebated, (*b*) the interest rate, if rates of 4% to 15% are used, or (*c*) the toilet life, if lives of 2 to 20 years are used?

5. What other factors might be important to (*a*) the participants and (*b*) the city in evaluating whether or not the program is a success?

PROBLEMS

16.1 Fixed costs for Universal Exports are $600,000 annually. Its main-line export is sold at a revenue of $2.10 per unit and has $1.50 variable costs. (*a*) Compute the annual breakeven quantity. (*b*) Plot the revenue and total cost relations, and estimate from your graph the annual profit if 1.3 million units are sold and if 1.8 million units are sold.

16.2 Define the term *average cost per unit,* AC, as total cost divided by volume. (*a*) Derive a relation for AC in terms of fixed and variable costs. Plot AC versus Q for a fixed cost of $60,000 and variable costs of $1.50 per unit. At what sales volume is a $3-per-unit average cost justified? (*b*) If the fixed cost in 1 year increases to $100,000, plot the new AC curve on the same graph and estimate the sales volume needed to justify a $3-per-unit average cost.

16.3 A company which sells reverse osmosis water purifiers has the following fixed- and variable-cost components for its product over a 1-year period.

Fixed Costs, $		Variable Costs, $ per Unit	
Administrative	10,000	Materials	5
Lease cost	20,000	Labor	3
Insurance	7,000	Indirect labor	5
Utilities	3,000	Other overhead	20
Taxes	10,000		
Other operations	50,000		

(*a*) Determine the revenue per unit required to break even if the domestic sales volume is estimated to be 5000 units.

(*b*) If foreign sales of 3000 units can be added to the 5000-unit domestic sales, determine the total profit provided the revenue per unit determined in part (*a*) is realized.

16.4 (*a*) The manufacturer of a toy called Willy Wax has a capacity of 200,000 units annually. If the fixed cost of the production line is $300,000 with a variable cost of $3 and a revenue of $5 per unit, find the percent of present capacity that must be used to break even. (*b*) The toy manufacturer plans to utilize an equivalent of 100,000 units of the 200,000-unit capacity for another product line. This is expected to reduce the fixed cost for Willy Wax to $150,000. What variable cost per unit will make 100,000 units the new breakeven volume? How does this compare with the current variable cost?

16.5 Nonlinear breakeven analysis often uses quadratic relations as the general form of total cost, TC, and revenue, R, relations, that is,

$$aQ^2 + bQ + c$$

where a, b, and c are unknown constants. Since profit is calculated as $R - TC$, a general form of R, TC, and the profit relations may be written.

$$R = dQ^2 + eQ$$

$$TC = fQ^2 + gQ + h$$

$$\text{Profit} = R - TC = aQ^2 + bQ + c \qquad \text{[16.9]}$$

where a through h are now constants to be determined.

As indicated in Figure 16–3, maximum profit occurs at a quantity Q_p where the greatest distance between R and TC occurs. This maximum profit point can be found using calculus; it occurs at the quantity

$$Q_p = \frac{-b}{2a} \qquad \text{[16.10]}$$

Substitution of Q_p into Equation [16.9] yields a maximum profit estimate.

$$\text{Maximum profit} = \frac{-b^2}{4a} + c \qquad \text{[16.11]}$$

(a) Verify the relations in Equations [16.10] and [16.11].

(b) Using the following revenue and total-cost relations, determine the maximum profit amount and the quantity at which it occurs. Indicate these values on a plot of the relations.

$$R = -0.005Q^2 + 25Q$$

$$TC = 0.002Q^2 + 3Q + 2$$

16.6 A consultant to the Weiner Corporation has determined that the total cost can best be described by the relation $TC = 0.002Q^2 + 3Q + 2$ and that revenue is linear with $r = \$25$ per unit. (a) Plot the profit function and determine the value of the maximum profit and the quantity at which it occurs. (b) What is the difference in the quantity at which maximum profit occurs between this solution and that in Problem 16.5?

16.7 Two pumps can be used for pumping a corrosive liquid. A pump with a brass impeller costs $800 and is expected to last 3 years. A pump with a stainless-steel impeller will cost $1900 and last 5 years. An overhaul costing $300 will be required after 2000 operating hours for the brass impeller pump, while an overhaul costing $700 will be required for the stainless-steel pump after 9000 hours. If the operating cost of each pump is $0.50 per hour, how many hours per year must the pump be required to justify the purchase of the more expensive pump? Use an interest rate of 10% per year.

16.8 A corporate headquarters manager has received two proposals from contractors to improve the staff parking areas. Proposal *A* includes filling, grading, and paving at an initial cost of $50,000. The life of the parking lot constructed in this manner is expected to be 4 years with annual costs for maintenance and repainting of strips of $3000. Proposal *B* provides a higher-quality pavement with an expected life of 16 years. The annual maintenance cost will be negligible for the paved parking area, but the markings will have to be repainted every 2 years at a cost of $5000. If the company's current MARR is 12% per year, how much can it afford to spend for the paving contract now so the proposals would just break even?

16.9 Quality Construction is considering the purchase of a small-load tractor for dirt scraping and leveling. The equipment has the following estimates: initial cost of $75,000, a life of 15 years, a $5000 salvage value, an operating cost of $30 a day, and an annual maintenance cost of $6000.

Alternatively, Quality can lease the same tractor and a driver as needed for $210 per day. If the company's minimum attractive rate of return is 12% per year, how many days per year must the scraper be required to justify its purchase?

16.10 Mr. and Mrs. Smith-James have started a home-based direct mail service in antiques. They have collected cost information for the computer system and Internet connection they plan to use for this specialized business. Compare the lease and purchase options at the market rate of return of 10% for the next 4 years. Determine *T*, the number of connect hours per month, necessary to justify purchase of the computer.

Lease alternative	
Computer and peripherals	$800 per month
System software use	$100 per month flat charge
Internet access to specialized databases	Free up to 200 connect hours per month; $1.50 above 200 connect hours
Purchase alternative	
Computer and peripherals	$P = \$10{,}000$, SV $= \$500$, $n = 4$ years
System software use	$100 per month flat charge
Internet access to specialized databases	$4 per connect hour per month

16.11 A textile company is evaluating the purchase of an automatic cloth-cutting machine. The machine will have a first cost of $22,000, a life of 10 years, and a $500 salvage value. The annual maintenance cost of the machine is expected to be $2000 per year. The machine will require one operator at a total cost of $24 an hour. Approximately 1500 yards of material can be cut each hour with the machine. Alternatively, if human labor is used, five workers, each earning $10 an hour, can cut 1000 yards per hour. If the company's MARR is 8% per year, how many yards per year of material must be cut to justify the purchase of the automatic machine? Assume an 8-hour workday and 2000 workdays per year.

16.12 The Johnsons have the opportunity to buy a fire-damaged house in a rural town for what they believe to be a bargain price of $58,000. They estimate that remodeling the house now will cost $12,000, that annual taxes will be approximately $1800 per year, that utilities will cost $1500 per year, and that the house must be repainted every 3 years at a cost of $400. At the present time, resale houses in the town are closing for $40 per square foot, but they expect this price to increase by $1.50 per square foot per year in the foreseeable future. They expect to continually lease the house for $5500 per year from this year on until it is sold. The house has 2500 square feet, and the Johnsons want to make 8% per year on their investment. (*a*) How long must they own and lease the house to break even? (*b*) What should the selling price be at breakeven time?

16.13 The Rawhide Tanning Company wishes to evaluate the economics of an in-house water-testing laboratory instead of sending samples to independent laboratories for analysis. If the lab were equipped so that all tests could be conducted in-house, the initial cost would be $25,000. A part-time technician would be employed at a salary of $13,000 per year. Cost of utilities, chemicals, etc., is estimated at $5 per sample. If the lab is partially equipped, the initial cost will be lower at $10,000. The part-time technician will have an annual salary of $5000. The cost of in-house sample analysis will be only $3 per sample, but since all tests cannot be conducted in-house, outside testing will be required at a cost of $20 per sample. Any laboratory equipment purchased will have a life of 12 years.

  If Rawhide Tanning continues to outsource its sample testing, the cost will average $55 per sample. If the MARR is 10% per year, how many samples must be tested each year in order to justify (*a*) the complete laboratory and (*b*) the partial laboratory? (*c*) If the company realistically expects to test 175 samples per year, which of the three options do you recommend?

16.14 A city engineer is considering two methods for lining water-holding tanks. A bituminous coating can be applied at a cost of $2000. If the coating is touched up after 4 years at a cost of $600, its life can be extended another 2 years. As an alternate, a plastic lining may be installed with a useful life of 15 years. If the city wants a rate of return of 5% per year on municipal capital, how much money can be spent for the plastic lining so that the two methods just break even?

16.15 A family plans to build a new house. It must decide between purchasing a lot in the city or in the rural suburbs. A 1000-square-meter lot in the city will cost $100,000 in the area in which they want to buy. If they purchase a lot outside the city limits, a similar parcel will cost only $20,000. For the house they plan to build, they expect annual taxes to amount to $3200 per year if they build in the city and only $1500 per year in the suburbs. If they purchase the rural property, they will have to drill a well for $4000. With their own well, they will save $150 per year in water charges, but they expect the city to provide water to their area in 5 years, after which time they will purchase the city water. They estimate that the increased travel distance will cost $325 the first year, $335 the second year, and amounts increasing by $10 per year. Using a 25-year analysis period and an interest rate of 6% per year, how much extra could the family afford to spend on a house in the rural suburbs and still have the same total investment? Assume that the city lot or rural parcel can be sold at the same price as its initial cost.

16.16 The Rite family will insulate the attic of their home to prevent heat loss. They are considering R-11 and R-19 insulation. They can install R-11 for $1600 and R-19 for $2400. They expect to save $150 per year in heating and cooling expenses if R-11 is installed. How much extra must they save in utility expenses per year in order to justify the R-19 insulation, if they expect to recover the extra cost in 7 years? The current interest rate is 6%.

16.17 A waste-holding lagoon situated near the main plant receives sludge on a daily basis. When the lagoon is full, it is necessary to remove the sludge to a site located 4.95 kilometers from the main plant. Currently, when the lagoon is full, the sludge is removed by pumping it into a tank truck and hauling it away. This process requires the use of a portable pump that initially cost $800 and has an 8-year life. The company supplies the individual to operate the pump at a cost of $25 per day, but the truck and driver must be rented for $110 per day.

Alternatively, the company has a proposal to install a pump and pipeline to the remote site. The pump would have an initial cost of $600, a life of 10 years, and a cost of $3 per day to operate. The company's MARR is 15%. (*a*) If the pipeline will cost $3.52 per meter to construct, how many days per year must the lagoon require pumping in order to justify construction of the pipeline? (*b*) If the company expects to pump the lagoon one time per week, how much money could it afford to spend on the pipeline in order to just break even? Assume a pipeline life of 10 years.

16.18 A building contractor is evaluating two alternatives for improving the exterior appearance of a small commercial building that he is renovating. The building can be completely painted at a cost of $2800. The paint is expected to remain attractive for 4 years, at which time repainting will be necessary. Every time the building is repainted, the cost will increase by 20% over the previous time.

As an alternative, the building can be sandblasted now and every 10 years at a cost 40% greater than the previous time. The remaining life of the building is expected to be 38 years. If the company's MARR is 10% per year, what is the maximum amount that could be spent now on the sandblasting alternative so that the two alternatives will just break even? Use present-worth analysis to solve this problem.

16.19 Why is it incorrect to use payback-period analysis as the primary measure of worth to choose between two alternatives?

16.20 (*a*) Use spreadsheet analysis to determine the number of years that an investor must retain ownership of a commercial property to make a current market return of 8% per year. The purchase price is $60,000 with taxes of $1800 the first year, increasing by $100 each year until sold. Assume the property must be retained for at least 2 years and that the sales price is estimated at $90,000 for year 3 and beyond. (*b*) If the property is not sold during the year determined above, find the year in which a $120,000 sales price must be obtained to just return the 8%.

16.21 (*a*) Determine the payback period for an asset that initially costs $8000, has a salvage value of $500 when sold, and generates a cash flow of $900 per year. The required return is 8% per year. (*b*) If the asset will be in service for an estimated 12 years, should it be purchased?

16.22 Darrell Enterprises expects to use newly constructed additions to its main building between 20 and 40 years. Determine the number of years that the two additions described below must be retained in service to make a 10%-per-year return. The estimated annual extra revenue is $6700, and the estimated salvage values apply for all years.

	Addition 1	Addition 2
First cost, $	30,000	5,000
Annual expenses, $	1,000	2,000
Salvage value, $	5,000	−200
Maximum life, years	40	20

16.23 The Sundance Detective Agency has purchased new surveillance equipment with the following estimates. The year index is $k = 1, 2, 3, \ldots$

$$\text{First cost} = \$1050$$

$$\text{Annual maintenance cost} = 70 + 5k$$

$$\text{Extra annual revenue} = 200 + 50k$$

$$\text{Salvage value} = \$600 \text{ for all years}$$

(*a*) Calculate the payback period to make a return of 10% per year.

(*b*) Should the equipment be purchased if the actual useful life is 7 years?

16.24 The annual lease cost of a delivery truck for Rundell R$_x$ Drugs is quoted as $3300 per year payable in full at the beginning of each year. This quote is good for the foreseeable future; however, no refund is given for partial-year leases. If Rundell purchases the same truck, it will cost $1700 now with a monthly payment of $300 for 4 years. If purchased, the truck can be sold for an estimated average of $1200 regardless of the length of time of ownership. The truck is expected to increase net revenue by $400 per month. The drug company must pay operating, maintenance, and insurance expenses for a leased or purchased truck, so these costs are equal for both alternatives. Apply a nominal return of 12% per year and determine

(*a*) For the purchase alternative and the lease plan separately, how many months are required to make the return.

(*b*) If a truck has an expected life of 6 years, whether it should be leased or purchased.

16.25 In Section 16.3 we learned that a shorter-lived alternative may be selected over a longer-lived one if only payback analysis is used to make the economic decision. Choose between the following two alternatives using payback-period analysis first and then present-worth analysis. Explain how and why it is possible to make different decisions. List any assumptions you had to make to complete these analyses.

A new college graduate has inherited $600,000. The family investment manager has presented two options. In both cases, the entire $600,000 is invested immediately. The first—option 1—is expected to yield $100,000 per year after taxes, while the second—option 2—will start low at a $15,000 after-tax yield next year which is expected to grow at the rate of 20% per year. A 6% after-tax return is desired. Option 2 has a maximum investment time of 16 years. Currently, option 1 is expected to produce the yield for 8 years only.

chapter

17

Capital Rationing among Independent Proposals

A company or business always has limited capital funds that may be divided between several proposed projects with the anticipation that the present value or rate of return of cash flows for the selected projects will be maximized. The proposals are commonly considered to be independent of each other; that is, if a particular proposal is selected, it does not prohibit any other proposal from selection. This is a basic departure from the situation in all the earlier chapters where the alternatives are mutually exclusive; that is, selection of one alternative precludes selection of other alternatives.

Project selection under capital rationing is referred to as the capital-budgeting problem. In this chapter, we will learn the present-worth approach to proposal selection under simple and quite specific capital-rationing conditions. More advanced texts illustrate additional characteristics of capital rationing.

Purpose: Select from several independent proposals when there is a stated capital investment limit.

Capital budgeting

Independent proposals with equal lives

Independent proposals with unequal lives

Linear program model

This chapter will help you:

1. Describe the basic characteristics of a capital-budgeting problem.

2. Use PW analysis to select from several different-life independent proposals.

3. Use PW analysis to select from several different-life independent proposals.

4. Understand the solution approach for the capital-budgeting problem using linear programming (optional section).

17.1   A BASIC OVERVIEW OF CAPITAL BUDGETING

For the evaluation methods we learned in Chapters 5 through 9 (PW, AW, ROR, and B/C), the alternatives were *mutually exclusive* of each other. For example, two trucks, two milling machines, three conveyor systems, four different bridge designs, etc., were compared with the intent to select one of the alternatives. (For a review, see Sections 8.1 through 8.4 for ROR analysis of mutually exclusive alternatives.) In order to differentiate, in this chapter on capital budgeting we use the terms projects or proposals (not alternatives) and each proposal describes an *independent* opportunity to invest capital. Selection of one independent project does not eliminate other projects from consideration. (In some cases, there may be *contingent projects*—the case where one project among independent projects may be selected in lieu of another specified project—and *dependent projects*—when selection of one project forces the inclusion of another project. These complicating situations are beyond our introductory scope. In practice, these complications can often be circumvented by forming a *package of related projects* which is evaluated as independent along with the other proposals.)

Most organizations have the opportunity to select from several independent capital-investment projects from time to time. Naturally, however, the funds available for investment are limited to an amount which is generally predetermined. Selection of projects may occur annually, quarterly, or on an ongoing basis. This is the most basic *capital-budgeting* problem or situation, and it has the following characteristics:

1. Several independent proposed projects are identified.
2. Each proposal is either selected entirely or not selected; that is, partial investment in a project is not possible.
3. A stated budgetary constraint restricts the total amount invested. Budget constraints may be present for the first year only or for several years.
4. The objective is to maximize the time value of the investments using some measure of worth such as PW or ROR.

By nature, independent projects are usually quite different from one another. For example, a city government may develop several projects to choose from: drainage project, city park, street-widening project, and an upgraded public bus system. A company may have several proposals: new warehousing facility, new information system, and acquisition of another firm. The typical capital-budgeting problem is illustrated in Figure 17–1. For each independent proposal there is an initial investment (the *capital expense*, as we called it in Chapter 15), project life, and estimated net cash flows, which can include a salvage value. Selection using the best total PW value of selected proposals evaluated at the MARR is the most common criterion, but ROR works equally well.

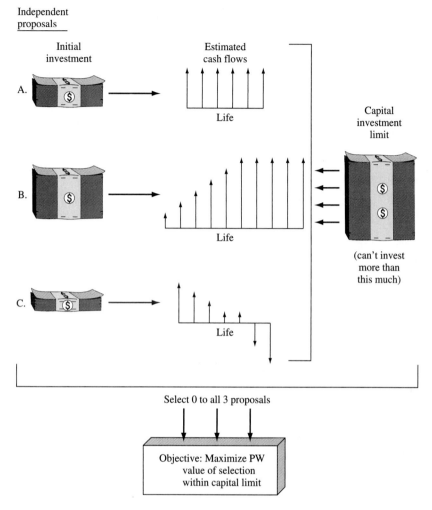

Figure 17–1   Basic characteristics of a capital-budgeting problem.

In order to solve a capital-budgeting problem, we will usually determine PW values at the MARR. We will use the inflation-adjusted MARR, as discussed in Section 12.1, for all capital-budgeting situations. Since capital funds are limited, some projects will likely not be funded. The return on a rejected project represents an opportunity that will not be realized. This *opportunity cost,* expressed as a percent return or interest rate, is the amount forgone, since the capital, and possibly other resources, are not available to fund one project because they have already been committed to some other project(s) which earn at least the MARR. As an illustration, if $500,000 is available and all of it is committed to projects estimated to earn 10% or more, and if one more project is left unfunded and is expected to return 9%, the opportunity cost is 9%. For limited capital conditions, as discussed in this chapter, the inflation-adjusted MARR is assumed to be the opportunity cost.

The discussion above assumes that all estimates are made with certainty. However, in reality most proposals have nondeterministic cash flows, involve different risk levels, and have investment requirements for more than the initial year. For simplicity, we assume the proposals are of equal risk, have only initial-year investments, and will generate the estimated (deterministic) cash flows.

Problem 17.1

17.2 CAPITAL RATIONING USING PW ANALYSIS FOR EQUAL-LIFE PROPOSALS

To select from proposals that all have the same expected life and to invest no more than a stated amount of capital, b, it is necessary to initially formulate all *mutually exclusive bundles*—one proposal at a time, two at a time, etc. Each feasible bundle must have a total investment that does not exceed b. One of these bundles is the do-nothing bundle which includes no proposals at all. Then the PW of each bundle is determined at the MARR. The bundle with the largest PW value is selected. The do-nothing bundle will always have initial investment and PW values of zero.

To illustrate the development of mutually exclusive bundles, consider these four proposals.

Proposal	Initial Investment
A	$10,000
B	5,000
C	8,000
D	15,000

If the capital-budget limit is $b = \$25,000$, there are 12 feasible bundles to be evaluated. The total number of bundles for m proposals is calculated using the relation 2^m. The number increases rapidly with m. For $m = 4$, there are $2^4 = 16$ bundles, but, in this example, four of them (*ABD*, *ACD*, *BCD*, and *ABCD*) have investment totals which exceed $25,000. The acceptable bundles are

Proposals	Total Initial Investment	Proposals	Total Initial Investment
A	$10,000	AD	$25,000
B	5,000	BC	13,000
C	8,000	BD	20,000
D	15,000	CD	23,000
AB	15,000	ABC	23,000
AC	18,000	Do nothing	0

The procedure to solve a capital-budgeting problem using PW analysis is

1. Develop all feasible mutually exclusive bundles which have a total initial investment that does not exceed the capital limit constraint b.
2. Estimate the net cash-flow sequence NCF_{jt} for each bundle j and each year t from 1 to the expected project life n_j. Refer to the initial investment for bundle j at time $t = 0$ as NCF_{j0}.
3. Compute the present-worth value, PW_j, for each bundle at the MARR using

$$PW_j = \text{PW of bundle net cash flows} - \text{initial investment}$$

$$= \sum_{t=1}^{t=n_j} NCF_{jt}(P/F,i,t) - NCF_{j0} \qquad \text{[17.1]}$$

4. Select the bundle with the largest PW_j value.

Selecting the maximum bundle PW_j value means that this bundle produces a return larger than any other bundle at the MARR. Any bundle with $PW_j < 0$ is discarded, because it does not produce a return greater than the MARR. This procedure is illustrated next for equal-life proposals.

Example 17.1

A&A Printing has $20,000 to invest next year on any or all of the five proposals in Table 17–1. Each proposal has an expected life of 9 years. Select from these independent proposals if a 15% return is expected and as much of the $20,000 as possible should be invested.

Solution

Use the procedure above with $b = \$20,000$ to select one bundle which maximizes present worth. A value $NCF_{jt} > 0$ is a net cash inflow and $NCF_{jt} < 0$ is a net cash outflow.

1. There are $2^5 = 32$ possible bundles. The eight bundles which require no more than $20,000 initial investments are identified in columns 2 and 3 of Table 17–2. The $21,000 investment for proposal E eliminates it.

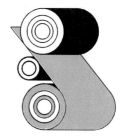

Table 17–1	Five equal-life independent proposals considered for investment		
Proposal	**Initial Investment**	**Annual Net Cash Flow**	**Proposal Life, Years**
A	$10,000	$2,870	9
B	15,000	2,930	9
C	8,000	2,680	9
D	6,000	2,540	9
E	21,000	9,500	9

Table 17–2	Summary of present-worth analysis of equal-life independent proposals			
(1)	**(2)**	**(3)**	**(4)**	**(5)**
Bundle, j	**Proposals Included**	**Initial Investment, NCF_{j0}**	**Annual Net Cash Flow, NCF_j**	**Present Worth, PW_j**
1	A	$-10,000	$2,870	$+3,694.49
2	B	-15,000	2,930	-1,019.21
3	C	-8,000	2,680	+4,787.89
4	D	-6,000	2,540	+6,119.86
5	AC	-18,000	5,550	+8,482.38
6	AD	-16,000	5,410	+9,814.36
7	CD	-14,000	5,220	+10,907.75
8	Do nothing	0	0	0

2. The annual net cash flows, column 4, are the sum of individual proposal net cash flows in a bundle for the 9 years.

3. Use Equation [17.1] to compute the present worth of each bundle. Here $n_j = 9$. Since the annual net cash flows are the same for a bundle, simplify the symbol NCF_{jt} to NCF_j (except for the initial investment NCF_{j0}). This reduces PW_j to

$$PW_j = NCF_j(P/A,15\%,9) - NCF_{j0}$$

4. Column 5 of Table 17–2 summarizes the PW_j values at $i = 15\%$. Note that bundle 2 does not return 15%, since $PW_2 < 0$. The largest measure of worth is $PW_7 = \$10,907.75$; therefore, invest $14,000 in proposals C and D. This leaves $6000 of uncommitted capital funds.

Comment

This analysis assumes that any funds not used by a bundle in the initial investment, for example, the $6000 uncommitted by selecting bundle 7, will return the MARR by placing it in some other unspecified investment opportunity.

Actually, the return on bundle 7 exceeds 15% per year, because $PW_7 > 0$. In fact, if we compute the actual rate of return i^*, using the relation $0 = -14,000 + 5220$ $(P/A,i^*,9)$, we obtain $i^* = 34.8\%$, which significantly exceeds MARR $= 15\%$.

Problems 17.2 to 17.4

17.3   CAPITAL RATIONING USING PW ANALYSIS FOR DIFFERENT-LIFE PROPOSALS

Usually independent proposals do not have the same expected life. When lives are different, solution of the capital-budgeting problem assumes that each proposal's initial investment will be made for the period of the longest-lived proposal, n_L, *with reinvestment of all positive cash flows at the (current, inflation-adjusted) MARR after year* n_j *through year* n_L. In capital-rationing problems, no computation is made for reinvestment in the identical or like-kind proposal at the end of the project life, as we did when evaluating mutually exclusive alternatives in previous chapters. Therefore, the use of the least common multiple (LCM) of proposal lives as the evaluation period for different-life proposals is not appropriate. If further investment opportunities are known to exist after a proposal's life is reached, the associated net cash-flow estimates should be explicitly considered in the economic analysis.

It is correct to use Equation [17.1] to select bundles by PW analysis for proposals of differing lives using the procedure in the previous section. The rationale is demonstrated after the next example, which illustrates independent proposal selection for different-life proposals.

Example 17.2

For a MARR $= 15\%$ per year and $b = \$20,000$, select from the following independent proposals.

Proposal	Initial Investment	Annual Net Cash Flow	Project Life, Years
A	$10,000	$2,870	6
B	15,000	2,930	9
C	8,000	2,680	5
D	6,000	2,540	4

Solution

The varying proposal life values make the net cash flows vary over a bundle's life, but the PW solution procedure is the same as that in Example 17.1. Of the

$2^4 = 16$ bundles, eight are economically feasible here since $b = \$20,000$. Their PW values by Equation [17.1] are summarized in column 6 of Table 17–3. As an illustration, for bundle 7 the PW expression is

$$PW_7 = -14,000 + 5220 \, (P/A,15\%,4) + 2680 \, (P/F,15\%,5) = \$2235.60$$

Select bundle 7 (proposals C and D) for a $\$14,000$ investment, since PW_7 is the largest value.

Table 17–3   Present-worth analysis for different-life independent proposals, Example 17.2

(1) Bundle, j	(2) Proposals Included	(3) Initial Investment, NCF_{j0}	(4) Year, t	(5) NCF_{jt}	(6) Present Worth, PW_j
1	A	$\$-10,000$	1–6	\$2,870	$ +861.52
2	B	−15,000	1–9	2,930	−1,019.21
3	C	−8,000	1–5	2,680	+983.90
4	D	−6,000	1–4	2,540	+1,251.70
5	AC	−18,000	1–5	5,550	+1,845.41
			6	2,870	
6	AD	− 16,000	1–4	5,410	+2,113.43
			5–6	2,870	
7	CD	−14,000	1–4	5,220	+2,235.60
			5	2,680	
8	Do nothing	0		0	0

Comment

It is only coincidental that this selection of different-life proposals is the same as that in Example 17.1 where all lives are equal at 9 years. It is the combination of net cash-flow estimates *and* the number of years over which they are received that determines the best economic bundle.

**Example 17.2
(Spreadsheet)**

For MARR = 15% per year and $b = \$20,000$, select from the following independent proposals using a spreadsheet analysis.

Proposal	Initial Investment	Annual Net Cash Flow	Project Life, Years
A	$10,000	$2,870	6
B	15,000	2,930	9
C	8,000	2,680	5
D	6,000	2,540	4

Solution

Figure 17–2 presents a spreadsheet solution with fundamentally the same information as in Table 17–3. Bundle 7 (projects C and D) has the largest PW value (row 14 cells). The present worth or net present value spreadsheet function is used to determine the PW for each bundle. Though the computation of PW is much easier using a spreadsheet system, it is still necessary for the analyst to initially develop the mutually exclusive bundles and associated net cash flows to load the spreadsheet.

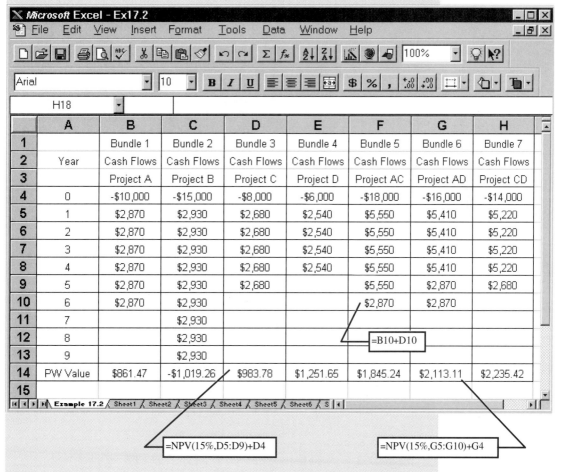

Figure 17–2   Spreadsheet-based selection from independent different-life proposals, Example 17.2.

Actually, it is easy to demonstrate why PW evaluation using Equation [17.1] for different-life proposals is correct. Refer to Figure 17–3 which uses the general layout of a two-proposal bundle for illustration. Let's assume each bundle has the same net cash flows each year so we can use the P/A factor for PW computation. Define n_L as the life of the longest-lived proposal. At the end of each shorter-lived project in the bundle, the bundle has a total future worth of $NCF_j(F/A,MARR,n_j)$ as determined for each project. Now, assume reinvestment from year n_{j+1} through year n_L (that's a total of n_L-n_j years) at the MARR. The assumption of the return at the MARR is important; this PW approach does not necessarily select the correct proposals if the return is not at MARR. The results are the two future-worth arrows in year n_L in Figure 17–3. Finally, compute the bundle PW value in the initial year. In Figure 17–3 this is the bundle $PW = PW_A + PW_B$. In general form, the bundle j present worth is

$$PW_j = NCF_j(F/A,MARR,n_j)(F/P,MARR,n_L-n_j)(P/F,MARR,n_L) \quad \text{[17.2]}$$

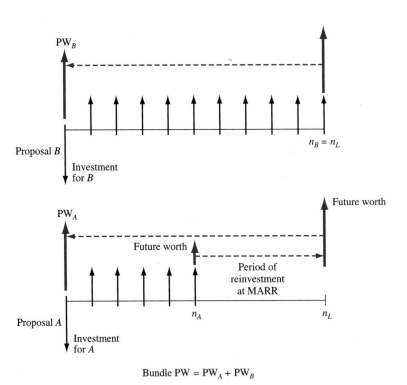

Figure 17–3   Representative cash flows used to compute PW for a bundle of two independent different-life proposals using Equation [17.1].

Let's substitute the symbol i for the MARR and use the factor formulas (inside the front cover) to simplify to

$$PW_j = NCF_j \frac{(1 + i)^{n_j} - 1}{i}(1 + i)^{n_L - n_j}\frac{1}{(1 + i)^{n_L}}$$

$$= NCF_j \left[\frac{(1+i)^{n_j} - 1}{i(1+i)^{n_j}} \right] \qquad [17.3]$$

$$= NCF_j(P/A,i,n_j)$$

Since the bracketed expression in Equation [17.3] is the $(P/A,i,n_j)$ factor, computation of PW_j of each bundle for n_j years assumes reinvestment at the MARR of all positive net cash flows until the longest-lived proposal is completed in year n_L.

To demonstrate numerically, consider bundle $j = 7$ in Example 17.2. The evaluation is in Table 17–3, and the net cash flow is pictured in Figure 17–4. At 15% the future worth in year 9, life of the longest-lived bundle, is

$$F = 5220(F/A,15\%,4)(F/P,15\%,5) + 2680(F/P,15\%,4) = \$57,111.36$$

The present worth at the initial investment time is

$$PW = -14,000 + 57,111.36(P/F,15\%,9) = \$2236$$

Allowing for the slight round-off error, we find that the PW value is the same as PW_7 in Table 17–3 and Figure 17–2. This demonstrates the reinvestment assumption in Equation [17.3]. So, indeed, it is not necessary to use the LCM

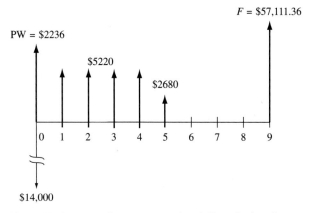

Figure 17–4   Initial investment and cash flows for bundle 7, proposals C and D, Example 17.2.

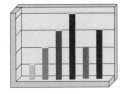

of bundles of independent projects to solve a capital-budgeting problem, provided the company is able to (or decides to) reinvest all positive net cash flows at the same MARR used to select from the current proposals. If this is not the case, the PW analysis must be conducted using the *LCM of all proposal lives.*

Proposal selection can be accomplished equally well using the *incremental-rate-of-return* procedure detailed in Section 8.6. Procedurally, once all feasible bundles are developed, they are ordered by increasing initial investment. The do-nothing bundle is always the first one to be listed. Determine the incremental rate of return i^* on the first bundle relative to the do-nothing bundle and the return for each incremental investment and incremental net cash-flow sequence on all other bundles. If any bundle has an incremental return less than the MARR, it is removed from consideration. The last justified increment indicates the best bundle. This approach will result in the same answer as a present-worth procedure.

There are a number of incorrect ways to apply the rate-of-return method to the capital-budgeting problem, but this procedure of incremental analysis of mutually exclusive bundles ensures a correct result—one that is the same as the PW selection. Advanced texts and current journal articles offer more detail on the incremental rate-of-return method and its uses.

Problems 17.5 to 17.11

17.4   CAPITAL BUDGETING PROBLEM FORMULATION USING LINEAR PROGRAMMING (OPTIONAL SECTION)

If you have studied the fundamentals of operations research or math programming in another course, you may be interested in learning how the basic capital-budgeting problem is stated in the form of the linear programming model. The problem is actually formulated using the integer linear programming (ILP) model, which means simply that all relations are linear and that the unknown variable, x, can take on only integer values. In fact, in this case, the variables can only take on the values 0 or 1, which makes it a special case called the 0-or-1 ILP model. The formulation in words follows.

Maximize: Sum of PW of net cash flows of independent proposals.

Constraints:

- Capital investment constraint, that is, the sum of initial investments in independent proposals, must not exceed a specified limit.
- Each proposal is completely selected or not selected at all.

For the math formulation, define b as the capital investment limit, as before, and let x_k ($k = 1$ to m proposals) be the variables to be determined. If $x_k = 1$, the proposal k is completely selected; if $x_k = 0$, proposal k is not selected. Note that the subscript k represents each *independent proposal,* not a mutually exclusive bundle.

If we refer to the sum of PW of the net cash flows as Z, the ILP or math programming formulation is

Maximize: $$\sum_{k=1}^{k=m} PW_k x_k = Z$$

Constraints: $$\sum_{k=1}^{k=m} NCF_{k0}\, x_k \le b \qquad\qquad \text{[17.4]}$$

$$x_k = 0 \text{ or } 1 \qquad \text{for } k = 1, 2, \ldots, m \text{ proposals}$$

The present worth of each proposal, PW_k, is calculated using Equation [17.1] at a MARR $= i$. That is,

$$PW_k = PW \text{ of proposal net cash flows for } n_k \text{ years}$$

$$= \sum_{t=1}^{t=n_k} NCF_{kt}(P/F,i,t) - NCF_{k0} \qquad\qquad \text{[17.5]}$$

The restriction that each $x_k = 0$ or 1 is the proposal indivisibility constraint. It is what makes this a 0-or-1 ILP formulation. Solution is best accomplished by a linear programming software package which treats the ILP model. However, it is possible to develop the ILP formulation and understand how the proposal selection works, as illustrated in the next example.

Example 17.3

For Example 17.2, (a) formulate the capital-budgeting problem using the math programming model presented in Equation [17.4], and (b) insert the solution (select proposals C and D) into the model to verify that it does indeed maximize present worth.

Solution

(a) Define the subscript $k = 1$ through 4 for the four proposals A through D, respectively, and use $b = \$20,000$ in the general model of Equation [17.4].

Maximize: $$\sum_{k=1}^{k=4} PW_k x_k = Z$$

Constraints: $$\sum_{k=1}^{k=4} NCF_{k0} x_k \le 20,000$$

$$x_k = 0 \text{ or } 1 \qquad \text{for } k = 1 \text{ through } 4$$

Calculate the PW_k for the estimated net cash flows using $i = 15\%$ and Equation [17.5].

Proposal, k	Net Cash Flow, NCF_{kt}	Life, n_k	Factor, $(P/A,15\%,n_k)$	Initial Investment, NCF_{k0}	Proposal PW, PW_k
A	$2,870	6	3.7845	$10,000	$ 861.52
B	2,930	9	4.7716	15,000	−1,019.21
C	2,680	5	3.3522	8,000	983.90
D	2,540	4	2.8550	6,000	1,251.70

Now, substitute the PW_k values for each proposal into the model, and put the proposal investments in the budget constraint. We have the complete 0-or-1 ILP formulation.

Maximize: $861.52x_1 - 1019.21x_2 + 983.90x_3 + 1251.70x_4 = Z$

Constraints: $10,000x_1 + 15,000x_2 + 8000x_3 + 6000x_4 \quad < 20,000$

$x_1, x_2, x_3,$ and $x_4 = 0$ or 1

(b) The conclusion of Example 17.2—select proposals C and D—is written as

$$x_1 = 0 \qquad x_2 = 0 \qquad x_3 = 1 \qquad x_4 = 1$$

for a PW value of $2235.60. There are larger PW values possible, but none less than $20,000 for the total investment. So this solution maximizes the value of Z while meeting the constraints on b and the x variables.

Problems 17.12 to 17.15

CHAPTER SUMMARY

Capital is always a scarce resource, and it must be rationed among several competing proposals using specific economic and noneconomic criteria. Capital budgeting involves proposed projects, each with an initial investment and cash flows estimated over the life of the project. The lives may be the same or different among proposals. The fundamental capital-budgeting problem has some specific characteristics (Figure 17–1).

- Selection is made from among independent proposals.
- Each proposal must be selected entirely or not at all.
- Maximizing the present worth of the cash flows is the objective.
- The total initial investment is limited to a specified (capital budget) amount.

The present worth is commonly used for evaluation, but it is first necessary to formulate the proposals into mutually exclusive bundles, including the do-nothing bundle. The initial investment for each bundle cannot exceed the capital-budget limit. There are a maximum of 2^m bundles for m proposals, but, depending on how much capital is available, some bundles may exceed the investment limitation. Calculate the present worth of each bundle, and select the one bundle with the largest present-worth value. This indicates the proposals to fund, provided there are no other conditions imposed.

This chapter concludes with an optional section on the formulation of the capital-budgeting problem using the linear programming model. Specifically the 0-or-1 integer linear programming model allows us to use the proposal estimates directly to formulate the problem; that is, mutually exclusive bundles need not be developed.

PROBLEMS

17.1 Describe the differences and similarities between decision making for mutually exclusive alternatives and for independent proposals. Give an example of each type of decision-making situation from your own life experiences.

17.2 (*a*) Determine which of the following independent proposals should be selected for investment if $30,000 is available and the inflation-adjusted MARR is 10% per year. Use the present-worth method to make your decision.

Proposal	Initial Investment	Net Cash Flow	Life, Years
A	$10,000	$3,950	8
B	12,000	2,400	8
C	18,000	5,750	8
D	22,000	3,530	8
E	32,000	8,200	8

(*b*) If these were mutually exclusive alternatives, how would you perform the present-worth analysis? Which alternative would be selected?

17.3 The capital fund for new investment at The Systems Corporation is limited to $100,000 for next year. Select any or all of the following proposals if a MARR of 15% is established by the board of directors.

Proposal	Initial Investment	Annual Net Cash Flow	Life, Years	Salvage Value
I	$25,000	$ 6,000	4	$ 4,000
II	30,000	9,000	4	−1,000
III	50,000	15,000	4	20,000

17.4 A pension fund manager must determine how to invest a total of $100,000 in the following independent proposals. Use a spreadsheet-based PW analysis and a 15% return requirement to make the decision from a purely economic perspective.

Proposal	Initial Investment	Annual Net Cash Flow	Life, Years	Salvage Value
A	$25,000	$ 6,000	4	$ 4,000
B	20,000	9,000	4	0
C	50,000	15,000	4	20,000

17.5 Resolve Problem 17.2 using a spreadsheet system. Retain the total investment limit of $30,000, but utilize the fact that further estimation has shown the proposal lives to be different from each other, as follows. Is the selection of proposals the same as that for equal lives?

Proposal	A	B	C	D	E
Life, Years	3	8	5	12	8

17.6 The management of East-West Music Productions has decided it has capital funds to invest in any three of four independent proposals. Each proposal has an initial investment of $10,000 and the 18% present-worth value shown below. Select the three proposals which offer the best investment opportunity.

Proposal	Life, Years	Proposal PW at 18%
1	13	$−1,820
2	5	375
3	10	1,820
4	8	25

17.7 Use the present-worth procedure to solve the following capital-budgeting problem: Select up to three of the four proposals at the top of the next page if the current MARR after taxes is 10% and the available capital budget is (*a*) $16,000 and (*b*) $24,000. The last year in which an after-tax NCF value is presented indicates the life estimate of the proposal. Note that the NCF patterns vary considerably from one proposal to another.

		NCF After Taxes				
Proposal	Investment	1	2	3	4	5
1	$ 6,000	$1,000	$1,700	$2,400	$ 3,100	$ 3,800
2	10,000	500	500	500	500	10,500
3	8,000	5,000	5,000	2,000		
4	10,000	0	0	0	15,000	

17.8 In the context of the four proposals in Problem 17.7, use proposals 3 and 4 to demonstrate numerically that equalization of the evaluation period is not necessary to make a correct decision with PW analysis of independent different-life proposals, if reinvestment is assumed at the MARR (10% in this case) through the life of the longest-lived bundle.

17.9 Rework Example 17.1 by considering the rate of return on the incremental net cash flows.

17.10 Use (a) the rate-of-return method and (b) the present-worth method to solve the following capital-budgeting problem for a budget limit of $5000 and a MARR of 14%. You may wish to utilize your spreadsheet system to determine the i^* and PW values once the mutually exclusive bundles are formulated.

Proposal	Required Investment	Estimated Net Cash Flow	Expected Life, Years
1	$3,000	$1,000	5
2	4,500	1,800	5
3	2,000	900	5

17.11 Use rate-of-return analysis to select from the independent proposals in Problem 17.3 using a minimum return criterion of MARR = 15%.

17.12 Set up the capital-budgeting problem for Example 17.1 using integer linear programming. Solve the problem manually or using a canned math programming software package.

17.13 Set up the linear programming formulation for Problem 17.2(a).

17.14 (a) Use the information in Problem 17.3 to develop the integer linear programming model which will solve the capital-budgeting problem. (b) What change is necessary in the model formulation if management has decided to allow partial investment in the proposals, assuming that cash flows can also be partitioned?

17.15 Set up the math programming model and the solution values for the variables x and the value of Z for Problem 17.7.

Determining a Minimum Attractive Rate of Return

Project risky or not?

Until now we have provided the minimum attractive rate of return (MARR) for alternative and proposal evaluation. In this chapter we learn more about the MARR and how to determine its value by using estimates of the actual cost (in terms of interest rate) of the capital funds employed by the corporation for its project operations, capital investments, and other activities. Additionally, some of the fundamental reasons that the MARR varies over time and for different types of alternatives are discussed.

In this chapter, we add to the basics of debt and equity financing introduced in Section 15.1. We learn about the advantages of each type of financing, why a balance between internal and external financing is important, and how each type of financing contributes to the overall cost of capital and how it impacts the MARR. We will further discuss after-tax analysis and how interest paid on loans and bonds is important, since it is tax-deductible in the United States.

The cost of capital is a complex estimate developed separately for debt and equity-financed funds and then combined to determine a weighted average cost of capital (WACC). Then, the WACC is translated into the MARR. Some aspects of project risk are discussed here; more are presented in the following chapters.

LEARNING OBJECTIVES

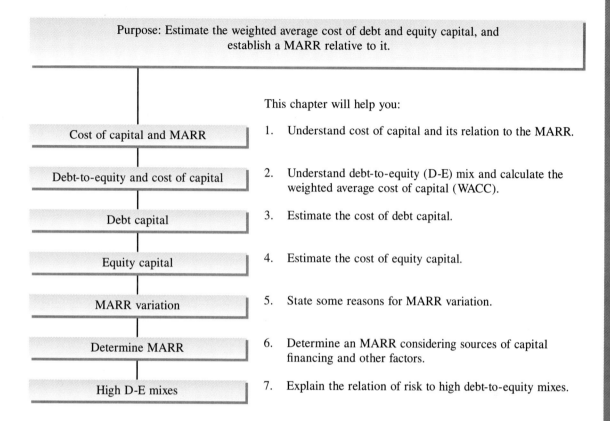

Purpose: Estimate the weighted average cost of debt and equity capital, and establish a MARR relative to it.

This chapter will help you:

Cost of capital and MARR

1. Understand cost of capital and its relation to the MARR.

Debt-to-equity and cost of capital

2. Understand debt-to-equity (D-E) mix and calculate the weighted average cost of capital (WACC).

Debt capital

3. Estimate the cost of debt capital.

Equity capital

4. Estimate the cost of equity capital.

MARR variation

5. State some reasons for MARR variation.

Determine MARR

6. Determine an MARR considering sources of capital financing and other factors.

High D-E mixes

7. Explain the relation of risk to high debt-to-equity mixes.

18.1   RELATING COST OF CAPITAL AND MARR

To determine a realistic MARR, the cost of each type of capital financing is initially computed separately and then the proportion from debt and equity sources is weighted in order to estimate the average interest rate paid for investment capital made available. This percentage is called the *cost of capital*. The MARR is then set equal to this cost, and sometimes higher depending upon the perceived risk inherent to the area where the capital may be invested, the *financial health* of the corporation, and many other factors active when a MARR is determined. If no specific MARR is established as a guideline by which alternatives are accepted or rejected, the *de facto* MARR is effectively set by the project net cash-flow estimates and limits on capital funds, as discussed in Chapter 17 on capital budgeting. That is, the MARR is, in reality, the opportunity cost, which is the $i*$ of the *first-rejected project due to limited capital funds*. Reread Section 17.1 to review the interpretation of opportunity cost.

To understand cost of capital, let's first review the two primary sources of corporate capital that we introduced in Section 15.1.

Debt financing represents borrowing outside of company resources, with the principal to be paid at a stated interest rate following a specified time schedule. Debt financing includes borrowing via *bonds, loans,* and *mortgages.* The lender takes no direct risk concerning the principal repayment and interest, nor does the lender share in the profits made using the funds. The amount of outstanding debt financing is indicated in the liabilities section of the corporate balance sheet. (If necessary, refer to Appendix B for a review of the balance sheet.)

Equity financing represents the use of corporate money comprised of the funds of owners and retained earnings. Owners' funds are further classified as common and preferred stock sale proceeds for a public corporation or owners' capital for a private (non-stock-issuing) company. Retained earnings are funds previously retained in the corporation for capital investment. The amount of equity is indicated in the net-worth section of the corporate balance sheet.

To illustrate the relation between cost of capital and the MARR, assume an extensive computer system project will be completely financed by a $5,000,000 bond issue (100% debt financing) and assume the dividend rate on the bonds is 8%. Therefore, the cost of debt capital is 8% as shown in Figure 18–1. This 8% is the fundamental MARR. Only if necessary should management increase this MARR in increments which reflect its desire for added return and its perception of risk. For example, management may add to this MARR an amount it believes necessary for all investments or capital commitments in a specific area. Suppose this amount is 2%. This increases the expected return to 10% (Figure 18–1). Also, if the risk associated with this computer system investment is considered

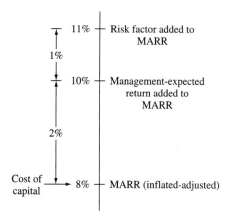

Figure 18–1   Fundamental relation between cost of capital and MARR.

substantial enough to warrant an additional 1% return requirement, the MARR will now be 11%.

The recommended approach is to utilize the cost of capital, 8% in this illustration, as the MARR, and calculate the alternative's $i*$ value using estimated net cash flows over the expected life. Suppose the computer system alternative is estimated to yield a return of 11%. Then, any anticipated return and risk factors are considered in order to determine if the 3% above the MARR of 8% is sufficient to justify the capital investment. After these considerations, if the project is rejected, the effective MARR is now 11%. This is the opportunity cost concept discussed several times previously—the rejected project $i*$ has established the actual MARR to be 11%, not 8%, for this computer system alternative.

Problems 18.1 to 18.3

18.2   DEBT-EQUITY MIX AND WEIGHTED AVERAGE COST OF CAPITAL

The *debt-to-equity (D-E) mix* identifies the percentages of debt and equity financing for a corporation. A company with a 40–60 D-E mix has 40% of its capital originating from debt capital sources (bonds, loans and mortgages) and 60% derived from equity sources (stocks and retained earnings).

Most projects are funded with a combination of debt and equity capital made available specifically for the project or taken from a corporate *pool of capital*. The *weighted average cost of capital (WACC)* of the pool is estimated by the relative fractions (or percentages) from debt and equity sources. If the fractions are known exactly, these fractions are used to estimate WACC; otherwise the historical fractions for each source are used in the relation:

$$\text{WACC} = \text{(equity fraction)(cost of equity capital)}$$
$$+ \text{ (debt fraction)(cost of debt capital)} \qquad [18.1]$$

The two *cost* terms are expressed as percentage interest rates.

Since virtually all public corporations have a mixture of capital sources, the WACC is a value between the debt and equity costs of capital. If the fraction of each type of equity financing—common stock, preferred stock, and retained earnings—is known, Equation [18.1] is expanded to include each equity component.

$$\text{WACC} = \text{(common stock fraction)(cost of common stock capital)}$$
$$+ \text{(preferred stock fraction)(cost of preferred stock capital)} \qquad [18.2]$$
$$+ \text{(retained earnings fraction)(cost of retained earnings capital)}$$
$$+ \text{(debt fraction)(cost of debt capital)}$$

The WACC value can be computed using before-tax or after-tax values for cost of capital; however, using the after-tax method is the correct one since debt financing has a distinct tax advantage.

Figure 18–2 indicates the usual shape of cost-of-capital curves. If 100% of the capital is derived from equity or from debt sources, the WACC equals the

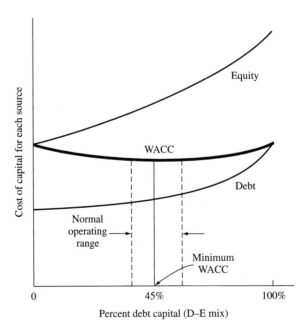

Figure 18–2   General shape of different cost-of-capital curves.

cost of capital of the source of funds. There is virtually always a mixture of capital sources involved for any capitalization program. As an illustration, Figure 18–2 indicates a minimum WACC at about 45% debt capital. Most firms operate over a range of D-E mixes. For example, a range of 30% to 50% debt financing for some companies may be very acceptable to lenders with no increases in risk or MARR. Yet, another company may be considered "risky" with only 20% debt capital. It takes knowledge about and trust in a management's ability, and knowledge about the current projects, to determine a reasonable operating range for the D-E mix for a particular company.

Example 18.1

A new program in genetics engineering will require $1 million in capital. The chief financial officer (CFO) has estimated the following amounts of financing at the indicated after-tax interest rates:

Common stock sales	$500,000 at 13.7%
Use of retained earnings	$200,000 at 8.9%
Debt financing through bonds	$300,000 at 7.5%

Historically, this company has financed projects using a D-E mix of 40% from debt sources costing 7.5%, and 60% from equity sources costing 10.0%. Compare the historical WACC value with that for this current genetics program.

Solution

Equation [18.1] is used to estimate the historical WACC.

$$\text{WACC} = 0.6(10) + 0.4(7.5) = 9.0\%$$

For the current program, the equity financing is comprised of 50% common stock ($500,000 out of $1.0 million) and 20% retained earnings, with the remaining 30% from debt sources. The program WACC by Equation [18.2] is

$$\text{WACC} = \text{stock portion} + \text{retained earnings portion} + \text{debt portion}$$

$$= 0.5(13.7) + 0.2(8.9) + 0.3(7.5) = 10.88\%$$

The current program is estimated to have a higher WACC than historically experienced (10.88% versus 9.0%).

Estimates of after-tax or before-tax cost of capital may be made using the tax rate, T, in the relation

$$\text{After tax cost} = (\text{before tax cost})(1-T)$$

This relation may be used for estimating equity capital cost or debt cost separately or for the WACC rate.

Problems 18.4 to 18.8

18.3 COST OF DEBT CAPITAL

Debt financing includes borrowing via bonds, loans, and mortgages. The dividend or interest paid is tax-deductible, so it reduces taxable income and therefore decreases U.S. federal taxes, as discussed in Chapter 15. The annual net cash flow (NCF) after-tax estimates are used to estimate the value of $i*$, which is the cost of debt capital. (Use the internal rate-of-return (IRR) function on your spreadsheet system to determine $i*$.)

**Example 18.2
(Spreadsheet)**

A total of $500,000 debt capital will be raised by issuing five hundred $1000 8%-per-year 10-year bonds. If the effective tax rate of the company is 50% and the bonds are discounted 2% for quick sale, compute the cost of debt capital (*a*) before taxes and (*b*) after taxes from the company perspective.

Solution

(*a*) The annual bond dividend is $1000(0.08) = $80, and the 2% discounted sales price is $980 now. Using the company perspective, find the $i*$ at which

$$0 = 980 - 80(P/A,i*,10) - 1000(P/F,i*,10)$$

The before-tax cost of debt capital is $i* = 8.3\%$, which is slightly higher than the 8% dividend, because of the 2% sales discount.

(*b*) With the allowance to reduce taxes by deducting the interest on borrowed money, a tax savings of $80(0.5) = $40 per year is realized. The effective bond dividend to substitute into the equation in part (*a*) for $i*$ is now $80 − $40 = $40. Solving for $i*$ after-taxes reduces the after-tax cost of debt capital by nearly one-half to 4.27%.

Comment

Figure 18–3 presents a spreadsheet solution to this example for both before-tax (column B) and after-tax analysis (column C).

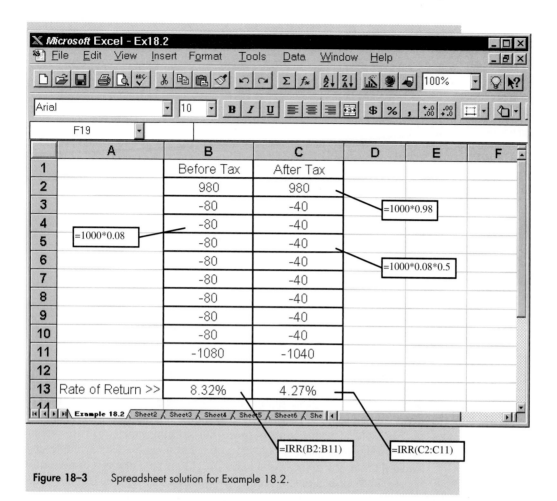

Figure 18–3   Spreadsheet solution for Example 18.2.

Example 18.3

The Roundtree Company plans to purchase a $20,000 10-year-life asset. Company owners have decided to put $10,000 down now and borrow $10,000 at an interest rate of 6% on the unpaid balance. The simplified repayment scheme will be $600 in interest each year with the entire $10,000 principal paid in year 10. What is the after-tax cost of debt capital if the effective tax rate is 50%?

Solution

The cash outflow for interest on the $10,000 loan includes an annual tax-deductible amount of 0.5($600) = $300 and the loan repayment of $10,000 in year 10. The following relation is developed to estimate a cost of debt capital of 3.0%.

$$0 = 10,000 - 300(P/A,i^*,10) - 10,000(P/F,i^*,10)$$

> **Comment**
> Note that the 6% annual interest on the $10,000 loan is not the WACC because 6% is paid only on the borrowed funds and 50% of the capital is from equity financing. Nor is the 3% determined here the WACC, since it is only the cost of debt capital.

Problems 18.9 to 18.12

18.4   COST OF EQUITY CAPITAL

Equity capital is usually obtained from the following sources:

   Sale of preferred stock
   Sale of common stock
   Use of retained earnings

The cost of each type of financing is estimated separately and entered into the WACC computation. A summary of one commonly accepted way to estimate each source's cost of capital is presented here. There are additional methods for estimating the cost of equity capital via common stock.

Issuance of *preferred stock* carries with it a commitment to pay a stated dividend annually. The cost of capital is the stated dividend percentage, for example, 10%, or the dividend amount divided by the price of the stock, that is, a $20 dividend paid on a $200 share is a 10% cost of equity capital. Preferred stock may be sold at a discount to speed the sale, in which case the actual proceeds from the stock should be used as the denominator. For example, if a 10% dividend preferred stock with a value of $200 is sold at a 5% discount for $190 per share, there is a cost of equity capital of $10\%/0.95 = 10.53\%$. This value may also be calculated as $(\$20/\$190) \times 100\% = 10.53\%$.

Estimating the cost of equity capital for *common stock* is more involved. The dividends paid are not a true indication of what the stock issue will actually cost the corporation in the future. Usually a valuation of the common stock is used to estimate the cost. If R_e is the equity cost of capital (in decimal form)

$$R_e = \frac{\text{first-year dividend}}{\text{price of stock}} + \text{expected dividend growth rate}$$

$$= \frac{DV_1}{P} + g \qquad\qquad [18.3]$$

The growth rate g is an estimate of the return that the shareholders expect to receive from owning stock in the company. Stated another way, it is the compound growth rate on dividends that the company believes is required to attract stockholders. For example, assume a multinational corporation plans to raise capital through its U.S. subsidiary for a new plant in South America by selling $2,500,000 worth of common stock valued at $20 each. If a 5% or $1 dividend is planned for the first year and an appreciation of 4% per year is anticipated for future dividends, the cost of capital for this common-stock issue from Equation [18.3] is 9%.

$$R_e = \frac{1}{20} + 0.04 = 0.09$$

The *retained-earnings* cost of equity capital is usually set equal to the common-stock cost, since it is the shareholders who will realize any returns from projects in which retained earnings are invested.

Once the cost of capital for all planned equity sources is estimated, the WACC is calculated using Equation [18.2]. Dividends paid on preferred and common stocks are not tax-deductible.

A second and optional method to estimate the cost of common-stock capital uses the *capital asset pricing model (CAPM)* in lieu of Equation [18.3]. Because of the fluctuations in stock prices and the higher return demanded by some corporation's stocks compared to others, this valuation technique has gained popularity. The cost of equity capital from common stock, R_e, using CAPM is

$$R_e = \text{risk-free return} + \text{premium above risk-free return}$$
$$= R_f + \beta(R_m - R_f) \qquad\qquad \text{[18.4]}$$

where β = volatility of a company's stock relative to other stocks in the market ($\beta = 1.0$ is the norm)

R_m = return on stocks in a defined market portfolio measured by a prescribed index

The term R_f in Equation [18.4] is usually the quoted U.S. Treasury bill rate, since it is considered a 'safe investment'. The term $(R_m - R_f)$ is the premium paid above the safe or risk-free rate. The coefficient β (beta) indicates how the stock is expected to vary compared to a selected portfolio of stocks in the same general market area, often the Standard and Poor's 500 stock index. If $\beta < 1.0$, the stock is less volatile, so the resulting premium can be smaller; when $\beta > 1.0$, larger price movements are expected, so the premium is increased.

Security is a word which identifies a stock, bond, or any other instrument used to develop capital. Figure 18–4 is a plot of a market security line, which is a linear fit via regression analysis to indicate the expected return for different β values. When $\beta = 0$, the risk-free return R_f is acceptable (no premium). As β increases, the premium return requirement grows. Beta values are published periodically for most stock-issuing corporations. Once complete, this estimated cost of equity capital can be included in the WACC computation in Equation [18.2].

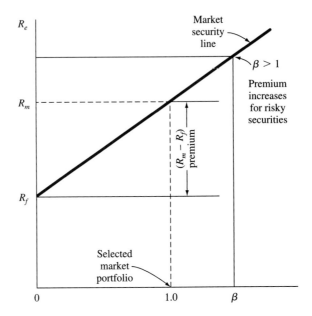

Figure 18–4   Expected return on common-stock issue using CAPM.

Example 18.4

MegaComputers plans to develop and market new software technology. A common-stock issue is a possibility if the cost of equity capital is below 15%. MegaComputers, which has a historical beta value of 1.7, uses CAPM to determine the premium of its stock compared to other public software corporations. The security market line indicates that a 5% premium above the risk-free rate is desirable. If U.S. Treasury bills are paying 7%, estimate the cost of common-stock capital.

Solution

The premium of 5% represents the term $(R_m - R_f)$ in Equation [18.4]. Then,

$$R_e = 7.0 + 1.7(5.0) = 15.5\%$$

Since this cost slightly exceeds 15%, MegaComputers should possibly use some mix of debt and equity financing for the new venture.

Problems 18.13 to 18.18

18.5 VARIATIONS IN MARR

Before reading this section, please refer back to Figure 1–6, where we first introduced the MARR, also called the *hurdle rate* for a project. As mentioned previously, the MARR is set relative to the cost of capital. However, setting the MARR is not a very exact process, because the debt and equity capital mix changes over time and from project to project. Yet, in all cases some MARR should be established for comparison with the estimated ROR of proposed projects. Also, the MARR is not a nonvarying value established corporatewide. Rather, it is altered for different opportunities and types of projects. For example, a corporation may use a MARR of 10% for depreciable assets and a MARR of 20% for diversification investments, such as purchasing smaller companies, land, etc.

The MARR varies from one project to another and through time because of factors such as the following:

Project risk.   Where there is more risk (perceived or actual) associated with an area of proposed projects, the tendency is to set a higher MARR. This is encouraged by the higher cost of debt capital commonly experienced in obtaining loans for projects considered risky, which usually means there is concern that the venture won't be able to fully realize its projected revenue requirements.

Investment opportunity.   If management is determined to diversify or invest in a certain area, the MARR may be lowered to encourage investment with the hope of recovering lost revenue or profit in other areas. This common reaction to investment opportunity can create havoc when the guidelines developed in texts such as this are strictly applied in an economy study. Flexibility becomes very important.

Tax structure.   If corporate taxes are rising (due to increased profits, capital gains, local taxes, etc.), pressure to increase the MARR is present. Use of after-tax analysis may assist in eliminating this reason for a fluctuating MARR, since accompanying business expenses will tend to decrease taxes and, therefore, reduce after-tax costs.

Limited capital.   As debt and equity capital become limited, the MARR is increased and management begins to look closely at the life of the project. As the demand for limited capital exceeds supply (capital budgeting), the MARR may tend to be set higher. The opportunity cost has a large role in determining the MARR actually used to make accept/reject decisions.

Market rates at other corporations.   If the rates increase at other firms that are used for comparison, a company may alter its MARR upward in response. These variations are often based on changes in market interest rates, which directly impact the cost of capital. If we consider the government as a 'corporation,' a typical standard is the current rates charged by a particular government agency.

Example 18.5

Shaun, who owns a small business in residential yard and landscape services, plans to purchase some new equipment. He has gone to his regular bank to apply for a small-business equipment loan. Identify some factors that might change the MARR that the bank would expect of Shaun's company. Base the comments on the five

factors mentioned in this section and indicate the possible direction of MARR change.

Solution

There are many considerations, but some typical ones are

Project risk. The MARR may be increased if there has been a severe drought in the area for the last 2 years, thus reducing the need for Shaun's services. Revenue generation capability is at risk in this situation.

Opportunity. The MARR increases if, for example, several other firms offering similar services have applied for a loan at the same bank during the last 6 months.

Taxes. If the local community had just removed personal and household services from the list of items subject to sales tax, Shaun would need less revenue, so the MARR may be lowered slightly.

Capital limitation. If all other equipment owned by Shaun's company was bought with his own funds and there are no outstanding loans, but additional equity capital is not available, the MARR should tend to be lower now for debt capital.

Market rates. If Shaun's local bank obtains its small-business loan money from a large regional bank, and the market rate to the local bank had just been increased substantially, the MARR will rise due to a *tighter money* situation caused by the higher market rate.

Problems 18.19 and 18.20

18.6   ESTABLISHING A MARR VALUE FOR DECISION MAKING

A correctly performed engineering economy study uses an MARR equal to the cost of the capital committed to the alternative. If a corporation's own funds are used exclusively, the cost of equity capital should be the MARR interest rate. If a combination of debt and equity capital is applied, the calculated WACC should be the MARR. Also, the risks associated with an alternative should be treated separately from the MARR determination with computations for the measure of worth under risk, as discussed in Chapter 20. This all leads to the guideline that the MARR should not be arbitrarily increased to account for the various types of risk associated with the cash-flow estimates. However, the MARR is often set above the WACC or the relevant cost of capital because *management does try to account for risk by increasing the MARR.*

We will use the terminology and definitions developed in this and previous chapters to identify general guidelines for establishing MARR values based on equity, debt, and combinations of the two sources of capital.

i_e = interest rate associated with the corporation's own funds (this is the cost of equity capital or the return on equity funds available and in the investment pool)

i_d = cost of debt capital from all sources

WACC = after-tax weighted average cost of capital

It is usual that $i_d > i_e$ because it costs more to use someone else's funds than your own. However, this may not be the case if, depending on the timing, debt capital is cheaper than the estimated cost of acquiring equity funds. See Example 18.1, for example. It is possible to make the determination of the MARR (without risk considerations) quite complex, but the simple approach is to establish the *MARR between* i$_e$ *and WACC*. The following situations describe how to establish the MARR more precisely for most of the *capital-available* and *capital limited* environments you are likely to experience.

Capital available.   This describes the situation when one mutually exclusive alternative *must* be selected from two or more, or when independent proposals with $i^* \geq$ MARR may be accepted without a strictly adhered to capital budget.

- If financing arrangements are specifically known, use MARR = i_e. (The debt capital cost is specifically known here, so it is not considered again when determining the MARR.)
- If financing specifics are unknown and/or financing is from a combination of equity and debt capital, use MARR = WACC.

Capital limited.   This describes the situation when a strict investment limit exists for mutually exclusive alternative selection, or when a capital budget is set for independent proposal selection (Chapter 17).

- If only corporate-owned (equity) funds are committed, use MARR = i_e.
- If a combination of debt and equity funds are needed, use MARR = WACC.

Of course, MARR values between these guideline boundaries are determined based on current circumstances. If the simple approach is used, that is, $i_e \leq$ MARR $\leq$ WACC is established as the operating range, then noneconomic factors, including project risk, are considered separately as each economy study is performed.

As indicated in Section 18.1, the pressure to impute risk into the MARR is strong in practice, and often some multiplier times the relevant cost-of-capital value is used, such as MARR = 1.1 (WACC) or MARR = 1.8 (i_e). Again, this is not a recommended method because it compromises the accuracy and effort taken to determine how much capital is actually costing. But the fact remains that some managers find this a convenient basis for decisions without doing additional analysis under risk of cash-flow estimates. (See Chapter 20 for more on this.)

Example 18.6

Bring-Hum Enterprises has two mutually exclusive alternatives A and B with incremental IRR values of $i^*_A = 9.2\%$ and $i^*_B = 5.9\%$. The financing scenario is yet unsettled, but it will be one of the following: Plan 1—Use all equity funds, which are currently earning about 8% for the corporation, Plan 2—Use funds from the corporate capital pool which is 25% debt capital costing 14.5% and the remainder from the same equity funds mentioned above. Debt capital is currently high because the company has not made its projected revenue on common stock for the last five quarters, and banks have increased the borrowing rate for Bring-Hum.

Make the economic decision on alternative A versus B under each financing scenario.

Solution

The capital is available for one of the two mutually exclusive alternatives. For Plan 1, 100% equity, the financing is specifically known, so the cost of equity capital is the MARR, that is, MARR $= i_e = 8\%$. Only alternative A is acceptable; alternative B is not since the estimated return of 5.9% does not exceed this MARR.

Under financing Plan 2, with a D-E mix of 25-75,

$$\text{WACC} = 0.25(14.5) + 0.75(8.0) = 9.625\%$$

Now, neither alternative is acceptable since both i^* values are less than MARR = WACC = 9.625%. The selected alternative should be do-nothing, unless one alternative absolutely must be selected in which case noneconomic factors must also be considered. Alternative A probably is more favorable since it returns 9.2%.

Comment

If the value of stock is the primary consideration, only i_e is used for the MARR, even if the financing specifically includes debt financing. Provided the company is not too heavily debt laden, this is a good way to measure the return on equity estimated for after-tax net cash flows. The analysis in this example would not change. However, if Plan 1 for specific financing included some percentage of debt capital, the 8% cost of equity capital would still be the established MARR.

Problems 18.21 and 18.22

18.7 EFFECT OF DEBT-EQUITY MIX ON INVESTMENT RISK

The D-E mix was introduced in Section 18.2. As the proportion of debt capital increases, the calculated cost of capital decreases based on tax advantages of debt capital. *However, the leverage offered by larger debt-capital percentages increases the risks taken by the company.* Then additional financing using debt (or equity) sources gets more difficult to justify and the corporation can be placed in a situation where it owns a smaller and smaller portion of itself. Inability to obtain operating and investment capital means increased difficulty for the company and its projects. Thus, a reasonable balance between debt and equity financing is important for the financial health of a corporation. Example 18.7 illustrates the disadvantages of unbalanced D-E mixes.

Example 18.7

Three bank-holding companies have the following debt and equity capital amounts and resulting D-E mixes. Assume all equity capital is in the form of common stock.

	Capital-Financing Source		
Bank	**Debt ($ in millions)**	**Equity ($ in millions)**	**D-E Mix (%–%)**
First National	10	40	20–80
United	20	20	50–50
Mercantile	40	10	80–20

Assume the annual revenue is $15 million for each bank and that, after interest on debt is considered, the net incomes are $14.4, $13.4, and $10.0 million for the three banks, respectively. Compute the return on common stock for each bank and comment on the returns for the different D-E mixes.

Solution

For each bank, divide the net income by the stock value to compute the common-stock return.

First National: $\quad\text{Return} = \dfrac{14.4}{40} = 0.36 \quad (36\%)$

United: $\quad\text{Return} = \dfrac{13.4}{20} = 0.67 \quad (67\%)$

Mercantile: $\quad\text{Return} = \dfrac{10.0}{10} = 1.00 \quad (100\%)$

As expected, the return for the highly leveraged Mercantile Bank is by far the largest, yet only 20% of the bank is in the hands of the ownership. The return is excellent, but the risk associated with this bank is high compared to First National, where the D-E mix is only 20% debt.

The use of large percentages of debt financing greatly *increases the risk* taken by lenders and stock owners. Confidence in the corporation becomes generally poor, no matter how large the short-term return on stock.

The leverage of large D-E mixes does increase the return on *equity capital*, as shown in previous examples; but it can also work against the owner or investor of funds. A small percent decrease in asset value will negatively affect a highly debt leveraged investment much more than one with small or no leveraging. Example 18.8 illustrates this fact. This is also the correct response to the question posed at the end of Additional Example 15.8, where we asked you to explain why a highly debt capital leveraged situation is not financially healthy.

Example 18.8

Two individuals placed $10,000 each in different investments. Marylynn invested $10,000 in silver mining stock and Carla leveraged the $10,000 by purchasing a $100,000 residence to be used as rental property. Compute the resulting value of the $10,000 equity capital if there is a 5% decrease in the value of the stock and residence. Do the same for a 5% increase. Neglect any dividend, income, or tax considerations.

Solution

The mining stock value decreases by $10,000(0.05) = \$500$ and the house value decreases by $100,000(0.05) = \$5000$. The effect upon the investment is a reduction in the amount of equity funds now available, if the investment is sold.

Marylynn's loss: $\dfrac{500}{10,000} = 0.05$ (5%)

Carla's loss: $\dfrac{5000}{10,000} = 0.50$ (50%)

The 10-to-1 leveraging by Carla gives her a 50% decrease in her equity position, while the 1-to-1 leveraging results in a 5% reduction for Marylynn.

The opposite is correct for a 5% increase; Carla would benefit by a 50% gain on her $10,000, while the 1-1 leverage would offer Marylynn only a 5% gain. The larger leverage is much more risky. It offers a much *higher return for an increase* in the value of the investment and a much *larger loss for a decrease* in the value of the investment.

Additional Example 18.9
Problems 18.23 to 18.28

ADDITIONAL EXAMPLE

Example 18.9

DEBT AND EQUITY FINANCING, SECTION 18.7   A pizza corporation plans to purchase 15 delivery vans for a total of $150,000. Each van will be depreciated over 10 years with SV = $1000 using the classical straight-line method. Debt financing for 50% of the capital and 100% equity financing are possible sources of the $150,000.

Equity financing requires the sale of $15-per-share common stock. The first dividend is expected to be $0.50 per share, and a 5% dividend growth rate is anticipated.

For debt financing, a $75,000 loan be taken. The loan arrangements will be 10 years at 8% compounded annually with equal end-of-year payments.

The MARR is determined to equal the cost of equity capital, or the WACC if debt and equity capital are combined. If the CFBT is expected to be $30,000 annually and a 35% effective tax rate applies, use an after-tax analysis to determine which financing method offers a better return relative to the MARR: 100% equity or 50% debt.

Solution

100% equity financing

Use Equation [18.3] to estimate the cost of equity capital.

$$R_e = \frac{0.5}{15} + 0.05 = 0.0833 \qquad (8.3\%)$$

Because there is no tax advantage for equity financing, and because the financing source is known specifically, the MARR is set at the equity cost of 8.3%.

Now, calculate the after-tax NCF estimates and return. Straight-line depreciation for the 10 trucks is $13,500 annually. Taxes and annual NCF estimates are

$$\text{Taxes} = 0.35(\text{TI}) = 0.35(30,000 - 13,500) = \$5775$$

$$\text{NCF} = \text{CFBT} - \text{taxes} = 30,000 - 5775 = \$24,225$$

The return is determined from the equation for annual cost.

$$0 = -150,000(A/P,i^*,10) + 15,000(A/F,i^*,10) + 24,225$$

The i^* is 10.7%, which exceeds the MARR of 8.3%; equity financing meets the MARR criterion.

50% debt–50% equity financing

Find the weighted average cost of capital for the 50-50 D-E mix. Equity cost is 8.3% as computed above. For the $75,000 10-year loan, we assume a uniform reduction of principal at the rate of $7500 annually. (This assumption is applied for simplification purposes only. Compound interest makes the actual contribution to principal reduction change each year.) The annual payment is

$$75,000(A/P,8\%,10) = \$11,177$$

The annual tax advantage of paying the interest is estimated as

$$(\text{Payment} - \text{principal portion})(\text{tax rate}) = \text{interest}(T)$$

$$(11,177 - 7500)(0.35) = 3677(0.35) = \$1287$$

The cost of debt capital is $i^* = 5.4\%$ from the AW relation

$$0 = 75,000(A/P,i^*,10) - (11,177 - 1287)$$

The WACC from Equation [18.1] is

$$\text{WACC} = 0.5(8.3) + 0.5(5.4) = 6.85\%$$

The MARR is set at the WACC of 6.85%.

To compute the return on the $75,000 of equity capital in the D-E mix of 50-50, first determine the annual NCF.

$$\text{Taxes} = \text{TI(tax rate)} = (30{,}000 - 13{,}500 - 3677)(0.35) = \$4488$$

$$\text{NCF} = \text{CFBT} - (\text{principal} + \text{interest}) - \text{taxes}$$

$$= 30{,}000 - 11{,}177 - 4488 = \$14{,}335$$

The return for the 50-50 D-E mix is $i^* = 15.2\%$ using the NCF value above in the AW relation.

$$0 = -75{,}000(A/P,i^*,10) + 15{,}000(A/F,i^*,10) + 14{,}335$$

The return of 15.2% is acceptable since it significantly exceeds MARR = WACC = 6.85%.

Both financing scenarios are acceptable, but the return on equity capital exceeds the MARR significantly more for the 50-50 D-E mix on the $75,000 equity capital.

Comment

The use of 50% debt financing increased the expected return on equity capital by 4.5% (15.2 versus 10.7) and decreased the MARR requirement by about 1.5% (8.3 versus 6.85%). However, the risk is slightly greater due to the loan payments incurred through debt financing.

CHAPTER SUMMARY

The mix and cost of debt and equity capital are combined to estimate the weighted average cost of capital (WACC), which is numerically between the cost of each source when calculated separately. Debt financing provides a federal tax advantage for the loan interest and bond dividends. There is no direct U.S. corporate tax advantage for using equity capital.

Estimating the cost of equity financing is often more involved than estimating the cost of debt financing. Equity funds come from three sources—preferred stock, common stock, and retained earnings. The cost of common-stock capital may be estimated by one of several methods. Two are summarized in this chapter: the valuation method and the CAPM model, which takes the security market and its variation into account.

The MARR will vary for several reasons: risk, investment opportunity, taxes, capital limitation, and market rates. Project risk is important but should not be considered directly in determining the MARR. Sources of funding and limitations on capital assist in determining the MARR, usually established between the cost of equity capital and the WACC.

For highly leveraged companies, where the debt percentage is large, it is more difficult to obtain new debt capital from lenders. Also, it is more risky to commit the corporation's limited equity funds. A high D-E mix makes a company and its projects more risky.

PROBLEMS

18.1 Mr. Richards, the franchise owner of a fast-food restaurant estimates his cost of new capital at 9%. He has established 12% per year as his MARR criterion to decide upon a new integrated order-taking, registering, and automated inventory system. (*a*) What return on the capital investment does he expect? (*b*) Mrs. Richards considers it necessary that risky ventures earn 3% over their cost of capital in addition to a return of 5%. What MARR does she recommend for use in this same evaluation, if this project is considered risk-neutral? Risky? (*c*) What is the recommended MARR in this situation, based upon what you have discovered in this chapter? How should the Richardes consider the required return and perceived risk factors when evaluating this capital project?

18.2 State whether each of the following involves debt financing or equity financing:

(*a*) Short-term loan of $8000 from a bank.

(*b*) $5000 taken from a co-owner's savings account to pay a company bill.

(*c*) $150,000 bond issue.

(*d*) Issue of preferred stock worth $355,000.

(*e*) Jane borrows $50,000 from her brother at 3% interest to operate her business. The brother is not a co-owner of the company.

18.3 Listed below are bundles for three independent proposals for which the PW and i^* have been estimated. (*a*) Select the acceptable projects if the capital budget limit is $40,000 and the MARR is the cost of capital, which is 9%. (*b*) What is the effective MARR for this situation using the opportunity cost concept?

Proposal Bundle	Initial Investment, $	Measure of Worth	
		PW @ 9%, $	Incremental i^*, %
1	10,000	3,000	10.5
2	25,000	−2,500	7.3
1,2	35,000	500	9.3
3	40,000	1,800	9.9

18.4   Randy and Jennifer plan to purchase a new car for $22,000 total cost. They plan to use $10,000 of their own funds which they will obtain by selling some common stock which has an average rate of return of 12% per year. The remainder of the purchase price will be borrowed at 8.6% per year from their credit union. Determine the D-E mix and WACC that this couple will experience for their car purchase.

18.5   A large company, Ben Products, plans to purchase a small business which has been a supplier for many years. A purchase price of $780,000 has been agreed to. The purchasing company does not know exactly how to finance the purchase to obtain a WACC as low as that for other ventures. The WACC is presently 10%.

   (*a*) Two possibilities for financing are available. The first requires that Ben invest 50% equity funds at 8% and borrow the balance at 11% per year. The second possibility requires only 25% equity funds and the balance borrowed at 9%. Which approach will result in the smaller overall cost of capital?

   (*b*) The finance committee at Ben Products has just decided that the WACC for the purchase must not exceed the historical average of 10% per year. What are the maximum costs of debt capital that can be incurred for each of the approaches outlined in part (*a*)?

18.6   Mariposa Pizza, S.A., intend to raise $100 million in new capital with 40% equity funds and 60% debt financing. The following percentages and rates are estimated. Compute the WACC.

Equity capital: 40% or $40 million
        Common stock sales: 25% at 11.5%
        Use of retained earnings: 75% at 9.0%

Debt capital: 60% or $60 million
        Short-term loans: 50% at 13.5%
        Long-term bonds: 50% at 9.0%

18.7   A friend employed in the finance department of a large corporation in which you own common stock tells you the company reported a WACC of 12.5% to bank auditors. The common stocks you own have averaged a total return of 10%-per-year over the last 5 years. The corporation states in its annual report that it uses 62% of its own funds for capitalization pro- jects. What is your best estimate of the cost of debt capital?

18.8   Georgia, a cost estimator with Sam-Young Aerospace, wants to know the debt-equity mix at which the WACC is most likely to have the lowest value. She has collected the information below for current capital-intensive projects. (*a*) Plot the curves for debt, equity, and weighted average costs of capital. (*b*) What debt-equity mix seems to historically offer the lower WACC values?

Project	Equity Capital		Debt Capital	
	Percent	Rate	Percent	Rate
1	80%	4.5%	20%	5.5%
2	66	12.7	34	5.5
3	56	10.9	44	8.2
4	35	9.3	65	9.8
5	20	10.0	80	13.8
6	100	13.2		

18.9 An automobile manufacturer requires the infusion of $2 million in new debt capital. If 12-year 8% bonds which pay semiannually are sold to a large brokerage firm at a 4% discount, (*a*) determine the total face value of the bonds required to provide the $2 million, and (*b*) compute the after-tax cost of debt capital if $T_e = 50\%$.

18.10 Barely Suits has to raise $500,000. Two methods of debt financing have been outlined. The first is to borrow it all from a bank. The company will pay an effective 8% compounded per year for 8 years to the bank. The principal on the loan will be reduced uniformly over the 8 years, with the remainder of each annual payment going toward interest. The second method is to issue five hundred $1000 10-year bonds which requires a 6% annual dividend payment. (*a*) Which method of financing would you recommend if the effective tax rate of 40% is considered? (*b*) Is the answer the same if a before-tax analysis is made? Why?

18.11 Harold is a new graduate working for Skim Milk Dairy. His boss tells him that a $75,000 purchase of new environmental-control equipment is planned. The equipment will last 5 years and has a salvage value of $15,000. The company has $25,000 in equity funds and expects to borrow the remainder for the 5-year period with uniform reduction in principal each year. The equipment is expected to increase cash flow before taxes by $18,000 per year. For analysis purposes, the depreciation model is classical straight line and an effective tax rate of 50% applies. Harold's boss, who estimates that the incremental taxes with the loan will be $1500 per year, asks him to (*a*) estimate the stated interest rate on the loan and (*b*) estimate the effective interest rate on the loan after taxes. Assume the current before-tax MARR is 10% per year. Help Harold develop his answers.

18.12 A tax consultant is working with an international firm, which can commit its own funds to a large capitalization project at an after-tax cost of 5.5% per year, or which can borrow from foreign sources through the issuance of $5 million worth of 20-year bonds which pay a 10%-per-year dividend on a semiannual basis. If the effective tax rate is 40%, which source should the consultant determine has the lower cost of capital?

18.13 The common stock for HiRiz Constructors can be evaluated using the dividend method or the CAPM. Last year, the first year for dividends, the stock paid $0.75 per share on an average price of $11.50 on the New York Stock Exchange. Management hopes to grow the dividend rate at 3% per year. HiRiz stock has a volatility which is higher than the norm at 1.3. If safe investments are returning 7.5% and the 3% growth rate on common stocks is also the premium above risk-free investments that HiRiz plans to pay, compare the cost of equity capital using the two estimates.

18.14 Describe the fundamental differences in the two approaches to estimate the cost of common-stock capital covered in Section 18.4.

18.15 A Fortune 500 corporation president expects to use a debt-equity mix of 40-60 to finance a $5 million venture. The after-tax cost of debt capital is 9.5%, but equity capital requires the sale of preferred and common stock, as well as the commitment of retained earnings. Use the following information to determine the WACC for the venture. All dollar quotes are per share.

> **Preferred stock:** $1 million to sell.
>   Face value = $30
>   Sales price = $27.60
>   Dividend rate = 8% per year
>
> **Common stock:** 50,000 shares to sell.
>   Price = $20
>   Initial dividend = $0.40
>   Dividend growth = 5% annually
>
> **Retained earnings:** Same cost of capital as for common stock.

18.16 The Murrays, owners of a quick-stop store, plan to construct a laundromat next door in what is currently a storage area. They will use 100% equity financing. It will cost $22,000 to build the facility. Depreciation over a 15-year life will be classical straight line with SV = $7000. The Murrays have the funds invested at the present time and make 10% per year. If the incremental annual cash flow before taxes is expected to be $5000 and the effective tax rate is 48%, is the venture projected to be profitable?

18.17 A man wants to buy lumber now for $600 in order to save a total of 25% during the next 6 months as prices rise, but he does not know how to finance the purchase. The financing plans he has developed are

> **Plan 1—all equity:** Take $600 from a savings account now and put $125 per month back in as it is available. This account pays 6%-per-year compounded quarterly.
>
> **Plan 2—all debt:** Borrow $600 now from the credit union at an effective 1% per month and repay the loan at $103.54 per month for 6 months. Then, put the difference between the payment and

the amount saved each month in the 6%-per-year compounded quarterly savings account.

Plan 3—50% debt–50% equity: Use $300 from the savings account and borrow $300 at 1% per month, and repay at the rate of $51.77 per month for 6 months. Again, the difference between the payments and the savings would be deposited at 6% per year compounded quarterly.

Perform a before-tax analysis to determine which financing plan has the lowest cost of capital.

18.18 The Holistic Drug Co. has a total of 153,000 shares of common stock outstanding at a market price of $28 per share. A before-tax cost of equity capital of 24% is incurred by these shares. Stocks fund 50% of the company's undertakings. The remaining investments are financed by bonds and short-term loans. It is known that 30% of the debt capital is from $1,285,000 worth of $10,000 6%-per-year 15-year bonds, which were sold at a 2% discount. The remaining 70% of debt capital is from loans repaid at an effective 17.3% before taxes. If the effective income tax rate is 48%, determine the weighted average cost of capital (*a*) before taxes and (*b*) after taxes.

18.19 Should each of the following tend to raise, lower, or not impact the MARR used to make the investment decision for a proposed capital project? Explain your answer.

(*a*) Investment in a chain of quick-food stores is contemplated, but the company president is very leery of such an undertaking due to the severe competition in this area of business.

(*b*) Finished Nail Construction built a 250-unit apartment house 3 years ago and still retains ownership. Because of the risk, when the project was undertaken, a 12%-per-year return was required; however, because of the favorable outcome management feels this is safer than some other types of investments.

(*c*) Income and other business taxes have increased by an average of 4% per year for the last 3 years.

(*d*) Federal government protection has just been announced for manufactured products which are threatened by competition from international firms, whose home governments subsidize them.

18.20 Each person has different impressions of risk and different ways to determine if an opportunity is financially risky or safe. Describe, as best you can, how you personally distinguish between risky and safe investment opportunities.

Now, assume you have $1000 to invest in some project. List two project opportunities in which you might invest the $1000. One of the opportunities should be an example of what you consider a risky venture; the second should be a safe venture.

18.21 For the following problems and added information, select the MARR using the guidelines in Section 18.6 and make the economic decision.

(a) Problem 15.19 (a) where the capital is available and the return on equity capital is 30% at this time.

(b) Problem 17.11, if capital is generated using 70% debt sources at 7% and 30% equity sources at 5.3%.

18.22 One view of government-imposed requirements upon business and industry in the areas of employee safety, environmental protection, noise control, etc., is that their compliance tends to decrease the return and/or increase the cost of capital to the corporation. These requirements cannot be evaluated like regular engineering economy alternatives. (a) Give your opinion of this view and describe one government-imposed requirement in industry that can substantiate your view. (b) Use your knowledge of engineering economic analysis to explain how a company should economically evaluate the way in which it will comply with imposed regulations.

18.23 Why is it unhealthy for a corporation to maintain a very high D-E mix over a long period of time?

18.24 Fairmont Industries uses 100% equity financing primarily. A good opportunity is now offered that will require $250,000 in capital. The Fairmont owner can supply the money from personal investments, which currently earn an average of 8½% per year. The annual net cash flow after taxes is estimated at $30,000 for the next 15 years. The effective tax rate is 50%.

Alternatively, 60% of the required amount can be borrowed at 5% compounded annually for 15 years. It is assumed that the principal is uniformly reduced and an average annual interest is paid. If the MARR is the WACC, determine which plan is better.

18.25 Suzanne has three plans for raising $50,000 for her company.

Type of Financing	Plan 1	Plan 2	Plan 3
Equity, %	90	60	20
Debt, %	10	40	80

At present, the before-tax cost for equity capital is 10% and for debt capital is 12%. If the project is expected to yield $10,000 annual net cash flow for 5 years, do a before-tax analysis to determine the return for each plan and identify the ones that are economically acceptable if the MARR is (*a*) the cost of equity capital, (*b*) the WACC, and (*c*) 1.5 times the WACC, which is an effort to consider the expected risk of this project.

18.26 Westfall Mattresses has an opportunity to invest $100,000 in a new style of mattress. Financing will be split between common-stock sales ($75,000) and a loan with an 8% after-tax interest rate. The estimated annual NCF after taxes is $18,900 for the next 7 years. The effective tax rate is 50%, and no depreciable assets are involved.

Westfall uses the capital asset pricing model for valuation of its common stock. Recent analysis shows that it has a volatility rating of 0.85 and is paying a premium of 5% common-stock dividend. Nationally, the safest investment is currently paying 8% per year. Is the venture finan-cially attractive if Westfall uses as the MARR its (*a*) equity cost of capital? (*b*) WACC?

18.27 The following information has been published by two companies. Total capitalization is the value of all assets for which the corporation has obtained investment capital in the past.

Corporation	Capitalization ($ in millions)	Percent Debt Capital
A	2.5	28
B	1.6	70

(*a*) What amount of total capitalization is from debt sources for each corporation?

(*b*) If a serious problem occurs in each corporation and the total capital-ization of the firm goes down by 15%, compute the new D-E mixes. (Remember that the amount of debt will remain constant even though the total value of assets in the firm decreases.)

18.28 Redraw the debt, equity, and WACC curves in Figure 18–2 under the condition that a high D-E mix has been present for some time. High D-E mixes cause the debt cost to increase substantially. This makes it harder to obtain equity funds, so the cost of equity capital also increases. Explain via your graph and words how the minimum WACC point will move for historically high D-E mixes relative to low D-E mixes.

Sensitivity Analysis and Expected-Value Decisions

This chapter includes three different, but related, topics about alternative selection. Since the sections naturally build upon the previous ones, you can include as much of this material as you have interest in and time to devote. The first three sections expand our capability to perform a *sensitivity analysis* for one estimate or an entire alternative. Then Sections 19.4 and 19.5 provide an elementary foundation of *expected-value* computations for cash-flow sequences.

After this very brief introduction to probability and the calculation of an expected value, a second method to consider variation and risk is discussed. This is the development and use of a *decision tree* to make a series of decisions considering multiple alternatives with estimated economic outcomes and probabilities.

LEARNING OBJECTIVES

Purpose: Perform a sensitivity analysis on one or more parameters using a selected measure of worth, and develop a decision tree to determine the most economic decision path for several related alternatives.

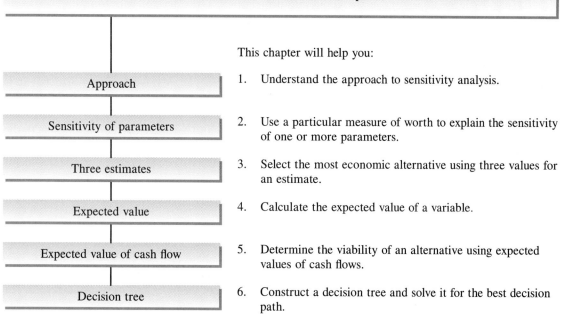

This chapter will help you:

Approach	1. Understand the approach to sensitivity analysis.
Sensitivity of parameters	2. Use a particular measure of worth to explain the sensitivity of one or more parameters.
Three estimates	3. Select the most economic alternative using three values for an estimate.
Expected value	4. Calculate the expected value of a variable.
Expected value of cash flow	5. Determine the viability of an alternative using expected values of cash flows.
Decision tree	6. Construct a decision tree and solve it for the best decision path.

19.1   THE APPROACH TO SENSITIVITY ANALYSIS

Economic analysis utilizes estimates of future happenings to assist decision makers. Since future estimates are always incorrect to some degree, inaccuracy is present in the economic projections. The effect of variation may be determined using sensitivity analysis. Some of the parameters or factors commonly evaluated for sensitivity are MARR, interest rates, life estimates, recovery period for tax purposes, all types of costs, sales, and many other factors. Usually, one factor at a time is varied and independence with other factors is assumed. This assumption is not completely correct in real-world situations, but it is practical since the ability to accurately account for actual dependencies is not generally possible.

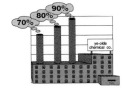

Sensitivity analysis, itself a study usually performed in conjunction with the engineering economy study, determines how a measure of worth—PW, AW, ROR, or B/C—and the alternative selected will be altered if a particular factor or parameter varies over a stated range of values. We use the term parameter, not factor, in this chapter. For example, variation in a parameter such as MARR would not alter the decision to select an alternative when all compared alternatives return more than the MARR; thus, the decision is relatively insensitive to the MARR. However, variation in the n value may indicate that the selection of al- ternatives is very sensitive to the estimate of asset life.

Usually the variations in life, annual costs, and revenues result from variations in selling price, operation at different levels of capacity, inflation, etc. For example, if an operating level of 90% of airline seating capacity is compared with 50% for a proposed new international route, operating cost and revenue per passenger mile will increase, but anticipated life will probably decrease only slightly. Usually several important parameters are studied to learn how the uncertainty of estimates affects the economic analysis.

Plotting the sensitivity of PW, AW, or ROR versus the parameter(s) studied is very helpful. Two alternatives can be compared with respect to a given parameter and the breakeven point computed. This is the value at which the two alternatives are economically equivalent. However, the breakeven chart commonly represents only one parameter per chart. Thus, several charts are constructed and independence of each parameter is assumed. (We will learn how to plot several parameters on one sensitivity chart in the next section.) In previous uses of breakeven analysis, we computed the measure of worth at only two values of a parameter and connected the points with a straight line. However, if the results are sensitive to the parameter value, several intermediate points should be used to better evaluate the sensitivity, especially if the relationships are not linear.

When several parameters are studied, a sensitivity analysis study may become quite complex. It may be performed one parameter at a time using a spreadsheet system, a specially prepared computer program, or manual computations. The computer facilitates comparison of multiple parameters and multiple measures of worth, and the software can rapidly plot the results.

19.2   DETERMINING SENSITIVITY TO PARAMETER ESTIMATES

There is a general procedure that you may follow when conducting a sensitivity analysis study. The steps are

1. Determine which parameter(s) of interest might vary from the most likely estimated value.
2. Select the probable range and increment of variation for each parameter.
3. Select the measure of worth to be calculated.
4. Compute the results for each parameter using the measure of worth as a basis.
5. To better interpret the results, graphically display the parameter versus the measure of worth.

This sensitivity analysis procedure should indicate the parameters which warrant closer study or require additional information collection. When there are two or more alternatives, it is better to use a monetary-based measure of worth (PW or AW) in step 3. If ROR is used, it requires the extra efforts of incremental analysis between alternatives. Example 19.1 illustrates sensitivity analysis for one project.

Example 19.1

Flavored Rice, Inc., is contemplating the purchase of a new asset for automated rice handling. Most likely estimates are a first cost of $80,000, zero salvage value, and a before-tax cash-flow relation of the form $NCF = \$27,000 - 2000t$ per year ($t = 1, 2, \ldots, n$). The MARR for the company varies from 10% to 25% per year for different types of asset investments. The economic life of similar machinery varies from 8 to 12 years. Evaluate the sensitivity of PW and AW by varying (a) the parameter MARR, while assuming a constant n value of 10 years and (b) the parameter n, while MARR is constant at 15% per year.

Solution

(a) Follow the procedure above.

Step 1. The MARR, i, is the parameter of interest.

Step 2. Select 5% increments to evaluate sensitivity to MARR; the range for i is 10% to 25%.

Step 3. The measures of worth are PW and AW.

Step 4. Set up the PW and AW relations. For example, at $i = 10\%$, use k values 1 to 10 for the cash flow,

$$PW = -80,000 + 25,000 \, (P/A,10\%,10) - 2000(P/G,10\%,10)$$

$$= \$27,830$$

$$AW = P(A/P,10\%,10) = \$4529$$

The measures of worth for all four *i* values at 5% intervals are

i	PW	AW
10%	$ 27,830	$ 4,529
15	11,512	2,294
20	−962	−229
25	−10,711	−3,000

Step 5.   A plot of MARR versus AW is shown in Figure 19–1. The steep negative slope indicates that the decision to accept the proposal based on AW is quite sensitive to variations in the MARR. If the MARR is established at the upper end of the range, the investment is not attractive.

(*b*)   **Step 1.**   Asset life *n* is the parameter.

Step 2.   Select 2-year increments to evaluate sensitivity to *n* over the range 8 to 12 years.

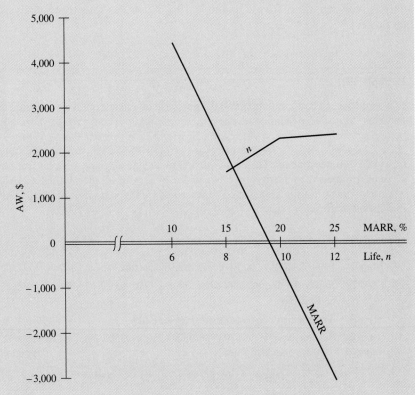

Figure 19–1   Plot of sensitivity of AW to MARR and life variation, Example 19.1.

Step 3.   The measures of worth are PW and AW.

Step 4.   Set up the same PW and AW relations as in part (a) for $i = 15\%$. Measures of worth results are

n	PW	AW
8	\$ 7,221	\$1,609
10	11,511	2,294
12	13,145	2,425

Step 5.   Figure 19–1 presents the nonlinear plot of AW versus n. This is a characteristic shape for sensitivity analysis of an n value. Since the PW and AW measures are positive for all values of n, the decision to invest is not materially affected by the estimated life.

Comment

Note that in part (b), the AW curve seems to level out above $n = 10$. This insensitivity to changes in cash flow in the distant future is an expected trait, because when discounted to time 0, the PW and AW values get smaller as n increases.

**Example 19.1
(Spreadsheet)**

Flavored Rice, Inc., is contemplating the purchase of a new asset for automated rice handling. Most likely estimates are a first cost of \$80,000, zero salvage value, and a before-tax cash-flow relation of the form CFBT = \$27,000 − 2000t per year ($t = 1, 2, \ldots, n$). The MARR for the company varies from 10% to 25% per year for different types of asset investments. The economic life of similar machinery varies from 8 to 12 years.

   Use a spreadsheet to evaluate the sensitivity of PW by varying the parameter MARR, while assuming a constant n value of 10 years, and by varying the parameter n, while the MARR is constant at 15% per year.

Solution

Figure 19–2 presents two spreadsheets for parts (a) and (b) and accompanying plots of PW versus i (fixed n) and PW versus n (fixed i). If we divide all values by \$1000 so no keystrokes for zeros are necessary, the general relation for cash-flow values in year t, CF_t, is

$$CF_t = \begin{cases} -80 & t = 0 \\ +27 - 2t & t = 1, \ldots \end{cases}$$

The spreadsheet was set up to calculate PW for i values from 10% to 25% and n values from 8 to 12 years. Sample spreadsheet cell functions are detailed in Figure 19–2 to calculate the PW [or net present value (NPV)]. As in the example above, PW is very sensitive to i, but not very sensitive to variations in n.

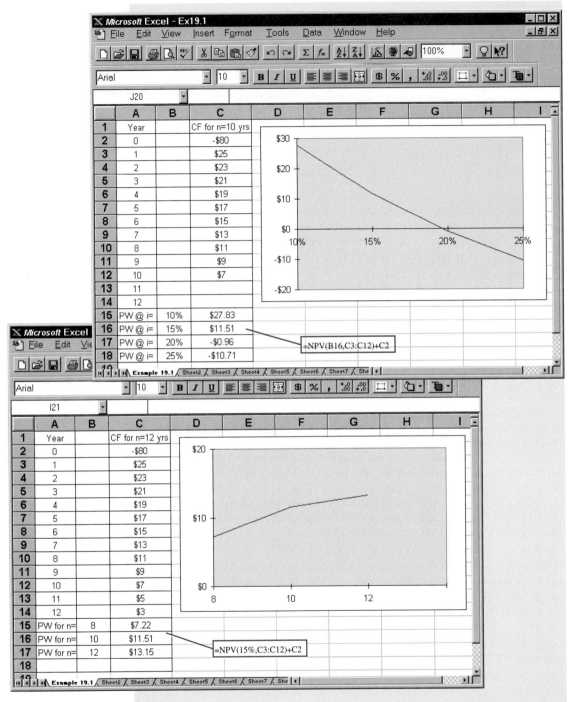

Figure 19–2   Spreadsheet approach to evaluating present-worth sensitivity to the parameters MARR and n, Example 19.1 (Spreadsheet).

When the sensitivity of *several parameters* is considered for *one alternative* using a *single measure of worth*, it is helpful to graph percentage change for each parameter versus the measure of worth. Figure 19–3 illustrates ROR versus six different parameters for one alternative. The variation in each parameter is indicated as a percentage deviation from the most likely estimate on the horizontal axis. To use the graph, select a parameter. If the ROR response curve is flat and approaches horizontal over the range of total variation graphed, there is little sensitivity of ROR to the parameter. This is the conclusion for indirect cost in Figure 19–3. On the other hand, ROR is very sensitive to sales price. A reduction of 30% from the expected sales price reduces the ROR from approximately 20% to −10%, whereas a 10% increase in price raises the ROR to about 30%.

If two *alternatives* are compared and the sensitivity to *one parameter* is sought, the graph may show quite nonlinear results. Observe the general shape of the sensitivity graphs in Figure 19–4. The curves are shown as linear

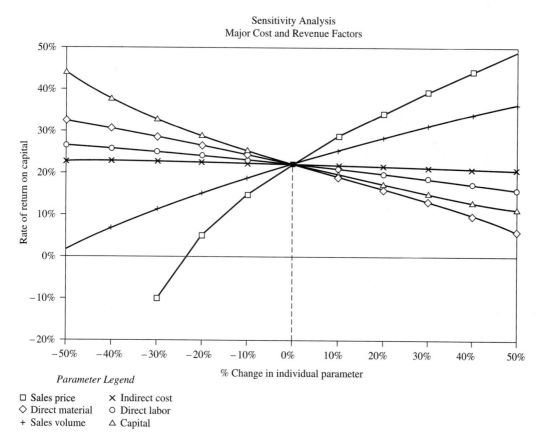

Figure 19–3   Sensitivity analysis graph of percent variation from the most likely estimate.

SOURCE: J. R. Heizer, "Sensitivity Analysis for Business Planning," Appendix D: *Engineering Economy* by L. T. Blank and A. J. Tarquin, New York, McGraw-Hill, 3d ed., 1909.

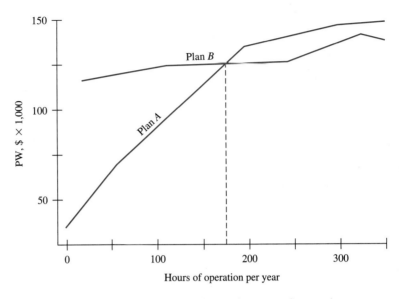

Figure 19–4   Sample PW sensitivity to hours of operation for two alternatives.

segments between specific computation points. We won't be concerned with the actual computations in this case. The graph indicates that the PW of each plan is a nonlinear function of hours of operation. Plan A is very sensitive in the range of 0 to 200 hours, but it is comparatively insensitive above 200 hours. Plan B is more attractive due to its relative insensitivity, provided that both plans A and B are justified, that is, the selected measure of worth indicates economic justification. This indicates the need to plot parameters versus the measure of worth at more frequent intermediate points to better understand the nature of the sensitivity.

Additional Example 19.8
Problems 19.1 to 19.16

19.3   SENSITIVITY ANALYSIS USING THREE ESTIMATES

We can thoroughly examine the economic advantages and disadvantages among two or more alternatives by borrowing from the field of project scheduling the concept of making three estimates for each parameter: *a pessimistic, a most likely, and an optimistic estimate.* You should realize that, depending upon the nature of a parameter, the pessimistic estimate may be the lowest value (alternative life is an example) or the largest value (such as asset first cost).

This approach allows us to study measure of worth and alternative selection sensitivity within a predicted range of variation for each parameter. Usually the most-likely estimate is used for all other parameters when the measure of worth is calculated for one particular parameter or one alternative. This approach, essentially the same as the one-parameter-at-a-time analysis of Section 19.2, is illustrated by the next example.

Example 19.2

You are an engineer evaluating three alternatives for which a management team has made three strategy estimates—pessimistic (P), most likely (ML), and optimistic (O)—for the life, salvage value, and annual operating costs. The estimates are presented on an alternative-by-alternative basis in Table 19–1. For example, alternative B has pessimistic estimates of SV = $500, AOC = $4000, and n = 2 years. The first costs are known, so they have the same value for all strategies. Perform a sensitivity analysis and try to determine the most economical alternative using AW analysis and a MARR of 12%.

Table 19-1	Competing alternatives with three estimates made for salvage value, AOC, and life parameters			
Strategy	**First Cost, $**	**Salvage Value, $**	**AOC, $**	**Life, n Years**
Alternative A				
Strategy P	−20,000	0	−11,000	3
Strategy ML	−20,000	0	−9,000	5
Strategy O	−20,000	0	−5,000	8
Alternative B				
Strategy P	−15,000	500	−4,000	2
Strategy ML	−15,000	1,000	−3,500	4
Strategy O	−15,000	2,000	−2,000	7
Alternative C				
Strategy P	−30,000	3,000	−8,000	3
Strategy ML	−30,000	3,000	−7,000	7
Strategy O	−30,000	3,000	−3,500	9

Solution

For each alternative description in Table 19–1, we calculate the AW of costs. For example, the AW relation for alternative A pessimistic estimates is

$$AW = -20,000(A/P,12\%,3) - 11,000 = \$-19,327$$

As in most sensitivity studies, a spreadsheet using the P, ML, and O parameter values will significantly reduce the computation time.

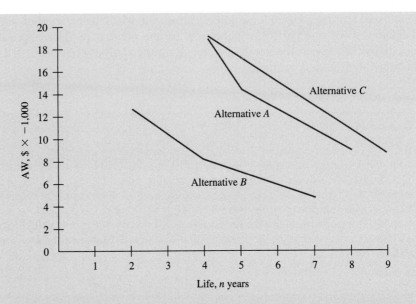

Figure 19-5   Plot of cost AW values for different-life estimates, Example 19.2.

Table 19–2 presents all AW values. Figure 19–5 is a plot of AW versus the three estimates of life for each alternative. Since the AW cost calculated using the ML estimates for alternative B ($\$-8229$) is economically better than even the optimistic AW value for alternatives A and C, alternative B is clearly favored.

Comment

While the alternative that should be selected in this example is quite obvious, this is not normally the case. For example, in Table 19–2 (omitting minus signs), if the pessimistic alternative B equivalent annual cost were much higher, say, $\$21,000$ per year (rather than $\$12,640$), and the optimistic AW costs for alternatives A and C were less than that for B ($\$5089$), the choice of B is not apparent or correct. In this case, it would be necessary to select one set of estimates (P, ML, or O) upon which to base the decision. Alternatively, the different estimates can be used in an expected-value analysis, which is introduced in the next section.

Table 19–2	Annual-worth values, Example 19.2		
	Alternative AW Values		
Strategy	**A**	**B**	**C**
P	$\$-19,327$	$\$-12,640$	$\$-19,601$
ML	$-14,548$	$-8,229$	$-13,276$
O	$-9,026$	$-5,089$	$-8,927$

Problems 19.17 to 19.21

19.4   ECONOMIC VARIABILITY AND THE EXPECTED VALUE

As engineers and economic analysts, we must deal with estimates about an uncertain *future* placing appropriate reliance on past data, if any exist. The use of probability and its basic computations by the engineering economist are not as common as they should be. The reason for this is not that the computations are difficult to perform or understand but that realistic probabilities associated with cash-flow estimates are difficult to make. Experience and judgment can often be used in conjunction with probabilities and expected values to evaluate the desirability of an alternative.

The *expected value* can be interpreted as a long-run average observable if the project is repeated many times. Since a particular alternative is evaluated or implemented only once, a *point estimate* of the expected value results. However, even for a single occurrence, the expected value is a meaningful number to know and utilize.

The expected value $E(X)$ is computed using the relation:

$$E(X) = \sum_{i=1}^{i=m} X_i P(X_i) \qquad \text{[19.1]}$$

where X_i = value of the variable X for i from 1 to m different values
          $P(X_i)$ = probability that a specific value of X will occur

Probabilities are always correctly stated in decimal form, but they are routinely spoken of in percentages and often referred to as *chance,* such as, *the chances are about 10%.* When placing the probability value in Equation [19.1] or any other relation, be sure to use the decimal equivalent of 10%, i.e., 0.1. In all probability statements the $P(X_i)$ values for a variable X must total to 1.0.

$$\sum_{i=1}^{i=m} P(X_i) = 1.0$$

We will commonly omit the subscript i on X for simplicity.

If X represents the estimated cash flows, some will be positive and some negative. If a cash-flow sequence includes revenues and costs and the present worth at the MARR is calculated, the result is the expected value of the discounted cash flows, $E(PW)$. If the expected value is negative, the overall outcome is expected to be a cash outflow. For example, if $E(PW) = \$-1500$, this indicates a proposal that is not expected to return the MARR.

Example 19.3

You expect to be mentioned in your favorite uncle's will. You believe that there is a 50% chance of being willed $5000 and a 45% chance of $50,000. And you believe there is a small chance (5%) of no inheritance at all. Compute your expected inheritance.

Solution

Let X be the inheritance values in dollars, and let $P(X)$ represent the associated probabilities. Your expected inheritance using Equation [19.1] is $25,000 based on your current estimates.

$$E(X) = 5000(0.5) + 50,000(0.45) + 0(0.05) = \$25,000$$

The 'no-inheritance' possibility is included because it makes the probability values sum to 1.0, and it makes the computation complete.

Problems 19.22 to 19.27

19.5 EXPECTED-VALUE COMPUTATIONS FOR ALTERNATIVES

The expected-value computation $E(X)$ is utilized in a variety of ways. Two ways are: (1) to prepare information for incorporation into a more complete engineering economy analysis, and (2) to evaluate the expected viability of a fully formulated alternative. Example 19.4 illustrates the first situation, and Example 19.5 determines the expected PW when the cash-flow sequence and probabilities are es- timated for an asset.

Example 19.4

An electric utility is experiencing a difficult time obtaining natural gas for electric generation. Fuels other than natural gas are purchased at an extra cost, which is transferred to the customer base. Total monthly fuel expenses are now averaging $7,750,000. An engineer with this city-owned utility has calculated the average revenue for the past 24 months using three fuel-mix situations—gas plentiful, less than 30% other fuels purchased, and 30% or more other fuels. Table 19–3 indicates the number of months that each fuel-mix situation occurred. Can the utility expect to

Table 19–3	Revenue and fuel-mix data, Example 19.4	
Fuel-Mix Situation	**Months in Past 24**	**Average Revenue, $ per Month**
Gas plentiful	12	5,270,000
< 30% other	6	7,850,000
≥ 30% other	6	12,130,000

meet future monthly expenses based on the 24 months of data, if a similar fuel-mix pattern continues?

Solution

Using the 24 months of data, a probability for each fuel mix is estimated.

Fuel-Mix Situation	Probability of Occurrence
Gas plentiful	12/24 = 0.50
< 30% other	6/24 = 0.25
≥ 30% other	6/24 = 0.25

Let the variable X represent average monthly revenue. Use Equation [19.1] to determine expected revenue per month.

$$E(\text{revenue}) = 5{,}270{,}000(0.50) + 7{,}850{,}000(0.25) + 12{,}130{,}000(0.25)$$

$$= \$7{,}630{,}000$$

With expenses averaging $7,750,000, the average monthly revenue shortfall is $7,750,000 − 7,630,000 = $120,000. To break even other sources of revenue must be generated, or, only if necessary, the additional costs may be transferred to the customer base in the form of a rate increase.

Example 19.5

The Tule Company has a substantial investment in automatic reaming equipment. A new piece of equipment costs $5000 and has a life of 3 years. Estimated annual cash flows are shown in Table 19–4 depending on economic conditions classified as receding, stable, or expanding. A probability is estimated that each of the economic conditions will prevail during the 3-year period. Apply expected value and PW analysis to determine if the equipment should be purchased. Use a MARR = 15%.

Table 19–4	Equipment cash-flow and probabilities, Example 19.5		
	Economic Condition		
Year	**Receding (Prob. = 0.2)**	**Stable (Prob. = 0.6)**	**Expanding (Prob. = 0.2)**
	Annual cash flow estimates, $		
0	$−5000	$−5000	$−5000
1	+2500	+2000	+2000
2	+2000	+2000	+3000
3	+1000	+2000	+3500

Solution

First determine the PW of the cash flows in Table 19–4 for each economic condition and then calculate $E(\text{PW})$ using Equation [19.1]. Define subscripts R for receding economy, S for stable, and E for expanding. The PW values for the three scenarios are

$$\text{PW}_R = -5000 + 2500(P/F,15\%,1) + 2000(P/F,15\%,2) + 1000(P/F,15\%,3)$$

$$= -5000 + 4344 = \$-656$$

$$\text{PW}_S = -5000 + 4566 = \$-434$$

$$\text{PW}_E = -5000 + 6309 = \$+1309$$

Only in an expanding economy will the cash flows return the 15% and justify the investment. The expected present worth is

$$E(\text{PW}) = \sum_{j=R,S,E} \text{PW}_j[P(j)]$$

$$= -656(0.2) - 434(0.6) + 1309(0.2)$$

$$= \$-130$$

At 15%, $E(\text{PW}) < 0$, so the reaming equipment is not justified using expected-value analysis.

Comment

It is correct to calculate the $E(\text{cash flow})$ for each year and then determine PW of the $E(\text{cash flow})$ series, because the PW computation is a linear function of cash flows. Computing $E(\text{cash flow})$ first may be easier in that it reduces the number of PW computations. In this example, calculate $E(\text{CF}_t)$ for each year $t = 0, 1, 2, 3$ from Table 19–4. Then determine the present worth, which is $E(\text{PW})$.

$$E(\text{CF}_0) = \$-5000$$

$$E(\text{CF}_1) = 2500(0.2) + 2000(0.6) + 2000(0.2) = \$2100$$

$$E(\text{CF}_2) = \$2200$$

$$E(\text{CF}_3) = \$2100$$

$$E(\text{PW}) = -5000 + 2100(P/F,15\%,1) + 2200(P/F,15\%,2) + 2100(P/F,15\%,3)$$

$$= \$-130$$

Additional Example 19.9
Problems 19.28 to 19.33

19.6 ALTERNATIVE SELECTION USING DECISION TREES

Alternative evaluation may require a series of decisions where the outcome from one stage is important to the next stage of decision making. When you can clearly define each economic alternative and you want to *explicitly account for risk,* it is helpful to perform the evaluation using a *decision tree.* A decision tree includes

- More than one stage of alternative selection.
- Selection of an alternative at one stage leads to another stage.
- Expected results from a decision at each stage.
- Probability estimates for each outcome.
- Estimates of economic value (cost or revenue) for each outcome.
- Measure of worth as the selection criterion, such as E(PW).

The decision tree is constructed left to right and includes each possible decision and outcome. A square represents a *decision node* with the possible alternatives indicated on the *branches* from the decision node (Figure 19–6a). A circle represents a *probability node* with the possible outcomes and estimated probabilities on the branches (Figure 19.6b). Since outcomes always follow decisions, the treelike structure in Figure 19–6c results as the entire situation is defined.

Usually each branch of a decision tree has some associated economic value in cost or in revenue or benefit terms (often referred to as *payoff*). These cash flows are expressed in terms of PW, AW, or FW values and are shown to the right

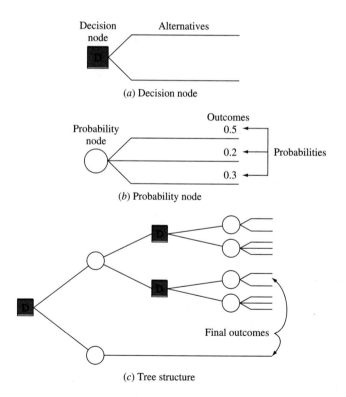

(a) Decision node

(b) Probability node

(c) Tree structure

Figure 19–6   Decision and probability nodes used to construct a decision tree.

of each final outcome branch. The cash-flow and probability values on each outcome branch are used in calculating the expected economic value of each decision branch. This process, called *solving the tree* or *foldback,* is explained after Example 19.6, which illustrates the construction of a decision tree.

Example 19.6

Jerry Hill is president and CEO of an American-based food-processing company, Hill Products and Services. He was recently approached by a large Indonesian-based supermarket chain which wants to market in-country its own brand of frozen, low-fat, midrange-calorie, full-taste microwaveable dinners. The offer made to Jerry by the supermarket corporation requires that a series of two decisions be made, now and 2 years hence. The current decision involves two alternatives: (1) *Lease* the facility from the supermarket chain, which has agreed to convert a current processing facility for immediate use by Jerry's company, or (2) *build and own* a processing and packaging facility in Indonesia. Possible outcomes of this first decision stage are good market or poor market depending upon the public's response.

The decision choices 2 years hence are dependent upon the lease-or-own decision made now. If Hill *decides to lease,* good market response means that the future decision alternatives are to produce at twice, equal-to, or one-half the original volume. This will be a mutual decision between the supermarket chain and Jerry's company. A poor market response will indicate a one-half level of production, or complete removal from the Indonesian market. Outcomes for the future decisions are, again, good and poor market responses.

As agreed by the supermarket company, the current decision for Jerry *to own* the facility will allow him to set the production level 2 years hence. If market response is good, the decision alternatives are four or two times original levels. The reaction to poor market response will be production at the same level or no production at all.

Construct the tree of decisions and outcomes for Hill Products and Services.

Solution

Identify the initial decision nodes and branches, and then develop the tree using the branches and the outcomes of good and poor market for each decision. Figure 19–7 details the first decision stage (D1) and outcome branches.

Decision now:
    Label it D1
    Alternatives: lease (L) and own (O)
    Outcomes: good and poor markets

Decision choices 2 years hence:
    Label them D2 through D5
    Outcomes: good market, poor market, and out-of-business

Choice of production levels for D2 through D5:
    Quadruple production ($4\times$); double production ($2\times$); level production ($1\times$); one-half production ($0.5\times$); stop production ($0\times$)

The alternatives for future production levels (D2 through D5) are added to the tree and followed by the market responses of good and poor (Figure 19–7). If the

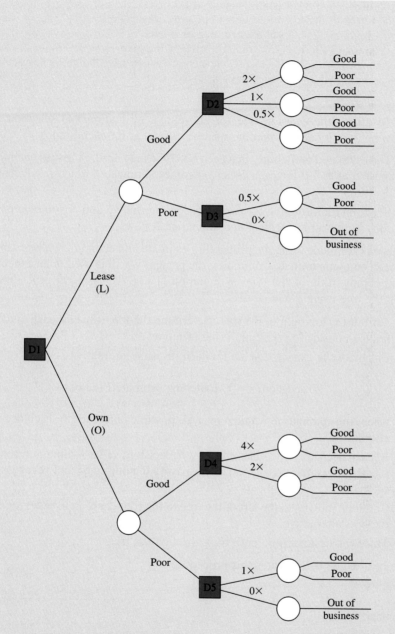

Figure 19–7   A two-stage decision tree identifying decision alternatives and possible outcomes.

stop-production (0✕) decision is made at D3 or D5, the only outcome is out-of-business. This completes the decision tree as presented in Figure 19–7. This is called a two-stage tree, since there are two decision points.

The size of the tree grows rapidly. A tree with as few as 10 decision nodes may have hundreds of final outcomes. Computer analysis to solve the tree and select the best decision path rapidly becomes essential.

To utilize the decision tree for alternative evaluation and selection, the following additional information must be estimated for each branch:

- The estimated probability that each outcome may occur. These probabilities must sum to 1.0 for each set of outcomes (branches) which result from a decision.
- Economic information for each decision alternative and possible outcome, such as, initial investment and annual cash flows.

Decisions are made using the probability estimate and economic value estimate for each outcome branch. Commonly the present worth is used in an expected-value computation of the type in Equation [19.1]. The general procedure to solve the tree using PW analysis is

1. Start at the top right of the tree. Determine the PW value for each outcome branch considering the time value of money.
2. Calculate the expected value for each decision alternative.

$$E(\text{decision}) = \sum (\text{outcome estimate})P(\text{outcome}) \qquad [19.2]$$

where the summation is taken over all possible outcomes for each decision alternative.

3. At each decision node, select the best E(decision) value—minimum cost for a cost-only situation, or maximum payoff if both costs and revenues are estimated.
4. Continue moving to the left of the tree to the root decision in order to select the best alternative.
5. Trace the best decision path back through the tree.

The next example illustrates this procedure.

Example 19.7

A decision is needed to either market or sell a new invention. If the product is marketed, the next decision is to take it international or national. Assume the details of the outcome branches result in the decision tree of Figure 19–8. The probabilities for each outcome and PW of costs and benefits (payoff in $ millions) are indicated. Determine the best decision at the decision node D1.

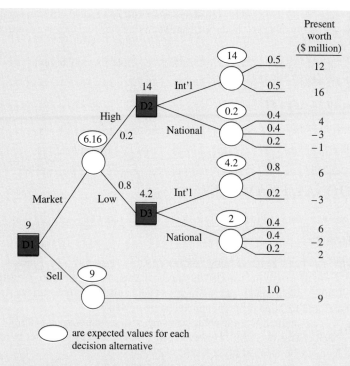

<table>
<tr><td></td><td></td><td></td><td></td><td></td><td>Present
worth
($ million)</td></tr>
</table>

Figure 19–8 are expected values for each decision alternative

Figure 19–8   Solution of a decision tree with stated present-worth values.

Solution

Use the procedure above to determine that the D1 decision alternative to sell the invention should maximize payoff.

1. Present worth of payoff is supplied in this example.

2. Calculate the expected PW payoff for alternatives from nodes D2 and D3 using Equation [19.2]. In Figure 19–8, to the right of decision node D2, the expected values of 14 and 0.2 in ovals are determined as

$$E(\text{international decision}) = 12(0.5) + 16(0.5) = 14$$

$$E(\text{national decision}) = 4(0.4) - 3(0.4) - 1(0.2) = 0.2$$

The expected PW payoffs of 4.2 and 2 for D3 are calculated in a similar fashion.

3. Select the larger expected payoff at each decision node. These are 14 (international) at D2 and 4.2 (international) at D3.

4. Calculate the expected payoff for the two D1 branches.

$$E(\text{market decision}) = 14(0.2) + 4.2(0.8) = 6.16$$

$$E(\text{sell decision}) = 9(1.0) = 9$$

The expected value for the sell decision is simple since the one outcome has a payoff of 9. The D1 node alternative sell yields the larger expected payoff of 9.

5. The largest expected PW payoff path is to select the sell branch at D1 for a guaranteed $9,000,000.

Problems 19.34 to 19.41

ADDITIONAL EXAMPLES

Example 19.8

SENSITIVITY OF TWO ALTERNATIVES, SECTION 19.2   The City of Blarney has a 0.3-mile stretch of heavily traveled highway to resurface. The Ajax Construction Company offers two methods of resurfacing. The first is a concrete surface for a cost of $150,000 and an annual maintenance charge of $1000.

The second method is an asphalt covering with a first cost of $100,000 and a yearly service charge of $2000. However, Ajax would also request that every third year the highway be touched up at a cost of $7500.

The city uses the interest rate on revenue bonds, 6% in this case, as the MARR. (*a*) Determine the breakeven number of years of the two methods. If the city expects an interstate to replace this stretch of highway in 20 years, which method should be selected? (*b*) If the touch-up cost increases by $500 per 0.1 mile every 3 years, is the decision sensitive to this cost?

Solution

(*a*) Set up the AW equations and determine the breakeven *n* value.

$$\text{AW of concrete} = \text{AW of asphalt}$$

$$-150{,}000(A/P,6\%,n) - 1000 = -100{,}000(A/P,6\%,n) - 2000$$

$$- 7500\left[\sum_{j} (P/F,6\%,j)\right](A/P,6\%,n)$$

where $j = 3, 6, 9, \ldots, n$. This relation is rewritten as

$$-50{,}000(A/P,6\%,n) + 1000 + 7500\left[\sum_{j}(P/F,6\%,j)\right](A/P,6\%,n) = 0 \qquad \textbf{[19.3]}$$

A value of $n = 39.6$ satisfies the equation, so a life of approximately 40 years is required to break even at 6%. Since the road is required for only 20 years, the asphalt surface should be constructed.

(b) The total touch-up cost will increase by $1500 every 3 years. Equation [19.3] is now

$$-50,000(A/P,6\%,n) + 1000$$
$$+ \sum_j \left[7500 + 1500\frac{j-3}{3}(P/F,6\%,j) \right](A/P,6\%,n) = 0$$

where $j = 3, 6, 9, \ldots, n$. The breakeven point is approximately 21 years which still favors the asphalt surface but only by a 1-year margin. The decision is technically insensitive to the stated increase in the estimated touch-up cost of $500 per 0.1 mile. However, intangible factors could alter the decision, as always.

Example 19.9

EXPECTED VALUES FOR A PROPOSAL, SECTION 19.5 The Holdar-XS Construction Company plans to build an apartment complex close to the edge of a partially leveled hill. Support for the soil on one side of the complex should ensure that no damage or injury will occur to the buildings or occupants during the rainy season. The amount of rainfall experienced at any one time can potentially cause varying amounts of damage. Table 19–5 itemizes the probability of certain rainfalls (inches) within a period of a few hours and the first cost to construct a support wall to ensure protection when the corresponding amount of rain occurs. Support wall construction will be financed by a 30-year loan with repayment at 9% per year compound interest.

Records indicate that an average of $20,000 in damage has occurred during heavy rains for similar complexes. Without taking the important intangibles of human safety, insurance requirements and income taxes into account, determine how much to spend on the support wall. Assume the MARR is established at the cost of debt capital, that is, 9% for the loan to construct the wall.

Table 19–5   Rainfall and support wall cost, Example 19.9

Rainfall, Inches	Probability of Greater Rainfall Occurring	Cost of Support Wall, $
2.0	0.3	10,000
2.5	0.1	15,000
3.0	0.05	22,000
3.5	0.01	30,000
4.0	0.005	42,000

Solution

Use AW analysis to find the most economic plan. Initially, determine the AW relation for loan payment and expected damage cost for each rainfall level. The $20,000 damage figure is used since it represents the best damage cost experience available.

AW = annual loan repayment + expected cost of annual damage

= wall cost$(A/P,9\%,30)$ + $(-20,000)$ (probability of greater rainfall)

For illustration, the AW at 3.0 inches is

$$AW = -22,000(A/P,9\%,30) - 20,000(0.05) = \$-3141$$

The resulting AW values in Table 19–6 indicate that the most economic choice is the $30,000 wall to protect against a 3.5-inch rainfall. The $22,000 wall for a 3.0-inch rain is a very close second.

Table 19–6	Annual worth for different support walls			
Rainfall, Inches	Support Wall Cost, $	Annual Loan Cost, $	Expected Annual Damage, $	AW, $
2.0	−10,000	−973	−6,000	−6,973
2.5	−15,000	−1,460	−2,000	−3,460
3.0	−22,000	−2,141	−1,000	−3,141
3.5	−30,000	−2,920	−200	−3,120
4.0	−42,000	−4,088	−100	−4,188

Comment

A rather large safety factor is commonly included when people are potentially endangered. So the economic analysis result is not used solely as the decision maker. By building to protect to a greater degree than actually needed on the average, the probabilities of damage are lowered and the actual costs are increased.

CHAPTER SUMMARY

In this chapter the emphasis is on determining the sensitivity of alternative selection to variation in one or more parameters or factors using a specific measure of worth. When two alternatives are compared, compute and graph the measure of worth for different values of the parameter to determine when each alternative is better.

When several parameters are expected to vary over a predictable range, the measure of worth can be plotted versus each parameter in terms of a percentage variation from the most likely estimate. This indicates at a glance where the decision is insensitive to a parameter (approximate horizontal plot) and where there is high sensitivity (larger slopes and nonlinear plots). Also, the use of three estimates for a parameter—most likely, pessimistic, and optimistic—can be made to determine which alternative is best among several. Independence between parameters is assumed in all these analyses.

The combination of estimates and probability allows us to calculate the expected value or anticipated long-run average. The simple expected-value relation

$$E(X) = \sum XP(X)$$

may be used to calculate amounts such as E(revenue), E(cost), E(cash flow), E(PW), and $E(i)$ for the entire cash-flow sequence of an alternative.

Decision trees are used to make a series of alternative selections. This is a way to explicitly take risk into account. It is necessary to make several types of estimates for a decision tree: outcomes for each possible decision, cash flows, and probabilities. Expected-value computations are coupled with measure-of-worth calculations to solve the tree and trace the best decision path.

CASE STUDY #7, CHAPTER 19: SENSITIVITY ANALYSIS OF WATER SUPPLY PLANS

Introduction

One of the most basic services provided by municipal governments is the delivery of a safe, reliable water supply. As cities grow and extend their boundaries to outlying areas, they often inherit water systems which were not constructed according to city codes. The upgrading of these systems is sometimes more expensive than installing one correctly in the first place. To avoid these problems, forward-thinking city officials sometimes install water systems beyond the existing city limits in anticipation of future growth. This case study was extracted from such a countywide water and waste-water management plan and is limited to only some of the water supply alternatives.

Procedure

From about a dozen suggested plans, five methods were developed by an executive committee as alternative ways for providing water to the study area. These methods were then subjected to a preliminary evaluation to identify the most promising alternatives. Six factors used in the initial rating were:

ability to serve area, relative cost, engineering feasibility, institutional issues, environmental considerations, and lead time requirement. Each factor carried the same weighting and had values ranging from 1 to 5, with 5 being best. After the top three alternatives were identified, each was subjected to a detailed economic evaluation for selection of the best alternative. These detailed evaluations included an estimate of the capital cost of each alternative amortized over 20 years at 8%-per-year interest and the annual maintenance and operation (M&O) costs. The annual cost (an AW value) was then divided by the population served to arrive at a monthly cost per household.

Results of Preliminary Screening

Table CS7–1 presents the results of the screening using the six rating factors on a scale of 1 to 5. Alternatives 1A, 3, and 4 were determined to be the three best and were chosen for further evaluation.

Detailed Cost Estimates for Selected Alternatives

Alternative 1A

Capital cost	
Land with water rights: 1720 hectares @ $5000 per hectare	$8,600,000
Primary treatment plant	2,560,000
Booster station at plant	221,425
Reservoir at booster station	50,325
Site cost	40,260
Transmission line from river	3,020,000
Transmission line right of way	23,350
Percolation beds	2,093,500
Percolation bed piping	60,400
Production wells	510,000
Well field gathering system	77,000
Distribution system	1,450,000
Additional distribution system	3,784,800
Reservoirs	250,000
Reservoir site, land and development	17,000
Subtotal	22,758,060
Engineering and contingencies	5,641,940
Total capital cost	$28,400,000

Table CS7-1   Results of rating using six factors for each alternative, Case Study #7

Alternative Name	Description	Ability to Supply Area	Relative Cost	Engineering Feasibility	Institutional Issues	Environmental Considerations	Lead-Time Requirement	Total
		Rating Factors						
1A	Receive city water and recharge wells	5	4	3	4	5	3	24
3	Joint city and county plant	5	4	4	3	4	3	23
4	County treatment plant	4	4	3	3	4	3	21
8	Desalt groundwater	1	2	1	1	3	4	12
12	Develop military water	5	5	4	1	3	1	19

Maintenance and operation costs (annual)
Pumping 9,812,610 KWH per year
@ $0.08 per KWH $785,009
Fixed operating cost   180,520
Variable operating cost   46,730
Taxes for water rights   48,160
       Total annual M&O cost $1,060,419

$$\text{Total annual cost} = \text{amortized capital cost} + \text{M\&O cost}$$
$$= 28,400,000(A/P,8\%,20) + 1,060,419$$
$$= 2,892,540 + 1,060,419$$
$$= \$3,952,959$$

Average monthly household cost to serve 95% of 4980 households is

$$\text{Household cost} = (3,952,959)\,\frac{1}{12}\,\frac{1}{4980}\,\frac{1}{0.95}$$
$$= \$69.63 \text{ per month}$$

Alternative 3

$$\text{Total capital cost} \quad = \$29,600,000$$
$$\text{Total annual M\&O cost} = \$867,119$$
$$\text{Total annual cost} \quad = \$3,881,879$$
$$\text{Household cost} \quad = \$68.38 \text{ per month}$$

Alternative 4

$$\text{Total capital cost} \quad = \$29,000,000$$
$$\text{Total annual M\&O cost} = \$1,063,449$$
$$\text{Total annual cost} \quad = 29,000,000(A/P,8\%,20) + 1,063,449$$
$$= \$4,017,099$$

$$\text{Household cost} \quad = (4,017,099)\,\frac{1}{12}\,\frac{1}{4980}\,\frac{1}{0.95}$$
$$= \$70.76 \text{ per month}$$

Conclusion

On the basis of the lowest monthly household cost, alternative 3 (joint city and county plant) is the most attractive.

Questions to Consider

1. If the environmental considerations factor was to have a weighting of twice as much as any of the other five factors, what would its percentage weighting be?

2. If the ability-to-supply-area and relative-cost factors were each weighted 20% and the other four factors 15% each, which alternatives would be ranked in the top three?

3. By how much would the capital cost of alternative 4 have to decrease in order to make it more attractive than alternative 3?

4. If alternative 1A served 100% of the households instead of 95%, by how much would the monthly household cost decrease?

CASE STUDY #8, CHAPTER 19: **SELECTING FROM SEQUENTIAL ALTERNATIVES USING A DECISION TREE**

Jerry Hill, president of Hill Products and Services, has asked a team of three colleagues, including you, to answer the questions about leasing or owning a building posed in Example 19.6. In a brief meeting with your team, Jerry is about to provide the CEO's view of the options described in the decision tree of Figure 19–7.

In the meeting at 2:00 on Thursday, after some introductory comments, Mr. Hill says: "Assume that the lease arrangement with the Indonesians will cost at least $175,000 per year for 4 years. If no dinners are made in years 3 and 4, the lease cost is still incurred. The build-and-own decision will cost approximately $800,000 for the initial investment in a building with an estimated market value of about $400,000 after 4 years." Jerry says: "The building will not be sold immediately if the stop-production decision $(0\times)$ is made for years 3 and 4."

Jerry also provides in a paper handout (see Table CS8–1), with no discussion, his staff's conclusions of the revenue and probability estimates for the tree in Figure 19–7 (decision nodes D1 through D5).

Hill Products and Services routinely uses a MARR of 15% and PW analysis for its major decisions. Use your group's discussion results, engineering economic analysis techniques, and a decision-tree approach to determine whether the lease or own option is better now and to determine the planned level of production for years 3 and 4 using the estimates provided by Jerry.

If requested to do so, prepare a 5- to 10-minute presentation of your team's approach and conclusions to be discussed at a class meeting.

Table CS8–1	Jerry Hill's handout of decisions and outcomes, Case Study #8		
Decision D1	**Estimate**	**Outcomes—Years 1 and 2**	
Lease or own		**Good**	**Poor**
	Revenue	$300,000	$150,000
	Probability	0.6	0.4
Decision D2	**Estimate**	**Outcomes—Years 3 and 4**	
2×		**Good**	**Poor**
	Revenue	$600,000	$300,000
	Probability	0.8	0.2
1×		**Good**	**Poor**
	Revenue	$300,000	$150,000
	Probability	0.8	0.2
0.5×		**Good**	**Poor**
	Revenue	$150,000	$75,000
	Probability	0.2	0.8
Decision D3	**Estimate**	**Outcomes—Years 3 and 4**	
0.5×		**Good**	**Poor**
	Revenue	$150,000	$75,000
	Probability	0.5	0.5
0×		**Out of Business**	
	Revenue	0	
	Probability	1.0	
Decision D4	**Estimate**	**Outcomes—Years 3 and 4**	
4×		**Good**	**Poor**
	Revenue	$1,200,000	$600,000
	Probability	0.8	0.2
2×		**Good**	**Poor**
	Revenue	$600,000	$300,000
	Probability	0.8	0.2
Decision D5	**Estimate**	**Outcomes—Years 3 and 4**	
1×		**Good**	**Poor**
	Revenue	$300,000	$150,000
	Probability	0.5	0.5
0×		**Out of Business**	
	Revenue	0	
	Probability	1.0	

PROBLEMS

Note: Once you have developed the engineering economy relations for a problem, it may be advantageous to use your spreadsheet system to determine the measure of worth for the resulting cash-flow sequence. This will significantly reduce the time needed to perform the computations and to plot the sensitivity results.

19.1 Metal Recyclers needs to purchase a new magnetic pickup device for moving scrap metal around the yard. The complete device will cost $62,000 and have an 8-year life and a salvage value of $1500. Annual maintenance, fuel, and overhead costs are estimated at $0.50 per metric ton moved. Labor cost will be $8 per hour for regular wages and $12 for overtime. A total of 25 tons can be moved in an 8-hour period. The salvage yard has handled in the past anywhere from 10 to 30 tons of scrap per day. If the company uses a MARR of 10%, plot the sensitivity of present worth of costs to the annual volume moved. Assume the operator is paid for 200 days of work per year. Use a 5-metric-ton increment for the graph.

19.2 An equipment alternative is being economically evaluated by three engineers separately. The first cost will be $77,000, and the life is estimated at 6 years with a salvage value of $10,000. The engineers disagree, however, on the estimated new revenue (gross income) the equipment will generate. Engineer Joe has made an estimate of $14,000 per year. Engineer Jane states that this is too low and estimates $16,000, while Engineer Carlos estimates $18,000 per year. If the before-tax MARR is 8%, use PW as the measure of worth to determine if these different estimates will change the decision to purchase the equipment.

19.3 Perform the same analysis as in Problem 19.2, except make it an after-tax consideration using MACRS depreciation and a 40% effective tax rate. Use estimated annual expenses of $1000 and an effective after-tax MARR estimated from the before-tax rate of 8%.

19.4 A rural-based manufacturing company needs 1000 square meters of storage space for 3 years. Purchasing land for $8000 and erecting a temporary metal building at $70 per square meter is one option. The president expects to sell the land for $9000 and the building for $12,000 after the 3 years. Another option is to lease space for $1.50 per square meter per month payable at the beginning of each year. The company president's MARR is 20%.

Use a present-worth analysis between building and leasing storage space to determine the sensitivity of the decision to the following situations: (*a*) construction costs go up 10% and lease costs go down to $1.25 per square meter per month; (*b*) lease costs remain at $1.50 per square meter per month, but construction costs vary from $50 to $90 per square meter.

19.5 For the data below, plot the sensitivity of the rate of return to the amount of the revenue gradient only for values from $500 to $1500. If the company would like a return of at least 20% per year, would this variation in the revenue gradient affect the decision to buy the truck?

> **Truck:** $P = \$74,000$; $n = 10$ years; SV = $15,000
> **Expenses:** $36,000 first year, increasing $3000 per year thereafter
> **Revenue:** $66,000 first year, decreasing $1000 per year thereafter

19.6 Consider the two air-conditioning systems detailed below.

	System 1	System 2
First cost, $	10,000	17,000
Annual operating cost, $	600	150
Salvage value, $	−100	−300
New compressor and motor cost at midlife, $	1,750	3,000
Life, years	8	12

Use AW analysis to determine the sensitivity of the economic decision to MARR values of 4, 6, 8, and 10%. Plot the sensitivity curve of AW for each air-conditioning system alternative to the MARR.

19.7 Reread Problem 19.6. If the MARR is 10%, plot the AW for each system for estimated life values from 4 to 8 years for system 1 and 6 to 12 years for system 2. Assume that the salvage values and annual operating costs retain the same estimated amounts for each life value. Further, assume that the compressor and motor are replaced at midlife. Plot AW for even-numbered years only. Which AW is more sensitive to a varying-life estimate?

19.8 Bonnie and John Bickers, who live in an urban area, tentatively plan to purchase a place in the woods for a weekend home. Alternatively, they have thought of buying a travel trailer and a four-wheel-drive vehicle to pull the trailer for vacations. The Bickerses have found a 5-acre tract with a small house 25 miles from their home. It will cost them $130,000, and they estimate they can sell the place for $145,000 in 10 years when their children are grown. The insurance and upkeep costs are estimated at $1500 per year, but this weekend site is expected to save the family $150 every day they don't go on a traveling vacation. The couple estimates that, even though the cabin is only 25 miles from home, they will travel 50 miles a day when at the cabin while working on it and visiting neighbors and local events. Their car averages 30 miles per gallon of gas.

The trailer and vehicle combination would cost $75,000 and could be sold for $20,000 in 10 years. Insurance and operating costs will average $1750 per year, but this alternative is expected to save $125 per vacation day. On a normal vacation, the Bickerses travel 300 miles each day. Mileage per gallon for the vehicle and trailer is estimated at 60% that of the family car. Assume gas costs $1.20 per U.S. gallon. The money earmarked for this purchase is currently invested and earns an average of 10% per year.

(a) Compute the breakeven number of days per year for the two plans.

(b) Plot the sensitivity of AW for each plan if the Bickerses' vacation time in the past has ranged from 10 to 22 days per year. Use 4-day increments.

19.9 For Problem 19.8, determine the sensitivity of the breakeven point to the cost of the trailer-and-vehicle alternative. The Bickerses have obtained quotes that range from $60,000 to $100,000 from different dealers.

19.10 (a) Manually calculate and plot the sensitivity of rate of return versus the life of a 5% $50,000 bond that is offered for $43,500 with bond interest paid quarterly. Consider bond lives of 10, 12, 15, 18, and 20 years. (b) Use spreadsheet analysis to solve this problem.

19.11 Leona has been offered an investment opportunity that will require a cash outlay of $30,000 now for a cash inflow of $3500 for each year of investment. However, she must state now the number of years she plans to retain the investment. Additionally, if the investment is retained for 6 years, $25,000 will be returned to investors, but after 10 years the return is anticipated to be only $15,000 and after 12 years it is estimated to be $8000. If money is currently worth 8% per year, is the decision sensitive to the retention period?

19.12 Determine the sensitivity of the economic service life of Problem 10.26, parts (a) and (b), to the cost gradient. Investigate the gradient values of $60 to $140 in increments of $20 and plot the results on the same graph. Spreadsheet analysis is recommended. Use $i = 5\%$ per year.

19.13 For plans A and B graph the sensitivity of PW values at 20% per year for the range -50% to $+100\%$ of the following single-point estimates for each of the factors: (a) first cost, (b) AOC, and (c) annual revenue.

	Plan *A*	Plan *B*
First cost, $	500,000	375,000
AOC, $	75,000	80,000
Annual revenue, $	150,000	130,000
Salvage value, $	50,000	37,000
Expected life, years	5	5

19.14 Use a spreadsheet system to determine and graph the sensitivity of the rate of return to a $\pm 25\%$ change in (*a*) purchase price and (*b*) selling price for the following investment. A family purchased an old house for $25,000 with the plan to make major improvements and sell it at a profit. Improvement expenses were $5000 the first year, $1000 the second year, and $800 the third year. They paid taxes of $500 per year and sold the house after 3 years for $35,000.

19.15 For process *M* estimates plot on one graph the sensitivity of AW for the following range of each factor.

Factor	Range
First Cost	−30% to +100%
AOC	−30% to +30%
Annual revenue	−50% to +25%

Process *M*	Estimate
First cost, $	80,000
Salvage value, $	10,000
Life, years	10
AOC, $	15,000
Annual revenue, $	39,000

19.16 Graph the sensitivity of what you should be willing to pay now for a 9% $10,000 bond due in 10 years if there is a $\pm 30\%$ change in (*a*) purchase price, (*b*) dividend rate, or (*c*) your desired nominal rate of return, which is 8% per year compounded semiannually. The bond dividends are paid semiannually.

19.17 An engineer must decide between two ways to pump concrete up to the top floors of a seven-story office building to be constructed. Plan 1 requires the purchase of equipment for $6000 which costs between $0.40 and $0.75 per metric ton to operate, with a most likely cost of $0.50 per metric ton. The asset is able to pump 100 metric tons per day. If purchased, the asset will last for 5 years, have no salvage value, and be used from 50 to 100 days per year. Plan 2 is an equipment-leasing option and is expected to cost the company $2500 per year for equipment with a low cost estimate of $1800 and a high estimate of $3200 per year. In addition, an extra $5-per-hour labor cost will be incurred for operating the leased equipment. Plot the AW of each plan versus total annual operating cost or lease cost at $i = 12\%$. Which plan should the engineer recommend if the most likely estimate of use is (*a*) 50 days per year? (*b*) 100 days per year?

19.18 Either a new or used machine may be purchased. The estimated costs are

	New Machine	Used Machine
First cost, $	44,000	23,000
Annual operating cost, $	7,000	9,000
Annual repair cost, $	210	350
Overhaul every 2 years, $	—	1,900
Overhaul every 5 years, $	2,500	—
Salvage value, $	4,000	3,000
Life, years	15	8

The time of overhaul can vary from 2 to 4 years for the used machine and from 4 to 6 years for the new machine. Plot the AW values for these three estimates at $i = 18\%$ per year, and determine if they will alter the decision of which machine to purchase.

19.19 A meat-packing plant must decide between two ways to cool cooked hams. Spraying cools to 30°C using approximately 80 liters of water for each ham. The immersion method uses only 16 liters per ham, but an extra initial cost for equipment of $2000 and extra maintenance costs of $100 per year for the 10-year life are estimated. Ten million hams per year are cooked, and water costs $0.12 per 1000 liters. Another cost is $0.04 per 1000 liters for wastewater treatment. The MARR is 15% per year.

If the spray method is selected, the amount of water used can vary from an optimistic value of 60 liters to a pessimistic value of 120 liters with 80 liters being the most likely amount. The immersion technique always takes 16 liters per ham. How will this varying use of water for the spray method affect the economic decision?

19.20 When the country's economy is expanding, the AB Investment Company is optimistic and expects a MARR of 15% for new investments. However, in a receding economy the expected return is 8%. Normally a 10% return is required. An expanding economy causes the estimates of asset life to go down about 20%, and a receding economy makes the n values increase about 10%. Plot the sensitivity of present worth versus (*a*) the MARR and (*b*) the life values for the two plans detailed below using the most likely estimates for the other factors. (*c*) Considering all your analyses, under which scenario, if any, should plan M or Q be rejected?

	Plan M	Plan Q
Initial investment, $	−100,000	−110,000
Annual NCF, $	+15,000	+19,000
Life, years	20	20

19.21 *As a team assignment, two or three students* may use the same scenario of estimates described in Problem 19.20, but for data use the estimates below for plans A and B. Be prepared to discuss your answers and conclusions in class.

	Plan A	Plan B Asset 1	Plan B Asset 2
First cost, $	10,000	30,000	5,000
Annual operating costs, $	500	100	200
Salvage value, $	1,000	5,000	−200
Estimated life, years	40	40	20

19.22 Lowell's father is drilling two oil wells. The estimated flow rate in barrels per day and probabilities that Lowell has recorded, based on several conversations, are listed below. Calculate the expected flow rate for each well. Comment on the relative size of Lowell's probability estimates for the three estimated flow rates.

	Expected Flow, Barrels/Day 100	200	300
Well #1	0.15	0.75	0.10
Well #2	0.45	0.10	0.45

19.23 You have an assignment to do for your engineering economy course. You ask four different people in your class how long it took for them to finish the assignment. Their estimates, in minutes, are: 2, 20, 30, and 120. (*a*) If you place equal weight on each of their answers in order to estimate your own time, what is the expected time for you to finish the assignment? (*b*) If you completely disregard the smallest and largest times because you don't believe them, estimate your expected time to completion. Did the two extreme estimates seem to significantly change the expected value?

19.24 The variable Y is defined as 2^n for $n = 1, 2, 3, 4$ with probabilities of 0.3, 0.4, 0.233, 0.067, respectively. Determine the expected value of Y.

19.25 The AOC value for an alternative is expected to be one of two values. Your office partner told you that the high value is $2800 per year. If her computations show a probability of 0.75 for the high value and an expected AOC of $2575, what is the low AOC value used in the computation of the average AOC value?

19.26 A total of 28 different proposals have been evaluated by your department during the past year. Their rate-of-return estimates are summarized below with the $i*$ values rounded to the nearest integer. If all proposals were accepted for investment, calculate the overall expected rate of return, $E(i)$.

Proposal Rate of Return, %	Number of Proposals
−8	1
−5	1
0	5
2	1
4	3
6	4
8	2
10	5
12	1
15	4
20	1
	28

19.27 Lunch-a-Bunch food service company has performed an economic analysis of proposed service in a new region of the country. The three-estimate approach to sensitivity analysis has been used with each of the optimistic and pessimistic estimates expected to occur with a 15% chance. Use the AW values to compute the expected AW for the new service proposal.

	Optimistic	Most Likely	Pessimistic
AW value, $	+150,000	+5,000	−275,000

19.28 (*a*) Determine the expected present worth of the following cash-flow series if each series may be realized with the probability shown at the head of each column. Assume $i = 20\%$.
(*b*) Determine the expected AW value for the same cash-flow series.

	Annual Cash Flows, $		
Year	Prob. = 0.5	Prob. = 0.2	Prob. = 0.3
0	−5000	−6000	−4000
1	1000	500	3000
2	1000	1500	1200
3	1000	2000	−800

19.29 The officers of a winter resort hotel are thinking of constructing an outdoor 18-hole miniature golf course for guest use. Because of the northerly location of the resort, there is a 60% chance of a 120-day season of good outdoor weather, a 20% chance of a 150-day season, and a 20% chance of a 165-day season. The miniature golf course will be used by an estimated 350 persons each day of the 4-month season, but by only 100 per day for each extra day the season lasts. The course will cost $375,000 to construct and will require a $25,000 rework cost after 4 years, and the annual maintenance cost will be $36,000. The green fees will be $4.25 per person. If a life of 10 years is anticipated before the course is replaced and a 12%-per-year return is desired, determine if the course should be constructed.

19.30 The owner of Ace Roofing may invest $100,000 in new equipment. A life of 6 years and a salvage value of 12% of first cost are anticipated. The annual extra revenue will depend upon the state of the housing and construction industry. The extra revenue is expected to be only $20,000 per year if the current slump in the industry continues. Real estate economists estimate a 50% chance of the slump lasting 3 years and give it a 20% chance of continuing for 3 additional years. However, if the depressed market does improve, either during the first or second 3-year period, the revenue of the investment is expected to increase $35,000 annually. Can the company expect to make a return of 8%-per-year on its investment? Use present-worth analysis.

19.31 Perform an after-tax analysis to determine for which rainfall level to construct an expected 25-year-life retaining wall and rainwater drainage system at a new townhouse complex. Use the criterion of lowest equivalent annual cost over a 25-year period to make the decision. Assume that the effective tax rate is 50% and that the construction cost will be provided by an 8%-per-year 25-year loan. Assume that the loan principal is reduced an equal amount each year with the remainder of the payment applied to interest.

The probability of a 2-hour rain shower greater than a specific amount, construction costs, and the expected damage cost are summarized below.

2-Hour Rainfall, centimeters	Probability of Greater Rainfall	Construction Cost to Carry Rainfall, $	Expected Annual Damage Cost for 2-Hour Rainfall Level, $
1.0	0.6	15,000	1,000
2.0	0.3	16,000	1,500
2.5	0.1	18,000	2,000
3.0	0.02	21,000	5,000
3.5	0.005	28,000	9,000
4.0	0.001	35,000	14,000

19.32 Jeremy has $5000 to invest. If he puts the money in a certificate of deposit (CD), he is assured of receiving an effective 6.35% per year for 5 years. If he invests the money in stocks, he has a 50/50 chance of one of the following cash-flow sequences for the next 5 years.

	Annual Cash Flows, $	
	Prob. = 0.5	**Prob. = 0.5**
Year	**Stock 1**	**Stock 2**
0	−5000	−5000
1–4	+250	+600
5	+6800	+5400

Finally, Jeremy can invest his $5000 in real estate for the 5 years with the following cash flow and probability estimates.

	Annual Cash Flows, $		
Year	**Prob. = 0.3**	**Prob. = 0.5**	**Prob. = 0.2**
0	−5000	−5000	−5000
1	−425	0	+500
2	−425	0	+600
3	−425	0	+700
4	−425	0	+800
5	+9500	+7200	+5200

Which of the three investment opportunities offers the best expected rate of return for Jeremy?

19.33 The California Company has $1 million in an investment pool which the board of directors plans to place in projects with different D-E mixes varying from 20-80 to 80-20. To assist with the decision, the plot shown below, prepared by the chief financial officer, of currently estimated annual equity rates of return (*i* on equity capital) versus various D-E mixes will be used. All investments will be for 10 years with no intermediate cash flows in or out of the projects. The motion passed by the board is to invest as follows:

D-E mix	20–80	50–50	80–20
Percent of pool, %	30	50	20

(a) What is the current estimate of the expected annual rate of return on the company's equity capital for the $1 million investments after 10 years?

(b) What is the actual amount of equity capital invested now, and what is the expected total amount after 10 years for the board-approved investment plan?

(c) If inflation is expected to average 4.5% per year over the next 10-year period, determine the real interest rates at which the equity investment funds will grow and determine the buying power in terms of today's dollars of the actual amount accumulated after the 10 years.

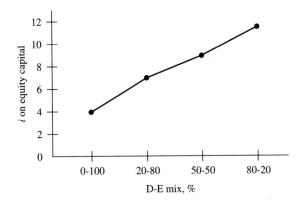

19.34 For the decision tree branch shown, determine the expected values of the two outcomes if decision D3 is already selected and the maximum outcome value is sought. (This decision branch is part of a larger tree.)

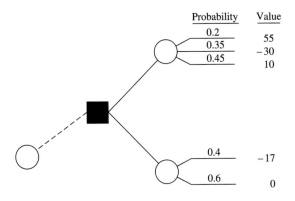

19.35 A large decision tree has an outcome branch that is detailed for this problem. If decisions D1, D2, and D3 are all options in a 1-year time period, find the decision path which maximizes the outcome value. There are specific dollar investments necessary for decision nodes D1, D2, and D3 as indicated on each branch.

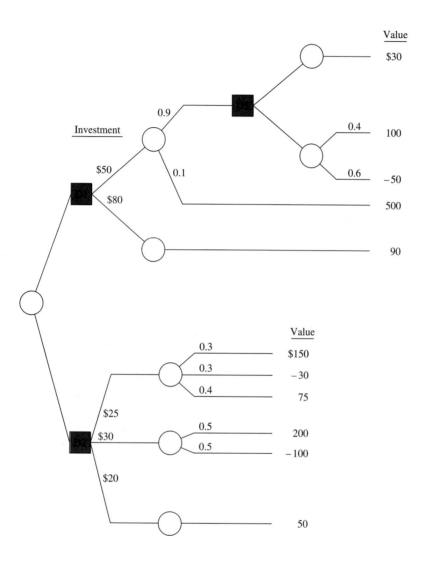

19.36 Decision D4, which has three possible outcomes—*x*, *y*, or *z*—must be made in year 3 of a 6-year study period in order to maximize the expected value of present worth. Using a rate of return of 15%, the investment required in year 3, and the estimated net cash flow for years 4 through 6, determine which decision should be made in year 3. (This decision node is part of a larger tree.)

	Investment Required, Year 3	NCF (× $1000) Year 4	NCF (× $1000) Year 5	NCF (× $1000) Year 6	Outcome Probability
High	$200,000	$50	$50	$50	0.7
Low		40	30	20	0.3
High	75,000	30	40	50	0.45
Low		30	30	30	0.55
High	350,000	190	170	150	0.7
Low		−30	−30	−30	0.3

19.37 A total of 5000 mechanical subassemblies are needed annually on a final assembly line. The subassemblies can be obtained in one of three ways: (1) *Make them* in one of three plants owned by the company; (2) *buy them off the shelf* from the one and only manufacturer; or (3) *contract to have them made* to specifications by a vendor.

The estimated annual cost for each alternative is dependent upon specific circumstances of the plant, producer, or contractor. The information below details the circumstance, a probability of occurrence, and the estimated annual cost. Construct and solve a decision tree to determine the least-cost alternative to provide the subassemblies.

Decision Alternative	Outcomes	Probability	Annual Cost for 5000 Units, $
(1) Make	Plant:		
	A	0.3	250,000
	B	0.5	400,000
	C	0.2	350,000
(2) Buy off shelf	Quantity:		
	< 5,000, pay premium	0.2	550,000
	5,000 available	0.7	250,000
	> 5,000, forced to buy	0.1	290,000
(3) Contract	Delivery:		
	Timely delivery	0.5	175,000
	Late delivery; then buy some off shelf	0.5	450,000

19.38 A friend at work developed the accompanying decision tree to determine if high- or low-volume production (node D1) will maximize the value of the outcomes. Assume the investments and outcome values are coded dollar values. Do not attempt to account for the time value of money. (*a*) Perform the "foldback" using probability and expected-value computations to determine if high or low production is better. (*b*) Would the decision at node D1 be different if the initial investment for high production were increased by 10?

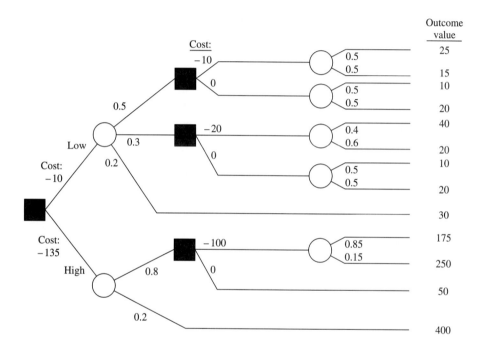

19.39 The president of ChemTech is trying to decide whether to start a new product line or purchase a small company. It is not financially possible to do both. To make the product for a 3-year period will require an initial investment of $250,000. The expected annual cash flows with probabilities in parentheses are: $75,000 (0.5), $90,000 (0.4), and $150,000 (0.1).

To purchase the small company will cost $450,000 now. Market surveys indicate a 55% chance of increased sales for the company and a 45% chance of severe decreases with an annual cash flow of $25,000. If decreases are experienced in the first year, the company will be sold immediately (during year 1) at a price of $200,000. Increased sales could be $100,000 the first 2 years. If this occurs, a decision to expand after

2 years at an additional investment of $100,000 will be considered. This expansion could generate cash flows with indicated probabilities as follows: $120,000 (0.3), $140,000 (0.3), and $175,000 (0.4). If expansion is not chosen, the current size will be maintained with anticipated sales to continue.

Assume there are no salvage values on any investments. Use the description above and a 15%-per-year return to do the following:

(*a*) Construct a decision tree with all values and probabilities shown.

(*b*) Determine the expected PW values at the 'expansion/no expansion' decision node after 2 years provided sales are up.

(*c*) Determine what decision should be made now to offer the greatest return possible for ChemTech.

(*d*) Explain in words what would happen to the expected values at each decision node if the planning horizon were extended beyond 3 years and all cash-flow values continued as forecasted in the description.

19.40 Select the decision path for the accompanying decision tree which will maximize expected revenue cash flow. The tree defines two options concerning an inspection task: the finished product can be inspected first (top branch), or it can be placed directly into service and checked during operation.

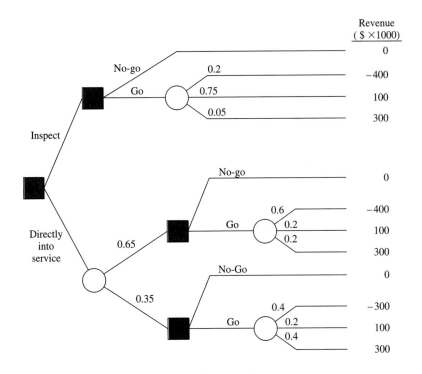

19.41 Think of your own decision making currently or in the recent past. Select a situation which requires some thought and that has several alternatives and possible outcomes. Describe the situation as a decision tree. To the degree possible, place values and probabilities on the outcome branches. Solve the tree and determine which decision path is best for you. Is this the decision you selected or want to select? Be as objective as you can in your appraisal of the situation. (This may also be useful as a team assignment for two or three students.)

chapter

20

More on Variation and Decision Making under Risk

This chapter expands our understanding about the variation of estimates, consideration of risk and probability, and decision making under risk that was introduced in earlier chapters. Fundamentals discussed include probability distributions, especially their graphs and properties of expected value and dispersion; random sampling; and the use of the simulation approach and multiple-criteria analysis to help account for variation in engineering economy studies.

Though elementary in its coverage of variation and probability, this chapter offers additional approaches to topics discussed in the first sections of Chapter 1: the role of engineering economy in decision making and the use of economic analysis in implementing the problem-solving process. These techniques are more time-consuming than using estimates made with certainty, so they should be used for parameters considered essential to decision making about alternatives.

Purpose: Learn to incorporate decision making under risk into an engineering economy analysis using the basics of probability distributions, simulation analysis, and multiple-criteria evaluation.

This chapter will help you:

Certainty, risk, and uncertainty	1. Understand the role of certainty, risk, and uncertainty in an engineering economy analysis.
Variables and distributions	2. Construct the probability distribution and cumulative distribution for a variable.
Random sample	3. Develop a random sample from the cumulative distribution of a variable.
Average and dispersion	4. Estimate the expected value and standard deviation of a population from a random sample.
Monte Carlo and simulation	5. Use Monte Carlo sampling and the simulation approach to select an alternative.
Multiple criteria	6. Use a multiple-criteria evaluation technique to select an alternative.

20.1   INTERPRETATION OF CERTAINTY, RISK, AND UNCERTAINTY

All things in the world vary—one from another, over time, and with different environments. We are guaranteed that variation will occur in engineering economy due to its emphasis on decision making for the future. Except for the use of breakeven analysis (Chapters 8 and 15), sensitivity analysis (Chapter 16), and a very brief introduction to expected values (Chapter 19), all of our numbers and estimates have been *certain;* that is, no variation in the amount has entered into the computations of PW, AW, ROR, or any relations used. For example, the estimate that cash flow next year will be +$4500 is one of certainty. Certainty is, of course, not present in the "real world" now and surely not in the future. We can observe outcomes with a high degree of certainty, but even this depends upon the accuracy and preciseness of the scale or measuring instrument.

To allow a parameter of an engineering economy study to vary implies that risk, and possibly uncertainty, are introduced.

Risk.   Engineering economic analysis should consider risk when there are anticipated to be two or more observable values for a parameter *and* it is possible to assign or estimate the chance that each value may occur. As an illustration, decision making under risk may be introduced when an annual cash-flow estimate has a 50-50 chance of being either −$1000 or +$500.

Uncertainty.   Decision making under uncertainty means there are two or more values observable, but the chances of their occurring cannot be estimated or no one is willing to assign the chances. The observable values in uncertainty analysis are often referred to as *states of nature.* For example, consider the states of nature to be the rate of national inflation in a particular country during the next 2 to 4 years: remain low, increase 5% to 10% annually, or increase 20% to 50% annually. If there is absolutely no indication that the three values are equally likely, or that one is more likely than the others, this is a statement that indicates decision making under uncertainty.

Example 20.1 explains how a parameter can be described and graphed to prepare for decision making under risk.

Example 20.1

Charles and Sue are both seniors in college and plan to be married next year. Based upon conversations with friends who have recently married, the couple decided to make separate estimates of what they each expected the ceremony to cost, with the chance that each estimate is actually observed expressed as a percentage. (*a*) Their

separate estimates are tabulated below. Construct two graphs: one of Charles's estimated costs versus his chance estimates, and one for Sue. Comment on the shape of
the plots relative to each other. (*b*) After some discussion, they decided the ceremony
should cost somewhere between $7500 and $10,000. All values in-between the two
limits are equally likely with a chance of 1 in 25. Plot these values versus chance.

Charles		Sue	
Estimated Cost, $	**Chance, %**	**Estimated Cost, $**	**Chance, %**
3,000	65	8,000	33
5,000	25	10,000	33
10,000	10	15,000	33

Solution

(*a*) Figure 20–1*a* presents the plot for Charles's and Sue's estimates, with the cost
scales aligned. Sue expects the cost to be considerably higher than Charles.
Additionally, Sue places equal (or uniform) chances on each value. Charles
places a much higher chance on lower-cost values; 65% of his chances are devoted to $3000, and only 10% to $10,000, which is Sue's middle cost estimate.
The plots clearly show the different perceptions about their estimated wedding
costs.

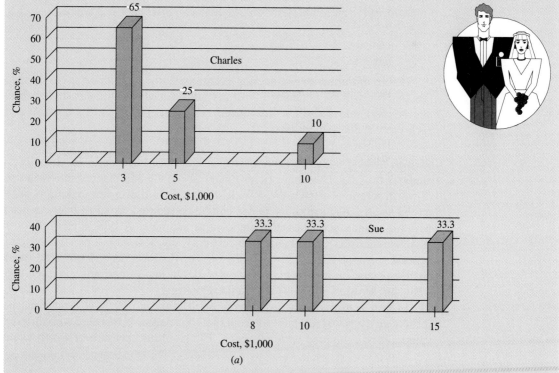

Figure 20–1a   Plots of cost estimate versus chance for (a) specified values.

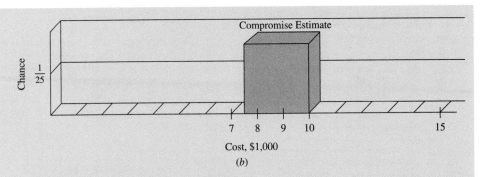

Figure 20–1b   (b) A continuous range of values, Example 20.1

(b) Figure 20–1b is the plot at a chance of 1 in 25 for the continuum of costs from $7500 through $10,000.

Comment

One significant difference between the cost estimates in parts (a) and (b) is that of discrete and continuous values. Charles and Sue first made specific, discrete estimates with chances associated with each value. The compromise estimate they reached is a continuous range of values from $7500 to $10,000 with some chance associated with every value in-between these limits. In the next section, we will introduce the term *variable* and define two types of variables—*discrete* and *continuous*—which have been illustrated here.

Before initiating an engineering economy study, it is important to decide if the analysis will be conducted with certainty for all parameters or if risk (or uncertainty) analysis will be introduced. A summary of the meaning and use for each type of analysis follows.

Decision Making under Certainty   This is what we have done in virtually all analyses thus far. Deterministic estimates are made and entered into measure-of-worth relations—PW, AW, FW, ROR, B/C—and decision making is based on the results. The values estimated can be considered the most likely to occur with all chance placed on the single-value estimate. A typical example is an asset's first cost estimate made with certainty, say, $50,000. A plot will have the general form of Figure 20–1a with one vertical bar at $50,000 and 100% chance placed on it. The term *deterministic*, in lieu of certainty, is often used when single-value estimates are used exclusively.

In fact, sensitivity analysis (Section 19.1 through 19.3) is simply another form of analysis with certainty, except that the analysis is repeated with different values, *each estimated with certainty*. The resulting measure-of-worth values are calculated and graphically portrayed to determine the decision's sensitivity to different estimates for one or more parameters.

Decision Making under Risk   Now the element of chance is formally taken into account. However, it is more difficult to make a clear decision because the analysis attempts to take *variation* into account. One or more parameters in an alternative will be allowed to vary. The estimates will be expressed as in Example 20.1 or in slightly more complex forms. Fundamentally, there are two ways to consider risk in an analysis:

*   **Expected-value analysis.** Use the chance and parameter estimates to calculate expected values, E(parameter), via formulas such as Equation [19.1]. Analysis results in series of E(cash flow), E(AOC), and the like, and the final result is the expected value for a measure of worth, such as E(PW), E(AW), E(ROR), E(B/C). To select the alternative, choose the most favorable expected value of the measure of worth. In an elementary form, this is what we learned in Sections 19.4 and 19.5 on expected values and then applied in Section 19.6 on decision trees. The computations can become more elaborate, but the principle is fundamentally the same.

*   **Simulation analysis.** Use the chance and parameter estimates to generate repeated computations of the measure-of-worth relation by randomly sampling from a plot for each varying parameter similar to those in Figure 20–1. When a representative and random sample is complete, an alternative decision is made utilizing a table or plot of the measure-of-worth results. Usually, graphics are an important part of decision making via simulation analysis. Basically, this is the approach discussed in the rest of this chapter.

Decision Making under Uncertainty   When chances are not known for the identified states of nature (or values) of the uncertain parameters, the use of expected-value-based decision making under risk as outlined above is not an option. In fact, it is difficult to determine what criterion to use to even make the decision. Note that if it is possible to agree that each state is equally likely, then all states have the same chance, and the situation reduces to one of decision making under risk because now expected values can be determined.

Because of the relatively inconclusive approaches necessary to incorporate decision making under uncertainty into an engineering economy study, the techniques can be quite useful but are beyond the intended scope of this text. Some of the various guidelines and rules—maximax, maximin, minimax regret, Hurwicz—are treated briefly in other engineering economy texts, such as Thuesen and Fabrycky (refer to the Bibliography at the end of the book).

The remainder of this chapter concentrates upon decision making under risk as applied in an engineering economy study. The next three sections provide foundation material necessary to design and correctly conduct a simulation analysis considering risk, as discussed in Section 20.5.

Problems 20.1 to 20.2

20.2   ELEMENTS IMPORTANT TO DECISION MAKING UNDER RISK

Some basics of probability and statistics are essential to correctly perform decision making under risk via expected-value or simulation analysis. These basics are explained here in very elementary terms. If you are already familiar with them, this section will provide a review.

Random Variable (or Variable)   A characteristic or parameter that can take on any one of several values. Variables are classified as *discrete* or *continuous*. Discrete variables have several specific, isolated values, while continuous variables can assume any value between two stated limits, called the range of the variable.

There is the estimated life of an asset is a discrete variable. As a simple illustration, n may be expected to have values of $n = 3, 5, 10,$ or 15 years, and no others. The rate of return is an example of a continuous variable. As we learned in Section 7.1, i can vary from -100% to ∞, that is, $-100\% \le i < \infty$. The range of possible values for n (discrete) and i (continuous) are shown as the x axes in Figure 20–2a. (In probability texts, capital letters symbolize a variable, say X, and small letters, x, identify a specific value of the variable. Though correct, this level of rigor in terminology is not included in this chapter.)

Probability   A number between 0 and 1.0 which expresses the chance in decimal form that a random variable (discrete or continuous) will take on any value from those identified for it. Probability is simply the amount of chance, discussed previously, divided by 100. Probabilities are commonly identified by $P(X_i)$ or $P(X = X_i)$, which is read as the probability that the variable X takes on the value X_i. (Actually, for a continuous variable, the probability at a single value is zero, as shown in a later example.) The sum of all $P(X_i)$ for a variable must be 1.0, a requirement we have already discussed in Section 19.4. The probability scale, like the percentage scale for chance in Figure 20–1, is indicated on the ordinate (y axis) of a graph. Figure 20–2b shows the 0 to 1.0 range for the random variables n and i.

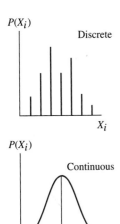

Probability Distribution   Describes how probability is distributed over the different values of a variable. Discrete variable distributions look significantly different from continuous variable distributions, as indicated by the inset at the left. The individual probability values are stated as

$$P(X_i) = \text{probability that } X \text{ equals } X_i \qquad \text{[20.1]}$$

The distribution may be developed in one of two ways: by listing each probability value for each possible variable value (see Example 20.2) or by a mathematical description or expression which states probability in terms of the possible variable values (Example 20.3).

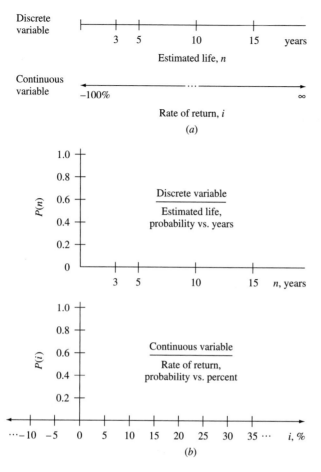

Figure 20–2 shows discrete variable and continuous variable scales with Estimated life, *n* and Rate of return, *i* labels, marked (*a*). Below are plots of P(n) vs. n, years labeled "Discrete variable / Estimated life, / probability vs. years" and P(i) vs. i, % labeled "Continuous variable / Rate of return, / probability vs. percent", marked (*b*).

Figure 20–2   (a) Discrete and continuous variable scales, and
(b) scales for a variable versus its probability.

Cumulative Distribution   Also called the cumulative probability distribution, this is the accumulation of probability over all values of a variable up to and including a specified value. Identified by $F(X_i)$, each cumulative value is calculated as

$$F(X_i) = \text{sum of all probabilities through the value } X_i$$

$$= P(X \le X_i) \qquad \text{[20.2]}$$

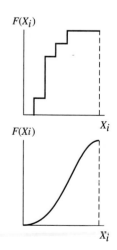

As with a probability distribution, cumulative distributions appear differently for discrete (stair-stepped) and continuous variables (smooth curve). Examples 20.2 and 20.3 illustrate cumulative distributions which correspond to specific probability distributions. We will use what we learn here about graphs of $F(X_i)$ in the next section to develop a random sample.

Example 20.2

Carl, a doctor who practices at Medical Center Hospital, knows of an antibiotic which may be prescribed for a patient with a particular infection expected to respond to the drug. Tests indicate the drug has been applied up to 6 times per day without harmful side effects. Also, if no drug is used, there is a positive probability that the infection will be reduced by a person's own system.

Published drug test results provide good probability estimates of positive reaction (that is, reduction in the infection count) within 48 hours for different numbers of treatments per day. Use the probabilities listed below to construct for Carl a probability distribution and a cumulative distribution for the number of treatments per day.

Number of Treatments per Day	Probability of Infection Reduction
0	0.07
1	0.08
2	0.10
3	0.12
4	0.13
5	0.25
6	0.25

Solution

Define the random variable T as the number of treatments per day. Since T can take on only seven different values, it is a discrete variable. The probability of infection count reduction is listed for each value in column 2 of Table 20–1. The cumulative probability $F(T_i)$ is determined using Equation [20.2] by adding all $P(T_i)$ values through T_i, as indicated in column 3 of Table 20–1.

Table 20–1	Probability distribution and cumulative distribution for Example 20.2	
(1)	**(2)**	**(3)**
		Cumulative
Number per Day,	**Probability,**	**Probability,**
T_i	$P(T_i)$	$F(T_i)$
0	0.07	0.07
1	0.08	0.15
2	0.10	0.25
3	0.12	0.37
4	0.13	0.50
5	0.25	0.75
6	0.25	1.00

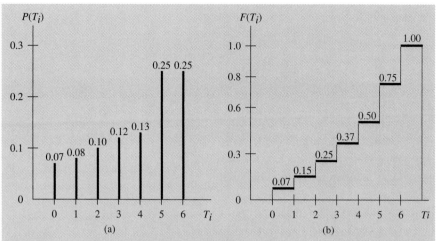

Figure 20-3   (a) Probability distribution, $P(T_i)$, and (b) cumulative distribution, $F(T_i)$, for Example 20.2.

Figure 20–3a and b shows plots of the probability distribution and cumulative distribution, respectively. The summing of probabilities to obtain $F(T_i)$ gives the cumulative distribution the stair-stepped appearance, and in all cases the final $F(T_i) = 1.0$, since the total of all $P(T_i)$ values must equal 1.0.

Comment

Rather than using a tabular form to state $P(T_i)$ and $F(T_i)$ values, it is common to express them for each value of the variable. For Example 20.2, the values are listed below, where $F(T_i) = P(X \le X_i)$ as in Equation [20.2].

$$P(T_i) = \begin{cases} 0.07 & T_1 = 0 \\ 0.08 & T_2 = 1 \\ 0.10 & T_3 = 2 \\ 0.12 & T_4 = 3 \\ 0.13 & T_5 = 4 \\ 0.25 & T_6 = 5 \\ 0.25 & T_7 = 6 \end{cases} \qquad F(T_i) = \begin{cases} 0.07 & T_1 = 0 \\ 0.15 & T_2 = 1 \\ 0.25 & T_3 = 2 \\ 0.37 & T_4 = 3 \\ 0.50 & T_5 = 4 \\ 0.75 & T_6 = 5 \\ 1.00 & T_7 = 6 \end{cases}$$

In basic engineering economy situations, the probability distribution for a continuous variable is commonly expressed as a mathematical function, such as a *uniform distribution,* a *triangular distribution* (both discussed in Example 20.3 in terms of cash flow), or the more complex, but commonly used, *normal distribution.* For continuous variable distributions, the symbol $f(X)$ is routinely used instead of $P(X_i)$ and $F(X)$ is used instead of $F(X_i)$, simply because the point probability for a continuous variable is zero. Thus, the $f(X)$ and $F(X)$ are continuous curves, not vertical bars and lines between point values, as is the case for discrete variables.

Example 20.3

As president of her own business, Sallie has observed the monthly cash flows into company accounts from two long-standing clients over the last 3 years. Remembering her college course in engineering economy, Sallie concludes the following about the distributions of these monthly cash flows:

Client 1

Estimated low cash flow: $10,000

Estimated high cash flow: $15,000

Most likely cash flow: same for all values

Distribution of probability: uniform

Client 2

Estimated low cash flow: $20,000

Estimated high cash flow: $30,000

Most likely cash flow: $28,000

Distribution of probability: mode at $28,000

The *mode* is the most frequently observed value for a variable. If Sallie assumes cash flow to be a continuous variable, referred to by C, and expresses all dollar values in $1000 increments, (*a*) write and graph the two probability distributions and cumulative distributions for monthly cash flow and (*b*) determine the probability that monthly cash flow is no more than $12 for client 1 and $25 for client 2.

Solution

Client 1: monthly cash-flow distribution.

(*a*) The distribution of cash flows for client 1, identified by the variable C_1, follows the *uniform distribution*. Probability and cumulative probability take the following general forms.

$$f(C_1) = \frac{1}{\text{high} - \text{low}} \qquad \text{low value} \leq C_1 \leq \text{high value}$$

$$= \frac{1}{H - L} \qquad L \leq C_1 \leq H \qquad \textbf{[20.3]}$$

$$F(C_1) = \frac{\text{value} - \text{low}}{\text{high} - \text{low}} \qquad \text{low value} \leq C_1 \leq \text{high value}$$

$$= \frac{C_1 - L}{H - L} \qquad L \leq C_1 \leq H \qquad \textbf{[20.4]}$$

For client 1, monthly cash flow is uniformly distributed with $L = \$10$, $H = \$15$, and $\$10 \leq C_1 \leq \$15$. (Remember all values are divided by $1000.) Figure 20–4 is a plot of $f(C_1)$ and $F(C_1)$ from Equations [20.3] and [20.4].

$$f(C_1) = \frac{1}{5} = 0.2 \qquad \$10 \leq C_1 \leq \$15$$

$$F(C_1) = \frac{C_1 - 10}{5} \qquad \$10 \leq C_1 \leq \$15$$

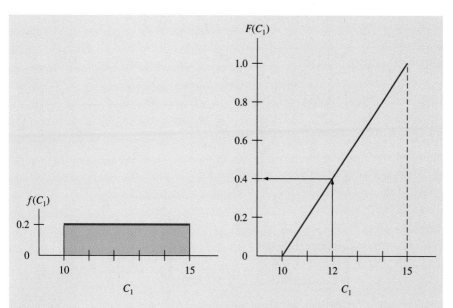

Figure 20–4   Uniform distribution for monthly cash flow, Example 20.3.

(b) The probability that client 1 has a monthly cash flow of less than $12 is easily determined from the $F(C_1)$ plot as 0.4, or a 40% chance.

$$F(\$12) = P(C_1 \leq \$12) = 0.4$$

Client 2: monthly cash-flow distribution.

(a) The distribution of cash flows for client 2, identified by the variable C_2, follows the *triangular distribution*. This probability distribution has the shape of an upward-pointing triangle with the peak at the mode, M, and downward sloping lines joining the x axis on either side at the low (L) and high (H) values. The mode of the triangular distribution has the maximum probability value.

$$f(\text{mode}) = f(M) = \frac{2}{H - L} \qquad [20.5]$$

The cumulative distribution is comprised of two curved line segments from 0 to 1 with a break point at the mode, where

$$F(\text{mode}) = F(M) = \frac{M - L}{H - L} \qquad [20.6]$$

For C_2, the low value is $L = \$20$, the high is $H = \$30$, and the most likely cash flow is the mode, $M = \$28$. The probability at M from Equation [20.5] is

$$f(28) = \frac{2}{30 - 20} = \frac{2}{10} = 0.2$$

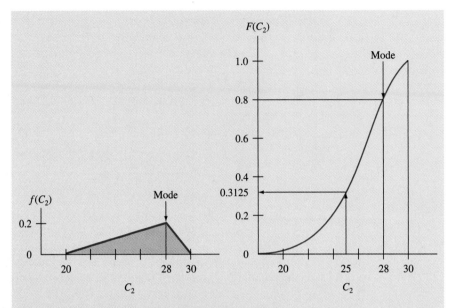

Figure 20–5   Triangular distribution for monthly cash flow, Example 20.3.

and the break point in the cumulative distribution occurs at $C_2 = 28$. Using Equation [20.6],

$$F(28) = \frac{28 - 20}{30 - 20} = 0.8$$

Figure 20–5 presents the plots for $f(C_2)$ and $F(C_2)$. Note that $f(C_2)$ is skewed, since the mode is not at the midpoint of the range $H - L$, and $F(C_2)$ is a smooth S-shaped curve with an inflection point at the mode.

(b) From the cumulative distribution in Figure 20–5, there is an estimated 31.25% chance that cash flow is $25 or less.

$$F(\$25) = P(C_2 \leq \$25) = 0.3125$$

Comment

Additional Example 20.9 provides further illustrations of developing probability statements about a variable from its probability distribution.

Additional Example 20.9
Problems 20.3 to 20.8

20.3   RANDOM SAMPLES

Estimating a parameter for the analysis in previous chapters is the equivalent of taking a *random sample of size one from an entire population* of possible values. If all values in the population were known, the probability distribution and cumulative distribution would be known. Then, the sample is not necessary. As an illustration, assume that estimates of first cost, annual operating cost, interest rate, and other parameters are used to compute one PW value in order to accept or reject an alternative. Each estimate is a sample of size one from an entire population of possible values for each parameter. Now, if a second estimate is made for each parameter and a second PW value is determined, a sample of size two has been taken from the PW population.

A *random sample of size* n is the selection in a random fashion of *n* values from the assumed or known probability distribution of the population, such that the values of the variable have the same chance of occurring in the sample as they are expected to occur in the population. For example, suppose Craig, a computer salesperson, estimates the number of months, *N*, that a new computer system he leases to clients will actually be used in the workplace before the clients decide it needs replacement and they enter into a new lease arrangement. Suppose that Craig identifies three *N* values—24, 30, and 36 months—because these are the only three options for lease length. Further, assume that nationwide 2000 systems were leased (the population) and that Craig knows the probability distribution of *N* to be

$$P(N = N_i) = \begin{cases} 0.20 & N_1 = 24 \text{ months} \\ 0.50 & N_2 = 30 \text{ months} \\ 0.30 & N_3 = 36 \text{ months} \end{cases} \qquad \text{[20.7]}$$

After 3 years, or 36 months, Craig samples his records for 100 (randomly selected) companies in two adjacent states to determine how many months each client retained the lease. If the sample is truly random, the actual lease periods for the 100 systems will be in approximately the same proportion as the population probabilities, that is, 20 leases for 24 months, etc. Since this is a sample, most likely the results won't be exactly the same. However, if they are relatively close, the experiment indicates that the sample from only two states may be useful in predicting actions of clients across the nation.

Whenever we perform an engineering economy study and utilize decision making under certainty, we use one estimate for each parameter to calculate a measure of worth. This is the equivalent of taking a (random) sample of size one for each parameter. The estimate is the most likely value, that is, one estimate of the expected value. We know that all parameters will vary somewhat, yet some are important enough, or will vary enough, that a probability distribution should be determined or assumed for it and the parameter treated as a random variable. This is using risk, and a sample from the parameter's probability distribution—$P(X)$ for discrete or $f(X)$ for continuous—helps formulate probability statements about the estimates. This approach complicates the analysis somewhat; however, it also provides a sense of confidence (or possibly a lack of confidence) about the decision made concerning the economic viability of the alternative based on the varying parameter. We will further discuss this aspect later in this chapter, after we learn how to correctly take a random sample from any probability distribution.

Table 20-2	Random digits clustered into two-digit numbers

51 82 88 18 19 81 03 88 91 46 39 19 28 94 70 76 33 15 64 20 14 52
73 48 28 59 78 38 54 54 93 32 70 60 78 64 92 40 72 71 77 56 39 27
10 42 18 31 23 80 80 26 74 71 03 90 55 61 61 28 41 49 00 79 96 78

45 44 79 29 81 58 66 70 24 82 91 94 42 10 61 60 79 30 01 26 31 42
68 65 26 71 44 37 93 94 93 72 84 39 77 01 97 74 17 19 46 61 49 67
75 52 14 99 67 74 06 50 97 46 27 88 10 10 70 66 22 56 18 32 06 24

To develop a random sample, use *random numbers (RN)* generated from a uniform probability distribution for the discrete numbers 0 through 9, that is,

$$P(X_i) = 0.1 \qquad \text{for } X_i = 0, 1, 2, \ldots, 9$$

In tabular form, the random digits so generated are commonly clustered in groups of two-digits, three-digits, or more. Table 20–2 is a sample of 264 random digits clustered into two-digit numbers. This format is very useful because the numbers 00 to 99 conveniently relate to the cumulative distribution values 0.01 to 1.00. This makes it easy to select a two-digit RN and enter $F(X)$ to determine a value of the variable with the same proportions as it occurs in the probability distribution. To apply this logic manually and develop a random sample of size n from a known discrete probability distribution $P(X)$ or a continuous variable distribution $f(X)$, the following procedure may be used.

1. Develop the cumulative distribution $F(X)$ from the probability distribution. Plot $F(X)$.

2. Assign the RN values from 00 to 99 to the $F(X)$ scale (the y axis) in the same proportion as the probabilities. For example, the probabilities from 0.0 to 0.15 are represented by the random numbers 00 to 14. Indicate the RNs on the graph.

3. To use a table of random numbers, determine the scheme or sequence of selecting RN values—down, up, across, diagonally. Any direction and pattern is acceptable, but the scheme should be used consistently for one entire sample.

4. Select the first number from the RN table, enter the $F(X)$ scale, and observe and record the corresponding variable value X. Repeat this step until there are n values of the variable which constitute the random sample.

5. Use the n sample values for analysis and decision making under risk. These may include;

   • Plotting the sample probability distribution.

   • Developing probability statements about the parameter.

   • Comparing sample results with the assumed population distribution.

- Determining sample statistics (next Section).
- Performing a simulation analysis (Section 20.5).

Example 20.4

Develop a random sample of size 10 for the variable N, the number of months a lease is active, as described by Equation [20.7]. Write the results in probability form and compare them with the original estimates.

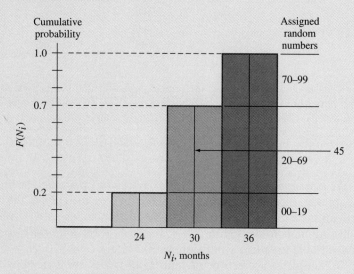

Figure 20–6   Cumulative distribution for a discrete variable with random number values assigned in proportion to probabilities, Example 20.4.

Solution

Apply the procedure above using the $P(N = N_i)$ values in Equation [20.7].

1. The cumulative distribution, Figure 20–6, is for the discrete variable N, which can assume three different values.
2. Assign 20 numbers (00 through 19) to $N_1 = 24$ months, where $P(N = 24) = 0.2$; 50 numbers to $N_2 = 30$; and 30 numbers to $N_3 = 36$.
3. Initially select any position in Table 20–2 and go across the row to the right and onto the row below toward the left. (Any routine can be developed, and a different sequence for each random sample may be used.)
4. Select the initial number 45 (4th row, 1st column), and enter Figure 20–6 in the RN range 20 to 69 to obtain $N = 30$ months.

5. Select and record the remaining nine values from Table 20–2 as shown below.

RN	45	44	79	29	81	58	66	70	24	82
N	30	30	36	30	36	30	30	36	30	36

Now, using the 10 values, develop the sample probabilities.

N, Months	Times in Sample	Sample Probability	Equation [20.7] Probability
24	0	0.00	0.2
30	6	0.60	0.5
36	4	0.40	0.3

With only 10 values, we can expect the sample probability estimates to be different from the values in Equation [20.7]. Only the value $N = 24$ months is significantly different, since no RN of 19 or less occurred. A larger sample will definitely make the probabilities closer to the original data.

To take a *random sample of size* n *for a continuous variable,* the procedure above is applied, except the random number values are assigned to the cumulative distribution on a continuous scale 00 to 99 corresponding to the $F(X)$ values. As an illustration, consider Figure 20–4, where C_1 is the *uniformly distributed* cash-flow variable for client 1 in Example 20.3. Here $L = \$10$, $H = \$15$, and $f(C_1) = 0.2$ for all values between L and H (all values are divided by 1000). The $F(C_1)$ is repeated as Figure 20–7 with the assigned random number values shown on the right scale. If the two-digit RN of 45 is chosen, the corresponding C_1 is graphically estimated to be $\$12.25$. It can also be linearly interpolated as $\$12.25 = 10 + (45/100)(15 - 10)$. See the Additional Example 20.10 for the next section to learn about sampling from the *normal distribution*.

For more accuracy when developing a random sample, especially for a continuous variable, some individuals like to use 3-, 4-, or 5-digit RNs. These can be developed from Table 20–2 simply by combining digits in the columns and rows or obtained from tables with RNs printed in larger clusters of digits. In computer-based sampling, most simulation software packages have an RN generator built-in which will generate values in the range 0 to 1 from a continuous variable uniform distribution, usually identified by the symbol $U(0, 1)$. The RN values, between 0.00000 and 0.99999 for example, are then used to sample directly from the cumulative distribution using essentially the same procedure we have learned here.

An initial question in random sampling is usually the *minimum size of* n required to ensure confidence in the results. Without detailing the mathematical

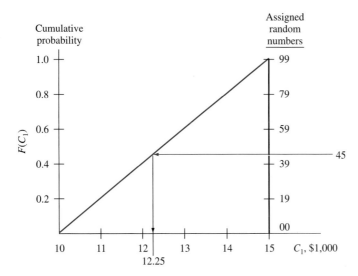

Figure 20-7   Random numbers assigned to the continuous variable of client 1 cash flows in Example 20.3.

logic, sampling theory, which is based upon the law of large numbers and the central-limit theorem (check a basic statistics book to learn about these), indicates that an n of 30 is sufficient. However, since reality does not follow theory exactly, and since engineering economy often deals with sketchy estimates, samples in the *range of 100 to 200* are the common practice. But, in many cases, samples as small as 10 to 25 provide a much better foundation for decision making under risk than the single-point (certainty) value made by using only one estimate for a parameter which is known to vary significantly.

Problems 20.9 to 20.12

20.4   EXPECTED VALUE AND STANDARD DEVIATION

Two very important measures or properties of a random variable are the expected value and standard deviation. (We introduced the expected value in Chapter 19.) If the entire population for a variable were known, these properties would be calculated directly. Since they are usually not known, random samples are commonly used to estimate them using the sample mean and the sample standard deviation, respectively. The following is a very brief introduction to the interpretation and calculation of these properties for a complete population and a random sample of size n from the population.

The usual symbols are Greek letters for the true population measures and English letters for the sample estimates.

	True Population Measure		Sample Estimate	
	Symbol	**Name**	**Symbol**	**Name**
Expected value	μ or $E(X)$	Mu or true mean	$\overline{X}$	Sample mean
Standard deviation	σ or $\sqrt{\mathrm{Var}(X)}$	Sigma or true standard deviation	s or $\sqrt{s^2}$	Sample standard deviation

The *expected value* is the long-run expected average to result if the variable is sampled many times. The population expected value is not known exactly, since the population itself is not known completely, so μ is estimated by $E(X)$ from a distribution or $\overline{X}$, the sample mean. Equation [19.1], repeated below as Equation [20.8], is used to compute the $E(X)$ of a probability distribution, and Equation [20.9] is the sample mean, also called the sample average.

Population: $\quad\quad\quad\quad\quad\quad\quad\quad \mu$

Probability distribution: $\quad E(X) = \sum X_i\, P(X_i)$

Sample: $\quad\quad\quad\quad\quad \overline{X} = \dfrac{\text{sum of sample values}}{\text{sample size}}$

$$= \frac{\Sigma X_i}{n} = \frac{\Sigma f_i X_i}{n} \quad\quad\quad \textbf{[20.9]}$$

The f_i in the second form of Equation [20.9] is the frequency of X_i, that is, the number of times each value occurs in the sample. The resulting $\overline{X}$ is not necessarily an observed value of the variable; it is the long-run average value and can take on any value within the range of the variable. We will omit the subscript i on X and f when there is no confusion introduced.

Example 20.5

Kayeu, an engineer with Worldwide Utilities, is planning to test several hypotheses about residential electricity bills in North American and Asian countries. The variable of interest is X, the monthly residential bill in U.S. dollars (rounded to the nearest dollar). Two small samples have been collected from different countries of North America and Asia. Estimate the population expected value. Do the samples (from a nonstatistical viewpoint) appear to be drawn from one population of electricity bills or from two different populations?

American, sample 1, $	40	66	75	92	107	159	275
Asian, sample 2, $	84	90	104	187	190		

Solution

Using Equation [20.9],

Sample 1: $n = 7, \quad \sum X_i = 814, \quad \bar{X} = \$116.29$

Sample 2: $n = 5, \quad \sum X_i = 655, \quad \bar{X} = \$131.00$

Based solely on the small-sample averages, the approximate $15 difference, which is less than 10% of the smaller average bill, does not seem sufficiently large to conclude that the two populations are different. There are several statistical tests available to determine if samples come from the same or different populations. Check a basic statistics text if you want to learn about them.

Comment

There are three commonly used measures of central tendency for data. The sample average is the most popular, but the *mode* and the *median* are also good measures. The mode, which is the most frequently observed value, was utilized in Example 20.3 for a triangular distribution. There is no specific mode in Kayeu's two samples, since all values are different. The median is the middle value of the sample. It is not biased by extreme sample values, as is the mean. The two medians in the samples above are $92 and $104. Based solely on the medians, the conclusion is still that the samples do not necessarily come from two populations of electricity bills.

The dispersion or spread of values about the expected value $E(X)$ or sample average $\bar{X}$ is commonly measured by the *standard deviation, s*. A probability distribution for data with strong central tendency is more closely clustered about the center of the data, and has a smaller s than a wider, more dispersed distribution (Figure 20–8). The sample with a larger s value has a flatter, wider probability distribution.

The sample *standard deviation s* estimates the property σ, which is the population measure of dispersion about the expected value of the variable.

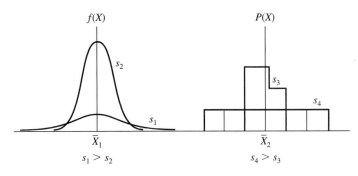

Figure 20–8   Sketches of distributions with different average and standard deviation values.

Actually, the variance, s^2, is often quoted as the measure of dispersion. The standard deviation is simply the square root of the variance, so either measure can be used. However, the s value is what we use routinely in making computations about risk and probability. Mathematically, the formulas and symbols for variance and standard deviation of a discrete variable and a random sample of size n from it are summarized here.

Population: $\sigma^2 \, \mathrm{Var}(X)$ and $\sigma = \sqrt{\sigma^2} = \sqrt{\mathrm{Var}(X)}$

Probability distribution: $\mathrm{Var}\,(X) = \sum [X_i - E(X)]^2 P(X_i)$ **[20.10]**

Sample: $s^2 = \dfrac{\text{sum of (sample value} - \text{sample average})^2}{\text{sample size} - 1}$

$$= \frac{\Sigma (X_i - \bar{X})^2}{n - 1} \qquad \textbf{[20.11]}$$

$s = \sqrt{s^2}$

Equation [20.11] for sample variance is usually applied in a more computationally convenient form.

$$s^2 = \frac{\Sigma X_i^2}{n-1} - \frac{n}{n-1}\bar{X}^2 = \frac{\Sigma f_i X_i^2}{n-1} - \frac{n}{n-1}\bar{X}^2 \qquad \textbf{[20.12]}$$

The standard deviation uses the sample average as a basis about which to measure the spread or dispersion of data via the calculation $(X - \bar{X})$, which can have a minus or plus sign. To accurately measure the dispersion in both directions from the average, $(X - \bar{X})$ is squared. In order to return to the dimension of the variable itself, the square root of Equation [20.11] is extracted. The term $(X - \bar{X})^2$ is called the *mean-squared deviation,* and s has historically also been referred to as the *root-mean-squared deviation.* The f_i in the second form of Equation [20.12] uses the frequency to calculate s^2.

One simple way to combine the average and standard deviation is to determine the percentage or fraction of the sample which is within ± 1, ± 2, or ± 3 standard deviations of the average, that is

$$\bar{X} \pm ts \qquad \text{for } t = 1, 2, \text{ or } 3 \qquad \textbf{[20.13]}$$

In probability terms, this is stated as

$$P(\bar{X} - ts \le X \le \bar{X} + ts) \qquad \textbf{[20.14]}$$

Virtually all the sample values will always be within the $\pm 3s$ range of $\bar{X}$, but the percent within $\pm 1s$ will vary depending on how the data points are distributed about $\bar{X}$. The following example illustrates the calculation of s to estimate σ, and incorporates s with the sample average using $\bar{X} \pm ts$.

Example 20.6

(*a*) Use the two samples of the previous example to estimate population variance and standard deviation for electricity bills. (*b*) Determine the percentages of each sample which are inside the ranges of 1 and 2 standard deviations from the mean.

Solution

(*a*) For illustration purposes only, we apply the two different relations to calculate s for the two samples. For sample 1 (American) with $n = 7$, we use X to identify the values, and Table 20–3 presents the computation of $\Sigma(X - \bar{X})^2$ for Equation [20.11], with $\bar{X} = \$116.29$. The resulting s^2 and s values are

$$s^2 = \frac{37743.40}{6} = 6290.57$$

$$s = \$79.31$$

For sample 2 (Asian), use Y to identify the values. With $n = 5$ and $\bar{Y} = 131$, Table 20–4 shows ΣY^2 for Equation [20.12]. Then,

$$s^2 = \frac{97,041}{4} - \frac{5}{4}(131)^2 = 42260.25 - 1.25(17,161) = 2809$$

$$s = \$53$$

The dispersion is smaller for the Asian sample (\$53) than for the American sample (\$79.31).

Table 20–3 Computation of standard deviation using Equation [20.11] with $\bar{X} = \$116.29$, Example 20.6

X	$(X - \bar{X})$	$(X - \bar{X})^2$
\$ 40	−76.29	5,820.16
66	−50.29	2,529.08
75	−41.29	1,704.86
92	−24.29	590.00
107	−9.29	86.30
159	+42.71	1,824.14
275	+158.71	25,188.86
\$814		\$37,743.40

Table 20–4	Computation of standard deviation using Equation [20.12] with $\overline{Y} = \$131$, Example 20.6	
Y	**Y²**	
$ 84	7,056	
90	8,100	
104	10,816	
187	34,969	
190	36,100	
$655	97,041	

(b) Equation [20.13] allows the determination of the ranges of $\overline{X} \pm 1s$ and $\overline{X} \pm 2s$. Count the number of sample data points between the limits, and calculate the corresponding percentage. See Figure 20–9 for a plot of the data and the standard deviation ranges.

American sample

$$\overline{X} \pm 1s = 116.29 \pm 79.31 \qquad \text{for a range of \$36.98 to \$195.60}$$

Six out of seven values are within this range, so the percentage is 85.7%.

$$\overline{X} \pm 2s = 116.29 \pm 158.62 \qquad \text{for a range of } -\$42.33 \text{ to } \$274.91$$

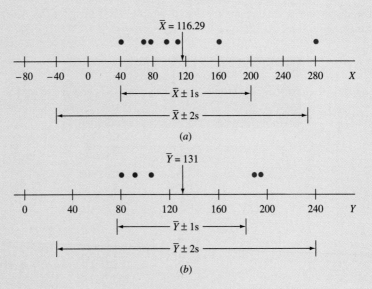

Figure 20–9   Values, averages, and standard deviation ranges for (a) American and (b) Asian samples, Example 20.6.

There are still six of the seven values within the $\pm 2s$ range. The limit $-\$42.33$ is meaningful only from the probabilistic perspective; from the practical viewpoint, use zero, that is, no amount billed.

Asian sample

$$\bar{X} \pm 1s = 131 \pm 53 \quad\quad \text{for a range of \$78 to \$184}$$

There are three of five values, or 60%, within the range.

$$\bar{X} \pm 2s = 131 \pm 106 \quad\quad \text{for a range of \$25 to \$237}$$

All five of the values are within the $\pm 2s$ range.

Comment

A second common measure of dispersion is the *range*. which is simply the largest minus the smallest sample values. In the two samples here, the range estimates are $235 and $106.

Before moving to simulation analysis in engineering economy, it may be of interest to summarize the expected value and standard deviation relations for a continuous variable, since Equations [20.8] through [20.12] address only discrete variables. The primary differences are that the sum symbol is replaced by the integral over the defined range of the variable, which we identify as R, and that $P(X)$ is replaced by the differential element $f(X)\, dX$. For a stated continuous probability distribution $f(X)$, the formulas are

Expected value:

$$E(X) = \int_R X f(X)\, dX \qquad\qquad \text{[20.15]}$$

Variance:

$$\text{Var}(X) = \int_R X^2 f(X)\, dX - [E(X)]^2 \qquad\qquad \text{[20.16]}$$

For a numerical example, again use the uniform distribution in Example 20.3 (Figure 20–4) over the range R from $10 to $15. If we identify the variable as X, rather than C_1, the following are correct.

$$f(X) = \frac{1}{5} = 0.2 \quad\quad \$10 \le X \le \$15$$

$$E(X) = \int_R X(0.2)\, dX = 0.1X^2 \Big|_{10}^{15} = 0.1(225 - 100) = \$12.5$$

$$\text{Var}(X) = \int_R X^2(0.2)\, dX - (12.5)^2 = \frac{0.2}{3}X^3 \Big|_{10}^{15} - (12.5)^2$$

$$= 0.06667(3375 - 1000) - 156.25 = 2.08$$

$$\sigma = \sqrt{2.08} = \$1.44$$

Therefore, the uniform distribution between $L = \$10$ and $H = \$15$ has an expected value of $\$12.5$ (the midpoint of the range, as expected), and a standard deviation of $\$1.44$. You should plot this probability distribution and the values of ± 1, ± 2, and ± 3 standard deviations from the expected value.

Additional Example 20.10
Problems 20.13 to 20.18

20.5   MONTE CARLO SAMPLING AND SIMULATION ANALYSIS

Up to this point, all decisions about alternatives have been made using estimates with certainty, possibly followed by some testing of the decision via sensitivity analysis or expected values. In this section, we will use a simulation approach which incorporates the material of the previous sections to facilitate the engineering economy decision about one alternative or between two or more alternatives.

The random sampling technique discussed in Section 20.3 is called *Monte Carlo sampling.* The general procedure outlined below uses Monte Carlo sampling to obtain samples of size n for selected parameters of formulated alternatives. These parameters, expected to vary according to a stated probability distribution, warrant decision making under risk. All other parameters in an alternative are considered certain; that is, they are known or they can be estimated with enough precision to consider them certain. An important assumption that is made, usually without realizing it, is that all parameters are independent; that is, one variable's distribution does not affect the value of any other variable of the alternative. This is referred to as the *property of independent random variables.*

The simulation approach to engineering economy analysis is summarized in the following basic steps.

Step 1: Formulate alternative(s).   Set up each alternative in the form to be considered using engineering economic analysis and select the measure of worth upon which to base the decision. Determine the form of the relation(s) to calculate the measure of worth.

Step 2: Parameters with variation.   Select the parameters in each alternative to be treated as random variables. Estimate values for all other (certain) parameters for the analysis.

Step 3: Determine probability distributions.   Determine whether each variable is discrete or continuous and describe a probability distribution for each variable in each alternative. Use standard distributions where possible to simplify the sampling process and to prepare for computer-based simulation.

Step 4: Random sampling.   Incorporate the random sampling procedure of Section 20.3 (the first four steps) into this procedure. This results in the cumulative distribution, assignment of RNs, selection of the RNs, and a sample of size n for each variable.

Step 5: Measure-of-worth calculation.   Compute n values of the selected measure of worth from the relation(s) determined in step 1. Use the estimates made with certainty and the n sample values for the varying parameters. (This is when the property of independent random variables is actually applied.)

Step 6: Measure-of-worth description.   Construct the probability distribution of the measure of worth using between 10 and 20 cells of data, and calculate measures such as X, s, $X \pm ts$, and relevant probabilities.

Step 7: Conclusions.   Make conclusions about each alternative and decide which one is to be implemented. If the alternative(s) have been previously evaluated under the assumption of complete certainty for all parameters, comparison of results may help determine the final decision.

Example 20.7 illustrates this procedure using an abbreviated manual simulation analysis.

Example 20.7

Yvonne Ramos is the CEO of Exercizer^{++}, a chain of 50 fitness centers in the United States and Canada. An equipment salesperson has offered Yvonne two long-term opportunities on new equipment which is charged to customers on a per-use basis on top of the monthly fees paid by customers. As an enticement, the sales offer includes a guarantee of net cash flow (revenue) for one of the systems for the first 5 years.

Since this is an entirely new and risky concept of revenue generation, Yvonne wants to do a careful analysis of each alternative. Details for the two systems follow:

System 1.   First cost is $P = \$12,000$ for a set period of $n = 7$ years with no salvage value. No guarantee for annual net revenue is offered.

System 2.   First cost is $P = \$8000$, no salvage value, and a guaranteed annual net revenue of $\$1000$ for the first 5 years, but after this period, there is no guarantee. The equipment, with updates, may be useful up to 15 years, but the exact number is not known. Cancellation any time after the initial 5 years is possible at no cost.

For either system, new versions of the equipment are installed upon release with no added costs. If an MARR of 15% is required, use PW analysis to determine if neither, one, or both of the systems should be installed.

Solution

Estimates which Yvonne makes to correctly use the simulation analysis procedure are included in the following steps.

Step 1: Formulate alternatives. Using PW analysis, the relations for system 1 and system 2 are developed including the parameters known with certainty. The symbol NCF identifies the annual net cash flows and NCF_G is the guaranteed NCF of $1000 for system 2.

$$PW_1 = -P_1 + NCF_1(P/A,15\%,n_1) \tag{20.17}$$

$$PW_2 = -P_2 + NCF_G(P/A,15\%,5) \tag{20.18}$$

$$+ NCF_2(P/A,15\%,n_2-5)(P/F,15\%,5)$$

Step 2: Parameters with variation. Yvonne summarizes the parameters estimated with certainty and makes distribution assumptions about three parameters treated as random variables.

System 1

Certainty. $P_1 = \$12,000$; $n_1 = 7$ years.
Variable. NCF_1 is a continuous variable, uniformly distributed between $L = -\$4000$ and $H = \$6000$ per year, because this is considered a high-risk venture.

System 2

Certainty. $P_2 = \$8000$; NCF = $1000 for first 5 years.
Variable. NCF_2 is a discrete variable, uniformly distributed over the values $L = \$1000$ to $H = \$6000$ only in $1000 increments, that is, $1000, $2000, etc.
Variable. n_2 is a continuous variable that is uniformly distributed between $L = 6$ and $H = 15$ years.

Now, rewrite Equations [20.17] and [20.18] to reflect the estimates made with certainty.

$$PW_1 = -12,000 + NCF_1(P/A,15\%,7)$$

$$= -12,000 + NCF_1(4.1604) \tag{20.19}$$

$$PW_2 = -8000 + 1000(P/A,15\%,5)$$

$$+ NCF_2(P/A,15\%,n_2-5)(P/F,15\%,5)$$

$$= -4648 + NCF_2(P/A,15\%,n_2-5)(0.4972) \tag{20.20}$$

Step 3: Determine probability distributions. Figure 20–10 (left side) shows the assumed probability distributions for NCF_1, NCF_2, and n_2.

Step 4: Random sampling. Yvonne decides on a sample of size 30 and applies the first four of the random sample steps in Section 20.3. Figure 20–10 (right side) shows the cumulative distributions (step 1) and assigns RNs to

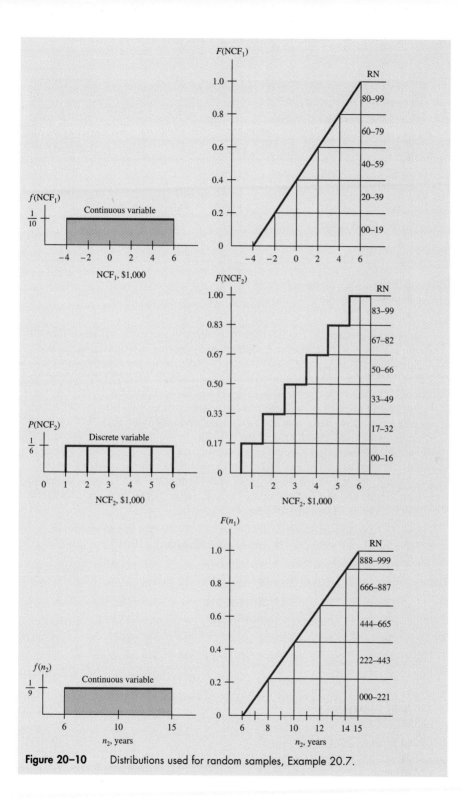

Figure 20–10 Distributions used for random samples, Example 20.7.

| Table 20–5 | Random numbers and variable values for NCF_1, NCF_2, and n_2, Example 20.7 |

NCF₁		NCF₂		n₂		
RN⁽¹⁾	**Value**	**RN⁽²⁾**	**Value**	**RN⁽³⁾**	**Value**	**Rounded⁽⁴⁾**
18	$-2,200	10	$1,000	586	11.3	12
59	+2,000	10	1,000	379	9.4	10
31	−1,100	77	5,000	740	12.7	13
29	−900	42	3,000	967	14.4	15
71	+3,100	55	4,000	144	7.3	8

[1] Randomly start with row 1, column 4 in Table 20–2.
[2] Start with row 6, column 14.
[3] Start with row 4, column 6.
[4] The n_2 value is rounded up.

each variable (step 2). The RNs for NCF_2 identify the x axis values so that all net cash flows will be in even $1000 amounts. For the continuous variable n_2, three-digit RN values are used in order to make the numbers come out evenly, and they are shown in cells only as 'indexers' for easy reference when an RN is used to find a variable value. However, we round the number to the next higher value of n_2 because it is likely the contract may be canceled on an anniversary date. Also, now the tabulated compound-interest factors for n_2-5 years can be used directly (see Table 20–5).

Once the first RN is selected randomly from Table 20–2, the sequence (step 3) used will be to proceed down the RN table column and then up the column to the left. Table 20–5 shows only the first five RN values selected for each sample and the corresponding variable values taken from the cumulative distributions in Figure 20–10 (step 4).

Step 5: Measure-of-worth calculation.   With the five sample values in Table 20–5, calculate the PW values using Equations [20.19] and [20.20].

1. $PW_1 = -12,000 + (-2200)(4.1604)$ $= \$-21,153$
2. $PW_1 = -12,000 + 2000(4.1604)$ $= \$-3679$
3. $PW_1 = -12,000 + (-1100)(4.1604)$ $= \$-16,576$
4. $PW_1 = -12,000 + (-900)(4.1604)$ $= \$-15,744$
5. $PW_1 = -12,000 + 3100(4.1604)$ $= \$+897$

1. $PW_2 = -4648 + 1000(P/A,15\%,7)(0.4972)$ $= \$-2579$
2. $PW_2 = -4648 + 1000(P/A,15\%,5)(0.4972)$ $= \$-2981$
3. $PW_2 = -4648 + 5000(P/A,15\%,8)(0.4972)$ $= \$+6507$
4. $PW_2 = -4648 + 3000(P/A,15\%,10)(0.4972)$ $= \$+2838$
5. $PW_2 = -4648 + 4000(P/A,15\%,3)(0.4972)$ $= \$-107$

Now, 25 more RNs are selected for each variable from Table 20–2 and the PW values are calculated.

Step 6: Measure-of-worth description.   Figure 20–11*a* and *b* presents the PW_1 and PW_2 probability distributions for the 30 samples with 14 and 15 cells, respectively, as well as the range of individual PW values and the $\overline{X}$ and *s* values.

PW_1.   Sample values range from $-\$24,481$ to $+\$12,962$. The calculated measures of the 30 values are

$$\overline{X}_1 = -\$7729$$

$$s_1 = \$10,190$$

PW_2.   Sample values range from $-\$3031$ to $+\$10,324$. The sample measures are

$$\overline{X}_2 = \$2724$$

$$s_2 = \$4336$$

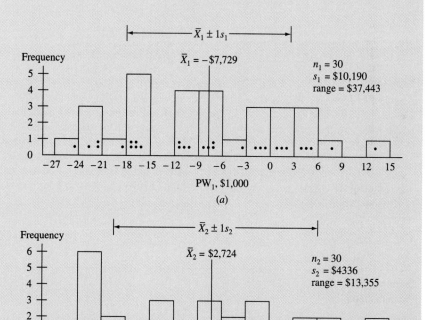

(a)

(b)

Figure 20–11   Probability distributions of PW values for a sample of size 30, Example 20.7.

Step 7: Conclusions.   Additional sample values will surely make the central tendency of the PW distributions more evident and may reduce the s values, which are quite large. Of course, many conclusions are possible once the PW distributions are known, but the following seem clear.

System 1.   Based on this small sample of 30 observations, *do not accept*. The likelihood of making the MARR $= 15\%$ is relatively small, since the sample indicates a probability of 0.27 (8 out of 30 values) that the PW will be positive, and the $\overline{X}_1$ is a large negative. Though appearing large, the standard deviation may be used to determine that about 20 of the 30 sample PW values (two-thirds) are within the limits $\overline{X} \pm 1s$, which are $-\$17,919$ and $\$2461$. A larger sample may alter this analysis somewhat.

System 2.   If Yvonne is willing to accept the longer-term commitment which may raise the NCF some years out, the sample of 30 observations indicates to *accept* this alternative. At a MARR of 15%, the simulation approximates the chance for a positive PW as 67% (20 of the 30 PW values in Figure 20–11b are positive). However, the probability of observing PW within the $\overline{X} \pm 1s$ limits ($\$-1612$ and $\$7060$) is 0.53 (16 of 30 sample values). This indicates that the PW sample distribution is more widely dispersed about its average than the system 1 PW sample.

Conclusion at this point.   Reject system 1; accept system 2; and carefully watch net cash flow, especially after the initial 5-year period.

Comment

A review of Example 16.5, where all estimates were made with certainty (NCF$_1$ = \$3000, NCF$_2$ = \$3000, and n_2 = 15 years), shows that the alternative machines, which have similar parameter characteristics, are evaluated by the payback period method at a MARR $= 15\%$, and the first alternative is selected. However, the subsequent AW analysis, which is equivalent to the PW analysis here, selected alternative 2 based, in part, upon the anticipated larger cash flow in the later years. You may find other comparative bases between these two examples if you increase the sample size of the simulation analysis.

**Example 20.7
(Spreadsheet)**

Help Yvonne Ramos set up an Excel spreadsheet simulation for the three random variables and PW analysis in Example 20.7 above. Does the PW distribution vary appreciably from that developed using manual simulation? Do the decisions to reject the system 1 proposal and accept the system 2 proposal still seem reasonable?

Solution

Figures 20–12 and 20–13 are spreadsheets which accomplish the simulation portion of the analysis described above in steps 3 (determine probability distribution) through 6 (measure-of-worth description). Excel, and most spreadsheet systems, are limited in the variety of distributions they can accept for sampling, but typical ones such as uniform and normal are available.

Figure 20–12 indicates the results of a small sample of 30 values (only a portion of the spreadsheet is printed here) from the three distributions using the RAND function.

NCF$_1$.   Continuous uniform from $-4000 to $6000. Cells in column B translate RN1 values (column A) into NCF1 amounts.

NCF$_2$.   Discrete uniform in $1000 increments from $1000 to $6000. Column D cells display NCF2 in the $1000 increments using the IF operator to translate from the RN2 values.

n_2.   Continuous uniform from 6 to 15 years. The cells in column F are integer values obtained using the INT function operating on the RN3 values.

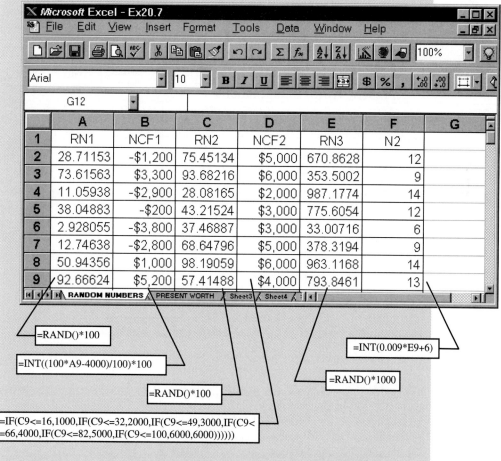

Figure 20–12   Sample values generated using spreadsheet simulation, Example 20.7.

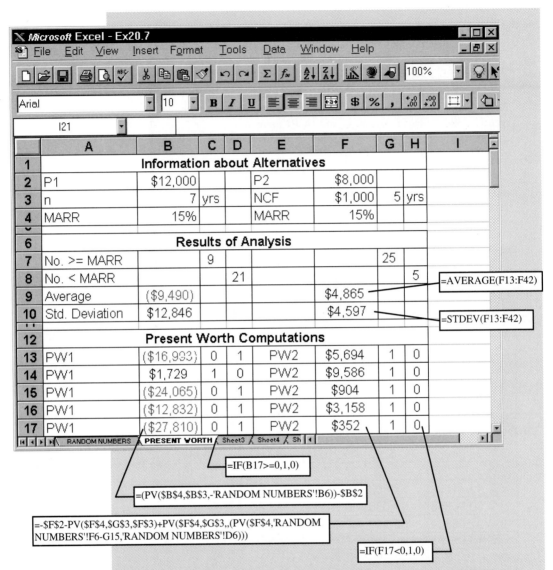

Figure 20–13   Spreadsheet simulation results for 30 sample PW values, Example 20.7.

Figure 20–13 presents the two alternatives' estimates in the top section. The PW1 and PW2 computations for the 30 repetitions of NCF1, NCF2, and n_2 are the spreadsheet equivalent of Equations [20.19] and [20.20]. The tabular approach used here tallies the number of PW values below zero ($0) and equal to or exceeding zero using the IF operator. For example, cell D13 contains a 1, indicating PW1 < 0 when NCF1 = $-1200 was used to calculate PW1 = $-16,993 by Equation [20.19].

Cells in rows 7 and 8 show the number of times in the 30 samples that system 1 and system 2 may return at least the MARR = 15% because the corresponding PW ≥ 0. Sample averages and standard deviations are also indicated.

Comparison between the manual and spreadsheet simulations is presented below.

	System 1 PW			System 2 PW		
	$\bar{X}$, $	s, $	No. of PW ≥ 0	$\bar{X}$, $	s, $	No. of PW ≥ 0
Manual	−7,729	10,190	8	2,724	4,336	20
Spreadsheet	−9,490	12,846	9	4,865	4,597	25

For the spreadsheet simulation, 9 (30%) of the PW1 values exceed zero, while the manual simulation included 8 (27%) positive values. These comparative results will change every time this spreadsheet is activated since the RAND function is set up (in this case) to produce a new RN each time. It is possible to define RAND to keep the same RN values. See the Excel User's Guide.

The conclusion to reject the system 1 proposal and accept system 2 is appropriate for the spreadsheet simulation as it is for the manual one, since there are comparable chances that PW ≥ 0.

Problems 20.19 and 20.20

20.6   MULTIPLE-CRITERIA EVALUATION

Besides the consideration of risk for economic factors, it is often important to include noneconomic factors when evaluating alternatives. In order to explicitly include these added factors in the evaluation, multiple-criteria decision making may be used. The general procedure is

1. Clearly define the alternatives.
2. Determine all the factors to be considered. These should include some form of an economic measure of worth we have already learned.
3. Utilize a multiple-criteria evaluation technique.
4. Choose the alternative with the best combined result.

These steps are not significantly different from those used thus far. However, the economic and noneconomic factors are combined to more closely reflect the individual style and interest of the decision maker. Several examples of noneconomic factors are

- Response time
- Cycle time

- Throughput rate
- Software availability
- Public reaction
- Litigation exposure
- Contract services
- Training requirements

Of the basic multiple-criteria evaluation techniques, most are scoring techniques, also called *rank and rate* techniques, the most popular of which is the weighted evaluation method. These techniques are especially helpful when there are three or more alternatives and the differences between them are not immediately obvious to the person doing the analysis.

Unweighted Evaluation Method   Once all the evaluation factors are determined, each alternative is rated on each factor using a preestablished scale. The alternative with the largest sum of ratings should be selected. Four of the possible rating scales for each factor are

$$-1, 0, +1 \qquad -2, -1, +1, +2$$

$$0 \text{ to } 1 \qquad\qquad 0 \text{ to } 100$$

If the 0-to-1 or 0-to-100 scale is used, the highest rating of 1 or 100 is assigned to the best alternative for each factor, and all other alternatives are rated relative to it.

Weighted Evaluation Method   Factors are rated as in the unweighted method, usually on a 0-to-1 or 0-to-100 scale. Additionally, each factor is rated for importance by the decision maker. For each alternative i, the total weighted value V_i is computed.

$$V_i = \sum_{j=1}^{j=n} w_j r_{ij} \qquad\qquad \text{[20.21]}$$

where w_j = weight for factor j $(j = 1, 2, \ldots, n)$
r_{ij} = rating for alternative i on factor j

Select the alternative with the largest V_i. The factor weights w_j are normalized by initially assigning the score of 100 to the most important one, then assigning relative scores to the other factors, and finally dividing each assigned weight by the total of all weights. The results are the w_j values in

Equation [20.21]. This process satisfies the requirement that the sum of all factor weights equal 1.0. Example 20.8 illustrates this method.

Example 20.8

An interactive information system used to dispatch trains has been in place for some years at Frontier Railroad. Computations and management discussions have led to the definition of three alternatives and six evaluation factors of differing importance.

Alternative 1.   Purchase new hardware and develop customized software.

Alternative 2.   Lease new hardware and use contract database services from an established vendor.

Alternative 3.   Keep the old hardware and upgrade the software.

Selected evaluation factors:

1. Payback period.
2. Initial investment requirement.
3. Response time.
4. User interface.
5. Software availability and maintainability.
6. Customer service.

Use the weighted evaluation method to determine which alternative is the best. Assume that the meaning of each factor has been established and that adequate information is available to rate each alternative.

Solution

Each factor and alternative is rated on a 0-to-100 scale. The factor 'initial investment requirement' is considered the most important since the current system (alternative 3) is already owned, so factor 2 is assigned a score of 100. Other factors are assigned values relative to the 100 score as shown in Table 20–6. The total for all factors (450) is divided into each score to compute the normalized weights w_j in the second column of Table 20–7.

The scores determined for the three alternatives (Table 20–6) are again relative to the best alternative for each factor using a 0-to-100 scale. For example, software availability and maintainability (factor 5) is judged best using a contract vendor (alternative 2); a score of 100 is assigned. However, the current system is very poor since it uses an outdated programming language; score assigned is 10.

Table 20–7 presents the weighted ratings and totals using Equation [20.21]. Alternative 2 is the clear choice.

Comment

Note how any economic measures can be incorporated into a multiple-criteria evaluation using this method. Of course, different measures of worth—PW, ROR, B/C, etc.—can also be included; however, their importance may vary relative to other noneconomic factors.

Table 20–6	Relative ratings for three alternatives using the weighted evaluation method			

Factor j	Factor Importance	Relative Importance (0 to 100), r_{ij}		
		Alternative 1	Alternative 2	Alternative 3
1	50	75	50	100
2	100	60	75	100
3	90	50	100	20
4	80	100	90	40
5	50	75	100	10
6	80	100	100	75
Total	450			

Table 20–7	Weighted values for three alternatives using the weighted evaluation method		

Factor j	Normalized Weight, w_j	Weighted Values, V_i		
		Alternative 1	Alternative 2	Alternative 3
1	0.11	8.3	5.5	11.0
2	0.22	13.2	16.5	22.0
3	0.20	10.0	20.0	4.0
4	0.18	18.0	16.2	7.2
5	0.11	8.3	11.0	1.1
6	0.18	18.0	18.0	13.5
Totals	1.00	75.8	87.2	58.8

Factors

1. Payback period.
2. Initial investment requirement.
3. Response time.
4. User interface.
5. Software availability and maintainability.
6. Customer service.

Problems 20.21 to 20.26

ADDITIONAL EXAMPLES

Example 20.9

PROBABILITY STATEMENTS, SECTION 20.2   Use the cumulative distribution for the variable C_1 in Figure 20–4 (Example 20.3, monthly cash flow for client 1) to determine the following probabilities:

(a) More than $14.

(b) Between $12 and $13.

(c) No more than $11 or more than $14.

(d) Exactly $12.

Solution

The shaded areas in Figure 20–14a through d indicate the points on the cumulative distribution $F(C_1)$ used to determine the probabilities.

(a) The probability of more than $14 per month is easily determined by subtracting the value of $F(C_1)$ at 14 from the value at 15. (Since the probability at a point is zero for a continuous variable, the equal sign does not change the value of the resulting probability.)

$$P(C_1 > 14) = P(C_1 \le 15) - P(C_1 \le 14)$$

$$= F(15) - F(14)$$

$$= 1.0 - 0.8$$

$$= 0.2 \quad (20\%)$$

(b) $\qquad P(12 \le C_1 \le 13) = P(C_1 \le 13) - P(C_1 \le 12)$

$$= 0.6 - 0.4$$

$$= 0.2 \quad (20\%)$$

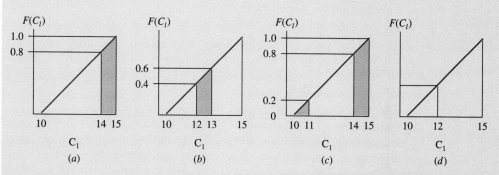

Figure 20–14   Calculation of probabilities from the cumulative distribution, Example 20.9.

(c) $P(C_1 \leq 11) + P(C_1 > 14) = [F(11) - F(10)] + [F(15) - F(14)]$

$$= (0.2 - 0) + (1.0 - 0.8)$$

$$= 0.2 + 0.2$$

$$= 0.4 \quad (40\%)$$

(d) $P(C_1 = 12) = F(12) - F(12) = 0.0$

There is no area under the cumulative distribution curve at a point for a continuous variable, as mentioned earlier. If two very closely placed points are used, it is possible to obtain a probability, for example, between 12.0 and 12.1 or between 12 and 13, as in part (b).

Example 20.10

THE NORMAL DISTRIBUTION, SECTION 20.4   Camilla is the regional manager for a large chain of franchise-based gasoline and quick-food stores. The home office in a large U.S. city has had many complaints and several legal actions from employees and customers about slips and falls due to liquids (water, oil, gas, soda, etc.) on concrete surfaces. Corporate management has authorized each regional manager to contract locally to apply to all exterior concrete surfaces a newly marketed product which absorbs up to 100 times its own weight in liquid and to charge a home office account for the installation. The authorizing letter to Camilla states that, based upon their simulation and random samples which assume a normal population, the cost of the locally arranged installation should be about $10,000 and almost always is within the range of $8000 to $12,000.

Camilla, a marketing major in college, asks you, TJ, an engineering technology graduate, to write a brief but thorough summary about the normal distribution, explain the $8000-to-$12,000 range statement, and explain the phrase "random samples which assume a normal population" in this situation.

Solution

Assume that you kept this book and a basic engineering statistics text when you graduated and that you have developed the following response to Camilla using them and the letter from the home office.

Camilla,

Here is a brief summary of how the home office appears to be using the normal distribution. As a refresher, I've included a summary of what the normal distribution is all about.

Normal distribution and probabilities

The normal distribution is also referred to as the bell-shaped curve, the Gaussian distribution, or the error distribution. It is, by far, the most commonly used probability distribution in all applications. It places exactly one-half the probability on either side of the mean or expected value. It is used for continuous variables over the entire range of numbers. The normal is found to accurately predict many types of outcomes, such as IQ values, manufacturing errors about a specified size, volume, weight, etc.;

and the distribution of sales revenues, costs, and many other business parameters around a specified mean, which is why it may apply in this situation.

The normal distribution, identified by the symbol $N(\mu,\sigma^2)$, where μ is the expected value or mean and σ^2 is the variance, or measure of spread, can be described as follows:

- The mean μ locates the probability distribution (Figure 20–15a), and the spread of the distribution varies with variance σ^2 (Figure 20–15b), growing wider and flatter for larger variance values.

- When a sample is taken, the estimates are identified as sample mean $\overline{X}$ for μ and sample standard deviation s for σ.

- The normal probability distribution $f(X)$ for a variable X is quite complicated, because its formula is

$$f(X) = \frac{1}{\sigma\sqrt{2\pi}} \exp - \left[\frac{(X - \mu)^2}{2\sigma^2} \right]$$

where exp represents the number $e = 2.71828+$ and it is raised to the power of the $-[\,]$ term. In short, if X is given different values, for a given mean μ and standard deviation σ, a curve looking like those in Figure 20–15a and b is developed.

Since $f(X)$ is so unwieldy, random samples and probability statements are developed using a transformation, called the *standard normal distribution (SND)*, which uses the μ and σ (population) or $\overline{X}$ and s (sample) to compute values of the variable Z.

Population: $$Z = \frac{\text{deviation from mean}}{\text{standard deviation}} = \frac{X - \mu}{\sigma} \qquad \textbf{[20.22]}$$

Sample: $$Z = \frac{X - \overline{X}}{s} \qquad \textbf{[20.23]}$$

The SND distribution for Z (Figure 20–15c) is the same as for X, except that it always has a mean of 0 and a standard deviation of 1, and it is identified by the symbol $N(0,1)$. Therefore, the probability values under the SND curve can be stated exactly. It is always possible to transfer back to the original values from sample data by solving Equation [20.22] for X:

$$X = Z\sigma + \mu \qquad \textbf{[20.24]}$$

Several probability statements for Z and X are summarized below and are shown for Z in Figure 20–15c.

Variable X Range	Probability	Variable Z Range
$\mu + 1\sigma$	0.3413	0 to +1
$\mu \pm 1\sigma$	0.6826	−1 to +1
$\mu + 2\sigma$	0.4773	0 to +2
$\mu \pm 2\sigma$	0.9546	−2 to +2
$\mu + 3\sigma$	0.4987	0 to +3
$\mu \pm 3\sigma$	0.9974	−3 to +3

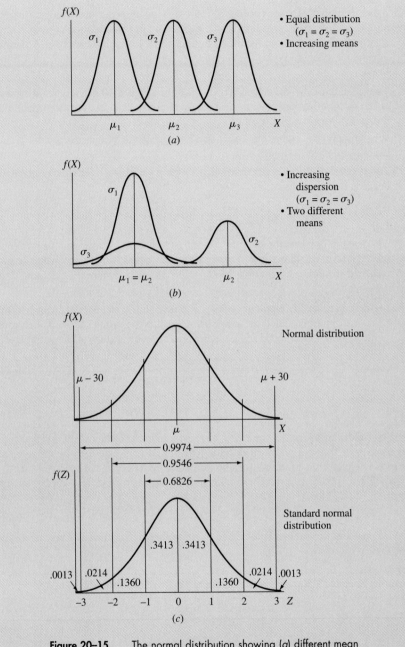

Figure 20–15   The normal distribution showing (a) different mean values μ; (b) different standard deviation values σ; and (c) relation of a normal for X and standard normal for $Z = (X-\mu)/\sigma$.

As an illustration, probability statements from this tabulation and Figure 20–15c for X and Z are

The probability that X is within 2σ of its mean is 0.9546.

The probability that Z is within 2σ of its mean, which is the same as between the values -2 and $+2$, is also 0.9546.

Random samples from a normal distribution

In order to take a random sample from a normal $N(\mu,\sigma^2)$ population, a specially prepared table of SND random numbers is used. (Tables of SND values are available in many statistics books.) The numbers are actually values from the Z or $N(0,1)$ distribution and have values such as -2.10, $+1.24$, etc. Translation from the Z value back to the sample values for X is via Equation [20.24].

The statement that virtually all the local contract amounts should be between \$8000 and \$12,000 may be interpreted as follows: A normal distribution is assumed with a mean of $\mu = \$10,000$, and a standard deviation of $\sigma = \$667$, or a variance of $\sigma^2 = (\$667)^2$; that is, an $N[\$10,000, (\$667)^2]$ distribution is assumed. The value $\sigma = \$667$ is calculated using the fact that virtually all the probability (99.74%) is within 3σ of the mean, as stated above. Therefore,

$$3\sigma = \$2000 \quad \text{and} \quad \sigma = \$667 \quad \text{(rounded off)}$$

As an illustration, if 6 SND random numbers are selected and used to take a sample of size 6 from the normal distribution $N[\$10,000, (\$667)^2]$, the results are as follows.

SND Random Number, Z	X Using Equation [20.24], $X = Z\sigma + \mu$
-2.10	$X = (-2.10)(667) + 10,000 = \$8,599$
$+3.12$	$X = (+3.12)(667) + 10,000 = \$12,081$
-0.23	$X = (-0.23)(667) + 10,000 = \$9,847$
$+1.24$	$X = (+1.24)(667) + 10,000 = \$10,827$
-2.61	$X = (-2.61)(667) + 10,000 = \$8,259$
-0.99	$X = (-0.99)(667) + 10,000 = \$9,340$

If we consider this a sample of six typical concrete surfacing contract amounts for sites in our region, the average is \$9825 and five of six values are within the range of \$8000 and \$12,000, with the sixth being only \$81 above the upper limit. So we should have no real problems, but it is important that we keep a close watch on the contract amounts, because the assumption of the normal distribution with a mean of about \$10,000 and virtually all contract amounts within $\pm\$2000$ of it may not prove to be correct for our region.

If you have any questions about this summary, please contact me.

TJ

CHAPTER SUMMARY

To perform decision making under risk implies that some parameters of an engineering alternative are treated as random variables. Assumptions about the shape of the variable's probability distribution are used to explain how the estimates of parameter values may vary. Additionally, measures such as the expected value and standard deviation describe the characteristic shape of the distribution. In this chapter, we learned several of the simple but useful, discrete and continuous population distributions used in engineering economy—uniform and triangular—as well as specifying our own distribution or assuming the normal distribution.

Since the population's probability distribution for a parameter is not fully known, a random sample of size n is usually taken, and its sample average and standard deviation are determined. The results are used to make probability statements about the parameter, which help make the final decision with risk considered.

The Monte Carlo sampling method is combined with engineering economy relations for a measure of worth such as PW to implement a simulation approach to risk analysis. The results of such an analysis can then be compared with decisions when parameter estimates are made with certainty.

The weighted evaluation method is a straightforward way to take several different factors into consideration, including noneconomic factors. This is another way to consider risk and to make the engineering economy analysis fulfill its role in decision making.

PROBLEMS

20.1 For each situation below, determine if the information involves certainty, risk, and/or uncertainty. If risk is involved, graph the information in the general form of Figure 20–1.

(*a*) A friend in real estate tells you the price per square foot for new houses will go up slowly or rapidly during the next 6 months.

(*b*) Your manager informs the staff there is an equal chance that sales will be between 500 and 550 units next month.

(*c*) You got paid yesterday and $320 was taken out in income taxes. The amount withheld next month will be different because of a pay raise you will receive.

(*d*) There is a 20% chance of rain and a 30% chance of snow today.

20.2 Describe cases of certainty, risk, and uncertainty from your own experiences which can be reduced to quantitative terms.

20.3 A survey of households included a question about the number of operating automobiles, N, currently owned by people living at the residence and the interest rate, i, on the lowest-rate loan for the cars. The results for 100 households are shown below.

Number of cars, N	0	1	2	3	≥ 4
Households	12	56	26	3	3

Loan rate, i	0.0–2	2.01–4	4.01–6	6.01–8	8.01–10	10.01–12
Households	22	10	12	42	8	6

(a) Determine whether each variable is discrete or continuous, and state why.

(b) Plot the probability distributions and cumulative distributions for N and i.

(c) From the data collected, what is the probability that a household has 1 or 2 cars? Three or more cars?

(d) Use the data for i to estimate the chances that the interest rate is between 6% and 9.75% per year.

20.4 An officer of the State Lottery Commission has sampled lottery ticket purchasers over a 1-week period at one location. The amounts distributed back to the purchasers and the associated probabilities for 5000 tickets are

Distribution, $	0	2	5	10	100
Probability	0.91	0.045	0.025	0.013	0.007

(a) Plot the cumulative distribution of winnings.

(b) Calculate the expected value of the distribution of dollars per ticket.

(c) If all tickets cost $2, what is the expected long-term income to the state per ticket, based upon this sample?

20.5 Bob is working on two separate probability-related projects. The first involves a variable N, which is the number of consecutively manufactured parts that weigh in above the weight specification limit. The variable N is described by the formula $(0.5)^N$ because each unit has a 50-50 chance of being below or above the limit. The second involves a battery life, L, which varies between 2 and 5 months. The probability distribution is triangular with the mode at 5 months, which is the design life. Some batteries fail early, but 2 months is the smallest life experienced thus far (a) Write out and plot the probability distributions and cumulative distributions for Bob. (*b*) Determine the probability of *N* being 1, 2, or 3 consecutive units above the weight limit.

20.6 An alternative to buy and an alternative to lease hydraulic lifting equipment have been formulated. Use the parameter estimates and assumed distribution data below to plot the probability distributions on one graph for corresponding parameters. Label the parameters carefully.

Purchase alternative

Parameter	Estimated Value High	Low	Assumed Distribution
First cost, $	25,000	20,000	Uniform; continuous
Salvage value, $	3,000	2,000	Triangular; mode at $2,500
Life, years	8	4	Triangular; mode at 6
AOC, $	9,000	5,000	Uniform; continuous

Lease alternative

Parameter	Estimated Value High	Low	Assumed Distribution
Lease first cost, $	2,000	1,800	Uniform; continuous
AOC, $	9,000	5,000	Triangular; mode at $7000
Lease term, years	2	2	Certainty

20.7 Carla is a statistician with a bank. She has collected debt-to-equity mix data on mature (*M*) and young (*Y*) companies. The debt percentages vary from 20% to 80% in her sample. Carla has defined D_M as a variable for the mature companies from 0 to 1, with $D_M = 0$ interpreted as the low of 20% debt and $D_M = 1.0$ as the high of 80% debt. The variable for young corporation debt percentages, D_Y, is similarly defined. The probability distributions used to describe D_M and D_Y are

$$f(D_M) = 3(1-D_M)^2 \qquad 0 \le D_M \le 1$$
$$f(D_Y) = 2D_Y \qquad 0 \le D_Y \le 1$$

(*a*) Use different values of the debt percentage between 20% and 80% to calculate values for the probability distributions and then plot them. (*b*) What can you comment about the probability that either of the companies will have a low debt percentage? A high debt percentage?

20.8 A discrete variable *X* can take on integer values of 1 to 10. A sample of size 50 results in the following probability estimates:

X_i	1	2	3	6	9	10
$P(X_i)$	0.2	0.2	0.2	0.1	0.1	0.2

(*a*) Write out and graph the cumulative distribution.

(*b*) Calculate the following probabilities using the cumulative distribution: X is between 6 and 10, and X has the values 4, 5, or 6.

(*c*) Use the cumulative distribution to show that $P(X = 7 \text{ or } 8) = 0.0$. Even though this probability is zero, the statement about X is that it can take on integer values of 1 to 10. How do you explain the apparent contradiction in these two statements?

20.9 Use the discrete variable probability distribution in Problem 20.8 to develop a sample of size 25. Estimate the probabilities for each value of X from your sample and compare them with those of the originating $P(X_i)$ values.

20.10 The interest rate paid on loans over a 1-year period varied from 5% to 10% in all cases. Because of the distribution of i values, the assumed probability distribution of i for the next year is

$$f(X) = 2X \qquad 0 \le X \le 1$$

where

$$X = \begin{cases} 0 & \text{when } i = 5\% \\ 1 & \text{when } i = 10\% \end{cases}$$

For a continuous variable the cumulative distribution $F(X)$ is the integral of $f(X)$ over the same range of the variable. In this case

$$F(X) = X^2 \qquad 0 \le X \le 1$$

(*a*) Graphically assign RNs to the cumulative distribution and take a sample of size 30 for the variable. Transform the X values into interest rates.

(*b*) Calculate the average i value from the sample.

20.11 Develop a probability distribution of your own, assign random numbers to $F(X)$, and take a sample from it. Now plot the probability values from the sample for each X value (or cluster of X values if you choose a continuous variable) and compare them with the original probability distribution.

20.12 Use the RAND or RANDBETWEEN function in Excel (or corresponding random number generator in your spreadsheet system) to generate 100 values from a $U(0,1)$ distribution.

(*a*) Calculate the average and compare it to 0.5, the expected value for a random sample between 0 and 1.

(*b*) Cluster the results into cells of 0.1 width, that is 0.0–0.1, 0.1–0.2, etc., where the upper-limit value is excluded from each cell. Determine the probability for each grouping from the results. Does your sample come close to having approximately 10% in each cell?

20.13 Carol sampled the monthly maintenance costs for automated soldering machines a total of 100 times during 1 year. She clustered the costs into $200 cells, for example, $500 to $700, with cell midpoints of $600, $800, $1000, etc., and she indicated the number of times (frequency) each cell value was observed. The costs and frequency data are as follows:

Cell Midpoint	Frequency
600	6
800	10
1000	9
1200	15
1400	28
1600	15
1800	7
2000	10

(a) Estimate the expected value and standard deviation of the maintenance costs the company should anticipate based on Carol's sample.

(b) What is the best estimate of the percentage of costs that will fall within 2 standard deviations of the mean?

(c) Develop a probability distribution of the monthly maintenance costs from Carol's sample and indicate the answers to the previous two questions on it.

20.14 (a) Determine the values of sample average and standard deviation of the data in Problem 20.8. (b) Determine the values 1 and 2 standard deviations from the mean. Of the 50 sample points, how many fall within these two ranges?

20.15 (a) Use the relations in Section 20.4 for continuous variables to determine the expected value and standard deviation for the distribution of $f(D_Y)$ in Problem 20.7 (b) It is possible to calculate the probability of a continuous variable X between two points (a, b) using the integral

$$P(a \le X \le b) = \int_a^b f(X)\, dx$$

Determine the probability that D_Y is within 2 standard deviations of the expected value.

20.16 Answer the questions in Problem 20.15 for the variable D_M in Problem 20.7.

20.17 Estimate the expected value for the variable *N* in Problem 20.5.

20.18 A newsstand manager is tracking *Y*, the number of weekly magazines left on the shelf when the new edition is delivered. Data collected over a 30-week period is summarized by the following probability distribution. Plot the distribution and the estimates for expected value and standard deviation of *Y* on it.

Y, copies	3	7	10	12
P(*Y*)	$1/3$	$1/4$	$1/3$	$1/12$

20.19 Carl, an engineering colleague, estimated net cash flows after taxes for the plan he is working on. He calculated the PW value at the company's MARR of 7% with the following results. (The second NCF in year 10 is for the sale of capital assets.)

Year	NCF, $
0	$-28{,}800$
1–6	5,400
7–10	2,040
10	2,800

$$\text{PW} = -28{,}800 + 5400(P/A,7\%,6) + 2040(P/A,7\%,4)(P/F,7\%,6)$$
$$+ 2800(P/F,7\%,10)$$
$$= \$2966$$

Carl expects the MARR to vary, as will the NCF, especially during the out years of 7 through 10. He is willing to accept the other estimates as certain. Use the following probability distribution assumptions for MARR and NCF to perform a simulation—manual or computer based.

MARR. Uniform distribution over the range 6% to 10%.

NCF, years 7 through 10. Uniform distribution over the range $1600 to $2400 for each year.

Plot the resulting PW distribution. Should the plan be accepted using decision making under certainty? under risk?

20.20 Repeat Problem 20.19 except use the normal distribution for the NCF in years 7 through 10 with an expected value of $2040 and a standard deviation of $500.

20.21 A team of three individuals developed the following statements about the factors to be used in a weighted evaluation method. Use the statements to determine the normalized factor weights to be used in Equation [20.21].

Factor	Comment
1. Flexibility	The most important factor
2. Safety	70% as important as uptime
3. Uptime	One-half as important as flexibility
4. Rate of return	Twice as important as safety

20.22 Your cousin plans to purchase a new car. For three different models she has evaluated the initial cost and estimated annual costs for fuel and maintenance. She also evaluated the styling of each car for her role as a 30-year-old engineering professional. List some additional factors (tangible and intangible) that she might use in the weighted evaluation method.

20.23 John, who works at Wrist Watches, has decided to use the weighted evaluation method to compare three methods of manufacturing a watchband. The president and vice president have rated each of three factors in terms of importance to them, and John has placed an evaluation from 0 to 100 on each alternative for the three factors. John's ratings are as follows:

	Alternative		
Factor	1	2	3
Economic return	50	70	100
System performance	100	60	30
Technological competitiveness	100	40	50

Use the president's and vice president's weights below to evaluate the alternatives. Are the results the same for the two ratings? Why?

Importance Factor	President	Vice President
Economic return	100	20
System performance	80	80
Technological competitiveness	20	100

20.24 In Problem 20.23 the president and vice president are not consistent in their rating of the three factors used to evaluate alternatives. Assume you are a consultant to John and are asked to assist him.

(*a*) What are some conclusions you can make about the weighted evaluation technique as an alternative selection method, given the alternative ratings and results in Problem 20.23?

(*b*) Use the new relative ratings below that you have developed yourself to select an alternative. Using the same ratings for the president and vice president given in Problem 20.23, comment on any differences in the alternative selected.

(c) What do your new factor ratings tell you about the selections based on the president and vice president importance ratings?

	Alternative		
Factor	1	2	3
Economic return	30	40	100
System performance	70	100	70
Technological competitiveness	100	80	90

20.25 For Example 20.8, use the unweighted evaluation method to choose the alternative with the largest total value. Did the weighting of factors in Example 20.8 change the selected alternative?

20.26 The Athlete's Shop has evaluated two proposals for weight-lifting and exercise equipment. A present-worth analysis at $i = 15\%$ of estimated incomes and costs resulted in $PW_A = \$420,500$ and $PW_B = \$392,800$. In addition to this economic measure, three additional factors were independently assigned a relative importance rating of 0 to 100 by the shop manager and the lead trainer.

	Importance of Factor	
Factor	Manager	Trainer
Economic measure	100	80
Durability	35	10
Flexibility	20	100
Maintainability	20	10

Separately, you have used the four factors to rate the two equipment proposals on a scale of 0.0 to 1.0. The economic measure factor was rated using the PW values.

Factor	Proposal A	Proposal B
Economic measure	1.00	0.90
Durability	0.35	1.00
Flexibility	1.00	0.90
Maintainability	0.25	1.00

Select the proposal using each of the following methods:

(a) PW measure of worth.

(b) Unweighted evaluation method.

(c) Weighted evaluations of the manager.

(d) Weighted evaluations of the lead trainer.

APPENDIX

A

BASICS OF USING MICROSOFT EXCEL©

*T*his appendix provides a brief summary of the procedures to use Microsoft Excel (hereafter called Excel) in general and the functions which are helpful in engineering economy. As more advanced software is available, the particular procedures explained here will change slightly and additional functions will become available. However, the usage will be fundamentally the same. Refer to your User's Guide for more details on capabilities and their application for your particular computer and version of Excel.

Other spreadsheet systems offer similar capabilities and can be used in equivalent ways to those discussed here to help with engineering economy computations.

A.1   INTRODUCTION TO USING EXCEL

Running Excel on Windows95©

1. After booting up the computer, click on the START button located on the lower left corner of the screen.

2. Move the mouse pointer to PROGRAMS, and pause for 1 second. Another submenu will appear on the right of the PROGRAMS menu.

3. Move to the Microsoft Excel icon, and left click the mouse to run Microsoft Excel.

4. If the Microsoft Excel icon is not on the PROGRAMS submenu list, move to the Office95 (or Microsoft Office) icon and pause for 1 second. Then move the mouse over to the right of the Office95 (or Microsoft Office) icon and highlight the Microsoft Excel icon. Left click the mouse to run Microsoft Excel.

Setting Up a Spreadsheet

Some example computations are detailed below. The $=$ sign is necessary to perform any formula or function computation in a cell.

1. Using the <arrow keys>, move to cell B4 and type =4+3 and hit Enter.
2. A number 7 appears in cell B4.
3. To edit, use your <arrow keys> to move to cell B4; hit <F2>.
4. Hit the <Backspace> key twice, to delete +3.
5. Type −3 and hit <Enter>.
6. The answer 1 appears in cell B4.
7. To delete the cell entirely, move to cell B4 and hit the <Delete> key once.
8. To exit, move the mouse pointer to the top left corner and left click on File in the top bar menu.
9. Move the mouse down the File submenu, highlight Exit, and left click.
10. When the "Save Changes" box appears, left click "No" to exit without saving.
11. If you wish to save your work, left click "Yes."
12. Type in a file name (e.g., calcs 1) and click on "Save."

Using Microsoft Excel Functions

1. Run Microsoft Excel.
2. Move to cell C3. (Move the mouse pointer to C3 and left click.)
3. Type =pv(5%,12,10) and hit <Enter>. This function will calculate the present value of 12 payments of $10 at a 5%-per-year interest rate.

Another use: To calculate the future value of 12 payments of $10 at 6%-per-year interest, do the following:

1. Move to cell B3, and type INTEREST.
2. Move to cell C3, and type 6/100 (to represent 6%).
3. Move to cell B4, and type PAYMENT.
4. Move to cell C4, and type 10 (to represent the amount of each payment).
5. Move to cell B5, and type NUMBER OF PAYMENTS.
6. Move to cell C5, and type 12 (to represent the number of payments).
7. Move to cell B7, and type FUTURE VALUE.
8. Move to cell C7, and type =fv(c3,c5,c4) and hit <Enter>. The answer will appear in cell C7.

To edit the values of each cell (this feature is used repeatedly in sensitivity analysis and breakeven analysis),

1. Move to cell C3 and type =5/100 (the previous value will be replaced).
2. The value in cell C7 will change its answer automatically.

Printing Your Spreadsheet

First define the portion (or all) of the spreadsheet to be printed.

1. Move the mouse pointer to the top left corner of your spreadsheet.
2. Hold the left click button down. (Do not release the left click button.)
3. Drag the mouse to the lower right corner of your spreadsheet or to wherever you want to stop printing.
4. Release the left click button. It is ready to print.
5. Left click the File top bar menu.
6. Move the mouse down to select Print and left click.
7. In the Print dialog box, left click the option Selection in the Print What box (or similar command).
8. Left click the OK button to start printing.

Depending on your computer environment, you may have to select a network printer and queue your printout through a server. Contact your network administrator if you are operating in a network environment.

Saving Your Spreadsheet

You can save your spreadsheet at any time during or after completing your work. It is recommended that you save your work regularly.

1. Left click the File top bar menu.
2. To save the spreadsheet the first time, left click the Save As . . . option.
3. Type the file name, e.g., calcs2, and left click the Save button.

To save the spreadsheet after it has been saved the first time, i.e., a file name has been assigned to it, left click the File top bar menu, move the mouse pointer down, and left click on Save.

Create a Bar Chart

1. Run Excel.
2. Move to cell A1 and type 1. Move down to cell A2 and type 2. Type 3 in cell A3, 4 in cell A4, and 5 in cell A5.
3. Move to cell B1 and type 4. Type 3.5 in cell B2, 5 in cell B3, 7 in cell B4, and 12 in cell B5.
4. Move the mouse pointer to cell A1, left click and hold, while dragging the mouse to cell B5. (All the cells with numbers should be highlighted.)
5. Left click Insert on the top bar menu. Select Chart and then select On This Sheet. (Cross hairs appear instead of the mouse pointer.)
6. Move the mouse (cross hair) to cell D2 and left click and hold, while dragging the mouse to cell H14.

7. The ChartWizard screen appears.

8. Since the data range was selected in step 4 above, left click Next on the ChartWizard 1 of 5 screen.

9. Left click the Column Bar Graph icon and then left click Next.

10. Left click the #7 icon, then left click Next.

11. Left click Next on the ChartWizard 4 of 5 screen.

12. Answer "Add a Legend?" by left clicking "No."

13. Left click the Title box and type Example 1.

14. Left click Category (X) box and type Year.

15. Left click Value (Y) box and type Rate of Return.

16. The bar chart appears on the spreadsheet.

17. To adjust the size of the spreadsheet, left click anywhere inside the chart. Locate the small dots on the sides and corners of the graphics box.

18. Move your mouse pointer to the lower right corner of the graphics box, left click, and hold the small dots and drag the corner to change the size of the graphics area.

Getting Help

1. To get general help information, left click on the Help top bar menu (top right corner).

2. Left click on Microsoft Excel Help Topics.

3. For example, if you want to know more about how to save a file, type the word save in box 1.

4. Select the appropriate matching words in box 2. You can browse through the selected words in box 2 by left clicking on suggested words.

5. Observe the listed topics in box 3.

6. If you find a topic listed in box 3 that matches what you are looking for, double left click the selected topic in box 3.

A.2   USING EXCEL FUNCTIONS PERTINENT TO ENGINEERING ECONOMY (alphabetical order)

DB (Declining Balance)

Calculates the depreciation amount for an asset for a specified period n using the fixed declining-balance method. The depreciation rate used in the computation is $1/n$.

Usage Syntax   =DB(cost, salvage, life, period, month)

cost   First cost of the asset.

salvage   Salvage value of the asset.

life   Depreciation life (recovery period).

period   The period, year, for which the depreciation is to be calculated.

month   This is an optional entry. If this entry is omitted, a full year is assumed for the first year.

Example   A new machine costs $100,000 and is expected to last 10 years. At the end of 10 years, the salvage value of the machine is $50,000. What is the depreciation of the machine in the first year and the fifth year?

Depreciation in the first year: =DB(100000,50000,10,1)

Depreciation for the fifth year: =DB(100000,50000,10,5)

DDB (Double-Declining Balance)

Calculates the depreciation of an asset for a specified period n using the double-declining-balance method. A factor can also be entered for some other declining-balance depreciation method by specifying a factor in the function.

Usage Syntax   =DDB(cost, salvage, life, period, factor)

cost   First cost of the asset.

salvage   Salvage value of the asset.

life   Depreciation life.

period   The period, year, for which the depreciation is to be calculated.

factor   This is an optional entry. If this entry is omitted, the function will use a double-declining method with 2 times the straight-line rate. If, for example, the entry is 1.5, the 150% declining-balance method will be used.

Example   A new machine costs $200,000 and is expected to last 10 years. The salvage value is $10,000. Calculate the depreciation of the machine for the first and the eighth year. Finally, calculate the depreciation for the fifth year using the 175% declining-balance method.

Depreciation for the first year: =DDB(200000,10000,10,1)

Depreciation for the eighth year: =DDB(200000,10000,10,8)

Depreciation for the fifth year using 175% DB: =DDB(200000,10000,10,5,1.75)

FV (Future Value)

Calculates the future value based on periodic payments at a specific interest rate.

Usage Syntax   =FV(rate, nper, pmt, pv, type)

rate	Interest rate per compounding period.
nper	Number of compounding periods.
pmt	Constant payment amount.
pv	The present value amount. If pv is not specified, the function will assume it to be 0.
type	Either 0 or 1. A 0 represents payments made at the end of the period, and 1 represents payments at the beginning of the period.

Example   Jack wants to start a savings account that can be increased as desired. He will deposit $12,000 to start the account and plans to add $500 to the account at the beginning of each month for the next 24 months. The bank pays 0.25% per month. How much will be in Jack's account at the end of 24 months?

Future value in 24 months: =FV(0.25%,24,500,12000,1)

IPMT (Interest Payment)

Calculates the interest accrued for a given period n based on constant periodic payments and interest rate.

Usage Syntax   =IPMT(rate, per, nper, pv, fv, type)

rate	Interest rate per compounding period.
per	Period over which interest is to be calculated.
nper	Number of compounding periods.
pv	Present value. If pv is not specified, the function will assume it to be 0.
fv	Future value. If fv is omitted, the function will assume it to be 0. The fv can also be considered a cash balance after the last payment is made.
type	Either 0 or 1. A 0 represents payments made at the end of the period, and 1 represents payments made at the beginning of the period.

Example   Calculate the interest due in the tenth month for a 48-month, $20,000 loan. The interest rate is 0.25% per month.

Interest due: =IPMT(0.25%,10,48,20000)

IRR (Internal Rate of Return)

Calculates the internal rate of return for a series of payments at regular periods.

Usage Syntax   =IRR(values, guess)

values   A set of numbers in a spreadsheet column (or row) for which the rate of return will be calculated. The set of numbers must consist of at least *one* positive and *one* negative number. Negative numbers denote a payment made or cash outflow, and positive numbers denote income or cash inflow.

guess   To reduce the number of iterations, a *guessed rate of return* can be added to the IRR function. In most cases, a guess is not required and a 10% rate of return is initially assumed. If the #NUM! Error appears, try using different values for guess. Inputting different guess values makes it possible to determine the multiple roots for the rate-of-return equation of a nonconventional (nonsimple) cash-flow sequence.

Example   John wants to start a printing business. He will need $25,000 in capital and anticipates that the business will generate the following incomes during the first 5 years. Calculate his rate of return after 3 years and after 5 years.

year 1 $5,000
year 2 $7,500
year 3 $8,000
year 4 $10,000
year 5 $15,000

Set up an array in the spreadsheet.

In cell A1, type −25000 (negative for payment).
In cell A2, type 5000 (positive for income).
In cell A3, type 7500.
In cell A4, type 8000.
In cell A5, type 10000.
In cell A6, type 15000.

Therefore, cells A1 through A6 contain the array of cash flows for the first 5 years, including the capital outlay. Note that any years with a zero cash flow must be left blank to ensure that the year value is correctly maintained for computation purposes.

To calculate the internal rate of return after 3 years, move to cell A7, and type =IRR(A1 : A4).

To calculate the business's internal rate of return after 5 years, move to cell A8, and type =IRR(A1 : A6,5%).

MIRR (Modified Internal Rate of Return)

Calculates the modified internal rate of return for a series of cash flows and reinvestment of income and interest at a stated rate.

Usage Syntax   =MIRR(values, finance_rate, reinvest_rate)

   values   Refers to an array of cells in the spreadsheet. Negative numbers represent payments, and positive numbers represent cash inflow or income. The series of payments and income must occur at regular periods and must contain at least *one* positive number (income) and *one* negative number (payment).

   finance_rate   Interest rate of income or money used in the cash flows.

   reinvest_rate   Interest rate for all reinvestments.

Example   Jane opened a hobby store 4 years ago. When she started the business, Jane borrowed $50,000 from a bank at 12% per year. Since then, the business has yielded $10,000 the first year, $15,000 the second year, $18,000 the third year, and $21,000 the fourth year. Jane reinvests her profits, earning 8% per year. What is the modified rate of return after 3 years and after 4 years?

   In cell A1, type −50000.

   In cell A2, type 10000.

   In cell A3, type 15000.

   In cell A4, type 18000.

   In cell A5, type 21000.

   To calculate the modified rate of return after 3 years, move to cell A6, and type =MIRR(A1:A4,12%,8%).

   To calculate the modified rate of return after 4 years, move to cell A7, and type =MIRR(A1:A5,12%,8%).

NPER (Number of Periods)

Calculates the number of periods for the present worth of an investment to equal the future value specified, based on uniform regular payments and a stated interest rate.

Usage Syntax   =NPER(rate, pmt, pv, fv, type)

   rate   Interest rate per compounding period.

   pmt   Amount paid during each compounding period.

   pv   Present value (lump-sum amount).

   fv   Future value or cash balance after the last payment. If fv is omitted, the function will assume a value of 0.

type   Enter 0 if payments are due at the end of the compounding period, and 1 if payments are due at the beginning of the period. If omitted, 0 is assumed.

Example Sally plans to open a savings account which pays 0.25% per month. Her initial deposit is $3000, and she plans to deposit $250 at the beginning of every month. How many payments does she have to make to accumulate $15,000 to buy a new car?

Number of payments : =NPER(0.25%,−250,−3000,15000,1)

NPV (Net Present Value)

Calculates the net present value of a series of future cash flows at a stated interest rate.

Usage Syntax   =NPV(rate, series)

rate   Interest rate per compounding period.

series   Series of payments (negative number) and incomes (positive number) set up in a range of cells in the spreadsheet.

Example Mark is considering buying a sports store for $100,000 and expects to receive the following income during the next 6 years of business: $25,000, $40,000, $42,000, $44,000, $48,000, $50,000. The interest rate is 8% per year.

In cell A1, type −100000.
In cell A2, type 25000.
In cell A3, type 40000.
In cell A4, type 42000.
In cell A5, type 44000.
In cell A6, type 48000.
In cell A7, type 50000.
In cell A8, type = NPV(8%,A2 : A7)+A1.

The cell A1 value is already a present value.

PMT (Payments)

Calculates payments based on present value and/or future value at a constant interest rate.

Usage Syntax   =PMT(rate, nper, pv, fv, type)

rate   Interest rate per compounding period.

nper   Total number of periods.

pv   Present value (or loan amount).

fv   Future value or future cash balance.

type   Enter 0 for payments due at the end of the compounding period, and 1 if payment is due at the start of the compounding period.

Example   Jim plans to take a $15,000 loan to buy a new car. The interest rate is 7%. He wants to pay the loan off in 5 years (60 months). What are his monthly payments?

Monthly payments: =PMT(7%/12,60,15000)

PPMT (Principal Payment)   Calculates the payment on the principal based on uniform payments at a specified interest rate.

Usage Syntax   =PPMT(rate, per, nper, pv, fv, type)

rate   Interest rate per compounding period.

per   Period for which the payment on the principal is required.

nper   Total number of periods.

pv   Present value.

fv   Future value.

type   Enter 0 for payments that are due at the end of the compounding period, and 1 if payments are due at the start of the compounding period.

Example   John is planning to invest $10,000 in equipment which is expected to last 10 years with no salvage value. The interest rate is 5%. What is the principal payment at the end of year 4 and year 8?

At the end of year 4: =PPMT(5%,4,10,−10000)

At the end of year 8: =PPMT(5%,8,10,−10000)

PV (Present Value)

Calculates the present value of a future series of constant-amount payments and a single lump sum in the last period at a constant interest rate.

Usage Syntax   =PV(rate, nper, pmt, fv, type)

rate   Interst rate per compounding period.

nper   Total number of payments.

pmt   Payment made or received at regular intervals during the life of the annuity. Negative numbers represent payments (cash outflows), and positive numbers represent income.

fv   Future value or cash balance at the end of the last payment.

type   Enter 0 if payments are due at the end of the compounding period, and 1 if payments are due at the start of each compounding period. If omitted, 0 is assumed.

The primary differences between the PV function and the NPV function are: PV allows for end or beginning of period cash flows, and PV requires that all amounts have the same value, whereas they may vary for the NPV function.

Example   John is considering leasing a car for $300 a month for 3 years (36 months). After the 36-month lease, he can purchase the car for $12,000. Using an interest rate of 8% per year, find the present value of this option.

Present value: =PV(8%/12,36,−300,−12000)

RAND (Random Number)

Returns an evenly distributed number that is either: (1) ≥ 0 and < 1; (2) ≥ 0 and < 100; or (3) between two specified numbers.

Usage Syntax

=RAND()	for range 0 to 1
=RAND()*100	for range 0 to 100
=RAND()*(b−a)+a	for range a to b

a = minimum integer to be generated
b = maximum integer to be generated.

The Excel function RANDBETWEEN(a,b) may also be used to obtain a random number between two values.

Example   Randi wants to generate random numbers between the limits of −10 and 25. What is the Excel function? The minimum and maximum values are a = −10 and b = 25.

Random number: =RAND()*(25−−10)+−10

RATE (Interest Rate)

Calculates the interest rate per compounding period for a series of payments or incomes.

Usage Syntax   =RATE(nper, pmt, pv, fv, type, guess)

nper	Total number of payments.
pmt	Payment amount made each compounding period.
pv	Present value.
fv	Future value.
type	Enter 0 for payments due at the end of the compounding period, and 1 if payments are due at the start of each compounding period.

guess   To minimize computing time, include a guessed interest rate. If a value of guess is not specified, the function will assume a rate of 10%. This function usually converges to a solution, if the rate is between 0% to 100%.

Example   Mary wants to start a savings account at a bank. She will make an initial deposit of $1000 to open the account and plans to deposit $100 at the beginning of each month. She plans to do this for the next 3 years (36 months). At the end of 3 years, she wants to have at least $5000. What is the minimum interest required to achieve this result.

Interest rate: =RATE(36,−100,−1000,5000,1)

SLN (Straight-Line Depreciation)

Calculates the straight-line depreciation of an asset for a given year.

Usage Syntax   =SLN(cost, salvage, life)

cost   First cost of the asset.

salvage   Salvage value.

life   Depreciation life.

Example   Maria purchased a printing machine for $100,000. The machine has an allowed depreciation life of 8 years and an estimated salvage value of $15,000. What is the depreciation each year?

Depreciation: =SLN(100000,15000,8)

SYD (Sum-of-Year-Digits Depreciation)

Calculates the sum-of-year-digits depreciation of an asset for a given year.

Usage Syntax   =SYD(cost, salvage, life, per)

cost   First cost of the asset.

salvage   Salvage value.

life   Depreciation life.

per   The year for which the depreciation is sought.

Example   Jack bought equipment for $100,000 which has a depreciation life of 10 years. The salvage value is $10,000. What is the depreciation for year 1 and year 9?

Depreciation for year 1: =SYD(100000,10000,10,1)

Depreciation for year 9: =SYD(100000,10000,10,9)

A.3   Error Messages

If Excel is unable to complete a formula or function computation, an error message is displayed. Some of the common messages are

#DIV/0!	Is trying to divide by zero.
#N/A	Refers to a value that is not available.
#NAME?	Uses a name that Excel doesn't recognize.
#NULL!	Specifies an invalid intersection of two areas.
#NUM!	Uses a number incorrectly.
#REF!	Refers to a cell that is not valid.
#VALUE!	Uses an invalid argument or operand.
#####	Produces a result, or includes a constant numeric value, that is too long to fit in the cell. (Widen the column.)

A.4   LIST OF EXCEL FINANCIAL FUNCTIONS

Here is a listing and brief description of the output of all Excel financial functions. Not all these functions are available on all versions of Microsoft Excel. The Add-ins command can help you determine if the function is available on the system you are using. See your User's Guide.

ACCRINT	Returns the accrued interest for a security that pays periodic interest.
ACCRINTM	Returns the accrued interest for a security that pays interest at maturity.
AMORDEGRC	Returns the depreciation for each accounting period.
AMORLINC	Returns the depreciation for each accounting period.
COUPDAYBS	Returns the number of days from the beginning of the coupon period to the settlement date.
COUPDAYS	Returns the number of days in the coupon period that contains the settlement date.
COUPDAYSNC	Returns the number of days from the settlement date to the next coupon date.
COUPNCD	Returns the next coupon date after the settlement date.
COUPNUM	Returns the number of coupons payable between the settlement date and maturity date.
COUPPCD	Returns the previous coupon date before the settlement date.

CUMIPMT	Returns the cumulative interest paid between two periods.
CUMPRINC	Returns the cumulative principal paid on a loan between two periods.
DB	Returns the depreciation of an asset for a specified period using the fixed declining-balance method.
DDB	Returns the depreciation of an asset for a specified period using the double-declining-balance method or some other method you specify.
DISC	Returns the discount rate for a security.
DOLLARDE	Converts a dollar price expressed as a fraction into a dollar price expressed as a decimal number.
DOLLARFR	Converts a dollar price expressed as a decimal number into a dollar price expressed as a fraction.
DURATION	Returns the annual duration of a security with periodic interest payments.
EFFECT	Returns the effective annual interest rate.
FV	Returns the future value of an investment.
FVSCHEDULE	Returns the future value of an initial principal after applying a series of compound interest rates.
INTRATE	Returns the interest rate for a fully invested security.
IPMT	Returns the interest payment for an investment for a given period.
IRR	Returns the internal rate of return for a series of cash flows.
MDURATION	Returns the Macauley modified duration for a security with an assumed par value of $100.
MIRR	Returns the internal rate of return where positive and negative cash flows are financed at different rates.
NOMINAL	Returns the annual nominal interest rate.
NPER	Returns the number of periods for an investment.
NPV	Returns the net present value of an investment based on a series of periodic cash flows and a discount rate.
ODDFPRICE	Returns the price per $100 face value of a security with an odd first period.
ODDFYIELD	Returns the yield of a security with an odd first period.
ODDLPRICE	Returns the price per $100 face value of a security with an odd last period.

ODDLYIELD	Returns the yield of a security with an odd last period.
PMT	Returns the periodic payment for an annuity.
PPMT	Returns the payment on the principal for an investment for a given period.
PRICE	Returns the price per $100 face value of a security that pays periodic interest.
PRICEDISC	Returns the price per $100 face value of a discounted security.
PRICEMAT	Returns the price per $100 face value of a security that pays interest at maturity.
PV	Returns the present value of an investment.
RATE	Returns the interest rate per period of an annuity.
RECEIVED	Returns the amount received at maturity for a fully invested security.
SLN	Returns the straight-line depreciation of an asset for one period.
SYD	Returns the sum-of-year-digits depreciation of an asset for a specified period.
TBILLEQ	Returns the bond-equivalent yield for a Treasury bill.
TBILLPRICE	Returns the price per $100 face value for a Treasury bill.
TBILLYIELD	Returns the yield for a Treasury bill.
VDB	Returns the depreciation of an asset for a specified or partial period using a declining-balance method.
XIRR	Returns the internal rate of return for a schedule of cash flows that is not necessarily periodic.
XNPV	Returns the net present value for a schedule of cash flows that is not necessarily periodic.
YIELD	Returns the yield on a security that pays periodic interest.
YIELDDISC	Returns the annual yield for a discounted security. For example, a Treasury bill.
YIELDMAT	Returns the annual yield of a security that pays interest at maturity.

There are many more functions available on Excel in the following and other areas: math and trig, statistical, date and time, database, logical, and information.

BASICS OF ACCOUNTING REPORTS AND BUSINESS RATIOS

his appendix provides a very fundamental description of financial statements. The documents discussed here may assist in reviewing or understanding basic financial statements and in gathering information useful in an engineering eco-nomy study.

LEARNING OBJECTIVES

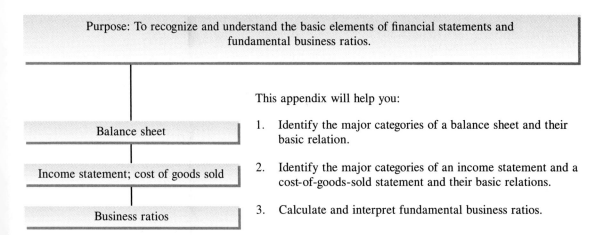

Purpose: To recognize and understand the basic elements of financial statements and fundamental business ratios.

Balance sheet	This appendix will help you: 1. Identify the major categories of a balance sheet and their basic relation.
Income statement; cost of goods sold	2. Identify the major categories of an income statement and a cost-of-goods-sold statement and their basic relations.
Business ratios	3. Calculate and interpret fundamental business ratios.

B.1   THE BALANCE SHEET

The fiscal year and the tax year are defined identically for a corporation or an individual—12 months in length. The fiscal year (FY) is commonly not the calendar year (CY) for a corporation. The U.S. government uses October through September as its FY. For example, October 1999 through September 2000 is FY2000. The fiscal or tax year is always the calendar year for an individual citizen.

At the end of each fiscal year, a company publishes a *balance sheet*. A sample balance sheet for the TeamWork Corporation is presented in Table B–1. This is a yearly presentation of the state of the firm at a particular time, for example, December 2001; however, a balance sheet is also usually prepared quarterly and monthly. Note that three main categories are used.

Assets.   A summary of all resources owned by or owed to the company. There are two main classes of assets. *Current assets* represent shorter-lived working capital (cash, accounts receivable, etc.), which are more easily converted to cash, usually within 1 year. Longer-lived assets are referred to as *fixed assets* (land, equipment, etc.). Conversion of these holdings to cash in a short period of time would require a major corporate reorientation.

Table B–1   Sample balance sheet

TEAMWORK CORPORATION
Balance Sheet
December 31, 2001

Assets		Liabilities	
Current			
Cash	$10,500	Accounts payable	$19,700
Accounts receivable	18,700	Dividends payable	7,000
Interest accrued receivable	500	Long-term notes payable	16,000
Inventories	52,000	Bonds payable	20,000
Total current assets	$81,700	Total liabilities	$62,700
		Net Worth	
Fixed			
Land	$25,000	Common stock	$275,000
Building and equipment	438,000	Preferred stock	100,000
Less: Depreciation allowance $82,000	356,000	Retained earnings	25,000
Total fixed assets	381,000	Total net worth	400,000
Total assets	$462,700	Total liabilities and net worth	$462,700

Liabilities. A summary of all financial obligations (debts, loans, etc.) of a corporation. Bond indebtedness is included here.

Net worth. Also called *owner's equity,* this is a summary of the financial value of ownership, including stocks issued and earnings retained by the corporation.

The balance sheet is constructed using the relation

$$\text{Assets} = \text{liabilities} + \text{net worth}$$

In Table B–1 each major category is further divided into standard categories. For example, current assets is comprised of cash, accounts receivable, etc. Each subdivision has a specific interpretation, such as, accounts receivable, which represents all money owed to the company by its customers.

Problem B.1

B.2 INCOME STATEMENT AND COST-OF-GOODS-SOLD STATEMENT

A second important financial statement is the *income statement* (Table B–2). The income statement summarizes the profits or losses of the corporation for a stated period of time. Income statements always accompany balance sheets.

Table B–2	Sample income statement	
TEAMWORK CORPORATION Income Statement Year Ended December 31, 2001		
Revenues		
Sales	$505,000	
Interest revenue	3,500	
Total revenues		$508,500
Expenses		
Cost of goods sold (from Table B–3)	$290,000	
Selling	28,000	
Administrative	35,000	
Other	12,000	
Total expenses		365,000
Income before taxes		143,500
Taxes for year		64,575
Net profit for year		**$ 78,925**

The major categories of an income statement are

Revenues.   All sales and interest revenue that the company has received in the past accounting period.

Expenses.   A summary of all expenses for the period. Some expense amounts are itemized in other statements, for example, cost of goods sold and income taxes.

The income statement, published at the same time as the balance sheet, uses the basic equation

$$\text{Revenues} - \text{expenses} = \text{profit (or loss)}$$

The *cost of goods sold* is an important accounting term. It represents the net cost of producing the product marketed by the firm. Cost of goods sold may also be called *factory cost.* A statement of the cost of goods sold, such as that shown in Table B–3, is useful in determining exactly how much it costs to make a particular product over a stated time period, usually a year. Note that the total of the cost-of-goods-sold statement is entered as an expense item on the income statement. This total is determined using the relations

$$\text{Prime cost} = \text{direct materials} + \text{direct labor}$$

$$\text{Cost of goods sold} = \text{prime cost} + \text{factory expense}$$

[B.1]

The item *factory expense* includes all indirect costs and overhead charged to a product. Indirect cost allocation methods are discussed in Chapter 12.

Problems B.2 and B.3

Table B–3	Sample cost-of-goods-sold statement

TEAMWORK CORPORATION
Statement of Cost of Goods Sold
Year Ended December 31, 2001

Materials		
Inventory, January 1, 2001	$ 54,000	
Purchases during year	174,500	
Total	$228,500	
Less: Inventory December 31, 2001	50,000	
Cost of materials		$178,500
Direct labor		110,000
Prime cost		288,500
Factory expense (indirect costs)		7,000
Factory cost		295,500
Less: Increase in finished goods inventory during year		5,500
Cost of goods sold (into Table B–2)		$290,000

B.3   BUSINESS RATIOS

Accountants and engineering economists frequently utilize business ratio analysis to evaluate the financial health (status) of a company over time and in relation to industry norms. Because the engineering economist must continually communicate with others, he or she should have a basic understanding of several ratios. For comparison purposes, it is necessary to compute the ratios for several companies in the same industry. Industrywide median ratio values are published annually by firms such as Dun and Bradstreet in *Industry Norms and Key Business Ratios*. The ratios are commonly classified according to their role in measuring the corporation.

> **Solvency ratios.**   An ability to meet short-term and long-term financial obligations.
>
> **Efficiency ratios.**   Measures which reflect management's ability to use and control assets.
>
> **Profitability ratios.**   Evaluate the ability to earn a return for the owners of the corporation.

Numerical data for several important ratios are discussed here and are extracted from the TeamWork balance sheet and income statement, Tables B–1 and B–2, respectively.

Current Ratio   This ratio is utilized to analyze the company's working-capital condition. It is defined as

$$\text{Current ratio} = \frac{\text{current assets}}{\text{current liabilities}}$$

Current liabilities include all short-term debts, such as accounts and dividends payable. Note that only balance-sheet data are utilized in the current ratio; that is, no association with revenues or expenses is made. For the balance sheet of Table B–1 current liabilities amount to $19,700 + $7,000 = $26,700 and

$$\text{Current ratio} = \frac{81,700}{26,700} = 3.06$$

Since current liabilities are those debts payable in the next year, the current ratio value of 3.06 means that the current assets would cover short-term debts approximately 3 times. Current-ratio values of 2 to 3 are common.

   The current ratio assumes that the working capital invested in inventory can be converted to cash quite rapidly. Often, however, a better idea of a company's *immediate* financial position can be obtained by using the acid-test ratio.

Acid-Test Ratio (Quick Ratio)   This ratio is

$$\text{Acid-test ratio} = \frac{\text{quick assets}}{\text{current liabilities}}$$

$$= \frac{\text{current assets} - \text{inventories}}{\text{current liabilities}}$$

It is meaningful for the emergency situation when the firm must cover short-term debts using its readily convertible assets. For the TeamWork Corporation,

$$\text{Acid-test ratio} = \frac{81,700 - 52,000}{26,700} = 1.11$$

Comparison of this and the current ratio shows that approximately 2 times the current debts of the company are invested in inventories. However, an acid-test ratio of approximately 1.0 is generally regarded as a strong current position, regardless of the amount of assets in inventories.

Equity Ratio   This ratio has historically been a measure of financial strength since it is defined as

$$\text{Equity ratio} = \frac{\text{total net worth}}{\text{total assets}}$$

For the TeamWork Corporation,

$$\text{Equity ratio} = \frac{400,000}{462,700} = 0.865$$

TeamWork is 86.5% stockholder owned. The $25,000 retained earnings is also called equity since it is actually owned by the stockholders, not the corporation. An equity ratio in the range 0.80 to 1.0 usually indicates a sound financial condition, with little fear of forced reorganization because of unpaid liabilities. However, a company with virtually no debts, that is, one with a very high equity ratio, may not have a promising future, because of its inexperience in dealing with short-term and long-term debt financing. The debt-equity (D-E) mix is another measure of financial strength as discussed in Chapter 18.

Return on Sales Ratio   This often-quoted ratio indicates the profit margin for the company. It is defined as

$$\text{Return on sales} = \frac{\text{net profit}}{\text{net sales}} (100\%)$$

Net profit is the after-tax value from the income statement. This ratio measures profit earned per sales dollar and indicates how well the corporation can sustain adverse conditions over time, such as falling prices, rising costs, and declining sales. For the TeamWork Corporation,

$$\text{Return on sales} = \frac{78{,}925}{505{,}000}\,(100\%) = 15.6\%$$

Corporations may point to small return-on-sales ratios, say 2.5% to 4.0%, as indications of sagging economic conditions. In truth, for a relatively large volume, high-turnover business, an income ratio of 3% is quite healthy. Of course, a steadily decreasing ratio indicates rising company expenses, which absorb net profit after taxes.

Return on Assets Ratio   This is the key indicator of profitability since it evaluates the ability of the corporation to transfer assets into operating profit. The definition and value for TeamWork are

$$\text{Return on assets} = \frac{\text{net profit}}{\text{total assets}}\,(100\%)$$

$$= \frac{78{,}925}{462{,}700}\,(100\%) = 17.1\%$$

Efficient use of assets indicates that the company should earn a high return, while low returns usually accompany lower values of this ratio compared to the industry group ratios.

Inventory Turnover Ratio   This ratio indicates the number of times the average inventory value passes through the operations of the company. If turnover of inventory to net sales is desired, the formula is

$$\text{Net sales to inventory} = \frac{\text{net sales}}{\text{average inventory}}$$

where average inventory is the figure recorded in the balance sheet. For the TeamWork Corporation this ratio is

$$\text{Net sales to inventory} = \frac{505{,}000}{52{,}000} = 9.71$$

which means that the average value of the inventory has been sold 9.71 times during the year. Values of this ratio vary greatly from one industry to another. If inventory turnover is related to cost of goods sold, the ratio to use is

$$\text{Cost of goods sold to inventory} = \frac{\text{cost of goods sold}}{\text{average inventory}}$$

where average inventory is computed as the average of the beginning and ending inventory values in the statement of cost of goods sold. This ratio is commonly used as a measure of the inventory turnover rate in manufacturing companies.

It varies with industries, but management likes to see it remain relatively constant as business increases. For TeamWork, using the values in Table B–3,

$$\text{Cost of goods sold to inventory} = \frac{290,000}{{}^1\!/_2(54,000 + 50,000)} = 5.58$$

There are, of course, many other ratios a company's personnel can use in various circumstances; however, the ones presented here are commonly used by both accountants and economic analysts.

Example B.1

The median values for three ratios of nationally surveyed companies are presented below. Compare the TeamWork Corporation ratios with these norms and comment on differences and similarities.

Ratio	House Furnishings (SIC 2392)*	Manufactured Ice (SIC 2097)*	Scheduled Air Transportation (SIC 4512)*
Current	2.5	1.6	1.3
Return on assets	5.9%	7.5%	1.8%
Net sales to inventory	7.9	39.7	25.8
Return on sales	2.5%	6.6%	1.8%

*SIC = Standard Industrial Code.
SOURCE: *Industry Norms and Key Business Ratios; Desk-top edition* (1994–95). Dun and Bradstreet, United States, 1995.

Solution

It is not appropriate to compare ratios for one company with those in different industry groups, that is, with different SIC codes. The comparison below is, therefore, made only for discussion purposes in this text.

The corresponding ratio values for TeamWork indicated above are

$$\text{Current ratio} = 3.06$$

$$\text{Return on assets} = 17.1\%$$

$$\text{Net sales to inventory} = 9.71$$

$$\text{Return on sales} = 15.6\%$$

If TeamWork makes furniture or ice or is an airline, its current ratio is higher than expected, since it can cover current liabilities over 3 times compared with the averages of 2.5 or less in the survey. If TeamWork makes ice or flies the commercial air lanes, its inventory turnover is too low compared with the 39.7 and 25.8 times published.

In this case, for all three SIC codes, TeamWork is substantially better, especially based on returns on assets and sales ratios.

Problems B.4 and B.5

PROBLEMS

The following financial data are for the month of July 20XX for Non-Stop, Inc. Use this information in solving Problems B.1 to B.5.

Present situation, July 31, 20XX

Account	Balance
Accounts payable	$35,000
Accounts receivable	29,000
Bonds payable (20 year)	110,000
Buildings (net value)	605,000
Cash on hand	17,000
Dividends payable	8,000
Inventory value (all inventories)	31,000
Land value	450,000
Long-term mortgage payable	450,000
Retained earnings	154,000
Stock value outstanding	375,000

Transactions for July 20XX

Category		Amount
Direct labor		$50,000
Expenses		
Insurance	$20,000	
Selling	62,000	
Rent and lease	40,000	
Salaries	110,000	
Other	62,000	
Total		294,000
Income taxes		20,000
Increase in finished goods inventory		25,000
Materials inventory, July 1, 20XX		46,000
Materials inventory, July 31, 20XX		25,000
Materials purchases		20,000
Overhead charges		75,000
Revenue from sales		500,000

B.1 Use the account summary information (*a*) to construct a balance sheet for Non-Stop as of July 31, 20XX, and (*b*) to determine the value of each term in the basic equation of the balance sheet.

B.2 What was the net change in materials inventory value during the month?

B.3 Use the summary information above to develop (*a*) an income statement for July 20XX and (*b*) the basic equation of the income statement. (*c*) What percent of revenue is reported as after-tax income?

B.4 (*a*) Compute the value of each business ratio that uses only balance-sheet information from the statement you constructed in Problem B.1. (*b*) What percent of the company's current debt is unavailable and in inventory?

B.5 (*a*) Compute the turnover of inventory ratio (based on net sales) for Non-Stop, Inc., and state its meaning. (*b*) What percent of each sales dollar can the company rely upon as profit? (*c*) If Non-Stop is an airline, how does its key profitability indicator compare with the median ratio value for its SIC?

C

FINAL ANSWERS TO SELECTED END-OF-CHAPTER PROBLEMS

Chapter 1

1.7 Use logic in Sections 1.2 and 1.3.

1.9 (*a*) Yes; (*c*) yes; (*d*) no and yes.

1.13 2, 3, 12 months.

1.15 $400, 12 months.

1.17 $500.

1.20 (*a*) $1224 and $1176.47; (*b*) $47.53, 3.96%.

1.22 15%.

1.23 Simple by $18.98.

1.25 (*a*) $33.75; (*b*) 2.25%.

1.28 $P = \$756.50$.

1.31 $n = 5.36$ years.

1.35 $P = \$1000$ each 2 years, $F = ?$, $i = 10\%$, $n = 9$ years.

1.37 $P = \$5000$, $F = \$10,000$, $i = 5.5\%$, $n = ?$

1.40 $36,834.

1.44

Year	0	2	4	5,7,9,11,12	6,8,10
$1,000	−5	−5	−5	+3	−2

1.47 Plan 1: $−350 in years 1 and 3, $F = ?$ in 6, $i = 8\%$.
Plan 2: $−125 in years 1 through 6, $F = ?$ in 6, $i = 8\%$.

1.50 $P = \$-10,000$ now, $F = ?$ in 10, $n = 10$, $i = 12\%$.

1.52 (*a*) 8.47; (*b*) 11.76; (*c*) 8.47 years.

1.53 (*a*) 20; (*b*) 14.4 years.

Chapter 2

2.2 (1) 14.4210; (2) 1.00000; (3) 0.30094; (4) 9.0770; (5) 0.0055.

2.6 $F/G = 59.365$.

2.9 (1) 30.7114 interpolation (I), 30.3762 by formula (F);
(2) 5.5029 by I, 5.4980 by F;
(3) 5.8602 by I, 5.8592 by F;
(4) 0.00027 by I, 0.00017 by F.

2.12 $P = \$22,696$.

2.15 $P_2 = \$12,021$.

2.18 $P = \$13,482$.

2.21 $F = \$11,384$.

2.24 $F = \$11,011$

2.27 $F = \$1,116,312$ (by formula).

2.30 (*b*) $F = \$1600$; (*c*) $P_G =$ year 0; (*d*) $n = 7$.

2.33 (*b*) $G = \$30$; (*c*) $\text{CF}_5 = \$650$; (*d*) $P = \$4454$.

2.36 $G = \$470.96$.

2.39 Factors = 0.9091, 1.7851, 2.6296.

2.42 $D = \$3831.63$.

2.45 $1112.50.

2.48 $i = 18.3\%$.

2.51 $i = 15.9\%$.

2.54 $8 < n < 9$: can withdraw for 8 years.

2.57 $7 < n < 8$: $n = 8$.

Chapter 3

3.3 (*a*) $r = 7.5\%$ per month; (*b*) $r = 3\%$ per month.

3.6 (*a*) Effective; (*b*) nominal; (*c*) nominal.

3.9 $i/\text{month} = 1.205\%$.

3.12 (a) $r = i = 1.098\%$ per month; (b) $r = 1.092\%$.

3.15 $i = 20\% > 19.72\%$.

3.18 $t = 2.5$ times per year.

3.21 $i = 22.17\%$ per year.

3.24 $P = \$10,731$ (by formula).

3.27 $x = \$515,100$ per year.

3.30 $x = 188$ pizzas per week.

3.33 $F = \$3006$.

3.36 $200/month.

Chapter 4

4.2 $P = \$16,831$.

4.5 $P = \$15,997$.

4.8 $P = \$19,481$.

4.11 $i = 12.68\%$ per year; amount $= \$7122$.

4.14 $A = \$4689$.

4.17 $P = \$4726$.

4.20 $x = \$1508$.

4.23 $i = 16.986\%$ per year; amount $= \$12,932$.

4.26 $P = \$20,684,000$.

4.29 $P = \$40,077$.

4.32 $i = 12.55\%$ per year; $x = \$12,169$.

4.35 $n = 33$; last receipt $= \$920$.

4.38 $G = \$3644$.

4.41 $P = \$147,325$.

4.44 $P = \$112,191$.

4.47 $5 < n < 6$: account depleted after sixth withdrawal.

4.50 $P = \$3332$; $A = \$542.28$.

Chapter 5

5.1. PW low $= \$-199.31$; PW ultra $= \$-200.45$.

5.4 PW stock $= \$48,950$; PW restaurant $= \$6772$.

5.7 PW gas $= \$-10,396$; PW dry $= \$-7984$.

5.10 PW new $= \$-93,194$; PW used $= \$-98,757$.

5.13 $PW_A = \$-11,829$; $PW_B = \$-28,378$.

5.16 PW spray $= \$-273,743$; PW truck $= \$-526,071$.

5.19 $PW_A = \$-65,533$; $PW_B = \$-55,765$.

5.22 (a) PW $= -\$293,333$; (b) $i = 17.227\%$, PW $= \$-230,121$.

5.25 $F = \$567,516$.

5.28 Cap cost $M = \$-189,856$; cap cost $N = \$-205,914$.

5.31 Capitalized cost $U = \$-17,738,000$; capitalized cost $W = \$-50,050,000$.

5.34 Capitalized cost $C = \$-1,315,300$; capitalized cost $D = \$-499,000$.

5.37 $PW_A = \$-834,750$; $PW_B = \$-819,630$; $PW_C = \$-921,690$; select B.

Chapter 6

6.1 (a) 6 years; (b) 3 years; (c) 6 years.

6.3 $i = 0.75\%$ per month; AW $= \$-2755$.

6.5 (a) AW $= \$-3852$; (b) AW $= \$-2755$.

6.7 $AW_x = \$-5252$; $AW_Y = \$-4766$.

6.10 $AW_G = \$-33,737$; $AW_H = \$-38,496$.

6.13 AW manual $= \$-22,212$; AW computer $= \$-34,672$.

6.16 AW manual $= \$-14,882$; AW computer $= \$-18,994$.

6.19 AW $= \$8757$.

6.22 PW $= \$-46,990$; AW $= \$-5639$.

6.25 (a) $AW_G = \$-10,121$, $AW_H = \$-25,000$; (b) $AW_G = \$-10,853$, $AW_H = \$-32,140$.

Chapter 7

7.3 $i = 1.5\%$ (by interpolation).

7.6 $i = 0.9\%$ per month.

7.9 $i = 1.14\%$ per year.

7.12 (a) $i = 4.84\%$ per year; (b) $i = 8.29\%$ per year.

7.15 (a) $i = 2.5\%$; (b) $i = 34.5\%$ per year.

7.18 (a) $i < 0\%$; (b) $i = 0.5\%$ per year.

7.21 $i = 5.25\%$ per quarter.

7.27 (a) two; (b) $i = 27.6\%$.

7.30 (a) one; (b) $i = -1.2\%$ per year.

Chapter 8

8.2 ROR = 12.75%.

8.5 Column totals: Alt P = $-81,000$; Alt Q = $-54,900$; $Q - P$ = $+26,100$.

8.9 i = 0.5% per month = 6% per year; they should rent.

8.12 i = 7.93%; select gravel.

8.15 i = -2.38%; select A.

8.17 i = 6.2%; select F.

8.21 Select method 5.

8.24 Select size 25.

8.27 (a) Select A, B, and C; (b) select project B.

8.30 (a) E versus G is 13.3%, E versus H is 18.7%, F versus G is -0.33%, F versus H is 16.5%, G versus H is 22.8%; (b) (DN = do nothing) DN versus E: eliminate DN; E versus F: eliminate E; F versus G: eliminate G; F versus H: eliminate F; select H; (c) F is best, H is second best.

Chapter 9

9.2 (a) B; (b) D; (c) D; (d) B; (e) C; (f) B; (g) D; (h) C.

9.5 (a) B/C = 0.85; (b) B/C = 0.79.

9.8 B/C = 0.89; do not install lab.

9.11 B/C north = 0.20; B/C east = 0.15; neither site is acceptable.

9.14 (a) Incremental B/C = 1.14; select short route; (b) modified B/C = 1.14.

9.17 Rank alternatives: 1, 2, 4, 3, 5; DN versus 1: eliminate 1; DN versus 2: eliminate DN; 2 versus 4: eliminate 2; 4 versus 3: eliminate 3; 4 versus 5: eliminate 4; select method 5.

9.20 B/C_1 = 1.18; B/C_2 = 1.28; B/C_3 = 1.84; B/C_4 = 1.02; B/C_5 = 2.05; B/C_6 = 1.94; B/C_7 = 1.01; accept all proposals.

9.23 Rank alternatives: 3, 1, 2; 3 versus 1: B/C = 2.3, eliminate 3; 1 versus 2: B/C negative, eliminate 2; select alternative 1.

Chapter 10

10.3 (a) Present: P = $115,000$, n = 10,

SV = 0, AOC = $4500; Urban: P = $7500, n = 10, SV = $7500, AOC = $16,600; (b) $5728.

10.5 (a) 0; (b) $4300.

10.7 (a) AW current = $-228,055$, AW new = $-379,023$; (b) $-88,252$.

10.8 No change, select current buses.

10.11 Select plan III (machine C) with AW = -6348.

10.13 P proposed = $-33,252$.

10.15 AW current = -6962, AW proposed = -8975; keep current.

10.18 Select plan I with AW_I = -5275.

10.20 $27,121.

10.23 Retain 9 years when AW = -2256 is lowest.

10.26 (a) 12.6 years; (b) 14 years with AW = -1879.

10.28 (a) $C_D(1)$ = -5000, select new sorter.

10.31 (a) $C_D(0)$ = -2040, yes; (b) keep equipment both years, $C_D(1)$ = -3860, $C_D(2)$ = -3090.

Chapter 11

11.4 I = $75 per quarter.

11.7 b = 6¼%.

11.10 I = $250 per month; PW = $26,260.

11.13 I = $675 per 6 months; i = 12.36% per 6 months; PW = $5552.

11.16 PW TIPS = $8146; PW corporate = $8754; buy corporate.

11.19 I = $75 per quarter; i = 0.33% per quarter.

11.22 I = $400 per 6 months; n = 17.8 semiannual periods or 9 years.

11.25 I = $440.62; b = 8.8%.

11.28 (a) I = $350,000 per 6 months, i = 6.25% per quarter; (b) ROR of 6.25% per 6 months > 10% per year; call bonds; (c) i = 7.7% per 6 months.

Chapter 12

12.1 $14,945.

12.4 i_f = 3.02% per quarter.

12.7 (a) $i = 13.21\%$ per year, $P = \$2927$; (b) $P = \$913$.

12.10 $i_f = 9.18\%$ per year; $PW_X = \$-128,340$; $PW_Y = \$-123,640$.

12.13 (a) $F = \$54,110$; (b) buying power = $\$27,767$.

12.16 (a) $i_f = 16.6\%$, $F = \$43,673$; (b) $F = \$34,308$.

12.19 (a) $F = \$21,003,000$; (b) buying power = $\$16,457,000$; (c) $i = 10.48\%$.

12.22 (a) Then dollars = $\$6,312,500$; (b) $A = \$1,360,200$.

12.25 $29,129.

12.28 $490,090.

12.31 $933.11.

12.34 $C_2 = \$2,203,000$.

12.37 Cost in 1970 = $\$22,440$; cost in 1996 = $\$76,890$.

12.39 $16.67, $50, $12.50, $12.50 per machine hour.

12.41 (a) $0.89 per square foot; (b) $16.35 per hour; (c) $4.60 per dollar.

12.43 (a) $12,500 for 1, $23,750 for 2, $188,715 for 5; (b) $9910 overallocation.

12.45 (a) Select to make, cost = $\$65,910$; (b) select to make, cost of new equipment is $73,891.

12.47 (a)

ABC allocation basis	A	B	C	D
Number of guests	$212,135	$242,440	$484,880	$60,610

(b)

ABC allocation basis	A	B	C	D
Number of guests nights	$297,885	$283,700	$283,700	$134,757

(c) Sites C and D change significantly.

12.49

State	(a)	(b)	(c)
CA	$128,700	$151,938	$149,080
AZ	71,500	48,262	51,120

Chapter 13

13.2 (a) $B = \$155,000$, $n = 10$, $SV = \$15,200$; (b) $n = 8$, market value = $\$43,000$, $BV = \$38,750$.

13.6 (a) $D_t = \$1250$, $BV_1 = \$10,750$, $BV_8 = \$2000$; (b) $d_t = 0.125$.

13.9 (a) and (b) DDB depreciation.

t	D_t	BV_t	d_t
1	$3,000.00	$9,000.00	0.2500
2	2,250.00	6,750.00	0.1875
3	1,687.50	5,062.50	0.1406
4	1,265.63	3,796.88	0.1055
5	949.22	2,847.66	0.0791
6	711.91	2,135.74	0.0593
7	135.74	2,000.00	0.0113
8	0	2,000.00	0

13.12

t	DDB	DB
2	$8099	$8048
10	3156	3159
18	1072	1240

13.17

t	MACRS D_t	DDB D_t
1	$4000	$4000
2	6400	3200
3	3840	2560
4	2304	2048
5	2304	1638
6	1152	1311
7	0	1048
8	0	195
9	0	0
10	0	0

13.19 (a) $n = 39$ years for real property; (b) $1,759.50 of depreciation moved to year 40.

13.20 (b) MACRS 7-year model: 56.27%, ADS 10-year model: 25%.

13.26 $PW_D = \$30,198$ with switch to SL in year 5.

13.28 $PW_D = \$64,211$ with switch to SL in year 6.

13.30   (*a*) No;   (*b*) $d < 0.04425$.

13.35 $3640 loss.

13.37   (*b*) MACRS: $PW_D = \$61,253$;   SL: $PW_D = \$56,915$.

13.39   (*a*) For 1998, allowance is $2700.

13.40 $112,500; $250,000; $173,250.

13.A3   (*a*) $4714;   (*b*) 1/7;   (*c*) $1457.

Chapter 14

14.1   (*a*) TI;   (*b*) operating loss;   (*c*) DR; (*d*) TI;   (*e*) DR;   (*f*) TI;   (*g*) GI and TI.

14.3

Part	Co. 1	Co. 2
(*a*)	$247,860	$61,250
(*b*)	16.5%	7.5%
(*c*)	0	−9.93%

14.6   (*a*) 19.3%;   (*b*) 25.76%;   (*c*) taxes = $20,608.

14.9   (*a*) Taxes = $14,530;   (*b*) 20.76%.

14.11   Taxes = $151,065.

14.15   Land: LTG = $5000;   machine 1: DR = $2000;   machine 2: LTG + DR = $1000 + $9000.

14.17

Year	1	2	3	4	5	6
CFAT, $	23,000	33,800	25,080	16,008	12,408	5,504

14.19 $T = 46.7\%$.

14.22   (*a*) *A*: PW tax = $41,358;   *B*: PW tax = $39,896.

14.25   PW tax = $12,266.

14.27   3 years: PW tax = $2591;   5 years: PW tax = $2694;   select 3 years.

14.30   *n* of 5: PW tax = $2481;   *n* of 8: PW tax = $4104;   select 5-year MACRS.

Chapter 15

15.3

t	0	1	2	3	4
NCF, $	−20,000	5,500	8,000	7,000	5,000

15.4   (*b*)

t	0	1	2	3	4
NCF, $	−20,000	6,333	9,945	5,981	4,241

15.8

t	0	1–5	6	
NCF, $	−10,000	3,560	4,199	Sum = $11,999

15.10

t	0	1–4	5
NCF, $	−5000	2187	4703

15.13 $PW_A = \$-4234$; $PW_B = \$-8021$; select *A*.

15.16 $AW_A = \$-2178$; $AW_B = \$-2124$; select *B*.

15.19   (*a*) $i^* = 39.13\%$;   (*b*) $i^* = 27.86\%$ (by Excel's IRR).

15.22   *A*: $i^* = 4.00\%$;   *B*: $i^* = 5.94\%$ (by Excel).

15.24   (*a*) 1.76%;   (*b*) 4.17%.

15.26   (*a*) $AW_D = \$-7794$; $AW_C = \$-14,096$ (opportunity cost approach);   keep defender;   (*b*) SV = $66,758.

15.28   (*a*) $AW_D = \$-95,294$; $AW_C = \$-226,040$ (cash-flow approach); (*b*) $AW_D = \$-49,667$; $AW_C = \$-90,307$; select defender.

Chapter 16

16.1   (*a*) 1 million units;   (*b*) $180,000, $480,000.

16.3   (*a*) *r* = $53/unit;   (*b*) $60,000.

16.6   (*a*) $Q_P = 5500$ units; profit = $60,498; (*b*) up 250%.

16.9   *x* = 94 days.

16.12   (*a*) *n* = 13.67 years;   (*b*) $151,263.

16.14 $P = \$-5100$.

16.17   (*a*) *n* = 25.9 days/year; (*b*) $P = \$-34,744$.

16.20   (*a*) *n* = 3.96 years; (*b*) 6.9 years.

16.23   (*a*) *n* = 3.3 years;   (*b*) yes.

16.25   Payback basis: select 1; PW basis: select 2.

Chapter 17

17.2   (*a*) *A* and *C* with PW = $23,749; (*b*) select *C*, $PW_E = \$11,746$.

17.4 B and C with PW = $9956.

17.6 2, 3, and 4 with PW = $2220.

17.7 (*a*) 1 and 3 with PW = $4774;
(*b*) 1, 3, and 4 with PW = $5019.

17.10 Select 2 with $i^* = 23.4\%$ over proposal 3.

17.15 (*a*) $X_1 = X_3 = 1$, $X_2 = X_4 = 0$,
$Z = \$4774$; (*b*) $X_1 = X_3 = X_4 = 1$,
$X_2 = 0$, $Z = \$5019$.

Chapter 18

18.1 (*a*) 3%;   (*b*) 14%, 17%;   (*c*) 9%.

18.2 (*a*) D;   (*b*) E;   (*c*) D;   (*d*) E;   (*e*) D.

18.5 (*a*) Approach 2 with WACC = 8.75%;
(*b*) 12% and 10.67%.

18.7 16.6%.

18.10 (*a*) i loan = 4.95%; i bonds = 3.6%, select
bonds;   (*b*) yes.

18.12 Equity at $i = 5.5\%$.

18.15 WACC = 8.34%.

18.17 $i_1 = 3.02\%$; $i_2 = 6.15\%$; $i_3 = 4.59\%$;
plan 1.

18.19 (*a*) Raise;   (*b*) lower;   (*c*) raise;
(*d*) lower.

18.24 i on 100% equity = 8.5%; i on 60-40
D-E = 15.9%;   use debt-equity financing.

18.26 (*a*) MARR = 12.25%, $i^* = 16.61\%$;
accept;   (*b*) MARR = 11.19%;
$i^* = 16.61\%$; accept.

Chapter 19

19.1

Tons	10	15	20	25	30
PW, $	−134,922	−137,589	−140,257	−142,924	−166,078

19.5

G, $	500	1000	1500	Yes
i^*, %	24.86	21.57	17.41	

19.7

	n				
	4	6	8	10	12
AW₁, $	−4232	−3211	−2707		
AW₂, $		−4610	−3746	−3239	−2907

19.11 $n = 6$: PW = $1935; $n = 10$: PW =
$433; $n = 12$: PW = $−447;   yes.

19.13

Estimate	%	PW_A	PW_B
(*a*) First cost	−50	$ −5,610	$ −23,100
	0	−255,610	−210,600
	+100	−755,610	−585,600
(*b*) AOC	−50	$−143,463	$ −90,976
	0	−255,610	−210,600
	+100	−479,905	−449,848
(*c*) Revenue	−50	$−479,905	$−404,989
	0	−255,610	−210,600
	+100	+192,980	+178,178

19.16 (*a*)

Purchase Price		
Variation	Amount	PW
−30%	$ 7,000	$ 9,311
−15	8,500	9,995
0	10,000	10,680
+15	11,500	11,365
+30	13,000	12,049

19.17 (*a*)

AOC, $/ton	AW₁	Lease	AW₂
0.40	$−3664	$1800	$−3800
0.50	−4164	2500	−4500
0.75	−5414	3200	−5200

19.22 $E(\text{flow}_1) = 195$; $E(\text{flow}_2) =$
200 barrels/day.

19.25 AOC = $1900.

19.27 $E(\text{AW}) = \$−15{,}250$.

19.30 $E(\text{PW}) = \$43{,}895$;   yes.

19.32 Stocks $E(i) = 10.7\%$;   real estate
$E(i) = 8.54\%$;   select stocks.

19.34 Top: $E(\text{value}) = 5.0$;   bottom:
$E(\text{value}) = -6.8$.

19.38 (*a*) High: $E(\text{value}) = \$14$;
low: $E(\text{value}) = \$8$;   (*b*) yes.

19.40 'Inspect' and 'Go' branches;
$E(\text{revenue}) = \$10{,}000$.

Chapter 20

20.3 (*a*) N is discrete, t is continuous;   (*c*) 0.82,
0.06;   (*d*) 50%.

20.5   (b) 0.875.

20.8   (b) $P(6 \leq X \leq 10) = 0.4$, $P(X = 4, 5$ or $6) = 0.1$.

20.10 (a)

X	0	0.2	0.4	0.6	0.8	1.0
F(X)	0	0.04	0.16	0.36	0.64	1.00

(b) One sample of 30 values had $\bar{i} = 6.3375\%$; yours will vary.

20.13 (a) $\bar{X} = 1344$; $s = 382.53$; (b) 578.94 and 2109.06.

20.16 (a) $E(D_M) = 0.25$; $\sigma = 0.1936$;   (b) 0.952.

20.18 $E(Y) = 7.083$; $\sigma = 3.227$.

20.21 0.39, 0.14, 0.20, 0.27.

20.23 President: $V_1 = 75$; vice president: $V_1 = 95$; same.

20.26 (a) A;   (b) B;   (c) $V_B = 0.92$; (d) $V_A = 0.93$.

Appendix B

B.1   (a) Current assets = $77,000, fixed assets = $1,055,000, liabilities = $603,000, equity = $529,000;   (b) $1,132,000 = 603,000 + 529,000.

B.3   (b) $500,000 − 455,000 = $45,000; (c) 9%.

B.5   (a) Inventory turnover = 16.13; income ratio = 9%;   (b) 9%;   (c) return on assets = 3.98%; better than SIC median.

BIBLIOGRAPHY

BOOKS

Au, T.; and T. P. Au. *Engineering Economics for Capital Investment Analysis,* 2nd ed. Prentice-Hall: Englewood Cliffs, NJ, 1992.

Barish, N. N.; and S. Kaplan. *Economic Analysis for Engineering and Management Decision Making,* 2nd ed. McGraw-Hill: New York, 1978.

Bussey, L. E.; and T. G. Eschenbach. *The Economic Analysis of Industrial Projects,* 2nd ed. Prentice-Hall: Englewood Cliffs, NJ, 1992.

Canada, J. R.; and J. A. White. *Capital Investment Decision Analysis for Management and Engineering,* Prentice-Hall: Englewood Cliffs, NJ, 1980.

Collier, C. A.; and W. B. Ledbetter. *Engineering Cost Analysis,* 2nd ed., Harper & Row: New York, 1988.

Degarmo, E. P.; W. Sullivan; J. Bontadelli; and E. Wicks. *Engineering Economy,* 10th ed. Prentice-Hall: Englewood Cliffs, NJ, 1997.

Fleischer, G. A. *Introduction to Engineering Economy,* PWS Publishing Company: Boston, 1994.

Grant, E. L.; W. G. Ireson; and R. S. Levenworth. *Principles of Engineering Economy,* 8th ed. John Wiley and Sons: New York, 1990.

Kleinfeld, I. H. *Engineering and Managerial Economics,* Holt, Rinehart and Winston: New York, 1986.

Newnan, D. G. *Engineering Economic Analysis,* 6th ed. Engineering Press: San Jose, CA, 1996.

Ostwald, P. F. *Engineering Cost Estimating,* 3rd ed. Prentice-Hall: Englewood Cliffs, NJ, 1992.

Park, C. S. *Contemporary Engineering Economics,* 2nd ed. Addison-Wesley: Menlo Park, CA, 1997.

Park, C. S. *Contemporary Engineering Economics: Case Studies,* Addison-Wesley: Menlo Park, CA, 1993.

Park, C. S. *Advanced Engineering Economics,* John Wiley and Sons: New York, 1990

Riggs, J. L.; D. D. Bedworth; and S. U. Randhawa. *Engineering Economics,* 4th ed. McGraw-Hill: New York, 1996.

Seldon, M. R. *Life Cycle Costing: A Better Method of Government Procurement.* Westview: Boulder, CO, 1979.

Smith, G. W. *Engineering Economy: Analysis of Capital Expenditures,* 4th ed. Iowa State University Press: Ames, IA, 1987.

Spraque, J. C.; and J. D. Whittaker. *Economic Analysis for Engineers and Managers.* Prentice-Hall: Englewood Cliffs, NJ, 1986.

Steiner, H. M. *Engineering Economy Principles,* 2nd ed. McGraw-Hill: New York, 1996.

Stevens, G. T., Jr. *The Economic Analysis of Capital Investments for Managers and Engineers.* Reston Publishing: Reston, VA, 1983.

Taylor, G. A. *Managerial and Engineering Economy: Economic Decision-Making,* 3rd ed. Van Nostrand: New York, 1980.

Thuesen, H. G.; and W. J. Fabrycky. *Engineering Economy,* 8th ed. Prentice-Hall: Englewood Cliffs, NJ, 1993.

White, J. A.; M. H. Agee; and K. E. Case. *Principles of Engineering Economic Analysis,* 3rd ed. John Wiley and Sons: New York, 1989.

SELECTED PERIODICALS AND ANNUAL PUBLICATIONS

American Journal of Agricultural Economics, American Agricultural Economics Association: Lexington, KY, 5 issues per year.

Engineering Economy Abstracts. Industrial Engineering Department, Iowa State University: Ames, IA.

Harvard Business Review. Harvard University Press: Boston, 6 issues per year.

Journal of Finance. American Finance Association: New York, 5 issues per year.

Journal of Financial and Quantitative Analysis. University of Washington: Seattle, WA, 5 issues per year.

Lasser, J. K. *Your Income Tax,* Prentice-Hall: New York, annually.

Public Utilities Fortnightly. Public Utilities Reports: Washington, DC, fortnightly.

Tax Information on Corporations. U.S. Internal Revenue Service Publication 542, Government Printing Office: Washington, DC, annually.

The Engineering Economist. Institute of Industrial Engineers: Norcross, GA, quarterly.

U.S. Master Tax Guide. Commerce Clearing House: Chicago, annually.

0.25%			**Table 1**	Discrete cash flow: compound interest factors				0.25%
	Single Payments		**Uniform-Series Payments**				**Uniform Gradient**	
	Compound Amount	Present Worth	Sinking Fund	Compound Amount	Capital Recovery	Present Worth	Gradient Present Worth	Gradient Annual Series
n	*F/P*	*P/F*	*A/F*	*F/A*	*A/P*	*P/A*	*P/G*	*A/G*
1	1.0025	0.9975	1.00000	1.0000	1.00250	0.9975		
2	1.0050	0.9950	0.49938	2.0025	0.50188	1.9925	0.9950	0.4994
3	1.0075	0.9925	0.33250	3.0075	0.33500	2.9851	2.9801	0.9983
4	1.0100	0.9901	0.24906	4.0150	0.25156	3.9751	5.9503	1.4969
5	1.0126	0.9876	0.19900	5.0251	0.20150	4.9627	9.9007	1.9950
6	1.0151	0.9851	0.16563	6.0376	0.16813	5.9478	14.8263	2.4927
7	1.0176	0.9827	0.14179	7.0527	0.14429	6.9305	20.7223	2.9900
8	1.0202	0.9802	0.12391	8.0704	0.12641	7.9107	27.5839	3.4869
9	1.0227	0.9778	0.11000	9.0905	0.11250	8.8885	35.4061	3.9834
10	1.0253	0.9753	0.09888	10.1133	0.10138	9.8639	44.1842	4.4794
11	1.0278	0.9729	0.08978	11.1385	0.09228	10.8368	53.9133	4.9750
12	1.0304	0.9705	0.08219	12.1664	0.08469	11.8073	64.5886	5.4702
13	1.0330	0.9681	0.07578	13.1968	0.07828	12.7753	76.2053	5.9650
14	1.0356	0.9656	0.07028	14.2298	0.07278	13.7410	88.7587	6.4594
15	1.0382	0.9632	0.06551	15.2654	0.06801	14.7042	102.2441	6.9534
16	1.0408	0.9608	0.06134	16.3035	0.06384	15.6650	116.6567	7.4469
17	1.0434	0.9584	0.05766	17.3443	0.06016	16.6235	131.9917	7.9401
18	1.0460	0.9561	0.05438	18.3876	0.05688	17.5795	148.2446	8.4328
19	1.0486	0.9537	0.05146	19.4336	0.05396	18.5332	165.4106	8.9251
20	1.0512	0.9513	0.04882	20.4822	0.05132	19.4845	183.4851	9.4170
21	1.0538	0.9489	0.04644	21.5334	0.04894	20.4334	202.4634	9.9085
22	1.0565	0.9466	0.04427	22.5872	0.04677	21.3800	222.3410	10.3995
23	1.0591	0.9442	0.04229	23.6437	0.04479	22.3241	243.1131	10.8901
24	1.0618	0.9418	0.04048	24.7028	0.04298	23.2660	264.7753	11.3804
25	1.0644	0.9395	0.03881	25.7646	0.04131	24.2055	287.3230	11.8702
26	1.0671	0.9371	0.03727	26.8290	0.03977	25.1426	310.7516	12.3596
27	1.0697	0.9348	0.03585	27.8961	0.03835	26.0774	335.0566	12.8485
28	1.0724	0.9325	0.03452	28.9658	0.03702	27.0099	360.2334	13.3371
29	1.0751	0.9301	0.03329	30.0382	0.03579	27.9400	386.2776	13.8252
30	1.0778	0.9278	0.03214	31.1133	0.03464	28.8679	413.1847	14.3130
36	1.0941	0.9140	0.02658	37.6206	0.02908	34.3865	592.4988	17.2306
40	1.1050	0.9050	0.02380	42.0132	0.02630	38.0199	728.7399	19.1673
48	1.1273	0.8871	0.01963	50.9312	0.02213	45.1787	1040.06	23.0209
50	1.1330	0.8826	0.01880	53.1887	0.02130	46.9462	1125.78	23.9802
52	1.1386	0.8782	0.01803	55.4575	0.02053	48.7048	1214.59	24.9377
55	1.1472	0.8717	0.01698	58.8819	0.01948	51.3264	1353.53	26.3710
60	1.1616	0.8609	0.01547	64.6467	0.01797	55.6524	1600.08	28.7514
72	1.1969	0.8355	0.01269	78.7794	0.01519	65.8169	2265.56	34.4221
75	1.2059	0.8292	0.01214	82.3792	0.01464	68.3108	2447.61	35.8305
84	1.2334	0.8108	0.01071	93.3419	0.01321	75.6813	3029.76	40.0331
90	1.2520	0.7987	0.00992	100.7885	0.01242	80.5038	3446.87	42.8162
96	1.2709	0.7869	0.00923	108.3474	0.01173	85.2546	3886.28	45.5844
100	1.2836	0.7790	0.00881	113.4500	0.01131	88.3825	4191.24	47.4216
108	1.3095	0.7636	0.00808	123.8093	0.01058	94.5453	4829.01	51.0762
120	1.3494	0.7411	0.00716	139.7414	0.00966	103.5618	5852.11	56.5084
132	1.3904	0.7492	0.00610	156.1582	0.00890	112.3121	6950.01	61.8813
144	1.4327	0.6980	0.00578	173.0743	0.00828	120.8041	8117.41	67.1949
240	1.8208	0.5492	0.00305	328.3020	0.00555	180.3109	19399	107.5863
360	2.4568	0.4070	0.00172	582.7369	0.00422	237.1894	36264	152.8902
480	3.6151	0.3016	0.00108	926.0595	0.00358	279.3418	53821	192.6699

| 0.5% | | | Table 2 | Discrete cash flow: compound interest factors | | | | 0.5% |

	Single Payments		Uniform-Series Payments				Uniform Gradient	
	Compound Amount	Present Worth	Sinking Fund	Compound Amount	Capital Recovery	Present Worth	Gradient Present Worth	Gradient Annual Series
n	F/P	P/F	A/F	F/A	A/P	P/A	P/G	A/G
1	1.0050	0.9950	1.00000	1.0000	1.00500	0.9950		
2	1.0100	0.9901	0.49875	2.0050	0.50375	1.9851	0.9901	0.4988
3	1.0151	0.9851	0.33167	3.0150	0.33667	2.9702	2.9604	0.9967
4	1.0202	0.9802	0.24813	4.0301	0.25313	3.9505	5.9011	1.4938
5	1.0253	0.9754	0.19801	5.0503	0.20301	4.9259	9.8026	1.9900
6	1.0304	0.9705	0.16460	6.0755	0.16960	5.8964	14.6552	2.4855
7	1.0355	0.9657	0.14073	7.1059	0.14573	6.8621	20.4493	2.9801
8	1.0407	0.9609	0.12283	8.1414	0.12783	7.8230	27.1755	3.4738
9	1.0459	0.9561	0.10891	9.1821	0.11391	8.7791	34.8244	3.9668
10	1.0511	0.9513	0.09777	10.2280	0.10277	9.7304	43.3865	4.4589
11	1.0564	0.9466	0.08866	11.2792	0.09366	10.6770	52.8526	4.9501
12	1.0617	0.9419	0.08107	12.3356	0.08607	11.6189	63.2136	5.4406
13	1.0670	0.9372	0.07464	13.3972	0.07964	12.5562	74.4602	5.9302
14	1.0723	0.9326	0.06914	14.4642	0.07414	13.4887	86.5835	6.4190
15	1.0777	0.9279	0.06436	15.5365	0.06936	14.4166	99.5743	6.9069
16	1.0831	0.9233	0.06019	16.6142	0.06519	15.3399	113.4238	7.3940
17	1.0885	0.9187	0.05651	17.6973	0.06151	16.2586	128.1231	7.8803
18	1.0939	0.9141	0.05323	18.7858	0.05823	17.1728	143.6634	8.3658
19	1.0994	0.9096	0.05030	19.8797	0.05530	18.0824	160.0360	8.8504
20	1.1049	0.9051	0.04767	20.9791	0.05267	18.9874	177.2322	9.3342
21	1.1104	0.9006	0.04528	22.0840	0.05028	19.8880	195.2434	9.8172
22	1.1160	0.8961	0.04311	23.1944	0.04811	20.7841	214.0611	10.2993
23	1.1216	0.8916	0.04113	24.3104	0.04613	21.6757	233.6768	10.7806
24	1.1272	0.8872	0.03932	25.4320	0.04432	22.5629	254.0820	11.2611
25	1.1328	0.8828	0.03765	26.5591	0.04265	23.4456	275.2686	11.7407
26	1.1385	0.8784	0.03611	27.6919	0.04111	24.3240	297.2281	12.2195
27	1.1442	0.8740	0.03469	28.8304	0.03969	25.1980	319.9523	12.6975
28	1.1499	0.8697	0.03336	29.9745	0.03836	26.0677	343.4332	13.1747
29	1.1556	0.8653	0.03213	31.1244	0.03713	26.9330	367.6625	13.6510
30	1.1614	0.8610	0.03098	32.2800	0.03598	27.7941	392.6324	14.1265
36	1.1967	0.8356	0.02542	39.3361	0.03042	32.8710	557.5598	16.9621
40	1.2208	0.8191	0.02265	44.1588	0.02765	36.1722	681.3347	18.8359
48	1.2705	0.7871	0.01849	54.0978	0.02349	42.5803	959.9188	22.5437
50	1.2832	0.7793	0.01765	56.6452	0.02265	44.1428	1035.70	23.4624
52	1.2961	0.7716	0.01689	59.2180	0.02189	45.6897	1113.82	24.3778
55	1.3156	0.7601	0.01584	63.1258	0.02084	47.9814	1235.27	25.7447
60	1.3489	0.7414	0.01433	69.7700	0.01933	51.7256	1448.65	28.0064
72	1.4320	0.6983	0.01157	86.4089	0.01657	60.3395	2012.35	33.3504
75	1.4536	0.6879	0.01102	90.7265	0.01602	62.4136	2163.75	34.6679
84	1.5204	0.6577	0.00961	104.0739	0.01461	68.4530	2640.66	38.5763
90	1.5666	0.6383	0.00883	113.3109	0.01383	72.3313	2976.08	41.1451
96	1.6141	0.6195	0.00814	122.8285	0.01314	76.0952	3324.18	43.6845
100	1.6467	0.6073	0.00773	129.3337	0.01273	78.5426	3562.79	45.3613
108	1.7137	0.5835	0.00701	142.7399	0.01201	83.2934	4054.37	48.6758
120	1.8194	0.5496	0.00610	163.8793	0.01110	90.0735	4823.51	53.5508
132	1.9316	0.5177	0.00537	186.3226	0.01037	96.4596	5624.59	58.3103
144	2.0508	0.4876	0.00476	210.1502	0.00976	102.4747	6451.31	62.9551
240	3.3102	0.3021	0.00216	462.0409	0.00716	139.5808	13416	96.1131
360	6.0226	0.1660	0.00100	1004.52	0.00600	166.7916	21403	128.3236
480	10.9575	0.0913	0.00050	1991.49	0.00550	181.7476	27588	151.7949

0.75%			Table 3	Discrete cash flow: compound interest factors				0.75%
	Single Payments		**Uniform-Series Payments**				**Uniform Gradient**	
n	Compound Amount F/P	Present Worth P/F	Sinking Fund A/F	Compound Amount F/A	Capital Recovery A/P	Present Worth P/A	Gradient Present Worth P/G	Gradient Annual Series A/G
1	1.0075	0.9926	1.00000	1.0000	1.00750	0.9926		
2	1.0151	0.9852	0.49813	2.0075	0.50563	1.9777	0.9852	0.4981
3	1.0227	0.9778	0.33085	3.0226	0.33835	2.9556	2.9408	0.9950
4	1.0303	0.9706	0.24721	4.0452	0.25471	3.9261	5.8525	1.4907
5	1.0381	0.9633	0.19702	5.0756	0.20452	4.8894	0.7058	1.9851
6	1.0459	0.9562	0.16357	6.1136	0.17107	5.8456	14.4866	2.4782
7	1.0537	0.9490	0.13967	7.1595	0.14717	6.7946	20.1808	2.9701
8	1.0616	0.9420	0.12176	8.2132	0.12926	7.7366	26.7747	3.4608
9	1.0696	0.9350	0.10782	9.2748	0.11532	8.6716	34.2544	3.9502
10	1.0776	0.9280	0.09667	10.3443	0.10417	9.5996	42.6064	4.4384
11	1.0857	0.9211	0.08755	11.4219	0.09505	10.5207	51.8174	4.9253
12	1.0938	0.9142	0.07995	12.5076	0.08745	11.4349	61.8740	5.4110
13	1.1020	0.9074	0.07352	13.6014	0.08102	12.3423	72.7632	5.8954
14	1.1103	0.9007	0.06801	14.7034	0.07551	13.2430	84.4720	6.3786
15	1.1186	0.8940	0.06324	15.8137	0.07074	14.1370	96.9876	6.8606
16	1.1270	0.8873	0.05906	16.9323	0.06656	15.0243	110.2973	7.3413
17	1.1354	0.8807	0.05537	18.0593	0.06287	15.9050	124.3887	7.8207
18	1.1440	0.8742	0.05210	19.1947	0.05960	16.7792	139.2494	8.2989
19	1.1525	0.8676	0.04917	20.3387	0.05667	17.6468	154.8671	8.7759
20	1.1612	0.8612	0.04653	21.4912	0.05403	18.5080	171.2297	9.2516
21	1.1699	0.8548	0.04415	22.6524	0.05165	19.3628	188.3253	9.7261
22	1.1787	0.8484	0.04198	23.8223	0.04948	20.2112	206.1420	10.1994
23	1.1875	0.8421	0.04000	25.0010	0.04750	21.0533	224.6682	10.6714
24	1.1964	0.8358	0.03818	26.1885	0.04568	21.8891	243.8923	11.1422
25	1.2054	0.8296	0.03652	27.3849	0.04402	22.7188	263.8029	11.6117
26	1.2144	0.8234	0.03498	28.5903	0.04248	23.5422	284.3888	12.0800
27	1.2235	0.8173	0.03355	29.8047	0.04105	24.3595	305.6387	12.5470
28	1.2327	0.8112	0.03223	31.0282	0.03973	25.1707	327.5416	13.0128
29	1.2420	0.8052	0.03100	32.2609	0.03850	25.9759	350.0867	13.4774
30	1.2513	0.7992	0.02985	33.5029	0.03735	26.7751	373.2631	13.9407
36	1.3086	0.7641	0.02430	41.1527	0.03180	31.4468	524.9924	16.6946
40	1.3483	0.7416	0.02153	46.4465	0.02903	34.4469	637.4693	18.5058
48	1.4314	0.6986	0.01739	57.5207	0.02489	40.1848	886.8404	22.0691
50	1.4530	0.6883	0.01656	60.3943	0.02406	41.5664	953.8486	22.9476
52	1.4748	0.6780	0.01580	63.3111	0.02330	42.9276	1022.59	23.8211
55	1.5083	0.6630	0.01476	67.7688	0.02226	44.9316	1128.79	25.1223
60	1.5657	0.6387	0.01326	75.4241	0.02076	48.1734	1313.52	27.2665
72	1.7126	0.5839	0.01053	95.0070	0.01803	55.4768	1791.25	32.2882
75	1.7514	0.5710	0.00998	100.1833	0.01748	57.2027	1917.22	33.5163
84	1.8732	0.5338	0.00859	116.4269	0.01609	62.1540	2308.13	37.1357
90	1.9591	0.5104	0.00782	127.8790	0.01532	65.2746	2578.00	39.4946
96	2.0489	0.4881	0.00715	139.8562	0.01465	68.2584	2853.94	41.8107
100	2.1111	0.4737	0.00675	148.1445	0.01425	70.1746	3040.75	43.3311
108	2.2411	0.4462	0.00604	165.4832	0.01354	73.8394	3419.90	46.3154
120	2.4514	0.4079	0.00517	193.5143	0.01267	78.9417	3998.56	50.6521
132	2.6813	0.3730	0.00446	224.1748	0.01196	83.6064	4583.57	54.8232
144	2.9328	0.3410	0.00388	257.7116	0.01138	87.8711	5169.58	58.8314
240	6.0092	0.1664	0.00150	667.8869	0.00900	111.1450	9494.12	85.4210
360	14.7306	0.0679	0.00055	1830.74	0.00805	124.2819	13312	107.1145
480	36.1099	0.0277	0.00021	4681.32	0.00771	129.6409	15513	119.6620

| 1% | | | | Table 4 | Discrete cash flow: compound interest factors | | | 1% | |

	Single Payments		Uniform-Series Payments				Uniform Gradient	
n	Compound Amount F/P	Present Worth P/F	Sinking Fund A/F	Compound Amount F/A	Capital Recovery A/P	Present Worth P/A	Gradient Present Worth P/G	Gradient Annual Series A/G
1	1.0100	0.9901	1.00000	1.0000	1.01000	0.9901		
2	1.0201	0.9803	0.49751	2.0100	0.50751	1.9704	0.9803	0.4975
3	1.0303	0.9706	0.33002	3.0301	0.34002	2.9410	2.9215	0.9934
4	1.0406	0.9610	0.24628	4.0604	0.25628	3.9020	5.8044	1.4876
5	1.0510	0.9515	0.19604	5.1010	0.20604	4.8534	9.6103	1.9801
6	1.0615	0.9420	0.16255	6.1520	0.17255	5.7955	14.3205	2.4710
7	1.0721	0.9327	0.13863	7.2135	0.14863	6.7282	19.9168	2.9602
8	1.0829	0.9235	0.12069	8.2857	0.13069	7.6517	26.3812	3.4478
9	1.0937	0.9143	0.10674	9.3685	0.11674	8.5660	33.6959	3.9337
10	1.1046	0.9053	0.09558	10.4622	0.10558	9.4713	41.8435	4.4179
11	1.1157	0.8963	0.08645	11.5668	0.09645	10.3676	50.8067	4.9005
12	1.1268	0.8874	0.07885	12.6825	0.08885	11.2551	60.5687	5.3815
13	1.1381	0.8787	0.07241	13.8093	0.08241	12.1337	71.1126	5.8607
14	1.1495	0.8700	0.06690	14.9474	0.07690	13.0037	82.4221	6.3384
15	1.1610	0.8613	0.06212	16.0969	0.07212	13.8651	94.4810	6.8143
16	1.1726	0.8528	0.05794	17.2579	0.06794	14.7179	107.2734	7.2886
17	1.1843	0.8444	0.05426	18.4304	0.06426	15.5623	120.7834	7.7613
18	1.1961	0.8360	0.05098	19.6147	0.06098	16.3983	134.9957	8.2323
19	1.2081	0.8277	0.04805	20.8109	0.05805	17.2260	149.8950	8.7017
20	1.2202	0.8195	0.04542	22.0190	0.05542	18.0456	165.4664	9.1694
21	1.2324	0.8114	0.04303	23.2392	0.05303	18.8570	181.6950	9.6354
22	1.2447	0.8034	0.04086	24.4716	0.05086	19.6604	198.5663	10.0998
23	1.2572	0.7954	0.03889	25.7163	0.04889	20.4558	216.0660	10.5626
24	1.2697	0.7876	0.03707	26.9735	0.04707	21.2434	234.1800	11.0237
25	1.2824	0.7798	0.03541	28.2432	0.04541	22.0232	252.8945	11.4831
26	1.2953	0.7720	0.03387	29.5256	0.04387	22.7952	272.1957	11.9409
27	1.3082	0.7644	0.03245	30.8209	0.04245	23.5596	292.0702	12.3971
28	1.3213	0.7568	0.03112	32.1291	0.04112	24.3164	312.5047	12.8516
29	1.3345	0.7493	0.02990	33.4504	0.03990	25.0658	333.4863	13.3044
30	1.3478	0.7419	0.02875	34.7849	0.03875	25.8077	355.0021	13.7557
36	1.4308	0.6989	0.02321	43.0769	0.03321	30.1075	494.6207	16.4285
40	1.4889	0.6717	0.02046	48.8864	0.03046	32.8347	596.8561	18.1776
48	1.6122	0.6203	0.01633	61.2226	0.02633	37.9740	820.1460	21.5976
50	1.6446	0.6080	0.01551	64.4632	0.02551	39.1961	879.4176	22.4363
52	1.6777	0.5961	0.01476	67.7689	0.02476	40.3942	939.9175	23.2686
55	1.7285	0.5785	0.01373	72.8525	0.02373	42.1472	1032.81	24.5049
60	1.8167	0.5504	0.01224	81.6697	0.02224	44.9550	1192.81	26.5333
72	2.0471	0.4885	0.00955	104.7099	0.01955	51.1504	1597.87	31.2386
75	2.1091	0.4741	0.00902	110.9128	0.01902	52.5871	1702.73	32.3793
84	2.3067	0.4335	0.00765	130.6723	0.01765	56.6485	2023.32	35.7170
90	2.4486	0.4084	0.00690	144.8633	0.01690	59.1609	2240.57	37.8724
96	2.5993	0.3847	0.00625	159.9273	0.01625	61.5277	2459.43	39.9727
100	2.7048	0.3697	0.00587	170.4814	0.01587	63.0289	2605.78	41.3426
108	2.9289	0.3414	0.00518	192.8926	0.01518	65.8578	2898.42	44.0103
120	3.3004	0.3030	0.00435	230.0387	0.01435	69.7005	3334.11	47.8349
132	3.7190	0.2689	0.00368	271.8959	0.01368	73.1108	3761.69	51.4520
144	4.1906	0.2386	0.00313	319.0616	0.01313	76.1372	4177.47	54.8676
240	10.8926	0.0918	0.00101	989.2554	0.01101	90.8194	6878.60	75.7393
360	35.9496	0.0278	0.00029	3494.96	0.01029	97.2183	8720.43	89.6995
480	118.6477	0.0084	0.00008	11765	0.01008	99.1572	9511.16	95.9200

| 1.25% | | | **Table 5** | Discrete cash flow: compound interest factors | | | | 1.25% |

	Single Payments		Uniform-Series Payments				Uniform Gradient	
	Compound Amount F/P	Present Worth P/F	Sinking Fund A/F	Compound Amount F/A	Capital Recovery A/P	Present Worth P/A	Gradient Present Worth P/G	Gradient Annual Series A/G
n								
1	1.0125	0.9877	1.00000	1.0000	1.01250	0.9877		
2	1.0252	0.9755	0.49680	2.0125	0.50939	1.9631	0.9755	0.4969
3	1.0380	0.9634	0.32920	3.0377	0.34170	2.9265	2.9023	0.9917
4	1.0509	0.9515	0.24536	4.0756	0.25786	3.8781	5.7569	1.4845
5	1.0641	0.9398	0.19506	5.1266	0.20756	4.8178	9.5160	1.9752
6	1.0774	0.9282	0.16153	6.1907	0.17403	5.7460	14.1569	2.4638
7	1.0909	0.9167	0.13759	7.2680	0.15009	6.6627	19.6571	2.9503
8	1.1045	0.9054	0.11963	8.3589	0.13213	7.5681	25.9949	3.4348
9	1.1183	0.8942	0.10567	9.4634	0.11817	8.4623	33.1487	3.9172
10	1.1323	0.8832	0.09450	10.5817	0.10700	9.3455	41.0973	4.3975
11	1.1464	0.8723	0.08537	11.7139	0.09787	10.2178	49.8201	4.8758
12	1.1608	0.8615	0.07776	12.8604	0.09026	11.0793	59.2967	5.3520
13	1.1753	0.8509	0.07132	14.0211	0.08382	11.9302	69.5072	5.8262
14	1.1900	0.8404	0.06581	15.1964	0.07831	12.7706	80.4320	6.2982
15	1.2048	0.8300	0.06103	16.3863	0.07353	13.6005	92.0519	6.7682
16	1.2199	0.8197	0.05685	17.5912	0.06935	14.4203	104.3481	7.2362
17	1.2351	0.8096	0.05316	18.8111	0.06566	15.2299	117.3021	7.7021
18	1.2506	0.7996	0.04988	20.0462	0.06238	16.0295	130.8958	8.1659
19	1.2662	0.7898	0.04696	21.2968	0.05946	16.8193	145.1115	8.6277
20	1.2820	0.7800	0.04432	22.5630	0.05682	17.5993	159.9316	9.0874
21	1.2981	0.7704	0.04194	23.8450	0.05444	18.3697	175.3392	9.5450
22	1.3143	0.7609	0.03977	25.1431	0.05227	19.1306	191.3174	10.0006
23	1.3307	0.7515	0.03780	26.4574	0.05030	19.8820	207.8499	10.4542
24	1.3474	0.7422	0.03599	27.7881	0.04849	20.6242	224.9204	10.9056
25	1.3642	0.7330	0.03432	29.1354	0.04682	21.3573	242.5132	11.3551
26	1.3812	0.7240	0.03279	30.4996	0.04529	22.0813	260.6128	11.8024
27	1.3985	0.7150	0.03137	31.8809	0.04387	22.7963	279.2040	12.2478
28	1.4160	0.7062	0.03005	33.2794	0.04255	23.5025	298.2719	12.6911
29	1.4337	0.6975	0.02882	34.6954	0.04132	24.2000	317.8019	13.1323
30	1.4516	0.6889	0.02768	36.1291	0.04018	24.8889	337.7797	13.5715
36	1.5639	0.6394	0.02217	45.1155	0.03467	28.8473	466.2830	16.1639
40	1.6436	0.6084	0.01942	51.4896	0.03192	31.3269	559.2320	17.8515
48	1.8154	0.5509	0.01533	65.2284	0.02783	35.9315	759.2296	21.1299
50	1.8610	0.5373	0.01452	68.8818	0.02702	37.0129	811.6738	21.9295
52	1.9078	0.5242	0.01377	72.6271	0.02627	38.0677	864.9409	22.7211
55	1.9803	0.5050	0.01275	78.4225	0.02525	39.6017	646.2277	23.8936
60	2.1072	0.4746	0.01129	88.5745	0.02379	42.0346	1084.84	25.8083
72	2.4459	0.4088	0.00865	115.6736	0.02115	47.2925	1428.46	30.2047
75	2.5388	0.3939	0.00812	123.1035	0.02062	48.4890	1515.79	31.2605
84	2.8391	0.3522	0.00680	147.1290	0.01930	51.8222	1778.84	34.3258
90	3.0588	0.3269	0.00607	164.7050	0.01857	53.8461	1953.83	36.2855
96	3.2955	0.3034	0.00545	183.6411	0.01795	55.7246	2127.52	38.1793
100	3.4634	0.2887	0.00507	197.0723	0.01757	56.9013	2242.24	39.4058
108	3.8253	0.2614	0.00442	226.0226	0.01692	59.0865	2468.26	41.7737
120	4.4402	0.2252	0.00363	275.2171	0.01613	61.9828	2796.57	45.1184
132	5.1540	0.1940	0.00301	332.3198	0.01551	64.4781	3109.35	48.2234
144	5.9825	0.1672	0.00251	398.6021	0.01501	66.6277	3404.61	51.0990
240	19.7155	0.0507	0.00067	1497.24	0.01317	75.9423	5101.53	67.1764
360	87.5410	0.0114	0.00014	6923.28	0.01264	79.0861	5997.90	75.8401
480	388.7007	0.0026	0.00003	31016	0.01253	79.7942	6284.74	78.7619

Table 6 Discrete cash flow: compound interest factors

	Single Payments		Uniform-Series Payments				Uniform Gradient	
	Compound Amount F/P	Present Worth P/F	Sinking Fund A/F	Compound Amount F/A	Capital Recovery A/P	Present Worth P/A	Gradient Present Worth P/G	Gradient Annual Series A/G
n								
1	1.0150	0.9852	1.00000	1.0000	1.01500	0.9852		
2	1.0302	0.9707	0.49628	2.0150	0.51128	1.9559	0.9707	0.4963
3	1.0457	0.9563	0.32838	3.0452	0.34338	2.9122	2.8833	0.9901
4	1.0614	0.9422	0.24444	4.0909	0.25944	3.8544	5.7098	1.4814
5	1.0773	0.9283	0.19409	5.1523	0.20909	4.7826	9.4229	1.9702
6	1.0934	0.9145	0.16053	6.2296	0.17553	5.6972	13.9956	2.4566
7	1.1098	0.9010	0.13656	7.3230	0.15156	6.5982	19.4018	2.9405
8	1.1265	0.8877	0.11858	8.4328	0.13358	7.4859	25.6157	3.4219
9	1.1434	0.8746	0.10461	9.5593	0.11961	8.3605	32.6125	3.9008
10	1.1605	0.8617	0.09343	10.7027	0.10843	9.2222	40.3675	4.3772
11	1.1779	0.8489	0.08429	11.8633	0.09929	10.0711	48.8568	4.8512
12	1.1956	0.8364	0.07668	13.0412	0.09168	10.9075	58.0571	5.3227
13	1.2136	0.8240	0.07024	14.2368	0.08524	11.7315	67.9454	5.7917
14	1.2318	0.8118	0.06472	15.4504	0.07972	12.5434	78.4994	6.2582
15	1.2502	0.7999	0.05994	16.6821	0.07494	13.3432	89.6974	6.7223
16	1.2690	0.7880	0.05577	17.9324	0.07077	14.1313	101.5178	7.1839
17	1.2880	0.7764	0.05208	19.2014	0.06708	14.9076	113.9400	7.6431
18	1.3073	0.7649	0.04881	20.4894	0.06381	15.6726	126.9435	8.0997
19	1.3270	0.7536	0.04588	21.7967	0.06088	16.4262	140.5084	8.5539
20	1.3469	0.7425	0.04325	23.1237	0.05825	17.1686	154.6154	9.0057
21	1.3671	0.7315	0.04087	24.4705	0.05587	17.9001	169.2453	9.4550
22	1.3876	0.7207	0.03870	25.8376	0.05370	18.6208	184.3798	9.9018
23	1.4084	0.7100	0.03673	27.2251	0.05173	19.3309	200.0006	10.3462
24	1.4295	0.6995	0.03492	28.6335	0.04992	20.0304	216.0901	10.7881
25	1.4509	0.6892	0.03326	30.0630	0.04826	20.7196	232.6310	11.2276
26	1.4727	0.6790	0.03173	31.5140	0.04673	21.3986	249.6065	11.6646
27	1.4948	0.6690	0.03032	32.9867	0.04532	22.0676	267.0002	12.0992
28	1.5172	0.6591	0.02900	34.4815	0.04400	22.7267	284.7958	12.5313
29	1.5400	0.6494	0.02778	35.9987	0.04278	23.3761	302.9779	12.9610
30	1.5631	0.6398	0.02664	37.5387	0.04164	24.0158	321.5310	13.3883
36	1.7091	0.5851	0.02115	47.2760	0.03615	27.6607	439.8303	15.9009
40	1.8140	0.5513	0.01843	54.2679	0.03343	29.9158	524.3568	17.5277
48	2.0435	0.4894	0.01437	69.5652	0.02937	34.0426	703.5462	20.6667
50	2.1052	0.4750	0.01357	73.6828	0.02857	34.9997	749.9636	21.4277
52	2.1689	0.4611	0.01283	77.9249	0.02783	35.9287	796.8774	22.1794
55	2.2679	0.4409	0.01183	84.5296	0.02683	37.2715	868.0285	23.2894
60	2.4432	0.4093	0.01039	96.2147	0.02539	39.3803	988.1674	25.0930
72	2.9212	0.3423	0.00781	128.0772	0.02281	43.8447	1279.79	29.1893
75	3.0546	0.3274	0.00730	136.9728	0.02230	44.8416	1352.56	30.1631
84	3.4926	0.2863	0.00602	166.1726	0.02102	47.5786	1568.51	32.9668
90	3.8189	0.2619	0.00532	187.9299	0.02032	49.2099	1709.54	34.7399
96	4.1758	0.2395	0.00472	211.7202	0.01972	50.7017	1847.47	36.4381
100	4.4320	0.2256	0.00437	228.8030	0.01937	51.6247	1937.45	37.5295
108	4.9927	0.2003	0.00376	266.1778	0.01876	53.3137	2112.13	39.6171
120	5.9693	0.1675	0.00302	331.2882	0.01802	55.4985	2359.71	42.5185
132	7.1370	0.1401	0.00244	409.1354	0.01744	57.3257	2588.71	45.1579
144	8.5332	0.1172	0.00199	502.2109	0.01699	58.8540	2798.58	47.5512
240	35.6328	0.0281	0.00043	2308.85	0.01543	64.7957	3870.69	59.7368
360	212.7038	0.0047	0.00007	14114	0.01507	66.3532	4310.72	64.9662
480	1269.70	0.0008	0.00001	84580	0.01501	66.6142	4415.74	66.2883

2%			Table 7	Discrete cash flow: compound interest factors				2%
	Single Payments		**Uniform-Series Payments**				**Uniform Gradient**	
	Compound Amount F/P	Present Worth P/F	Sinking Fund A/F	Compound Amount F/A	Capital Recovery A/P	Present Worth P/A	Gradient Present Worth P/G	Gradient Annual Series A/G
n								
1	1.0200	0.9804	1.00000	1.0000	1.02000	0.9804		
2	1.0404	0.9612	0.49505	2.0200	0.51505	1.9416	0.9612	0.4950
3	1.0612	0.9423	0.32675	3.0604	0.34675	2.8839	2.8458	0.9868
4	1.0824	0.9238	0.24262	4.1216	0.26262	3.8077	5.6173	1.4752
5	1.1041	0.9057	0.19216	5.2040	0.21216	4.7135	9.2403	1.9604
6	1.1262	0.8880	0.15853	6.3081	0.17853	5.6014	13.6801	2.4423
7	1.1487	0.8706	0.13451	7.4343	0.15451	6.4720	18.9035	2.9208
8	1.1717	0.8535	0.11651	8.5830	0.13651	7.3255	24.8779	3.3961
9	1.1951	0.8368	0.10252	9.7546	0.12252	8.1622	31.5720	3.8681
10	1.2190	0.8203	0.09133	10.9497	0.11133	8.9826	38.9551	4.3367
11	1.2434	0.8043	0.08218	12.1687	0.10218	9.7868	46.9977	4.8021
12	1.2682	0.7885	0.07456	13.4121	0.09456	10.5753	55.6712	5.2642
13	1.2936	0.7730	0.06812	14.6803	0.08812	11.3484	64.9475	5.7231
14	1.3195	0.7579	0.06260	15.9739	0.08260	12.1062	74.7999	6.1786
15	1.3459	0.7430	0.05783	17.2934	0.07783	12.8493	85.2021	6.6309
16	1.3728	0.7284	0.05365	18.6393	0.07365	13.5777	96.1288	7.0799
17	1.4002	0.7142	0.04997	20.0121	0.06997	14.2919	107.5554	7.5256
18	1.4282	0.7002	0.04670	21.4123	0.06670	14.9920	119.4581	7.9681
19	1.4568	0.6864	0.04378	22.8406	0.06378	15.6785	131.8139	8.4073
20	1.4859	0.6730	0.04116	24.2974	0.06116	16.3514	144.6003	8.8433
21	1.5157	0.6598	0.03878	25.7833	0.05878	17.0112	157.7959	9.2760
22	1.5460	0.6468	0.03663	27.2990	0.05663	17.6580	171.3795	9.7055
23	1.5769	0.6342	0.03467	28.8450	0.05467	18.2922	185.3309	10.1317
24	1.6084	0.6217	0.03287	30.4219	0.05287	18.9139	199.6305	10.5547
25	1.6406	0.6095	0.03122	32.0303	0.05122	19.5235	214.2592	10.9745
26	1.6734	0.5976	0.02970	33.6709	0.04970	20.1210	229.1987	11.3910
27	1.7069	0.5859	0.02829	35.3443	0.04829	20.7069	244.4311	11.8043
28	1.7410	0.5744	0.02699	37.0512	0.04699	21.2813	259.9392	12.2145
29	1.7758	0.5631	0.02578	38.7922	0.04578	21.8444	275.7064	12.6214
30	1.8114	0.5521	0.02465	40.5681	0.04465	22.3965	291.7164	13.0251
36	2.0399	0.4902	0.01923	51.9944	0.03923	25.4888	392.0405	15.3809
40	2.2080	0.4529	0.01656	60.4020	0.03656	27.3555	461.9931	16.8885
48	2.5871	0.3865	0.01260	79.3535	0.03260	30.6731	605.9657	19.7556
50	2.6916	0.3715	0.01182	84.5794	0.03182	31.4236	642.3606	20.4420
52	2.8003	0.3571	0.01111	90.0164	0.03111	32.1449	678.7849	21.1164
55	2.9717	0.3365	0.01014	98.5865	0.03014	33.1748	733.3527	22.1057
60	3.2810	0.3048	0.00877	114.0515	0.02877	34.7609	823.6975	23.6961
72	4.1611	0.2403	0.00633	158.0570	0.02633	37.9841	1034.06	27.2234
75	4.4158	0.2265	0.00586	170.7918	0.02586	38.6771	1084.64	28.0434
84	5.2773	0.1895	0.00468	213.8666	0.02468	40.5255	1230.42	30.3616
90	5.9431	0.1683	0.00405	247.1567	0.02405	41.5869	1322.17	31.7929
96	6.6929	0.1494	0.00351	284.6467	0.02351	42.5294	1409.30	33.1370
100	7.2446	0.1380	0.00320	312.2323	0.02320	43.0984	1464.75	33.9863
108	8.4883	0.1178	0.00267	374.4129	0.02267	44.1095	1569.30	35.5774
120	10.7652	0.0929	0.00205	488.2582	0.02205	45.3554	1710.42	37.7114
132	13.6528	0.0732	0.00158	632.6415	0.02158	46.3378	1833.47	39.5676
144	17.3151	0.0578	0.00123	815.7545	0.02123	47.1123	1939.79	41.1738
240	115.8887	0.0086	0.00017	5744.44	0.02017	49.5686	2374.88	47.9110
360	1247.56	0.0008	0.00002	62328	0.02002	49.9599	2482.57	49.7112
480	13430	0.0001			0.02000	49.9963	2498.03	49.9643

	Single Payments		Uniform-Series Payments				Uniform Gradient	
	Compound Amount F/P	Present Worth P/F	Sinking Fund A/F	Compound Amount F/A	Capital Recovery A/P	Present Worth P/A	Gradient Present Worth P/G	Gradient Annual Series A/G
n								
1	1.0300	0.9709	1.00000	1.0000	1.03000	0.9709		
2	1.0609	0.9426	0.49261	2.0300	0.52261	1.9135	0.9426	0.4926
3	1.0927	0.9151	0.32353	3.0909	0.35353	2.8286	2.7729	0.9803
4	1.1255	0.8885	0.23903	4.1836	0.26903	3.7171	5.4383	1.4631
5	1.1593	0.8626	0.18835	5.3091	0.21835	4.5797	8.8888	1.9409
6	1.1941	0.8375	0.15460	6.4684	0.18460	5.4172	13.0762	2.4138
7	1.2299	0.8131	0.13051	7.6625	0.16051	6.2303	17.9547	2.8819
8	1.2668	0.7894	0.11246	8.8923	0.14246	7.0197	23.4806	3.3450
9	1.3048	0.7664	0.09843	10.1591	0.12843	7.7861	29.6119	3.8032
10	1.3439	0.7441	0.08723	11.4639	0.11723	8.5302	36.3088	4.2565
11	1.3842	0.7224	0.07808	12.8078	0.10808	9.2526	43.5330	4.7049
12	1.4258	0.7014	0.07046	14.1920	0.10046	9.9540	51.2482	5.1485
13	1.4685	0.6810	0.06403	15.6178	0.09403	10.6350	59.4196	5.5872
14	1.5126	0.6611	0.05853	17.0863	0.08853	11.2961	68.0141	6.0210
15	1.5580	0.6419	0.05377	18.5989	0.08377	11.9379	77.0002	6.4500
16	1.6047	0.6232	0.04961	20.1569	0.07961	12.5611	86.3477	6.8742
17	1.6528	0.6050	0.04595	21.7616	0.07595	13.1661	96.0280	7.2936
18	1.7024	0.5874	0.04271	23.4144	0.07271	13.7535	106.0137	7.7081
19	1.7535	0.5703	0.03981	25.1169	0.06981	14.3238	116.2788	8.1179
20	1.8061	0.5537	0.03722	26.8704	0.06722	14.8775	126.7987	8.5229
21	1.8603	0.5375	0.03487	28.6765	0.06487	15.4150	137.5496	8.9231
22	1.9161	0.5219	0.03275	30.5368	0.06275	15.9369	148.5094	9.3186
23	1.9736	0.5067	0.03081	32.4529	0.06081	16.4436	159.6566	9.7093
24	2.0328	0.4919	0.02905	34.4265	0.05905	16.9355	170.9711	10.0954
25	2.0938	0.4776	0.02743	36.4593	0.05743	17.4131	182.4336	10.4768
26	2.1566	0.4637	0.02594	38.5530	0.05594	17.8768	194.0260	10.8535
27	2.2213	0.4502	0.02456	40.7096	0.05456	18.3270	205.7309	11.2255
28	2.2879	0.4371	0.02329	42.9309	0.05329	18.7641	217.5320	11.5930
29	2.3566	0.4243	0.02211	45.2189	0.05211	19.1885	229.4137	11.9558
30	2.4273	0.4120	0.02102	47.5754	0.05102	19.6004	241.3613	12.3141
31	2.5001	0.4000	0.02000	50.0027	0.05000	20.0004	253.3609	12.6678
32	2.5751	0.3883	0.01905	52.5028	0.04905	20.3888	265.3993	13.0169
33	2.6523	0.3770	0.01816	55.0778	0.04816	20.7658	277.4642	13.3616
34	2.7319	0.3660	0.01732	57.7302	0.04732	21.1318	289.5437	13.7018
35	2.8139	0.3554	0.01654	60.4621	0.04654	21.4872	301.6267	14.0375
40	3.2620	0.3066	0.01326	75.4013	0.04326	23.1148	361.7499	15.6502
45	3.7816	0.2644	0.01079	92.7199	0.04079	24.5187	420.6325	17.1556
50	4.3839	0.2281	0.00887	112.7969	0.03887	25.7298	477.4803	18.5575
55	5.0821	0.1968	0.00735	136.0716	0.03735	26.7744	531.7411	19.8600
60	5.8916	0.1697	0.00613	163.0534	0.03613	27.6756	583.0526	21.0674
65	6.8300	0.1464	0.00515	194.3328	0.03515	28.4529	631.2010	22.1841
70	7.9178	0.1263	0.00434	230.5941	0.03434	29.1234	676.0869	23.2145
75	9.1789	0.1089	0.00367	272.6309	0.03367	29.7018	717.6978	24.1634
80	10.6409	0.0940	0.00311	321.3630	0.03311	30.2008	756.0865	25.0353
84	11.9764	0.0835	0.00273	365.8805	0.03273	30.5501	784.5434	25.6806
85	12.3357	0.0811	0.00265	377.8570	0.03265	30.6312	791.3529	25.8349
90	14.3005	0.0699	0.00226	443.3489	0.03226	31.0024	823.6302	26.5667
96	17.0755	0.0586	0.00187	535.8502	0.03187	31.3812	858.6377	27.3615
108	24.3456	0.0411	0.00129	778.1863	0.03129	31.9642	917.6013	28.7072
120	34.7110	0.0288	0.00089	1123.70	0.03089	32.3730	963.8635	29.7737

3% Table 8 Discrete cash flow: compound interest factors 3%

4%			Table 9	Discrete cash flow: compound interest factors				4%
	Single Payments		**Uniform-Series Payments**				**Uniform Gradient**	
n	Compound Amount **F/P**	Present Worth **P/F**	Sinking Fund **A/F**	Compound Amount **F/A**	Capital Recovery **A/P**	Present Worth **P/A**	Gradient Present Worth **P/G**	Gradient Annual Series **A/G**
1	1.0400	0.9615	1.00000	1.0000	1.04000	0.9615		
2	1.0816	0.9246	0.49020	2.0400	0.53020	1.8861	0.9246	0.4902
3	1.1249	0.8890	0.32035	3.1216	0.36035	2.7751	2.7025	0.9739
4	1.1699	0.8548	0.23549	4.2465	0.27549	3.6299	5.2670	1.4510
5	1.2167	0.8219	0.18463	5.4163	0.22463	4.4518	8.5547	1.9216
6	1.2653	0.7903	0.15076	6.6330	0.19076	5.2421	12.5062	2.3857
7	1.3159	0.7599	0.12661	7.8983	0.16661	6.0021	17.0657	2.8433
8	1.3686	0.7307	0.10853	9.2142	0.14853	6.7327	22.1806	3.2944
9	1.4233	0.7026	0.09449	10.5828	0.13449	7.4353	27.8013	3.7391
10	1.4802	0.6756	0.08329	12.0061	0.12329	8.1109	33.8814	4.1773
11	1.5395	0.6496	0.07415	13.4864	0.11415	8.7605	40.3772	4.6090
12	1.6010	0.6246	0.06655	15.0258	0.10655	9.3851	47.2477	5.0343
13	1.6651	0.6006	0.06014	16.6268	0.10014	9.9856	54.4546	5.4533
14	1.7317	0.5775	0.05467	18.2919	0.09467	10.5631	61.9618	5.8659
15	1.8009	0.5553	0.04994	20.0236	0.08994	11.1184	69.7355	6.2721
16	1.8730	0.5339	0.04582	21.8245	0.08582	11.6523	77.7441	6.6720
17	1.9479	0.5134	0.04220	23.6975	0.08220	12.1657	85.9581	7.0656
18	2.0258	0.4936	0.03899	25.6454	0.07899	12.6593	94.3498	7.4530
19	2.1068	0.4746	0.03614	27.6712	0.07614	13.1339	102.8933	7.8342
20	2.1911	0.4564	0.03358	29.7781	0.07358	13.5903	111.5647	8.2091
21	2.2788	0.4388	0.03128	31.9692	0.07128	14.0292	120.3414	8.5779
22	2.3699	0.4220	0.02920	34.2480	0.06920	14.4511	129.2024	8.9407
23	2.4647	0.4057	0.02731	36.6179	0.06731	14.8568	138.1284	9.2973
24	2.5633	0.3901	0.02559	39.0826	0.06559	15.2470	147.1012	9.6479
25	2.6658	0.3751	0.02401	41.6459	0.06401	15.6221	156.1040	9.9925
26	2.7725	0.3607	0.02257	44.3117	0.06257	15.9828	165.1212	10.3312
27	2.8834	0.3468	0.02124	47.0842	0.06124	16.3296	174.1385	10.6640
28	2.9987	0.3335	0.02001	49.9676	0.06001	16.6631	183.1424	10.9909
29	3.1187	0.3207	0.01888	52.9663	0.05888	16.9837	192.1206	11.3120
30	3.2434	0.3083	0.01783	56.0849	0.05783	17.2920	201.0618	11.6274
31	3.3731	0.2965	0.01686	59.3283	0.05686	17.5885	209.9596	11.9371
32	3.5081	0.2851	0.01595	62.7015	0.05595	17.8736	218.7924	12.2411
33	3.6484	0.2741	0.01510	66.2095	0.05510	18.1476	227.5634	12.5396
34	3.7943	0.2636	0.01431	69.8579	0.05431	18.4112	236.2607	12.8324
35	3.9461	0.2534	0.01358	73.6522	0.05358	18.6646	244.8768	13.1198
40	4.8010	0.2083	0.01052	95.0255	0.05052	19.7928	286.5303	14.4765
45	5.8412	0.1712	0.00826	121.0294	0.04826	20.7200	325.4028	15.7047
50	7.1067	0.1407	0.00655	152.6671	0.04655	21.4822	361.1638	16.8122
55	8.6464	0.1157	0.00523	191.1592	0.04523	22.1086	393.6890	17.8070
60	10.5196	0.0951	0.00420	237.9907	0.04420	22.6235	422.9966	18.6972
65	12.7987	0.0781	0.00339	294.9684	0.04339	23.0467	449.2014	19.4909
70	15.5716	0.0642	0.00275	364.2905	0.04275	23.3945	472.4789	20.1961
75	18.9453	0.0528	0.00223	448.6314	0.04223	23.6804	493.0408	20.8206
80	23.0498	0.0434	0.00181	551.2450	0.04181	23.9154	511.1161	21.3718
85	28.0436	0.0357	0.00148	676.0901	0.04148	24.1085	526.9384	21.8526
90	34.1193	0.0293	0.00121	827.9833	0.04121	24.2673	540.7369	22.2826
96	43.1718	0.0232	0.00095	1054.30	0.04095	24.4209	554.9312	22.7236
108	69.1195	0.0145	0.00059	1702.99	0.04059	24.6383	576.8949	23.4146
120	110.6626	0.0090	0.00036	2741.56	0.04036	24.7741	592.2428	23.9057
144	283.6618	0.0035	0.00014	7066.55	0.04014	24.9119	610.1055	24.4906

	Single Payments		Uniform-Series Payments				Uniform Gradient	
n	Compound Amount F/P	Present Worth P/F	Sinking Fund A/F	Compound Amount F/A	Capital Recovery A/P	Present Worth P/A	Gradient Present Worth P/G	Gradient Annual Series A/G
1	1.0500	0.9524	1.00000	1.0000	1.05000	0.9524		
2	1.1025	0.9070	0.48780	2.0500	0.53780	1.8594	0.9070	0.4878
3	1.1576	0.8638	0.31721	3.1525	0.36721	2.7232	2.6347	0.9675
4	1.2155	0.8227	0.23201	4.3101	0.28201	3.5460	5.1028	1.4391
5	1.2763	0.7835	0.18097	5.5256	0.23097	4.3295	8.2369	1.9025
6	1.3401	0.7462	0.14702	6.8019	0.19702	5.0757	11.9680	2.3579
7	1.4071	0.7107	0.12282	8.1420	0.17282	5.7864	16.2321	2.8052
8	1.4775	0.6768	0.10472	9.5491	0.15472	6.4632	20.9700	3.2445
9	1.5513	0.6446	0.09069	11.0266	0.14069	7.1078	26.1268	3.6758
10	1.6289	0.6139	0.07950	12.5779	0.12950	7.7217	31.6520	4.0991
11	1.7103	0.5847	0.07039	14.2068	0.12039	8.3064	37.4988	4.5144
12	1.7959	0.5568	0.06283	15.9171	0.11283	8.8633	43.6241	4.9219
13	1.8856	0.5303	0.05646	17.7130	0.10646	9.3936	49.9879	5.3215
14	1.9799	0.5051	0.05102	19.5986	0.10102	9.8986	56.5538	5.7133
15	2.0789	0.4810	0.04634	21.5786	0.09634	10.3797	63.2880	6.0973
16	2.1829	0.4581	0.04227	23.6575	0.09227	10.8378	70.1597	6.4736
17	2.2920	0.4363	0.03870	25.8404	0.08870	11.2741	77.1405	6.8423
18	2.4066	0.4155	0.03555	28.1324	0.08555	11.6896	84.2043	7.2034
19	2.5270	0.3957	0.03275	30.5390	0.08275	12.0853	91.3275	7.5569
20	2.6533	0.3769	0.03024	33.0660	0.08024	12.4622	98.4884	7.9030
21	2.7860	0.3589	0.02800	35.7193	0.07800	12.8212	105.6673	8.2416
22	2.9253	0.3418	0.02597	38.5052	0.07597	13.1630	112.8461	8.5730
23	3.0715	0.3256	0.02414	41.4305	0.07414	13.4886	120.0087	8.8971
24	3.2251	0.3101	0.02247	44.5020	0.07247	13.7986	127.1402	9.2140
25	3.3864	0.2953	0.02095	47.7271	0.07095	14.0939	134.2275	9.5238
26	3.5557	0.2812	0.01956	51.1135	0.06956	14.3752	141.2585	9.8266
27	3.7335	0.2678	0.01829	54.6691	0.06829	14.6430	148.2226	10.1224
28	3.9201	0.2551	0.01712	58.4026	0.06712	14.8981	155.1101	10.4114
29	4.1161	0.2429	0.01605	62.3227	0.06605	15.1411	161.9126	10.6936
30	4.3219	0.2314	0.01505	66.4388	0.06505	15.3725	168.6226	10.9691
31	4.5380	0.2204	0.01413	70.7608	0.06413	15.5928	175.2333	11.2381
32	4.7649	0.2099	0.01328	75.2988	0.06328	15.8027	181.7392	11.5005
33	5.0032	0.1999	0.01249	80.0638	0.06249	16.0025	188.1351	11.7566
34	5.2533	0.1904	0.01176	85.0670	0.06176	16.1929	194.4168	12.0063
35	5.5160	0.1813	0.01107	90.3203	0.06107	16.3742	200.5807	12.2498
40	7.0400	0.1420	0.00828	120.7998	0.05828	17.1591	229.5452	13.3775
45	8.9850	0.1113	0.00626	159.7002	0.05626	17.7741	255.3145	14.3644
50	11.4674	0.0872	0.00478	209.3480	0.05478	18.2559	277.9148	15.2233
55	14.6356	0.0683	0.00367	272.7126	0.05367	18.6335	297.5104	15.9664
60	18.6792	0.0535	0.00283	353.5837	0.05283	18.9293	314.3432	16.6062
65	23.8399	0.0419	0.00219	456.7980	0.05219	19.1611	328.6910	17.1541
70	30.4264	0.0329	0.00170	588.5285	0.05170	19.3427	340.8409	17.6212
75	38.8327	0.0258	0.00132	756.6537	0.05132	19.4850	351.0721	18.0176
80	49.5614	0.0202	0.00103	971.2288	0.05103	19.5965	359.6460	18.3526
85	63.2544	0.0158	0.00080	1245.09	0.05080	19.6838	366.8007	18.6346
90	80.7304	0.0124	0.00063	1594.61	0.05063	19.7523	372.7488	18.8712
95	103.0347	0.0097	0.00049	2040.69	0.05049	19.8059	377.6774	19.0689
96	108.1864	0.0092	0.00047	2143.73	0.05047	19.8151	378.5555	19.1044
98	119.2755	0.0084	0.00042	2365.51	0.05042	19.8323	380.2139	19.1714
100	131.5013	0.0076	0.00038	2610.03	0.05038	19.8479	381.7492	19.2337

6%			Table 11	Discrete cash flow: compound interest factors				6%
	Single Payments		Uniform-Series Payments				Uniform Gradient	
n	Compound Amount F/P	Present Worth P/F	Sinking Fund A/F	Compound Amount F/A	Capital Recovery A/P	Present Worth P/A	Gradient Present Worth P/G	Gradient Annual Series A/G
1	1.0600	0.9434	1.00000	1.0000	1.06000	0.9434		
2	1.1236	0.8900	0.48544	2.0600	0.54544	1.8334	0.8900	0.4854
3	1.1910	0.8396	0.31411	3.1836	0.37411	2.6730	2.5692	0.9612
4	1.2625	0.7921	0.22859	4.3746	0.28859	3.4651	4.9455	1.4272
5	1.3382	0.7473	0.17740	5.6371	0.23740	4.2124	7.9345	1.8836
6	1.4185	0.7050	0.14336	6.9753	0.20336	4.9173	11.4594	2.3304
7	1.5036	0.6651	0.11914	8.3938	0.17914	5.5824	15.4497	2.7676
8	1.5938	0.6274	0.10104	9.8975	0.16104	6.2098	19.8416	3.1952
9	1.6895	0.5919	0.08702	11.4913	0.14702	6.8017	24.5768	3.6133
10	1.7908	0.5584	0.07587	13.1808	0.13587	7.3601	29.6023	4.0220
11	1.8983	0.5268	0.06679	14.9716	0.12679	7.8869	34.8702	4.4213
12	2.0122	0.4970	0.05928	16.8699	0.11928	8.3838	40.3369	4.8113
13	2.1329	0.4688	0.05296	18.8821	0.11296	8.8527	45.9629	5.1920
14	2.2609	0.4423	0.04758	21.0151	0.10758	9.2950	51.7128	5.5635
15	2.3966	0.4173	0.04296	23.2760	0.10296	9.7122	57.5546	5.9260
16	2.5404	0.3936	0.03895	25.6725	0.09895	10.1059	63.4592	6.2794
17	2.6928	0.3714	0.03544	28.2129	0.09544	10.4773	69.4011	6.6240
18	2.8543	0.3503	0.03236	30.9057	0.09236	10.8276	75.3569	6.9597
19	3.0256	0.3305	0.02962	33.7600	0.08962	11.1581	81.3062	7.2867
20	3.2071	0.3118	0.02718	36.7856	0.08718	11.4699	87.2304	7.6051
21	3.3996	0.2942	0.02500	39.9927	0.08500	11.7641	93.1136	7.9151
22	3.6035	0.2775	0.02305	43.3923	0.08305	12.0416	98.9412	8.2166
23	3.8197	0.2618	0.02128	46.9958	0.08128	12.3034	104.7007	8.5099
24	4.0489	0.2470	0.01968	50.8156	0.07968	12.5504	110.3812	8.7951
25	4.2919	0.2330	0.01823	54.8645	0.07823	12.7834	115.9732	9.0722
26	4.5494	0.2198	0.01690	59.1564	0.07690	13.0032	121.4684	9.3414
27	4.8223	0.2074	0.01570	63.7058	0.07570	13.2105	126.8600	9.6029
28	5.1117	0.1956	0.01459	68.5281	0.07459	13.4062	132.1420	9.8568
29	5.4184	0.1846	0.01358	73.6398	0.07358	13.5907	137.3096	10.1032
30	5.7435	0.1741	0.01265	79.0582	0.07265	13.7648	142.3588	10.3422
31	6.0881	0.1643	0.01179	84.8017	0.07179	13.9291	147.2864	10.5740
32	6.4534	0.1550	0.01100	90.8898	0.07100	14.0840	152.0901	10.7988
33	6.8406	0.1462	0.01027	97.3432	0.07027	14.2302	156.7681	11.0166
34	7.2510	0.1379	0.00960	104.1838	0.06960	14.3681	161.3192	11.2276
35	7.6861	0.1301	0.00897	111.4348	0.06897	14.4982	165.7427	11.4319
40	10.2857	0.0972	0.00646	154.7620	0.06646	15.0463	185.9568	12.3590
45	13.7646	0.0727	0.00470	212.7435	0.06470	15.4558	203.1096	13.1413
50	18.4202	0.0543	0.00344	290.3359	0.06344	15.7619	217.4574	13.7964
55	24.6503	0.0406	0.00254	394.1720	0.06254	15.9905	229.3222	14.3411
60	32.9877	0.0303	0.00188	533.1282	0.06188	16.1614	239.0428	14.7909
65	44.1450	0.0227	0.00139	719.0829	0.06139	16.2891	246.9450	15.1601
70	59.0759	0.0159	0.00103	967.9322	0.06103	16.3845	253.3271	15.4613
75	79.0569	0.0126	0.00077	1300.95	0.06077	16.4558	258.4527	15.7058
80	105.7960	0.0095	0.00057	1746.60	0.06057	16.5091	262.5493	15.9033
85	141.5789	0.0071	0.00043	2342.98	0.06043	16.5489	265.8096	16.0620
90	189.4645	0.0053	0.00032	3141.08	0.06032	16.5787	268.3946	16.1891
95	253.5463	0.0039	0.00024	4209.10	0.06024	16.6009	270.4375	16.2905
96	268.7590	0.0037	0.00022	4462.65	0.06022	16.6047	270.7909	16.3081
98	301.9776	0.0033	0.00020	5016.29	0.06020	16.6115	271.4491	16.3411
100	339.3021	0.0029	0.00018	5638.37	0.06018	16.6175	272.0471	16.3711

7% 7%

Table 12 Discrete cash flow: compound interest factors

	Single Payments		Uniform-Series Payments				Uniform Gradient	
	Compound Amount F/P	Present Worth P/F	Sinking Fund A/F	Compound Amount F/A	Capital Recovery A/P	Present Worth P/A	Gradient Present Worth P/G	Gradient Annual Series A/G
n								
1	1.0700	0.9346	1.00000	1.0000	1.07000	0.9346		
2	1.1449	0.8734	0.48309	2.0700	0.55309	1.8080	0.8734	0.4831
3	1.2250	0.8163	0.31105	3.2149	0.38105	2.6243	2.5060	0.9549
4	1.3108	0.7629	0.22523	4.4399	0.29523	3.3872	4.7947	1.4155
5	1.4026	0.7130	0.17389	5.7507	0.24389	4.1002	7.6467	1.8650
6	1.5007	0.6663	0.13980	7.1533	0.20980	4.7665	10.9784	2.3032
7	1.6058	0.6227	0.11555	8.6540	0.18555	5.3893	14.7149	2.7304
8	1.7182	0.5820	0.09747	10.2598	0.16747	5.9713	18.7889	3.1465
9	1.8385	0.5439	0.08349	11.9780	0.15349	6.5152	23.1404	3.5517
10	1.9672	0.5083	0.07238	13.8164	0.14238	7.0236	27.7156	3.9461
11	2.1049	0.4751	0.06336	15.7836	0.13336	7.4987	32.4665	4.3296
12	2.2522	0.4440	0.05590	17.8885	0.12590	7.9427	37.3506	4.7025
13	2.4098	0.4150	0.04965	20.1406	0.11965	8.3577	42.3302	5.0648
14	2.5785	0.3878	0.04434	22.5505	0.11434	8.7455	47.3718	5.4167
15	2.7590	0.3624	0.03979	25.1290	0.10979	9.1079	52.4461	5.7583
16	2.9522	0.3387	0.03586	27.8881	0.10586	9.4466	57.5271	6.0897
17	3.1588	0.3166	0.03243	30.8402	0.10243	9.7632	62.5923	6.4110
18	3.3799	0.2959	0.02941	33.9990	0.09941	10.591	67.6219	6.7225
19	3.6165	0.2765	0.02675	37.3790	0.09675	10.3356	72.5991	7.0242
20	3.8697	0.2584	0.02439	40.9955	0.09439	10.5940	77.5091	7.3163
21	4.1406	0.2415	0.02229	44.8652	0.09229	10.8355	82.3393	7.5990
22	4.4304	0.2257	0.02041	49.0057	0.09041	11.0612	87.0793	7.8725
23	4.7405	0.2109	0.01871	53.4361	0.08871	11.2722	91.7201	8.1369
24	5.0724	0.1971	0.01719	58.1767	0.08719	11.4693	96.2545	8.3923
25	5.4274	0.1842	0.01581	63.2490	0.08581	11.6536	100.6765	8.6391
26	5.8074	0.1722	0.01456	68.6765	0.08456	11.8258	104.9814	8.8773
27	6.2139	0.1609	0.01343	74.4838	0.08343	11.9867	109.1656	9.1072
28	6.6488	0.1504	0.01239	80.6977	0.08239	12.1371	113.2264	9.3289
29	7.1143	0.1406	0.01145	87.3465	0.08145	12.2777	117.1622	9.5427
30	7.6123	0.1314	0.01059	94.4608	0.08059	12.4090	120.9718	9.7487
31	8.1451	0.1228	0.00980	102.0730	0.07980	12.5318	124.6550	9.9471
32	8.7153	0.1147	0.00907	110.2182	0.07907	12.6466	128.2120	10.1381
33	9.3253	0.1072	0.00841	118.9334	0.07841	12.7538	131.6435	10.3219
34	9.9781	0.1002	0.00780	128.2588	0.07780	12.8540	134.9507	10.4987
35	10.6766	0.0937	0.00723	138.2369	0.07723	12.9477	138.1353	10.6687
40	14.9745	0.0668	0.00501	199.6351	0.07501	13.3317	152.2928	11.4233
45	21.0025	0.0476	0.00350	285.7493	0.07350	13.6055	163.7559	12.0360
50	29.4570	0.0339	0.00246	406.5289	0.07246	13.8007	172.9051	12.5287
55	41.3150	0.0242	0.00174	575.9286	0.07174	13.9399	180.1243	12.9215
60	57.9464	0.0173	0.00123	813.5204	0.07123	14.0392	185.7677	13.2321
65	81.2729	0.0123	0.00087	1146.76	0.07087	14.1099	190.1452	13.4760
70	113.9894	0.0088	0.00062	1614.13	0.07062	14.1604	193.5185	13.6662
75	159.8760	0.0063	0.00044	2269.66	0.07044	14.1964	196.1035	13.8136
80	224.2344	0.0045	0.00031	3189.06	0.07031	14.2220	198.0748	13.9273
85	314.5003	0.0032	0.00022	4478.58	0.07022	14.2403	199.5717	14.0146
90	441.1030	0.0023	0.00016	6287.19	0.07016	14.2533	200.7042	14.0812
95	618.6697	0.0016	0.00011	8823.85	0.07011	14.2626	201.5581	14.1319
96	661.9766	0.0015	0.00011	9442.52	0.07011	14.2641	201.7016	14.1405
98	757.8970	0.0013	0.00009	10813	0.07009	14.2669	201.9651	14.1562
100	867.7163	0.0012	0.00008	12382	0.07008	14.2693	202.2001	14.1703

8%			Table 13	Discrete cash flow: compound interest factors				8%

	Single Payments		Uniform-Series Payments				Uniform Gradient	
n	Compound Amount F/P	Present Worth P/F	Sinking Fund A/F	Compound Amount F/A	Capital Recovery A/P	Present Worth P/A	Gradient Present Worth P/G	Gradient Annual Series A/G
1	1.0800	0.9259	1.00000	1.0000	1.08000	0.9259		
2	1.1664	0.8573	0.48077	2.0800	0.56077	1.7833	0.8573	0.4808
3	1.2597	0.7938	0.30803	3.2464	0.38803	2.5771	2.4450	0.9487
4	1.3605	0.7350	0.22192	4.5061	0.30192	3.3121	4.6501	1.4040
5	1.4693	0.6806	0.17046	5.8666	0.25046	3.9927	7.3724	1.8465
6	1.5869	0.6302	0.13632	7.3359	0.21632	4.6229	10.5233	2.2763
7	1.7138	0.5835	0.11207	8.9228	0.19207	5.2064	14.0242	2.6937
8	1.8509	0.5403	0.09401	10.6366	0.17401	5.7466	17.8061	3.0985
9	1.9990	0.5002	0.08008	12.4876	0.16008	6.2469	21.8081	3.4910
10	2.1589	0.4632	0.06903	14.4866	0.14903	6.7101	25.9768	3.8713
11	2.3316	0.4289	0.06008	16.6455	0.14008	7.1390	30.2657	4.2395
12	2.5182	0.3971	0.05270	18.9771	0.13270	7.5361	34.6339	4.5957
13	2.7196	0.3677	0.04652	21.4953	0.12652	7.9038	39.0463	4.9402
14	2.9372	0.3405	0.04130	24.2149	0.12130	8.2442	43.4723	5.2731
15	3.1722	0.3152	0.03683	27.1521	0.11683	8.5595	47.8857	5.5945
16	3.4259	0.2919	0.03298	30.3243	0.11298	8.8514	52.2640	5.9046
17	3.7000	0.2703	0.02963	33.7502	0.10963	9.1216	56.5883	6.2037
18	3.9960	0.2502	0.02670	37.4502	0.10670	9.3719	60.8426	6.4920
19	4.3157	0.2317	0.02413	41.4463	0.10413	9.6036	65.0134	6.7697
20	4.6610	0.2145	0.02185	45.7620	0.10185	9.8181	69.0898	7.0369
21	5.0338	0.1987	0.01983	50.4229	0.09983	10.0168	73.0629	7.2940
22	5.4365	0.1839	0.01803	55.4568	0.09803	10.2007	76.9257	7.5412
23	5.8715	0.1703	0.01642	60.8933	0.09642	10.3711	80.6726	7.7786
24	6.3412	0.1577	0.01498	66.7648	0.09498	10.5288	84.2997	8.0066
25	6.8485	0.1460	0.01368	73.1059	0.09368	10.6748	87.8041	8.2254
26	7.3964	0.1352	0.01251	79.9544	0.09251	10.8100	91.1842	8.4352
27	7.9881	0.1252	0.01145	87.3508	0.09145	10.9352	94.4390	8.6363
28	8.6271	0.1159	0.01049	95.3388	0.09049	11.0511	97.5687	8.8289
29	9.3173	0.1073	0.00962	103.9659	0.08962	11.1584	100.5738	9.0133
30	10.0627	0.0994	0.00883	113.2832	0.08883	11.2578	103.4558	9.1897
31	10.8677	0.0920	0.00811	123.3459	0.08811	11.3498	106.2163	9.3584
32	11.7371	0.0852	0.00745	134.2135	0.08745	11.4350	108.8575	9.5197
33	12.6760	0.0789	0.00685	145.9506	0.08685	11.5139	111.3819	9.6737
34	13.6901	0.0730	0.00630	158.6267	0.08630	11.5869	113.7924	9.8208
35	14.7853	0.0676	0.00580	172.3168	0.08580	11.6546	116.0920	9.9611
40	21.7245	0.0460	0.00386	259.0565	0.08386	11.9246	126.0422	10.5699
45	31.9204	0.0313	0.00259	386.5056	0.08259	12.1084	133.7331	11.0447
50	46.9016	0.0213	0.00174	573.7702	0.08174	12.2335	139.5928	11.4107
55	68.9139	0.0145	0.00118	848.9232	0.08118	12.3186	144.0065	11.6902
60	101.2571	0.0099	0.00080	1253.21	0.08080	12.3766	147.3000	11.9015
65	148.7798	0.0067	0.00054	1847.25	0.08054	12.4160	149.7387	12.0602
70	218.6064	0.0046	0.00037	2720.08	0.08037	12.4428	151.5326	12.1783
75	321.2045	0.0031	0.00025	4002.56	0.08025	12.4611	152.8448	12.2658
80	471.9548	0.0021	0.00017	5886.94	0.08017	12.4735	153.8001	12.3301
85	693.4565	0.0014	0.00012	8655.71	0.08012	12.4820	154.4925	12.3772
90	1018.92	0.0010	0.00008	12724	0.08008	12.4877	154.9925	12.4116
95	1497.12	0.0007	0.00005	18702	0.08005	12.4917	155.3524	12.4365
96	1616.89	0.0006	0.00005	20199	0.08005	12.4923	155.4112	12.4406
98	1885.94	0.0005	0.00004	23562	0.08004	12.4934	155.5176	12.4480
100	2199.76	0.0005	0.00004	27485	0.08004	12.4943	155.6107	12.4545

9%			Table 14	Discrete cash flow: compound interest factors				9%
	Single Payments		**Uniform-Series Payments**				**Uniform Gradient**	
n	Compound Amount F/P	Present Worth P/F	Sinking Fund A/F	Compound Amount F/A	Capital Recovery A/P	Present Worth P/A	Gradient Present Worth P/G	Gradient Annual Series A/G
1	1.0900	0.9174	1.00000	1.0000	1.09000	0.9174		
2	1.1881	0.8417	0.47847	2.0900	0.56847	1.7591	0.8417	0.4785
3	1.2950	0.7722	0.30505	3.2781	0.39505	2.5313	2.3860	0.9426
4	1.4116	0.7084	0.21867	4.5731	0.30867	3.2397	4.5113	1.3925
5	1.5386	0.6499	0.16709	5.9847	0.25709	3.8897	7.1110	1.8282
6	1.6771	0.5963	0.13292	7.5233	0.22292	4.4859	10.0924	2.2498
7	1.8280	0.5470	0.10869	9.2004	0.19869	5.0330	13.3746	2.6574
8	1.9926	0.5019	0.09064	11.0285	0.18067	5.5348	16.8877	3.0512
9	2.1719	0.4604	0.07680	13.0210	0.16680	5.9952	20.5711	3.4312
10	2.3674	0.4224	0.06582	15.1929	0.15582	6.4177	24.3728	3.7978
11	2.5804	0.3875	0.05695	17.5603	0.14695	6.8052	28.2481	4.1510
12	2.8127	0.3555	0.04965	20.1407	0.13965	7.1607	32.1590	4.4910
13	3.0658	0.3262	0.04357	22.9534	0.13357	7.4869	36.0731	4.8182
14	3.3417	0.2992	0.03843	26.0192	0.12843	7.7862	39.9633	5.1326
15	3.6425	0.2745	0.03406	29.3609	0.12406	8.0607	43.8069	5.4346
16	3.9703	0.2519	0.03030	33.0034	0.12030	8.3126	47.5849	5.7245
17	4.3276	0.2311	0.02705	36.9737	0.11705	8.5436	51.2821	6.0024
18	4.7171	0.2120	0.02421	41.3013	0.11421	8.7556	54.8860	6.2687
19	5.1417	0.1945	0.02173	46.0185	0.11173	8.9501	58.3868	6.5236
20	5.6044	0.1784	0.01955	51.1601	0.10955	9.1285	61.7770	6.7674
21	6.1088	0.1637	0.01762	56.7645	0.10762	9.2922	65.0509	7.0006
22	6.6586	0.1502	0.01590	62.8733	0.10590	9.4424	68.2048	7.2232
23	7.2579	0.1378	0.01438	69.5319	0.10438	9.5802	71.2359	7.4357
24	7.9111	0.1264	0.01302	76.7898	0.10302	9.7066	74.1433	7.6384
25	8.6231	0.1160	0.01181	84.7009	0.10181	9.8226	76.9265	7.8316
26	9.3992	0.1064	0.01072	93.3240	0.10072	9.9290	79.5863	8.0156
27	10.2451	0.0976	0.00973	102.7231	0.09973	10.0266	82.1241	8.1906
28	11.1671	0.0895	0.00885	112.9682	0.09885	10.1161	84.5419	8.3571
29	12.1722	0.0822	0.00806	124.1354	0.09806	10.1983	86.8422	8.5154
30	13.2677	0.0754	0.00734	136.3075	0.09734	10.2737	89.0280	8.6657
31	14.4618	0.0691	0.00669	149.5752	0.09669	10.3428	91.1024	8.8083
32	15.7633	0.0634	0.00610	164.0370	0.09610	10.4062	93.0690	8.9436
33	17.1820	0.0582	0.00556	179.8003	0.09556	10.4644	94.9314	9.0718
34	18.7284	0.0534	0.00508	196.9823	0.09508	10.5178	96.6935	9.1933
35	20.4140	0.0490	0.00464	215.7108	0.09464	10.5668	98.3590	9.3083
40	31.4094	0.0318	0.00296	337.8824	0.09296	10.7574	105.3762	9.7957
45	48.3273	0.0207	0.00190	525.8587	0.09190	10.8812	110.5561	10.1603
50	74.3575	0.0134	0.00123	815.0836	0.09123	10.9617	114.3251	10.4295
55	114.4083	0.0087	0.00079	1260.09	0.09079	11.0140	117.0362	10.6261
60	176.0313	0.0057	0.00051	1944.79	0.09051	11.0480	118.9683	10.7683
65	270.8460	0.0037	0.00033	2998.29	0.09033	11.0701	120.3344	10.8702
70	416.7301	0.0024	0.00022	4619.22	0.09022	11.0844	121.2942	10.9427
75	641.1909	0.0016	0.00014	7113.23	0.09014	11.0938	121.9646	10.9940
80	986.5517	0.0010	0.00009	10951	0.09009	11.0998	122.4306	11.0299
85	1517.93	0.0007	0.00006	16855	0.09006	11.1038	122.7533	11.0551
90	2335.53	0.0004	0.00004	25939	0.09004	11.1064	122.9758	11.0726
95	3593.50	0.0003	0.00003	39917	0.09003	11.1080	123.1287	11.0847
96	3916.91	0.0003	0.00002	43510	0.09002	11.1083	123.1529	11.0866
98	4653.68	0.0002	0.00002	51696	0.09002	11.1087	123.1963	11.0900
100	5529.04	0.0002	0.00002	61423	0.09002	11.1091	123.2335	11.0930

10%			Table 15		Discrete cash flow: compound interest factors			10%
	Single Payments		**Uniform-Series Payments**				**Uniform Gradient**	
n	Compound Amount F/P	Present Worth P/F	Sinking Fund A/F	Compound Amount F/A	Capital Recovery A/P	Present Worth P/A	Gradient Present Worth P/G	Gradient Annual Series A/G
1	1.1000	0.9091	1.00000	1.0000	1.10000	0.9091		
2	1.2100	0.8264	0.47619	2.1000	0.57619	1.7355	0.8264	0.4762
3	1.3310	0.7513	0.30211	3.3100	0.40211	2.4869	2.3291	0.9366
4	1.4641	0.6830	0.21547	4.6410	0.31547	3.1699	4.3781	1.3812
5	1.6105	0.6209	0.16380	6.1051	0.26380	3.7908	6.8618	1.8101
6	1.7716	0.5645	0.12961	7.7156	0.22961	4.3553	9.6842	2.2236
7	1.9487	0.5132	0.10541	9.4872	0.20541	4.8684	12.7631	2.6216
8	2.1436	0.4665	0.08744	11.4359	0.18744	5.3349	16.0287	3.0045
9	2.3579	0.4241	0.07364	13.5795	0.17364	5.7590	19.4215	3.3724
10	2.5937	0.3855	0.06275	15.9374	0.16275	6.1446	22.8913	3.7255
11	2.8531	0.3505	0.05396	18.5312	0.15396	6.4951	26.3963	4.0641
12	3.1384	0.3186	0.04676	21.3843	0.14676	6.8137	29.9012	4.3884
13	3.4523	0.2897	0.04078	24.5227	0.14078	7.1034	33.3772	4.6988
14	3.7975	0.2633	0.03575	27.9750	0.13575	7.3667	36.8005	4.9955
15	4.1772	0.2394	0.03147	31.7725	0.13147	7.6061	40.1520	5.2789
16	4.5950	0.2176	0.02782	35.9497	0.12782	7.8237	43.4164	5.5493
17	5.0545	0.1978	0.02466	40.5447	0.12466	8.0216	46.5819	5.8071
18	5.5599	0.1799	0.02193	45.5992	0.12193	8.2014	49.6395	6.0526
19	6.1159	0.1635	0.01955	51.1591	0.11955	8.3649	52.5827	6.2861
20	6.7275	0.1486	0.01746	57.2750	0.11746	8.5136	55.4069	6.5081
21	7.4002	0.1351	0.01562	64.0025	0.11562	8.6487	58.1095	6.7189
22	8.1403	0.1228	0.01401	71.4027	0.11401	8.7715	60.6893	6.9189
23	8.9543	0.1117	0.01257	79.5430	0.11257	8.8832	63.1462	7.1085
24	9.8497	0.1015	0.01130	88.4973	0.11130	8.9847	65.4813	7.2881
25	10.8347	0.0923	0.01017	98.3471	0.11017	9.0770	67.6964	7.4580
26	11.9182	0.0839	0.00916	109.1818	0.10916	9.1609	69.7940	7.6186
27	13.1100	0.0763	0.00826	121.0999	0.10826	9.2372	71.7773	7.7704
28	14.4210	0.0693	0.00745	134.2099	0.10745	9.3066	73.6495	7.9137
29	15.8631	0.0630	0.00673	148.6309	0.10673	9.3696	75.4146	8.0489
30	17.4494	0.0573	0.00608	164.4940	0.10608	9.4269	77.0766	8.1762
31	19.1943	0.0521	0.00550	181.9434	0.10550	9.4790	78.6395	8.2962
32	21.1138	0.0474	0.00497	201.1378	0.10497	9.5264	80.1078	8.4091
33	23.2252	0.0431	0.00450	222.2515	0.10450	9.5694	81.4856	8.5152
34	25.5477	0.0391	0.00407	245.4767	0.10407	9.6086	82.7773	8.6149
35	28.1024	0.0356	0.00369	271.0244	0.10369	9.6442	83.9872	8.7086
40	45.2593	0.0221	0.00226	442.5926	0.10226	9.7791	88.9525	9.0962
45	72.8905	0.0137	0.00139	718.9048	0.10139	9.8628	92.4544	9.3740
50	117.3909	0.0085	0.00086	1163.91	0.10086	9.9148	94.8889	9.5704
55	189.0591	0.0053	0.00053	1880.59	0.10053	9.9471	96.5619	9.7075
60	304.4816	0.0033	0.00033	3034.82	0.10033	9.9672	97.7010	9.8023
65	490.3707	0.0020	0.00020	4893.71	0.10020	9.9796	98.4705	9.8672
70	789.7470	0.0013	0.00013	7887.47	0.10013	9.9873	98.9870	9.9113
75	1271.90	0.0008	0.00008	12709	0.10008	9.9921	99.3317	9.9410
80	2048.40	0.0005	0.00005	20474	0.10005	9.9951	99.5606	9.9609
85	3298.97	0.0003	0.00003	32980	0.10003	9.9970	99.7120	9.9742
90	5313.02	0.0002	0.00002	53120	0.10002	9.9981	99.8118	9.9831
95	8556.68	0.0001	0.00001	85557	0.10001	9.9988	99.8773	9.9889
96	9412.34	0.0001	0.00001	94113	0.10001	9.9989	99.8874	9.9898
98	11389	0.0001	0.00001		0.10001	9.9991	99.9052	9.9914
100	13781	0.0001	0.00001		0.10001	9.9993	99.9202	9.9927

11%			Table 16	Discrete cash flow: compound interest factors				11%
	Single Payments		Uniform-Series Payments				Uniform Gradient	
	Compound Amount F/P	Present Worth P/F	Sinking Fund A/F	Compound Amount F/A	Capital Recovery A/P	Present Worth P/A	Gradient Present Worth P/G	Gradient Annual Series A/G
n								
1	1.1100	0.9009	1.00000	1.0000	1.11000	0.9009		
2	1.2321	0.8116	0.47393	2.1100	0.58393	1.7125	0.8116	0.4739
3	1.3676	0.7312	0.29921	3.3421	0.40921	2.4437	2.2740	0.9306
4	1.5181	0.6587	0.21233	4.7097	0.32233	3.1024	4.2502	1.3700
5	1.6851	0.5935	0.16057	6.2278	0.27057	3.6959	6.6240	1.7923
6	1.8704	0.5346	0.12638	7.9129	0.23638	4.2305	9.2972	2.1976
7	2.0762	0.4817	0.10222	9.7833	0.21222	4.7122	12.1872	2.5863
8	2.3045	0.4339	0.08432	11.8594	0.19432	5.1461	15.2246	2.9585
9	2.5580	0.3909	0.07060	14.1640	0.18060	5.5370	18.3520	3.3144
10	2.8394	0.3522	0.05980	16.7220	0.16980	5.8892	21.5217	3.6544
11	3.1518	0.3173	0.05112	19.5614	0.16112	6.2065	24.6945	3.9788
12	3.4985	0.2858	0.04403	22.7132	0.15403	6.4924	27.8388	4.2879
13	3.8833	0.2575	0.03815	26.2116	0.14815	6.7499	30.9290	4.5822
14	4.3104	0.2320	0.03323	30.0949	0.14323	6.9819	33.9449	4.8619
15	4.7846	0.2090	0.02907	34.4054	0.13907	7.1909	36.8709	5.1275
16	5.3109	0.1883	0.02552	39.1899	0.13552	7.3792	39.6953	5.3794
17	5.8951	0.1696	0.02247	44.5008	0.13247	7.5488	42.4095	5.6180
18	6.5436	0.1528	0.01984	50.3959	0.12984	7.7016	45.0074	5.8439
19	7.2633	0.1377	0.01756	56.9395	0.12756	7.8393	47.4856	6.0574
20	8.0623	0.1240	0.01558	64.2028	0.12558	7.9633	49.8423	6.2590
21	8.9492	0.1117	0.01384	72.2651	0.12384	8.0751	52.0771	6.4491
22	9.9336	0.1007	0.01231	81.2143	0.12231	8.1757	54.1912	6.6283
23	11.0263	0.0907	0.01097	91.1479	0.12097	8.2664	56.1864	6.7969
24	12.2392	0.0817	0.00979	102.1742	0.11979	8.3481	58.0656	6.9555
25	13.5855	0.0736	0.00874	114.4133	0.11874	8.4217	59.8322	7.1045
26	15.0799	0.0663	0.00781	127.9988	0.11781	8.4881	61.4900	7.2443
27	16.7386	0.0597	0.00699	143.0786	0.11699	8.5478	63.0433	7.3754
28	18.5799	0.0538	0.00626	159.8173	0.11626	8.6016	64.4965	7.4982
29	20.6237	0.0485	0.00561	178.3972	0.11561	8.6501	65.8542	7.6131
30	22.8923	0.0437	0.00502	199.0209	0.11502	8.6938	67.1210	7.7206
31	25.4104	0.0394	0.00451	221.9132	0.11451	8.7331	68.3016	7.8210
32	28.2056	0.0355	0.00404	247.3236	0.11404	8.7686	69.4007	7.9147
33	31.3082	0.0319	0.00363	275.5292	0.11363	8.8005	70.4228	8.0021
34	34.7521	0.0288	0.00326	306.8374	0.11326	8.8293	71.3724	8.0836
35	38.5749	0.0259	0.00293	341.5896	0.11293	8.8552	72.2538	8.1594
40	65.0009	0.0154	0.00172	581.8261	0.11172	8.9511	75.7789	8.4659
45	109.5302	0.0091	0.00101	986.6386	0.11101	9.0079	78.1551	8.6763
50	184.5648	0.0054	0.00060	1668.77	0.11060	9.0417	79.7341	8.8185
55	311.0025	0.0032	0.00035	2818.20	0.11035	9.0617	80.7712	8.9135
60	524.0572	0.0019	0.00021	4755.07	0.11021	9.0736	81.4461	8.9762
65	883.0669	0.0011	0.00012	8018.79	0.11012	9.0806	81.8819	9.0172
70	1488.02	0.0007	0.00007	13518	0.11007	9.0848	82.1614	9.0438
75	2507.40	0.0004	0.00004	22785	0.11004	9.0873	82.3397	9.0610
80	4225.11	0.0002	0.00003	38401	0.11003	9.0888	82.4529	9.0720
85	7119.56	0.0001	0.00002	64714	0.11002	9.0896	82.5245	9.0790

12%			Table 17	Discrete cash flow: compound interest factors				12%
	Single Payments		**Uniform-Series Payments**				**Uniform Gradient**	
	Compound Amount F/P	Present Worth P/F	Sinking Fund A/F	Compound Amount F/A	Capital Recovery A/P	Present Worth P/A	Gradient Present Worth P/G	Gradient Annual Series A/G
n								
1	1.1200	0.8929	1.00000	1.0000	1.12000	0.8929		
2	1.2544	0.7972	0.47170	2.1200	0.59170	1.6901	0.7972	0.4717
3	1.4049	0.7118	0.29635	3.3744	0.41635	2.4018	2.2208	0.9246
4	1.5735	0.6355	0.20923	4.7793	0.32923	3.0373	4.1273	1.3589
5	1.7623	0.5674	0.15741	6.3528	0.27741	3.6048	6.3970	1.7746
6	1.9738	0.5066	0.12323	8.1152	0.24323	4.1114	8.9302	2.1720
7	2.2107	0.4523	0.09912	10.0890	0.21912	4.5638	11.6443	2.5512
8	2.4760	0.4039	0.08130	12.2997	0.20130	4.9676	14.4714	2.9131
9	2.7731	0.3606	0.06768	14.7757	0.18768	5.3282	17.3563	3.2574
10	3.1058	0.3220	0.05698	17.5487	0.17698	5.6502	20.2541	3.5847
11	3.4785	0.2875	0.04842	20.6546	0.16842	5.9377	23.1288	3.8953
12	3.8960	0.2567	0.04144	24.1331	0.16144	6.1944	25.9523	4.1897
13	4.3635	0.2292	0.03568	28.0291	0.15568	6.4235	28.7024	4.4683
14	4.8871	0.2046	0.03087	32.3926	0.15087	6.6282	31.3624	4.7317
15	5.4736	0.1827	0.02682	37.2797	0.14682	6.8109	33.9202	4.9803
16	6.1304	0.1631	0.02339	42.7533	0.14339	6.9740	36.3670	5.2147
17	6.8660	0.1456	0.02046	48.8837	0.14046	7.1196	38.6973	5.4353
18	7.6900	0.1300	0.01794	55.7497	0.13794	7.2497	40.9080	5.6427
19	8.6128	0.1161	0.01576	63.4397	0.13576	7.3658	42.9979	5.8375
20	9.6463	0.1037	0.01388	72.0524	0.13388	7.4694	44.9676	6.0202
21	10.8038	0.0926	0.01224	81.6987	0.13224	7.5620	46.8188	6.1913
22	12.1003	0.0826	0.01081	92.5026	0.13081	7.6446	48.5543	6.3514
23	13.5523	0.0738	0.00956	104.6029	0.12956	7.7184	50.1776	6.5010
24	15.1786	0.0659	0.00846	118.1552	0.12846	7.7843	51.6929	6.6406
25	17.0001	0.0588	0.00750	133.3339	0.12750	7.8431	53.1046	6.7708
26	19.0401	0.0525	0.00665	150.3339	0.12665	7.8957	54.4177	6.8921
27	21.3249	0.0469	0.00590	169.3740	0.12590	7.9426	55.6369	7.0049
28	23.8839	0.0419	0.00524	190.6989	0.12524	7.9844	56.7674	7.1098
29	26.7499	0.0374	0.00466	214.5828	0.12466	8.0218	57.8141	7.2071
30	29.9599	0.0334	0.00414	241.3327	0.12414	8.0552	58.7821	7.2974
31	33.5551	0.0298	0.00369	271.2926	0.12369	8.0850	59.6761	7.3811
32	37.5817	0.0266	0.00328	304.8477	0.12328	8.1116	60.5010	7.4586
33	42.0915	0.0238	0.00292	342.4294	0.12292	8.1354	61.2612	7.5302
34	47.1425	0.0212	0.00260	384.5210	0.12260	8.1566	61.9612	7.5965
35	52.7996	0.0189	0.00232	431.6635	0.12232	8.1755	62.6052	7.6577
40	93.0510	0.0107	0.00130	767.0914	0.12130	8.2438	65.1159	7.8988
45	163.9876	0.0061	0.0074	1358.23	0.12074	8.2825	66.7342	8.0572
50	289.0022	0.0035	0.00042	2400.02	0.12042	8.3045	67.7624	8.1597
55	509.3206	0.0020	0.00024	4236.01	0.12024	8.3170	68.4082	8.2251
60	897.5969	0.0011	0.00013	7471.64	0.12013	8.3240	68.8100	8.2664
65	1581.87	0.0006	0.00008	13174	0.12008	8.3281	69.0581	8.2922
70	2787.80	0.0004	0.00004	23223	0.12004	8.3303	69.2103	8.3082
75	4913.06	0.0002	0.00002	40934	0.12002	8.3316	69.3031	8.3181
80	8658.48	0.0001	0.00001	72146	0.12001	8.3324	69.3594	8.3241
85	15259	0.0001	0.00001		0.12001	8.3328	69.3935	8.3278

	Single Payments		Uniform-Series Payments				Uniform Gradient	
	Compound Amount F/P	Present Worth P/F	Sinking Fund A/F	Compound Amount F/A	Capital Recovery A/P	Present Worth P/A	Gradient Present Worth P/G	Gradient Annual Series A/G
n								
1	1.1400	0.8772	1.00000	1.0000	1.14000	0.8772		
2	1.2996	0.7695	0.46729	2.1400	0.60729	1.6467	0.7695	0.4673
3	1.4815	0.6750	0.29073	3.4396	0.43073	2.3216	2.1194	0.9129
4	1.6890	0.5921	0.20320	4.9211	0.34320	2.9137	3.8957	1.3370
5	1.9254	0.5194	0.15128	6.6101	0.29128	3.4331	5.9731	1.7399
6	2.1950	0.4556	0.11716	8.5355	0.25716	3.8887	8.2511	2.1218
7	2.5023	0.3996	0.09319	10.7305	0.23319	4.2883	10.6489	2.4832
8	2.8526	0.3506	0.07557	13.2328	0.21557	4.6389	13.1028	2.8246
9	3.2519	0.3075	0.06217	16.0853	0.20217	4.9464	15.5629	3.1463
10	3.7072	0.2697	0.05171	19.3373	0.19171	5.2161	17.9906	3.4490
11	4.2262	0.2366	0.04339	23.0445	0.18339	5.4527	20.3567	3.7333
12	4.8179	0.2076	0.03667	27.2707	0.17667	5.6603	22.6399	3.9998
13	5.4924	0.1821	0.03116	32.0887	0.17116	5.8424	24.8247	4.2491
14	6.2613	0.1597	0.02661	37.5811	0.16661	6.0021	26.9009	4.4819
15	7.1379	0.1401	0.02281	43.8424	0.16281	6.1422	28.8623	4.6990
16	8.1372	0.1229	0.01962	50.9804	0.15962	6.2651	30.7057	4.9011
17	9.2765	0.1078	0.01692	59.1176	0.15692	6.3729	32.4305	5.0888
18	10.5752	0.0946	0.01462	68.3941	0.15462	6.4674	34.0380	5.2630
19	12.0557	0.0829	0.01266	78.9692	0.15266	6.5504	35.5311	5.4243
20	13.7435	0.0728	0.01099	91.0249	0.15099	6.6231	36.9135	5.5734
21	15.6676	0.0638	0.00954	104.7684	0.14954	6.6870	38.1901	5.7111
22	17.8610	0.0560	0.00830	120.4360	0.14830	6.7429	39.3658	5.8381
23	20.3616	0.0491	0.00723	138.2970	0.14723	6.7921	40.4463	5.9549
24	23.2122	0.0431	0.00630	158.6586	0.14630	6.8351	41.4371	6.0624
25	26.4619	0.0378	0.00550	181.8708	0.14550	6.8729	42.3441	6.1610
26	30.1666	0.0331	0.00480	208.3327	0.14480	6.9061	43.1728	6.2514
27	34.3899	0.0291	0.00419	238.4993	0.14419	6.9352	43.9289	6.3342
28	39.2045	0.0255	0.00366	272.8892	0.14366	6.9607	44.6176	6.4100
29	44.6931	0.0224	0.00320	312.0937	0.14320	6.9830	45.2441	6.4791
30	50.9502	0.0196	0.00280	356.7868	0.14280	7.0027	45.8132	6.5423
31	58.0832	0.0172	0.00245	407.7370	0.14245	7.0199	46.3297	6.5998
32	66.2148	0.0151	0.00215	465.8202	0.14215	7.0350	46.7979	6.6522
33	75.4849	0.0132	0.00188	532.0350	0.14188	7.0482	47.2218	6.6998
34	86.0528	0.0116	0.00165	607.5199	0.14165	7.0599	47.6053	6.7431
35	98.1002	0.0102	0.00144	693.5727	0.14144	7.0700	47.9519	6.7824
40	188.8835	0.0053	0.00075	1342.03	0.14075	7.1050	49.2376	6.9300
45	363.6791	0.0027	0.00039	2590.56	0.14039	7.1232	49.9963	7.0188
50	700.2330	0.0014	0.00020	4994.52	0.14020	7.1327	50.4375	7.0714
55	1348.24	0.0007	0.00010	9623.13	0.14010	7.1376	50.6912	7.1020
60	2595.92	0.0004	0.00005	18535	0.14005	7.1401	50.8357	7.1197
65	4998.22	0.0002	0.00003	35694	0.14003	7.1414	50.9173	7.1298
70	9623.64	0.0001	0.00001	68733	0.14001	7.1421	50.9632	7.1356
75	18530	0.0001	0.00001		0.14001	7.1425	50.9887	7.1388
80	35677				0.14000	7.1427	51.0030	7.1406
85	68693				0.14000	7.1428	51.0108	7.1416

15%		Table 19		Discrete cash flow: compound interest factors			15%
	Single Payments		**Uniform-Series Payments**				**Uniform Gradient**	
n	Compound Amount *F/P*	Present Worth *P/F*	Sinking Fund *A/F*	Compound Amount *F/A*	Capital Recovery *A/P*	Present Worth *P/A*	Gradient Present Worth *P/G*	Gradient Annual Series *A/G*
1	1.1500	0.8696	1.00000	1.0000	1.15000	0.8696		
2	1.3225	0.7561	0.46512	2.1500	0.61512	1.6257	0.7561	0.4651
3	1.5209	0.6575	0.28798	3.4725	0.43798	2.2832	2.0712	0.9071
4	1.7490	0.5718	0.20027	4.9934	0.35027	2.8550	3.7864	1.3263
5	2.0114	0.4972	0.14832	6.7424	0.29832	3.3522	5.7751	1.7228
6	2.3131	0.4323	0.11424	8.7537	0.26424	3.7845	7.9368	2.0972
7	2.6600	0.3759	0.09036	11.0668	0.24036	4.1604	10.1924	2.4498
8	3.0590	0.3269	0.07285	13.7268	0.22285	4.4873	12.4807	2.7813
9	3.5179	0.2843	0.05957	16.7858	0.20957	4.7716	14.7548	3.0922
10	4.0456	0.2472	0.04925	20.3037	0.19925	5.0188	16.9795	3.3832
11	4.6524	0.2149	0.04107	24.3493	0.19107	5.2337	19.1289	3.6549
12	5.3503	0.1869	0.03448	29.0017	0.18448	5.4206	21.1849	3.9082
13	6.1528	0.1625	0.02911	34.3519	0.17911	5.5831	23.1352	4.1438
14	7.0757	0.1413	0.02469	40.5047	0.17469	5.7245	24.9725	4.3624
15	8.1371	0.1229	0.02102	47.5804	0.17102	5.8474	26.6930	4.5650
16	9.3576	0.1069	0.01795	55.7175	0.16795	5.9542	28.2960	4.7522
17	10.7613	0.0929	0.01537	65.0751	0.16537	6.0472	29.7828	4.9251
18	12.3755	0.0808	0.01319	75.8364	0.16319	6.1280	31.1565	5.0843
19	14.2318	0.0703	0.01134	88.2118	0.16134	6.1382	32.4213	5.2307
20	16.3665	0.0611	0.00976	102.4436	0.15976	6.2593	33.5822	5.3651
21	18.8215	0.0531	0.00842	118.8101	0.15842	6.3125	34.6448	5.4883
22	21.6447	0.0462	0.00727	137.6316	0.15727	6.3587	35.6150	5.6010
23	24.8915	0.0402	0.00628	159.2764	0.15628	6.3988	36.4988	5.7040
24	28.6252	0.0349	0.00543	184.1678	0.15543	6.4338	37.3023	5.7979
25	32.9190	0.0304	0.00470	212.7930	0.15470	6.4641	38.0314	5.8834
26	37.8568	0.0264	0.00407	245.7120	0.15407	6.4906	38.6918	5.9612
27	43.5353	0.0230	0.00353	283.5688	0.15353	6.5135	39.2890	6.0319
28	50.0656	0.0200	0.00306	327.1041	0.15306	6.5335	39.8283	6.0960
29	57.5755	0.0174	0.00265	377.1697	0.15265	6.5509	40.3146	6.1541
30	66.2118	0.0151	0.00230	434.7451	0.15230	6.5660	40.7526	6.2066
31	76.1435	0.0131	0.00200	500.9569	0.15200	6.5791	41.1466	6.2541
32	87.5651	0.0114	0.00173	577.1005	0.15173	6.5905	41.5006	6.2970
33	100.6998	0.0099	0.00150	664.6655	0.15150	6.6005	41.8184	6.3357
34	115.8048	0.0086	0.00131	765.3654	0.15131	6.6091	42.1033	6.3705
35	133.1755	0.0075	0.00113	881.1702	0.15113	6.6166	42.3586	6.4019
40	267.8635	0.0037	0.00056	1779.09	0.15056	6.6418	43.2830	6.5168
45	538.7693	0.0019	0.00028	3585.13	0.15028	6.6543	43.8051	6.5830
50	1083.66	0.0009	0.00014	7217.72	0.15014	6.6605	44.0958	6.6205
55	2179.62	0.0005	0.00007	14524	0.15007	6.6636	44.2558	6.6414
60	4384.00	0.0002	0.00003	29220	0.15003	6.6651	44.3431	6.6530
65	8817.79	0.0001	0.00002	58779	0.15002	6.6659	44.3903	6.6593
70	17736	0.0001	0.00001		0.15001	6.6663	44.4156	6.6627
75	35673				0.15000	6.6665	44.4292	6.6646
80	71751				0.15000	6.6666	44.4364	6.6656
85					0.15000	6.6666	44.4402	6.6661

16%			Table 20	Discrete cash flow: compound interest factors				16%
	Single Payments		**Uniform-Series Payments**				**Uniform Gradient**	
n	Compound Amount F/P	Present Worth P/F	Sinking Fund A/F	Compound Amount F/A	Capital Recovery A/P	Present Worth P/A	Gradient Present Worth P/G	Gradient Annual Series A/G
1	1.1600	0.8621	1.00000	1.0000	1.16000	0.8621		
2	1.3456	0.7432	0.46296	2.1600	0.62296	1.6052	0.7432	0.4630
3	1.5609	0.6407	0.28526	3.5056	0.44526	2.2459	2.0245	0.9014
4	1.8106	0.5523	0.19738	5.0665	0.35738	2.7982	3.6814	1.3156
5	2.1003	0.4761	0.14541	6.8771	0.30541	3.2743	5.5858	1.7060
6	2.4364	0.4104	0.11139	8.9775	0.27139	3.6847	7.6380	2.0729
7	2.8262	0.3538	0.08761	11.4139	0.24761	4.0386	9.7610	2.4169
8	3.2784	0.3050	0.07022	14.2401	0.23022	4.3436	11.8962	2.7388
9	3.8030	0.2630	0.05708	17.5185	0.21708	4.6065	13.9998	3.0391
10	4.4114	0.2267	0.04690	21.3215	0.20690	4.8332	16.0399	3.3187
11	5.1173	0.1954	0.03886	25.7329	0.19886	5.0286	17.9941	3.5783
12	5.9360	0.1685	0.03241	30.8502	0.19241	5.1971	19.8472	3.8189
13	6.8858	0.1452	0.02718	36.7862	0.18718	5.3423	21.5899	4.0413
14	7.9875	0.1252	0.02290	43.6720	0.18290	5.4675	23.2175	4.2464
15	9.2655	0.1079	0.01936	51.6595	0.17936	5.5755	24.7284	4.4352
16	10.7480	0.0930	0.01641	60.9250	0.17641	5.6685	26.1241	4.6086
17	12.4677	0.0802	0.01395	71.6730	0.17395	5.7487	27.4074	4.7676
18	14.4625	0.0691	0.01188	84.1407	0.17188	5.8178	28.5828	4.9130
19	16.7765	0.0596	0.01014	98.6032	0.17014	5.8775	29.6557	5.0457
20	19.4608	0.0514	0.00867	115.3797	0.16867	5.9288	30.6321	5.1666
22	26.1864	0.0382	0.00635	157.4150	0.16635	6.0113	32.3200	5.3765
24	35.2364	0.0284	0.00467	213.9776	0.16467	6.0726	33.6970	5.5490
26	47.4141	0.0211	0.00345	290.0883	0.16345	6.1182	34.8114	5.6898
28	63.8004	0.0157	0.00255	392.5028	0.16255	6.1520	35.7073	5.8041
30	85.8499	0.0116	0.00189	530.3117	0.16189	6.1772	36.4234	5.8964
32	115.5196	0.0087	0.00140	715.7475	0.16140	6.1959	36.9930	5.9706
34	155.4432	0.0064	0.00104	965.2698	0.16104	6.2098	37.4441	6.0299
35	180.3141	0.0055	0.00089	1120.71	0.16089	6.2153	37.6327	6.0548
36	209.1643	0.0048	0.00077	1301.03	0.16077	6.2201	37.8000	6.0771
38	281.4515	0.0036	0.00057	1752.82	0.16057	6.2278	38.0799	6.1145
40	378.7212	0.0026	0.00042	2360.76	0.16042	6.2335	38.2992	6.1441
45	795.4438	0.0013	0.00020	4965.27	0.16020	6.2421	38.6598	6.1934
50	1670.70	0.0006	0.00010	10436	0.16010	6.2463	38.8521	6.2201
55	3509.05	0.0003	0.00005	21925	0.16005	6.2482	38.9534	6.2343
60	7370.20	0.0001	0.00002	46058	0.16002	6.2492	39.0063	6.2419

18%			**Table 21**	Discrete cash flow: compound interest factors				18%
	Single Payments		**Uniform-Series Payments**				**Uniform Gradient**	
n	Compound Amount *F/P*	Present Worth *P/F*	Sinking Fund *A/F*	Compound Amount *F/A*	Capital Recovery *A/P*	Present Worth *P/A*	Gradient Present Worth *P/G*	Gradient Annual Series *A/G*
1	1.1800	0.8475	1.00000	1.0000	1.18000	0.8475		
2	1.3924	0.7182	0.45872	2.1800	0.63872	1.5656	0.7182	0.4587
3	1.6430	0.6086	0.27992	3.5724	0.45992	2.1743	1.9354	0.8902
4	1.9388	0.5158	0.19174	5.2154	0.37174	2.6901	3.4828	1.2947
5	2.2878	0.4371	0.13978	7.1542	0.31978	3.1272	5.2312	1.6728
6	2.6996	0.3704	0.10591	9.4420	0.28591	3.4976	7.0834	2.0252
7	3.1855	0.3139	0.08236	12.1415	0.26236	3.8115	8.9670	2.3526
8	3.7589	0.2660	0.06524	15.3270	0.24524	4.0776	10.8292	2.6558
9	4.4355	0.2255	0.05239	19.0859	0.23239	4.3030	12.6329	2.9358
10	5.2338	0.1911	0.04251	23.5213	0.22251	4.4941	14.3525	3.1936
11	6.1759	0.1619	0.03478	28.7551	0.21478	4.6560	15.9716	3.4303
12	7.2876	0.1372	0.02863	34.9311	0.20863	4.7932	17.4811	3.6470
13	8.5994	0.1163	0.02369	42.2187	0.20369	4.9095	18.8765	3.8449
14	10.1472	0.0985	0.01968	50.8180	0.19968	5.0081	20.1576	4.0250
15	11.9737	0.0835	0.01640	60.9653	0.19640	5.0916	21.3269	4.1887
16	14.1290	0.0708	0.01371	72.9390	0.19371	5.1624	22.3885	4.3369
17	16.6722	0.0600	0.01149	87.0680	0.19149	5.2223	23.3482	4.4708
18	19.6733	0.0508	0.00964	103.7403	0.18964	5.2732	24.2123	4.5916
19	23.2144	0.0431	0.00810	123.4135	0.18810	5.3162	24.9877	4.7003
20	27.3930	0.0365	0.00682	146.6280	0.18682	5.3527	25.6813	4.7978
22	38.1421	0.0262	0.00485	206.3448	0.18485	5.4099	26.8506	4.9632
24	53.1090	0.0188	0.00345	289.4945	0.18345	5.4509	27.7725	5.0950
26	73.9490	0.0135	0.00247	405.2721	0.18247	5.4804	28.4935	5.1991
28	102.9666	0.0097	0.00177	566.4809	0.18177	5.5016	29.0537	5.2810
30	143.3706	0.0070	0.00126	790.9480	0.18126	5.5168	29.4864	5.3448
32	199.6293	0.0050	0.00091	1103.50	0.18091	5.5277	29.8191	5.3945
34	277.9638	0.0036	0.00065	1538.69	0.18065	5.5356	30.0736	5.4328
35	327.9973	0.0030	0.00055	1816.65	0.18055	5.5386	30.1773	5.4485
36	387.0368	0.0026	0.00047	2144.65	0.18047	5.5412	30.2677	5.4623
38	538.9100	0.0019	0.00033	2988.39	0.18033	5.5452	30.4152	5.4849
40	750.3783	0.0013	0.00024	4163.21	0.18024	5.5482	30.5269	5.5022
45	1716.68	0.0006	0.00010	9531.58	0.18010	5.5523	30.7006	5.5293
50	3927.36	0.0003	0.00005	21813	0.18005	5.5541	30.7856	5.5428
55	8984.84	0.0001	0.00002	49910	0.18002	5.5549	30.8268	5.5494
60	20555			114190	0.18001	5.5553	30.8465	5.5526

20%					Table 22	Discrete cash flow: compound interest factors		20%

	Single Payments		Uniform-Series Payments				Uniform Gradient	
n	Compound Amount *F/P*	Present Worth *P/F*	Sinking Fund *A/F*	Compound Amount *F/A*	Capital Recovery *A/P*	Present Worth *P/A*	Gradient Present Worth *P/G*	Gradient Annual Series *A/G*
1	1.2000	0.8333	1.00000	1.0000	1.20000	0.8333		
2	1.4400	0.6944	0.45455	2.2000	0.65455	1.5278	0.6944	0.4545
3	1.7280	0.5787	0.27473	3.6400	0.47473	2.1065	1.8519	0.8791
4	2.0736	0.4823	0.18629	5.3680	0.38629	2.5887	3.2986	1.2742
5	2.4883	0.4019	0.13438	7.4416	0.33438	2.9906	4.9061	1.6405
6	2.9860	0.3349	0.10071	9.9299	0.30071	3.3255	6.5806	1.9788
7	3.5832	0.2791	0.07742	12.9159	0.27742	3.6046	8.2551	2.2902
8	4.2998	0.2326	0.06061	16.4991	0.26061	3.8372	9.8831	2.5756
9	5.1598	0.1938	0.04808	20.7989	0.24808	4.0310	11.4335	2.8364
10	6.1917	0.1615	0.03852	25.9587	0.23852	4.1925	12.8871	3.0739
11	7.4301	0.1346	0.03110	32.1504	0.23110	4.3271	14.2330	3.2893
12	8.9161	0.1122	0.02526	39.5805	0.22526	4.4392	15.4667	3.4841
13	10.6993	0.0935	0.02062	48.4966	0.22062	4.5327	16.5883	3.6597
14	12.8392	0.0779	0.01689	59.1959	0.21689	4.6106	17.6008	3.8175
15	15.4070	0.0649	0.01388	72.0351	0.21388	4.6755	18.5095	3.9588
16	18.4884	0.0541	0.01144	87.4421	0.21144	4.7296	19.3208	4.0851
17	22.1861	0.0451	0.00944	105.9306	0.20944	4.7746	20.0419	4.1976
18	26.6233	0.0376	0.00781	128.1167	0.20781	4.8122	20.6805	4.2975
19	31.9480	0.0313	0.00646	154.7400	0.20646	4.8435	21.2439	4.3861
20	38.3376	0.0261	0.00536	186.6880	0.20536	4.8696	21.7395	4.4643
22	55.2061	0.0181	0.00369	271.0307	0.20369	4.9094	22.5546	4.5941
24	79.4968	0.0126	0.00255	392.4842	0.20255	4.9371	23.1760	4.6943
26	114.4755	0.0087	0.00176	567.3773	0.20176	4.9563	23.6460	4.7709
28	164.8447	0.0061	0.00122	819.2233	0.20122	4.9697	23.9991	4.8291
30	237.3763	0.0042	0.00085	1181.88	0.20085	4.9789	24.2628	4.8731
32	341.8219	0.0029	0.00059	1704.11	0.20059	4.9854	24.4588	4.9061
34	492.2235	0.0020	0.00041	2456.12	0.20041	4.9898	24.6038	4.9308
35	590.6682	0.0017	0.00034	2948.34	0.20034	4.9915	24.6614	4.9406
36	708.8019	0.0014	0.00028	3539.01	0.20028	4.9929	24.7108	4.9491
38	1020.67	0.0010	0.00020	5098.37	0.20020	4.9951	24.7894	4.9627
40	1469.77	0.0007	0.00014	7343.86	0.20014	4.9966	24.8469	4.9728
45	3657.26	0.0003	0.00005	18281	0.20005	4.9986	24.9316	4.9877
50	9100.44	0.0001	0.00002	45497	0.20002	4.9995	24.9698	4.9945
55	22645		0.00001		0.20001	4.9998	24.9868	4.9976

| 22% | | | | Table 23 | Discrete cash flow: compound interest factors | | | 22% | |

	Single Payments		Uniform-Series Payments				Uniform Gradient	
	Compound Amount F/P	Present Worth P/F	Sinking Fund A/F	Compound Amount F/A	Capital Recovery A/P	Present Worth P/A	Gradient Present Worth P/G	Gradient Annual Series A/G
n								
1	1.2200	0.8197	1.00000	1.0000	1.22000	0.8197		
2	1.4884	0.6719	0.45045	2.2200	0.67045	1.4915	0.6719	0.4505
3	1.8158	0.5507	0.26966	3.7084	0.48966	2.0422	1.7733	0.8683
4	2.2153	0.4514	0.18102	5.5242	0.40102	2.4936	3.1275	1.2542
5	2.7027	0.3700	0.12921	7.7396	0.34921	2.8636	4.6075	1.6090
6	3.2973	0.3033	0.09576	10.4423	0.31576	3.1669	6.1239	1.9337
7	4.0227	0.2486	0.07278	13.7396	0.29278	3.4155	7.6154	2.2297
8	4.9077	0.2038	0.05630	17.7623	0.27630	3.6193	9.0417	2.4982
9	5.9874	0.1670	0.04411	22.6700	0.26411	3.7863	10.3779	2.7409
10	7.3046	0.1369	0.03489	28.6574	0.25489	3.9232	11.6100	2.9593
11	8.9117	0.1122	0.02781	35.9620	0.24781	4.0354	12.7321	3.1551
12	10.8722	0.0920	0.02228	44.8737	0.24228	4.1274	13.7438	3.3299
13	13.2641	0.0754	0.01794	55.7459	0.23794	4.2028	14.6485	3.4855
14	16.1822	0.0618	0.01449	69.0100	0.23449	4.2646	15.4519	3.6233
15	19.7423	0.0507	0.01174	85.1922	0.23174	4.3152	16.1610	3.7451
16	24.0856	0.0415	0.00953	104.9345	0.22953	4.3567	16.7838	3.8524
17	29.3844	0.0340	0.00775	129.0201	0.22775	4.3908	17.3283	3.9465
18	35.8490	0.0279	0.00631	158.4045	0.22631	4.4187	17.8025	4.0289
19	43.7358	0.0229	0.00515	194.2535	0.22515	4.4415	18.2141	4.1009
20	53.3576	0.0187	0.00420	237.9893	0.22420	4.4603	18.5702	4.1635
22	79.4175	0.0126	0.00281	356.4432	0.22281	4.4882	19.1418	4.2649
24	118.2050	0.0085	0.00188	532.7501	0.22188	4.5070	19.5635	4.3407
26	175.9364	0.0057	0.00126	795.1653	0.22126	4.5196	19.8720	4.3968
28	261.8637	0.0038	0.00084	1185.74	0.22084	4.5281	20.0962	4.4381
30	389.7579	0.0026	0.00057	1767.08	0.22057	4.5338	20.2583	4.4683
32	580.1156	0.0017	0.00038	2632.34	0.22038	4.5376	20.3748	4.4902
34	863.4441	0.0012	0.00026	3920.20	0.22026	4.5402	20.4582	4.5060
35	1053.40	0.0009	0.00021	4783.64	0.22021	4.5411	20.4905	4.5122
36	1285.15	0.0008	0.00017	5837.05	0.22017	4.5419	20.5178	4.5174
38	1912.82	0.0005	0.00012	8690.08	0.22012	4.5431	20.5601	4.5256
40	2847.04	0.0004	0.00008	12937	0.22008	4.5439	20.5900	4.5314
45	7694.71	0.0001	0.00003	34971	0.22003	4.5449	20.6319	4.5396
50	20797		0.00001	94525	0.22001	4.5452	20.6492	4.5431
55	56207				0.22000	4.5454	20.6563	4.5445

| 24% | | | | Table 24 | | Discrete cash flow: compound interest factors | | | 24% |

	Single Payments		Uniform-Series Payments				Uniform Gradient	
	Compound Amount F/P	Present Worth P/F	Sinking Fund A/F	Compound Amount F/A	Capital Recovery A/P	Present Worth P/A	Gradient Present Worth P/G	Gradient Annual Series A/G
n								
1	1.2400	0.8065	1.00000	1.0000	1.24000	0.8065		
2	1.5376	0.6504	0.44643	2.2400	0.68643	1.4568	0.6504	0.4464
3	1.9066	0.5245	0.26472	3.7776	0.50472	1.9813	1.6993	0.8577
4	2.3642	0.4230	0.17593	5.6842	0.41593	2.4043	2.9683	1.2346
5	2.9316	0.3411	0.12425	8.0484	0.36425	2.7454	4.3327	1.5782
6	3.6352	0.2751	0.09107	10.9801	0.33107	3.0205	5.7081	1.8898
7	4.5077	0.2218	0.06842	14.6153	0.30842	3.2423	7.0392	2.1710
8	5.5895	0.1789	0.05229	19.1229	0.29229	3.4212	8.2915	2.4236
9	6.9310	0.1443	0.04047	24.7125	0.28047	3.5655	9.4458	2.6492
10	8.5944	0.1164	0.03160	31.6434	0.27160	3.6819	10.4930	2.8499
11	10.6571	0.0938	0.02485	40.2379	0.26485	3.7757	11.4313	3.0276
12	13.2148	0.0757	0.01965	50.8950	0.25965	3.8514	12.2637	3.1843
13	16.3863	0.0610	0.01560	64.1097	0.25560	3.9124	12.9960	3.3218
14	20.3191	0.0492	0.01242	80.4961	0.25242	3.9616	13.6358	3.4420
15	25.1956	0.0397	0.00992	100.8151	0.24992	4.0013	14.1915	3.5467
16	31.2426	0.0320	0.00794	126.0108	0.24794	4.0333	14.6716	3.6376
17	38.7408	0.0258	0.00636	157.2534	0.24636	4.0591	15.0846	3.7162
18	48.0386	0.0208	0.00510	195.9942	0.24510	4.0799	15.4385	37840
19	59.5679	0.0168	0.00410	244.0328	0.24410	4.0967	15.7406	38423
20	73.8641	0.0135	0.00329	303.6006	0.24329	4.1103	15.9979	3.8922
22	113.5735	0.0088	0.00213	469.0563	0.24213	4.1300	16.4011	3.9712
24	174.6306	0.0057	0.00138	723.4610	0.24138	4.1428	16.6891	4.0284
26	268.5121	0.0037	0.00090	1114.63	0.24090	4.1511	16.8930	4.0695
28	412.8642	0.0024	0.00058	1716.10	0.24058	4.1566	17.0365	4.0987
30	634.8199	0.0016	0.00038	2640.92	0.24038	4.1601	17.1369	4.1193
32	976.0991	0.0010	0.00025	4062.91	0.24025	4.1624	17.2067	4.1338
34	1500.85	0.0007	0.00016	6249.38	0.24016	4.1639	17.2552	4.1440
35	1861.05	0.0005	0.00013	7750.23	0.24013	4.1664	17.2734	4.1479
36	2307.71	0.0004	0.00010	9611.28	0.24010	4.1649	17.2886	4.1511
38	3548.33	0.0003	0.00007	14781	0.24007	4.1655	17.3116	4.1560
40	5455.91	0.0002	0.00004	22729	0.24004	4.1659	17.3274	4.1593
45	15995	0.0001	0.00002	66640	0.24002	4.1664	17.3483	4.1639
50	46890		0.00001		0.24001	4.1666	17.3563	4.1653
55					0.24000	4.1666	17.3593	4.1663

25%		Table 25		Discrete cash flow: compound interest factors.				25%
	Single Payments		**Uniform-Series Payments**				**Uniform Gradient**	
n	Compound Amount **F/P**	Present Worth **P/F**	Sinking Fund **A/F**	Compound Amount **F/A**	Capital Recovery **A/P**	Present Worth **P/A**	Gradient Present Worth **P/G**	Gradient Annual Series **A/G**
1	1.2500	0.8000	1.00000	1.0000	1.25000	0.8000		
2	1.5625	0.6400	0.44444	2.2500	0.69444	1.4400	0.6400	0.4444
3	1.9531	0.5120	0.26230	3.8125	0.51230	1.9520	1.6640	0.8525
4	2.4414	0.4096	0.17344	5.7656	0.42344	2.3616	2.8928	1.2249
5	3.0518	0.3277	0.12185	8.2070	0.37185	2.6893	4.2035	1.5631
6	3.8147	0.2621	0.08882	11.2588	0.33882	2.9514	5.5142	1.8683
7	4.7684	0.2097	0.06634	15.0735	0.31634	3.1611	6.7725	2.1424
8	5.9605	0.1678	0.05040	19.8419	0.30040	3.3289	7.9469	2.3872
9	7.4506	0.1342	0.03876	25.8023	0.28876	3.4631	9.0207	2.6048
10	9.3132	0.1074	0.03007	33.2529	0.28007	3.5705	9.9870	2.7971
11	11.6415	0.0859	0.02349	42.5661	0.27349	3.6564	10.8460	2.9663
12	14.5519	0.0687	0.01845	54.2077	0.26845	3.7251	11.6020	3.1145
13	18.1899	0.0150	0.05454	68.7596	0.26454	3.7801	12.2617	3.2467
14	22.7374	0.0440	0.01150	86.9495	0.26150	3.8241	12.8334	3.3559
15	28.4217	0.0352	0.00912	109.6868	0.25912	3.8593	13.3260	3.4530
16	35.5271	0.0281	0.00724	138.1085	0.25724	3.8874	13.7482	3.5366
17	44.4089	0.0225	0.00576	173.6357	0.25576	3.9099	14.1085	3.6084
18	55.5112	0.0180	0.00459	218.0446	0.25459	3.9279	14.4147	3.6698
19	69.3889	0.0144	0.00366	273.5558	0.25366	3.9424	14.6741	3.7222
20	86.7362	0.0115	0.00292	342.9447	0.25292	3.9539	14.8932	3.7667
22	135.5253	0.0074	0.00186	538.1011	0.25186	3.9705	15.2326	3.8365
24	211.7582	0.0047	0.00119	843.0329	0.25119	3.9811	15.4711	3.8861
26	330.8722	0.0030	0.00076	1319.49	0.25076	3.9879	15.6373	3.9212
28	516.9879	0.0019	0.00048	2063.95	0.25048	3.9923	15.7524	3.9457
30	807.7936	0.0012	0.00031	3227.17	0.25031	3.9950	15.8316	3.9628
32	1262.18	0.0008	0.00020	5044.71	0.25020	3.9968	15.8859	3.9746
34	1972.15	0.0005	0.00013	7884.61	0.25013	3.9980	15.9229	3.9828
35	2465.19	0.0004	0.00010	9856.76	.025010	3.9984	15.9367	3.9858
36	3081.49	0.0003	0.00008	12322	0.25008	3.9987	15.9481	3.9883
38	4814.82	0.0002	0.00005	19255	0.25005	3.9992	15.9651	3.9921
40	7523.16	0.0001	0.00003	30089	0.25003	3.9995	15.9766	3.9947
45	22959		0.00001	91831	0.25001	3.9998	15.9915	3.9980
50	70065				0.25000	3.9999	15.9969	3.9993
55					0.25000	4.0000	15.9989	3.9997

30%				Table 26	Discrete cash flow: compound interest factors				30%

	Single Payments		Uniform-Series Payments				Uniform Gradient	
	Compound Amount F/P	Present Worth P/F	Sinking Fund A/F	Compound Amount F/A	Capital Recovery A/P	Present Worth P/A	Gradient Present Worth P/G	Gradient Annual Series A/G
n								
1	1.3000	0.7692	1.00000	1.0000	1.30000	0.7692		
2	1.6900	0.5917	0.43478	2.3000	0.73478	1.3609	0.5917	0.4348
3	2.1970	0.4552	0.25063	3.9900	0.55063	1.8161	1.5020	0.8271
4	2.8561	0.3501	0.16163	6.1870	0.46163	2.1662	2.5524	1.1783
5	3.7129	0.2693	0.11058	9.0431	0.41058	2.4356	3.6297	1.4903
6	4.8268	0.2072	0.07839	12.7560	0.37839	2.6427	4.6656	1.7654
7	6.2749	0.1594	0.05687	17.5828	0.35687	2.8021	5.6218	2.0063
8	8.1573	0.1226	0.04192	23.8577	0.34192	2.9247	6.4800	2.2156
9	10.6045	0.0943	0.03124	32.0150	0.33124	3.0190	7.2343	2.3963
10	13.7858	0.0725	0.02346	42.6195	0.32346	3.0915	7.8872	2.5512
11	17.9216	0.0558	0.01773	56.4053	0.31773	3.1473	8.4452	2.6833
12	23.2981	0.0429	0.01345	74.3270	0.31345	3.1903	8.9173	2.7952
13	30.2875	0.0330	0.01024	97.6250	0.31024	3.2233	9.3135	2.8895
14	39.3738	0.0254	0.00782	127.9125	0.30782	3.2487	9.6437	2.9685
15	51.1859	0.0195	0.00598	167.2863	0.30598	3.2682	9.9172	3.0344
16	66.5417	0.0150	0.00458	218.4722	0.30458	3.2832	10.1426	3.0892
17	86.5042	0.0116	0.00351	285.0139	0.30351	3.2948	10.3276	3.1345
18	112.4554	0.0089	0.00269	371.5180	0.30269	3.3037	10.4788	3.1718
19	146.1920	0.0068	0.00207	483.9734	0.30207	3.3105	10.6019	3.2025
20	190.0496	0.0053	0.00159	630.1655	0.30159	3.3158	10.7019	3.2275
22	321.1839	0.0031	0.00094	1067.28	0.30094	3.3230	10.8482	3.2646
24	542.8008	0.0018	0.00055	1806.00	0.30055	3.3272	10.9433	3.2890
25	705.6410	0.0014	0.00043	2348.80	0.30043	3.3286	10.9773	3.2979
26	917.3333	0.0011	0.00033	3054.44	0.30033	3.3297	11.0045	3.3050
28	1550.29	0.0006	0.00019	5164.31	0.30019	3.3312	11.0437	3.3153
30	2620.00	0.0004	0.00011	8729.99	0.30011	3.3321	11.0687	3.3219
32	4427.79	0.0002	0.00007	14756	0.30007	3.3326	11.0845	3.3261
34	7482.97	0.0001	0.00004	24940	0.30004	3.3329	11.0945	3.3288
35	9727.86	0.0001	0.00003	32423	0.30003	3.3330	11.0980	3.3297

35%			Table 27	Discrete cash flow: compound interest factors				35%
	Single Payments		**Uniform-Series Payments**				**Uniform Gradient**	
n	Compound Amount F/P	Present Worth P/F	Sinking Fund A/F	Compound Amount F/A	Capital Recovery A/P	Present Worth P/A	Gradient Present Worth P/G	Gradient Annual Series A/G
1	1.3500	0.7407	1.00000	1.0000	1.35000	0.7407		
2	1.8225	0.5487	0.42553	2.3500	0.77553	1.2894	0.5487	0.4255
3	2.4604	0.4064	0.23966	4.1725	0.58966	1.6959	1.3616	0.8029
4	3.3215	0.3011	0.15076	6.6329	0.50076	1.9969	2.2648	1.1341
5	4.4840	0.2230	0.10046	9.9544	0.45046	2.2200	3.1568	1.4220
6	6.0534	0.1652	0.06926	14.4384	0.41926	2.3852	3.9828	1.6698
7	8.1722	0.1224	0.04880	20.4919	0.39880	2.5075	4.7170	1.8811
8	11.0324	0.0906	0.03489	28.6640	0.38489	2.5982	5.3515	2.0597
9	14.8937	0.0671	0.02519	39.6964	0.37519	2.6653	5.8886	2.2094
10	20.1066	0.0497	0.01832	54.5902	0.36832	2.7150	6.3363	2.3338
11	27.1439	0.0368	0.01339	74.6967	0.36339	2.7519	6.7047	2.4364
12	36.6442	0.0273	0.00982	101.8406	0.35982	2.7792	7.0049	2.5205
13	49.4697	0.0202	0.00722	138.4848	0.35722	2.7994	7.2474	2.5889
14	66.7841	0.0150	0.00532	187.9544	0.35532	2.8144	7.4421	2.6443
15	90.1585	0.0111	0.00393	254.7385	0.35393	2.8255	7.5974	2.6889
16	121.7139	0.0082	0.00290	344.8970	0.35290	2.8337	7.7206	2.7246
17	164.3138	0.0061	0.00214	466.6109	0.35214	2.8398	7.8180	2.7530
18	221.8236	0.0045	0.00158	630.9247	0.35158	2.8443	7.8946	2.7756
19	299.4619	0.0033	0.00117	852.7483	0.35117	2.8476	7.9547	2.7935
20	404.2736	0.0025	0.00087	1152.21	0.35087	2.8501	8.0017	2.8075
22	736.7886	0.0014	0.00048	2102.25	0.35048	2.8533	8.0669	2.8272
24	1342.80	0.0007	0.00026	3833.71	0.35026	2.8550	8.1061	2.8393
25	1812.78	0.0006	0.00019	5176.50	0.35019	2.8556	8.1194	2.8433
26	2447.25	0.0004	0.00014	6989.28	0.35014	2.8560	8.1296	2.8465
28	4460.11	0.0002	0.00008	12740	0.35008	2.8565	8.1435	2.8509
30	8128.55	0.0001	0.00004	23222	0.35004	2.8568	8.1517	2.8535
32	14814	0.0001	0.00002	42324	0.35002	2.8569	8.1565	2.8550
34	26999		0.00001	77137	0.35001	2.8570	8.1594	2.8559
35	36449		0.00001		0.35001	2.8571	8.1603	2.8562

40%			Table 28	Discrete cash flow: compound interest factors				40%
	Single Payments		**Uniform-Series Payments**				**Uniform Gradient**	
n	Compound Amount F/P	Present Worth P/F	Sinking Fund A/F	Compound Amount F/A	Capital Recovery A/P	Present Worth P/A	Gradient Present Worth P/G	Gradient Annual Series A/G
1	1.4000	0.7143	1.00000	1.0000	1.40000	0.7143		
2	1.9600	0.5102	0.41667	2.4000	0.81667	1.2245	0.5102	0.4167
3	2.7440	0.3644	0.22936	4.3600	0.62936	1.5889	1.2391	0.7798
4	3.8416	0.2603	0.14077	7.1040	0.54077	1.8492	2.0200	1.0923
5	5.3782	0.1859	0.09136	10.9456	0.49136	2.0352	2.7637	1.3580
6	7.5295	0.1328	0.06126	16.3238	0.46126	2.1680	3.4278	1.5811
7	10.5414	0.0949	0.04192	23.8534	0.44192	2.2628	3.9970	1.7664
8	14.7579	0.0678	0.02907	34.3947	0.42907	2.3306	4.4713	1.9185
9	20.6610	0.0484	0.02034	49.1526	0.42034	2.3790	4.8585	2.0422
10	28.9255	0.0346	0.01432	69.8137	0.41432	2.4136	5.1696	2.1419
11	40.4957	0.0247	0.01013	98.7391	0.41013	2.4383	5.4166	2.2215
12	56.6939	0.0176	0.00718	139.2348	0.40718	2.4559	5.6106	2.2845
13	79.3715	0.0126	0.00510	195.9287	0.40510	2.4685	5.7618	2.3341
14	111.1201	0.0090	0.00363	275.3002	0.40363	2.4775	5.8788	2.3729
15	155.5681	0.0064	0.00259	386.4202	0.40259	2.4839	5.9688	2.4030
16	217.7953	0.0046	0.00185	541.9883	0.40185	2.4885	6.0376	2.4262
17	304.9135	0.0033	0.00132	759.7837	0.40132	2.4918	6.0901	2.4441
18	426.8789	0.0023	0.00094	1064.70	0.40094	2.4941	6.1299	2.4577
19	597.6304	0.0017	0.00067	1491.58	0.40067	2.4958	6.1601	2.4682
20	836.6826	0.0012	0.00048	2089.21	0.40048	2.4970	6.1828	2.4761
22	1639.90	0.0006	0.00024	4097.24	0.40024	2.4985	6.2127	2.4866
24	3214.20	0.0003	0.00012	8033.00	0.40012	2.4992	6.2294	2.4925
25	4499.88	0.0002	0.00009	11247	0.40009	2.4994	6.2347	2.4944
26	6299.83	0.0002	0.00006	15747	0.40006	2.4996	6.2387	2.4959
28	12348	0.0001	0.00003	30867	0.40003	2.4998	6.2438	2.4977
30	24201		0.00002	60501	0.40002	2.4999	6.2466	2.4988
32	47435		0.00001		0.40001	2.4999	6.2482	2.4993
34	92972				0.40000	2.5000	6.2490	2.4996
35					0.40000	2.5000	6.2493	2.4997

50%			Table 29	Discrete cash flow: compound interest factors				50%
	Single Payments		**Uniform-Series Payments**				**Uniform Gradient**	
n	Compound Amount F/P	Present Worth P/F	Sinking Fund A/F	Compound Amount F/A	Capital Recovery A/P	Present Worth P/A	Gradient Present Worth P/G	Gradient Annual Series A/G
1	1.5000	0.6667	1.00000	1.0000	1.50000	0.6667		
2	2.2500	0.4444	0.40000	2.5000	0.90000	1.1111	0.4444	0.4000
3	3.3750	0.2963	0.21053	4.7500	0.71053	1.4074	1.0370	0.7368
4	5.0625	0.1975	0.12308	8.1250	0.62308	1.6049	1.6296	1.0154
5	7.5938	0.1317	0.07583	13.1875	0.57583	1.7366	2.1564	1.2417
6	11.3906	0.0878	0.04812	20.7813	0.54812	1.8244	2.5953	1.4226
7	17.0859	0.0585	0.03108	32.1719	0.53108	1.8829	2.9465	1.5648
8	25.6289	0.0390	0.02030	49.2578	0.52030	1.9220	3.2196	1.6752
9	38.4434	0.0260	0.01335	74.8867	0.51335	1.9480	3.4277	1.7596
10	57.6650	0.0173	0.00882	113.3301	0.50882	1.9653	3.5838	1.8235
11	86.4976	0.0116	0.00585	170.9951	0.50585	1.9769	3.6994	1.8713
12	129.7463	0.0077	0.00388	257.4927	0.50388	1.9846	3.7842	1.9068
13	194.6195	0.0051	0.00258	387.2390	0.50258	1.9897	3.8459	1.9329
14	291.9293	0.0034	0.00172	581.8585	0.50172	1.9931	3.8904	1.9519
15	437.8939	0.0023	0.00114	873.7878	0.50114	1.9954	3.9224	1.9657
16	656.8408	0.0015	0.00076	1311.68	0.50076	1.9970	3.9452	1.9756
17	985.2613	0.0010	0.00051	1968.52	0.50051	1.9980	3.9614	1.9827
18	1477.89	0.0007	0.00034	2953.78	0.50034	1.9986	3.9729	1.9878
19	2216.84	0.0005	0.00023	4431.68	0.50023	1.9991	3.9811	1.9914
20	3325.26	0.0003	0.00015	6648.51	0.50015	1.9994	3.9868	1.9940
22	7481.83	0.0001	0.00007	14962	0.50007	1.9997	3.9936	1.9971
24	16834	0.0001	0.00003	33666	0.50003	1.9999	3.9969	1.9986
25	25251		0.00002	50500	0.50002	1.9999	3.9979	1.9990
26	37877		0.00001	75752	0.50001	1.9999	3.9985	1.9993
28	85223		0.00001		0.50001	2.0000	3.9993	1.9997
30					0.50000	2.0000	3.9997	1.9998
32					0.50000	2.0000	3.9998	1.9999
34					0.50000	2.0000	3.9999	2.0000
35					0.50000	2.0000	3.9999	2.0000

INDEX

Glossary of Common Terms

Term	Symbol	Description (with initial section reference in parentheses)
Annual amount or worth	A or AW	Equivalent uniform annual worth of all cash inflows and outflows over estimated life (1.7, 6.1).
Annual operating cost	AOC	Estimated annual costs to maintain and support an alternative (1.3).
Benefit/cost ratio	B/C	Ratio of a project's benefits to costs expressed in PW, AW, or FW terms (9.1).
Breakeven point	Q_{BE}	Quantity at which revenues and costs are equal (16.1).
Book value	BV	Remaining capital investment in an asset after depreciation is accounted for (13.1).
Capital budget	b	Amount of money available for capital investment projects (1.1, 17.1).
Capital gain	CG	Monetary gain when a capital asset is sold for more than its purchase price (14.1).
Cash flow	CF	Actual cash amounts which are receipts (inflow) and disbursements (outflow) (1.9).
Cash flow before or after taxes	CFBT or CFAT	Cash-flow amount before relevant taxes or after these taxes are applied (14.4).
Cost of capital	i	Interest rate paid for the use of capital funds; includes both debt and equity funds (18.1).
Cumulative distribution	$F(X)$	Accumulation of probability up to and including a specified variable value (20.2).
Debt-equity mix	D-E	Percentages of debt and equity capital used by a corporation (18.2).
Depreciation	D	Reduction in the value of owned assets using specific models and rules; may not reflect actual usage of the asset (13.1).
Depreciation rate	d_t	Annual rate for reducing the value of owned assets using depreciation models (13.1).
Expected value (average)	$\overline{X}$, μ or $E(X)$	Long-run expected average to result if a random variable is sampled many times (19.4, 20.4).
First cost	P or B	Total initial cost for an alternative—purchase, construction, setup, etc. (1.3, 13.1).
Future amount or worth	F or FW	Amount at some future date considering time value of money (1.7).
Gradient	G	Uniform change ($+$ or $-$) in cash flow each time period; arithmetic (2.5) and percentage (2.6).
Gross income	GI	Income from all sources for corporations or individuals (14.1).
Inflation rate	f	Rate which reflects changes in the value of a currency over time (12.1).